HANDBOOK OF NANOCERAMIC AND NANOCOMPOSITE COATINGS AND MATERIALS

HANDBOOK OF NANOCERAMIC AND NANOCOMPOSITE COATINGS AND MATERIALS

Edited By

ABDEL SALAM HAMDY MAKHLOUF

DIETER SCHARNWEBER

ELSEVIER

AMSTERDAM • BOSTON • HEIDELBERG • LONDON
NEW YORK • OXFORD • PARIS • SAN DIEGO
SAN FRANCISCO • SINGAPORE • SYDNEY • TOKYO

Butterworth-Heinemann is an imprint of Elsevier

B H

Acquiring Editor: Steve Merken
Editorial Project Manager: Jeff Freeland
Project Manager: Priya Kumaraguruparan
Designer: Matthew Limbert

Butterworth Heinemann is an imprint of Elsevier
The Boulevard, Langford Lane, Kidlington, Oxford OX5 1GB, UK
225 Wyman Street, Waltham, MA 02451, USA

Library of Congress Cataloging-in-Publication Data

A catalog record for this book is available from the Library of Congress

British Library Cataloguing in Publication Data

A catalogue record for this book is available from the British Library

ISBN: 978-0-12-799947-0

For information on all Butterworth Heinemann publications visit our website at http://store.elsevier.com/

ABOUT THE EDITORS

Dr. Makhlouf is RGV STAR Professor in the Dept. of Manufacturing Engineering, University of Texas-Pan American (UTPA). He is the founder of Surface Engineering Laboratory and a leading faculty of the Rapid Response Manufacturing Center. His research interests include preparation and characterization of advanced coatings, corrosion, nano/biomaterials, renewable energy, and advanced materials and polymers. Dr. Makhlouf was able to make breakthroughs in several highly important areas of materials science and corrosion engineering.

Prof. Makhlouf is a multiple-award winner for his academic excellence: He received several prestigious awards in **Germany** (Humboldt Research Award for Experienced Scientists at Max Planck Institute); **USA** (Fulbright Visiting Scholar, NSF Fellow, and Dept. of Energy Fellow); **Belgium** (Belgian Federal Science Research Fellowship); **Arab League** (Arab Youth Excellence Award in Innovation 2013); **Jordan** (Abdul Hameed Shoman Award in Engineering Science 2012); **Egypt** (National Prize of Egypt in Advanced Science and Technology 2006; Egyptian Prize of Excellence in Surface Technology and Corrosion 2006; and Egyptian Prize of Excellence and Innovation in Materials Science and Their Applications 2009); and **Palestine** (An-Najah Prize for Research 2014). Makhlouf's biography was selected to be included in *Who's Who in the World*® 2006, 2007, and 2015.

Prof. Makhlouf is a persistent journal reviewer, advisor, and judge of the work of his peers. He is a referee for over 30 international journals of a high caliber and a continued board member of over 20 journals. He is also an experienced editor with board titles at journals published by Springer and Elsevier, an expert evaluator for the European Union's Framework Programme Seven, with an estimated budget of over €50.521 billion, DAAD and German Aerospace Center, the US Fulbright Commission, the Qatar National Research Fund, the Kuwait Foundation for the Advancement of Sciences, and a consultant at Innosquared GmbH, Germany. He is a reviewer/panelist for the NSF programs: Manufacturing Machines and Equipment, Materials Engineering and Processing, and CREST; with an estimated budget of over $7.6 billion.

Prof. Makhlouf has organized and served as a head speaker at numerous highly prestigious international symposiums and conferences over 30 times, including by way of example the "International Conference on Aerospace Science and Engineering" and "International Conference on Nanotechnologies and Biomedical Engineering." His publication list (+160) includes studies and review papers authored in journals from top publishers. He is the editor of 12 books, 17 book chapters, and has 2 US patents. One of his articles has been ranked the second among the *Top 25 Hottest Articles in Materials Science*, Elsevier, Oct-Dec 2006.

Prof. Makhlouf has a long-term experience in designing training programs and teaching courses for senior science teachers and mangers from high schools, chemists and engineers from different industrial sectors. He developed four courses at UTPA for master engineering students. His work as a professor has also brought him acclaim, with numerous appointments at outstanding institutions and universities in the United States, Germany, Italy, Egypt, and Asia, and a record of having supervised and graduated 11 PhD and Master's students, and 5 postdoctoral fellows.

Prof. Dr. Dieter Scharnweber is head of the Biomaterial-Development group at the Max Bergmann Center of Biomaterials at the Institute of Materials Science, Technische Universität Dresden (TU Dresden). He is local speaker of the DFG-Transregio 67 "Functional biomaterials for controlling healing processes in bone and skin—from material science to clinical application."

His major scientific interests are on biomaterials focusing on (i) composition-structure-property relationships of oxide layers on titanium-based materials; (ii) techniques for surface engineering of biomaterials; (iii) matrix engineering, that is, the design of artificial extracellular matrices mimicking the native cellular microenvironment; and (iv) the design and combination of defined biochemical and physical microenvironments for applications in tissue engineering and regenerative therapies.

Prof. Scharnweber has studied chemistry at the TU Dresden with a PhD in Physical Chemistry. Following his PhD, he worked on corrosion under hydrothermal conditions at the Central Institute for Nuclear Research of the Academy of Sciences of the GDR and returned to TU Dresden in 1993. With that, he changed his focus to biomaterials and surface engineering including electrochemical processing routes. After a habilitation on biomaterials in 2002, he is professor at the TU Dresden since 2008.

Prof. Scharnweber is author of almost 150 papers in peer-reviewed journals and book chapters and is inventor of more than 20 patents and patent applications.

CONTENTS

CONTRIBUTORS

Abdalla Abdal-hay
Department of Computer Science, Faculty of Engineering, Universidad de Cuenca, Cuenca, Ecuador; Department of Engineering Materials and Mechanical Design, Faculty of Engineering, South Valley of University, Qena, Egypt, and Department of Bionano System Engineering, College of Engineering, Chonbuk National University, Jeonju, Republic of Korea

Nedal Y. Abu-Thabit
Department of Chemical and Process Engineering Technology, Jubail Industrial College, Jubail Industrial City, Saudi Arabia

Saeed Bastani
Surface Coatings and Corrosion Department, Institute for Color Science and Technology, and Center of Excellence for Color Science and Technology, Tehran, Iran

Maria Benelmekki
Okinawa Institute of Science and Technology Graduate University, Onna-son, Japan

Robert Bogdanowicz
Department of Metrology and Optoelectronics, Faculty of Electronics, Telecommunications and Informatics, Gdansk University of Technology, Gdansk, Poland

Xue-Qian Fang
Department of Engineering Mechanics, Shijiazhuang Tiedao University, Shijiazhuang, PR China

Pierre Fauchais
SPCTS UMR 7315, CEC, University of Limoges, Limoges, France

Aleksandr Aleksandrovich Fomin
Saratov State Technical University, Saratov, Russia

Uwe Gbureck
Department for Functional Materials in Medicine and Dentistry, University of Würzburg, Würzburg, Germany

Martha Geffers
Department for Functional Materials in Medicine and Dentistry, University of Würzburg, Würzburg, Germany

Sascha Heinemann
Max Bergmann Center of Biomaterials and Institute for Materials Science, Technische Universität Dresden, Dresden, Germany

Alain Hoornaert
CHU Nantes, Nantes, France

Ahmad Hussain
Department of Biochemistry, National Dairy Research Institute (Deemed University), Karnal, India

Evgeni Ivanov
Open Laboratory on Experimental Micro- and Nano Mechanics, Institute of Mechanics, Bulgarian Academy of Sciences, Sofia, Bulgaria

Xudong Jiang
School of Physics and Technology, Wuhan University, and Hubei Provincial Museum, Wuhan, PR China

Balasubramanian Kandasubramanian
Department of Materials Engineering, Defence Institute of Advanced Technology (Deemed University), Pune, India

Hideyuki Kanematsu
Department of Materials Science and Engineering, Suzuka National College of Technology (SNCT), Suzuka, Japan

Kaviyarasu Kasinathan
Department of Physics, Sri Sankara Arts & Science College, Kanchipuram, Tamil Nadu, India

Gautam Kaul
Department of Biochemistry, National Dairy Research Institute (Deemed University), Karnal, India

Tahir I. Khan
Department of Mechanical & Industrial Engineering, Qatar University, Doha, Qatar

Dinkar Kisanrao Kharat
Department of Materials Engineering, Defence Institute of Advanced Technology (Deemed University), Pune, India

Rumiana Kotsilkova
Open Laboratory on Experimental Micro- and Nano Mechanics, Institute of Mechanics, Bulgarian Academy of Sciences, Sofia, Bulgaria

Pierre Layrolle
Inserm U957, Lab. Pathophysiology of bone resorption, University of Nantes, Nantes, France

Jin-Xi Liu
Department of Engineering Mechanics, Shijiazhuang Tiedao University, Shijiazhuang, PR China

Chandrashekhar Madhav Lonkar
Department of Materials Engineering, Defence Institute of Advanced Technology (Deemed University), Pune, India

Guy Louarn
Institute of Materials Jean Rouxel, CNRS, University of Nantes, Nantes, France

Abdel Salam Hamdy Makhlouf
Department of Manufacturing Engineering, College of Engineering and Computer Science, University of Texas Pan-American, Edinburg, Texas, USA

Vikas Mittal
Department of Chemical Engineering, The Petroleum Institute, Abu Dhabi, United Arab Emirates

Majid Mohseni
Surface Coatings and Corrosion Department, Institute for Color Science and Technology, Tehran, Iran

Claus Moseke
Department for Functional Materials in Medicine and Dentistry, University of Würzburg, Würzburg, Germany

Akira Otsuki
GeoRessources UMR 7359 CNRS-CREGU, ENSG, Université de Lorraine, Vandoeuvre-lés-Nancy, France

Chunxu Pan
School of Physics and Technology, Wuhan University, Wuhan, PR China

Sahab Prasad
Department of Materials Engineering, Defence Institute of Advanced Technology (Deemed University), Pune, India

Igor Vladimirovich Rodionov
Saratov State Technical University, Saratov, Russia

Laëtitia Salou
Inserm U957, Lab. Pathophysiology of bone resorption, University of Nantes; Biomedical Tissues SAS, IRSUN, and Institute of Materials Jean Rouxel, CNRS, University of Nantes, Nantes, France

Sandhya Sanand
Department of Biochemistry, National Dairy Research Institute (Deemed University), Karnal, India

Dieter Scharnweber
Technische Universität Dresden, Institute of Materials Science, Max Bergmann Center of Biomaterials, Dresden, Germany

Hanaa Soliman
Central Metallurgical Research and Development Institute, CMRDI, Cairo, Egypt

Mukhles Sowwan
Okinawa Institute of Science and Technology Graduate University, Onna-son, Japan

Tomáš Suchý
Department of Composites and Carbon Materials, Institute of Rock Structure and Mechanics, Academy of Sciences of the Czech Republic, Prague, Czech Republic

Monika Šupová
Department of Composites and Carbon Materials, Institute of Rock Structure and Mechanics, Academy of Sciences of the Czech Republic, Prague, Czech Republic

Pablo Vanegas
Department of Computer Science, Faculty of Engineering, Universidad de Cuenca, Cuenca, Ecuador

Armelle Vardelle
SPCTS UMR 7315, CEC, University of Limoges, Limoges, France

Michel Vardelle
SPCTS UMR 7315, CEC, University of Limoges, Limoges, France

Muthukumaraswamy Rangaraj Vengatesan
Department of Chemical Engineering, The Petroleum Institute, Abu Dhabi, United Arab Emirates

Elke Vorndran
Department for Functional Materials in Medicine and Dentistry, University of Würzburg, Würzburg, Germany

Cornelia Wolf-Brandstetter
Technische Universität Dresden, Institute of Materials Science, Max Bergmann Center of Biomaterials, Dresden, Germany

Michiko Yoshitake
MANA, National Institute for Materials Science (NIMS), Tsukuba, Japan

PART 1

Synthetic Approaches

CHAPTER 1

Polymer-Based Matrix Composites

Sascha Heinemann
Max Bergmann Center of Biomaterials and Institute for Materials Science, Technische Universität Dresden, Dresden, Germany

Contents

1. INTRODUCTION AND DEFINITIONS

In the last years of research and development in the fields of technical materials and especially biomaterials, it has become obvious that a wide range of demanding applications cannot be satisfied by conventional monophasic materials although they are well established for several applications. Based on findings that defined mixtures show combinations of materials properties to be more useful than monophases, the development of composites was strongly established. In general, a composite provides improved specific or synergistic characteristics not obtainable by any of the original phases alone. The first composites were generated in constructional and technical materials science. With some delay, composites found their way also into biomedical materials research. There is no universal accepted definition of the term composite. However, in the field of biomaterials, the statements commonly include (i) combinations of two or more heterogeneous materials, (ii) multiphase materials built up from a single phase that differ in composition and form, and (iii) phases retaining their identities and specific properties while forming an interface between each other [1,2].

Handbook of Nanoceramic and Nanocomposite Coatings and Materials
http://dx.doi.org/10.1016/B978-0-12-799947-0.00001-8

There are several ways of classifying composites, which will be presented in the following lines. Most of the composites consist of a structural phase (mostly inorganic particles, whiskers, fibers, lamellae, or meshes), embedded in a continuous matrix phase (mostly an organic substance). The function of the structural phase is usually to improve the strength of the composite, whereas the matrix phase acts as a binder for the inorganic building blocks and can provide elasticity and ductility. From a structural point of view, four types of composites can be distinguished: (i) fibrous (fibers in a matrix), (ii) laminar (layers of phases), (iii) particulate (particles or flakes in a matrix), and (iv) hybrids that represent combinations of any of the above three types. Literature related to biomaterials often makes use of the term *hybrid* or *hybrid material* instead of the term *composite*, which should not be mixed up with the above definition. In general, hybrid means that two phases are blended on the molecular scale while being characterized by the nature of the phases, interactions between the phases, and the resulting structure. When considering the nature of the matrix, crystalline/amorphous or organic/inorganic combinations are common. Building blocks can be molecules, macromolecules, particles, or fibers. Inorganic phases are mostly formed *in situ* by molecular precursors that often tend to form clusters or particles potentially templated by an organic phase. Based on the bonding characteristics, two classes of such hybrids can be defined [2]: Class I hybrids exhibit weak interactions between the phases caused by *van der Waals* forces, hydrogen bonds, and weak electrostatic interactions. An example is organic polymers with an entrapped inorganic phase lacking a strong interaction. Weak cross-linking can occur by the inorganic moieties or by the polymer matrix. When the inorganic phase forms its own network, additionally to the network of the organic phase, the term interpenetrating polymer networks is used. An example for that situation is the formation of an inorganic sol-gel phase in the presence of an organic network-forming phase. On the other hand, class II hybrids exhibit strong chemical interactions between the phases. In these cases, mostly discrete building blocks or polymers are linked to each other by forming covalent bonds.

Decreasing the total size of the building blocks—of the incorporated phase and of the matrix phase—and especially matching the sizes of the components to a common level drastically enhance the homogeneity of the composite and the range of adjustable properties. Especially, mechanical strength (tension, compression, and bending), often being a predominant characteristic of the material, can be dramatically improved by following this route [3]. In detail, improved characteristics mainly result from increased specific surface area and therefore enhanced interface area and cohesion between the phases. Based on this strategy, building blocks most frequently are of nanometer size, which means that at least one dimension of the basic unit is $\leq$100 nm. If these criteria are fulfilled, the composite can be termed a *nanocomposite* [4]. Since this group and the above hybrid materials make use of building blocks ranging in the nanometer size, the terms *nanocomposites* and *nanohybrids* are not clearly discriminated and often used as synonyms in the literature. Last but not least, the expression *hybrid nanocomposite* is valid and should be used for combinations of several individual nanocomposites forming one material [5].

Taking into account the large number of parameters, systematization of the class of composite materials is challenging and depends on the intended purpose. Since the focus is on biomaterials, this chapter will start from the classification of (i) organic/organic and (ii) inorganic/inorganic composites as a basis and will develop the focus on organic/inorganic composites. The discussion of that group is accompanied by the analysis of mammalian bone being a role model of naturally occurring polymer-based matrix composites. Furthermore, artificial analogs making use of a large number of natural or synthetic organic components or mixtures of that used as matrix phases and the addition of inorganic phases—often related to bioinspired and naturally occurring models—in order to specifically adjust material properties will be presented.

2. CONDITIONS DETERMINING POSSIBILITIES AND LIMITATIONS

During the conception of novel biomaterials—no matter if making use of starting from natural or synthetic components as a basis—many conditions have to be taken into account, for example, some related to processing limitations of the basic materials and some related to the purpose that the final product is intended to satisfy. The complete system determines the techniques, which can be used for all processing steps, and the range of processing conditions, which can be applied without facing unfavorable effects. By implication, the parameters identified to be applicable will determine the achievable range of material properties of the final product. The essential requirements defined for biomaterials are biocompatibility, degradability, porosity, mechanical strength, and manufacturing and commercial aspects.

In the following, these terms are explained in detail. Biocompatibility is the very fundamental requirement that characterizes the ability of a biomaterial to trigger a useful response of the host tissue [6]. At least, the biomaterial and the host tissue should coexist without having any undesirable effect on each other, including inflammation, immunogenicity, cytotoxicity, irritation, allergenic, or carcinogenic effects. Degradability includes the opportunity that the biomaterial gradually disappears and is replaced by a newly formed, healthy tissue [7]. Ideally, these two processes are linked to each other and proceed simultaneously—which is especially important when load-bearing tissue or tissue with essential functions is involved. On the one hand, the degradation process can be based on simple dissolution without any enzymatic or cellular action being involved. On the other hand, enzymes or cell-induced chemicals, for example, acids in the case of osteoclasts when resorbing the bone tissue, can be responsible for active resorption of the biomaterial. This circumstance should highlight the difference between the terms degradation (without cellular activity) and resorption (with cellular activity). During biodegradation, no toxic or otherwise harmful substances (residues from the degraded material) may be produced and the biomaterial should disappear in a similar time span as the new tissue is formed [4]. Furthermore, the biomaterial should be highly porous, which is the requirement for cells from the host tissue to invade the biomaterial, to form capillary for vascularization for oxygen and nutrient supply, and to dispatch degradation products. On the other hand, high porosity is useful for tissue engineering and biomolecular engineering techniques, where high specific surface area allows strong cell/material interaction and effectiveness of modifications. In general, the porosity should be homogeneously distributed, meaning also homogeneous occurrence of all cellular activity and degradation. However, also anisotropic orientation, for example, obtained by modified freeze-drying, can be useful for individual situations. The ideal parameters are dependent on the individual situation. Generally accepted is the advantage of about 90% overall porosity and interconnected pores with diameters of at least 100 μm [8]. Next, the biomaterial should be mechanically compatible with the host tissue. On the one hand, it must not collapse during the patient's normal activities, which is a topic of intensive research in the case of load-bearing implantation sites, for example, large bone defects. Unfortunately, high mechanical strength and high porosity are material parameters that in general are working in opposite directions. One route to overcome this limitation is the support of highly porous (and mechanically weak) materials with additional stabilization structures that can be removed later. Examples are metal parts like plates, nails, or screws during bone fracture treatment [9]. As a next point, biointegration, describing the bioadhesive contact between biomaterial and host tissue, must be established, allowing a fast integration and cellular and vascular ingrowth [10]. From a production point of view, some additional technical points should be taken into account. The manufacturing procedure and handling, including sterilization of the biomaterial, should be as simple as possible with parameters that can easily be checked and steered [11]. Finally, the manufacturing procedure must be cost-effective, which often depends on batch sizes, upscaling opportunities, automation, and individualization, for example [9].

All the points mentioned above are valid for each component of a biomaterial, making the situation much more complicated for composites when more than one component and additionally all interactions between components must be considered. On the other hand, composite formation can be a means of choice to be able to create innovative material property profiles.

The evolution of nature also made use of highly organized composite materials. In the following section, this is explained in detail for two exceptional examples: mammalian bone and siliceous spicules in sponges. Both structures are intended to achieve mechanical stability; both are based on an organic polymer that is mineralized to form a nanocomposite. The mimicry of both of these material families is the focus of extensive research and development.

3. NATURALLY OCCURRING COMPOSITES

Evolution developed a variety of polymer/mineral composites occurring in several species. Invertebrates often produce composites of calcium carbonate mineral particles and organic phases (e.g., nacre) or those consisting of amorphous silica and organic components (e.g., in glass sponge spicules) [12,13]. Further information on bionanoceramics and bionanocomposites can also be found in Chapter 3 of this book. Vertebrates developed nanocomposites with calcium phosphate mineral phases, for example, the extracellular matrix of the bone (including fish bone), dentin, and antler. Surprisingly, many of the composites mentioned are based on a common polymer matrix.

On the one hand, a natural composite material, evolutionarily much older than bone, was found within the highly flexible siliceous basal spicules of marine glass sponges (*Porifera*: *Hexactinellida*) [14]. After demineralization, a collagenous protein was identified, apparently serving as a template for silicification and from which the organic matrix was built up. As collagen

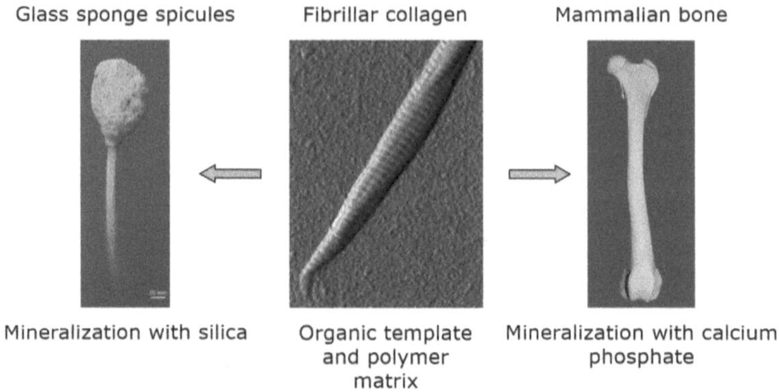

Figure 1 Fibrillar collagen fibrils (atomic force microscopy image, middle) act as templates for biomineralization with silica to form glass sponge spicules or with calcium phosphate phases to form mammalian bone.

also serves as a template for calcium phosphate and carbonate deposition in the bone, this suggests that silica and bone skeletons share a common origin with respect to collagen as a unified template for biomineralization (Figure 1). Understanding the composition, hierarchical structure, and resulting properties of the glass sponge spicules gives impetus for the development of equivalents designed *in vitro*. Outstanding examples for such artificial composites will be presented in the following sections.

In mammals, several types of bones are present or are formed temporarily during development or healing, mainly compact (cortical), trabecular (spongy), and woven bone. The major components are calcium phosphate mineral structurally similar to hydroxyapatite (HAP) and type I fibrous collagen, which form a composite from the nanoscale. Water is the third major component and is located within the fibril, in the gaps, between triple-helical collagen molecules, and between fibrils and between fibers [15]. Bone apatite crystals are plate-shaped, with average lengths and widths of 50×25 nm. The thickness varies from around 1.5 nm for mineralized tendon to 4 nm for some mature bone types. The organic component is type I collagen and a low content of noncollagenous proteins like osteocalcin and osteopontin, which together with other organic substances like proteoglycans seem to play an important role in controlling the HAP crystallization and therefore formation of the nanocomposite.

Interestingly, the three-dimensional packing of the triple-helical collagen molecules in the fibril still seems to be unclear, even though a variety of models have been proposed in the last decades [16]. Most of the experimental data were derived from transmission electron microscope experiments, revealing the nanoscopic dimensions of the HAP crystals and a partly observable patterned alignment that fits to the 67 nm spacing of the gap regions in the collagen fibrils. This was interpreted as a strong hint that the nucleation of the HAP platelets starts in the gap regions of the fibril.

The nanoscopic size of the mineral phase and their platelike morphology result in a very high surface area. The interaction between the mineral components and the fibrous collagen matrix is therefore very strong. The protein matrix does control not only the size and shape of the HAP crystals but also their crystallographic orientation. It could be shown that the *c*-axes of the hexagonal HAP platelets are well aligned with the collagen fibril long axis [15]. Collagen is deposited by osteoblasts, also expressing the enzyme alkaline phosphatase, mainly responsible for calcium phosphate mineral formation. With the exception of woven bone, collagen fibrils are deposited in a sheetlike manner and with a parallel fiber alignment into the free space, created by resorbing osteoclasts during bone remodeling. These lamellae form osteons in compact bone and trabeculae in spongy bone. These structural elements are responsible for the outstanding mechanical properties of the bone tissue and its perfect adaptation to the local force distribution [17]. The optimization for mechanical strength is one motivation for nature to develop hierarchical structures, taking advantage of the combination of nanometer-sized inorganic phases glued by organic phases. Gao et al. discussed the influence of the shape, size, and distribution of a mineral phase embedded in an organic matrix, present in many biological materials, with respect to optimized mechanical strength [18]. A large volume concentration and a large aspect ratio of the mineral building blocks result in a high stiffness of such composites. The optimal size of a single mineral crystal is limited to the nanometer scale due to the destabilizing effects of flaws.

The mechanical properties of bone vary in a wide range, which depends on bone type and testing procedure. Table 1 intends to highlight the composite formation issue, which of course hampers the development and processing of optimized artificial equipollents for bone substitution.

4. ARTIFICIAL MINERAL COMPOSITES

The combination of established inorganic phases in order to develop new materials for biomedical applications is a common strategy in materials science. Instances applying sintering and melting processes often yield modified but monophasic materials, which will not be considered as a nanocomposite in this chapter. Nanocomposites, consisting solely of inorganic components,

Table 1 Overview of some measurement results on mechanical testing of bone components individually and compositely

Sample	Mechanical test	Strength (MPa)	Young's modulus (GPa)	Reference
Collagen	Tension	100	1.5	[19]
HAP	Compression	500-1000	80-110	[20]
	Tension	9-120	80-117	
	Bending	115-200	-	
Cortical bone	Compression	170-193	14-20	[21]
	Compression	193	17-25	[22]
	Compression	134-167	11.9-17.3	[23]
	Tension	133-150	17-25	[22]
	Tension	116-151	11.9-17.3	[23]
	Bending	150-185	5.4-15.8	
	Torsion	53-70	3.3	[24]
Cancellous bone	Compression	7-10	0.05-0.5	[21]
	Compression	3.6-9.3	0.26-0.90	[22]
	Tension	2.42-2.63		[25]

are highly relevant for bone substitution applications. By combining different mineral phases, the solubility and therefore degradability of the resulting composite can be modified and adapted to the clinical needs. Such materials can be formed from two or more (nano)crystalline phases or by embedding crystalline phases in amorphous ones (e.g., silica). Both approaches have been successfully developed to manufacture nanocomposites with remarkable properties. Stoichiometric HAP is known to be the most stable and least soluble calcium phosphate phase in aqueous solutions under physiological conditions [26]. In order to overcome this disadvantage, HAP was combined with other more soluble calcium phosphate phases, for example, β-tricalcium phosphate (β-TCP). The combination in specific ratios results in biphasic calcium phosphates with controllable degradation behavior and bioactivity [27,28]. The higher release of calcium and phosphate ions compared to monophasic calcium phosphate phases after implantation also promotes new bone formation. It has been shown that a composition consisting of 20% HAP and 80% β-TCP is able to stimulate osteogenic differentiation of human mesenchymal stem cells, the precursors of bone-forming osteoblasts [29].

In most of these composites, HAP nanoparticles are used as one component. The methods for preparing can be used to determine the crystal morphology and include solid-state reactions, wet chemical syntheses/

Table 2 Examples for calcium phosphate mineral composite materials

Name	Producer	Composition
4-Bone SBS	MIS Germany	60% HAP + 40% TCP
Alaska	Argomedical	HAP + β-TCP
Artosal	aap Implantate	HAP + TCP
Bi-Ostetic	Berkeley Advanced Biomaterials, Inc.	HAP + TCP
BoneSave	Stryker	20% HAP + 80% TCP
Calciresorb-Cerapatite	Ceraver Osteal	65% HAP + 35% TCP
Ceraform	Teknimed	65% HAP + 35% TCP
Eurocer	Bioland	HAP + TCP
Ostilit	Stryker	20% HAP + 80% TCP
Straumann BoneCeramic	Straumann	60% HAP + 40% TCP
TricOs	Baxter	HAP + TCP
Triosite	Zimmer	60% HAP + 40% TCP

precipitation, hydrothermal and mechanochemical processes, pH shock wave, and microwave synthesis [1]. Some commercially established examples for calcium phosphate mineral composites are presented in Table 2.

Furthermore, silica and calcium phosphates (HAP and TCP) are combined to form composites that can be used as implant materials. Silicon is the second most abundant element in the Earth's crust. When associated with oxygen, it forms silica (SiO_2), which can be crystalline or amorphous. In the latter case, it is mainly found from biogenic origin. It is mostly formed by diatoms, glass sponges, higher plants, and insects in distinct biomineralization processes [12]. Silica is also known to take part in the cartilage and bone formation of higher animals. It has been found to be a major contributor to bone growth at the early stages of mineralization [30], which has led to the substitution of silicon for calcium in synthetic HAP and the development of different silica-based biomaterials including bioglasses [31], silica-containing cements [32], cyclosilicate-based silica-calcium phosphate nanocomposites [33], nanosilica-fused whiskers and silicon-substituted calcium phosphate powders [34], calcium phosphate HAP nanoparticles within organized silica structures [35], and composites containing silica gel and pure xerogel [36,37]. Currently, the use of silica in medical applications is mainly driven by the development of bioactive glasses that have been found to bond to living bone tissue [31,38]. Based on silica, these materials can be differentiated by specific amounts of CaO, P_2O_5, and Na_2O, which control their solubility and therefore degradation behavior. The components are mixed, melted, homogenized, and cast into molds. Some studies

Table 3 Research studies and commercial examples for mineral composite materials containing calcium phosphates and silica

Silica component	Calcium phosphate component	Reference/trade name
Calcium silicate	β-TCP	[41]
Silica	β-Rhenanite (β-$NaCaPO_4$), $Ca_2P_2O_7$	[42]
4% SiO_2	58% HAP, 38% β-TCP	[43,44]
26% SiO_2	74% HAP	[45]
30% SiO_2	42% HAP, 28% β-TCP	[45]
24% SiO_2	76% HAP	[46]
16% SiO_2	50.4% HAP, 33.6% β-TCP	[46]
1-1.7% SiO_2	98.3-99% HAP slurry	[47]
5-30% $CaSiO_3$	CPC	[48]
45.1-67.6% $CaSiO_3$	CPC	[49]
0.8% SiO_2	Synthetic HAP	Actifuse (ApaTech)
SiO_2-gel	Calcium phosphate shell	Biogran (Biomet)
13% SiO_2	52.2% HAP, 34.8% β-TCP	BONITmatrix (DOT)
24% SiO_2	76% HAP	NanoBone (Artoss)
SiO_2 matrix	HAP + TCP	Nanos (Dr. Ihde)
SiO_2	92% HAP, β-TCP	Ossa Nova (DOT)
13% SiO_2 matrix	87% HAP + β-TCP	ReBone (Schütz Dental)

report that silica present in calcium phosphate matrices influences and enhances the bioactivity [39,40]. Table 3 provides some research results and commercial examples for mineral composites containing calcium phosphates and silica.

5. ARTIFICIAL POLYMER/MINERAL COMPOSITES

Owing to the remarkable material properties of natural bone, the development of novel composites has focused on the combination of organic phases and inorganic mineral phases in order to obtain artificial equivalents (Table 4). In general, organic biopolymers (e.g., collagen and fibrin) or synthetic polymers (e.g., polylactic acid (PLA) and poly (glycolic acid) (PGA)) act as binding agents in the presence of inorganic components and improve bulk handling characteristics [4], or inorganic components added to organic matrices enhance mechanical stability. A large variety of preparation techniques including blending, extrusion, compounding, and compression

Table 4 Research studies and commercial examples for polymer/mineral composite materials containing calcium phosphate phases

Polymer	Mineral	Reference/trade name
Collagen	HAP	[1,50–62]
Collagen, PLA	HAP	[63]
Collagen, chitosan	HAP	[64–66]
Chitosan	HAP	[67–72]
Chitosan, carboxymethyl cellulose	HAP	[73]
Chitosan, PCL	HAP	[74]
Chitosan, PLA	HAP	[75]
Alginate	HAP	[76]
Alginate, gelatin	HAP	[77,78]
Gelatin	HAP	[79]
Gelatin, PCL	HAP	[80]
PLA	HAP	[81]
PLGA	HAP	[82]
PC (polycarbonate)	HAP	[83]
Cellulose	HAP	[84]
pHEMA, PCL	HAP	[85]
Chitosan	HAP and montmorillonite	[86]
PCL	TCP	[87]
9.5% Equine collagen, 2.5% chondroitin sulfate	88% HAP	Biostite (Acteon)
Equine collagen	Bone mineral	Bio-Gen Putty (Mectron)
Porcine collagen	Bio-Oss (10:90)	Bio-Oss Collagen (Geistlich Biomaterials)
Bovine collagen	HAP	Cerapatite-Collagen (Ceraver Osteal)
Bovine collagen	β-TCP	Calciresorb-Collagen (Ceraver Osteal)
Bovine collagen	HAP/TCP (65:35)	Collagraft (Zimmer)
Bovine collagen	HAP	Collapat II (Biomet)
Xenogenous collagen	HAP coating	Healos Fx (DePuy)

molding [88] have been qualified in order to obtain organic/inorganic composites for several biomaterial-related purposes, and finally, various material combinations were investigated that have been comprehensively reviewed [1,9,89]. In the following subsections, novel and recently published polymer/mineral composites are presented, which most often are intended to mimic hard tissue. These are divided in those making use of calcium and/or phosphate phases as the mineral phase and those making use of silica phases.

Composite materials of the first group consist of defined ratios of collagen and HAP, which from a chemical point of view comes close to the biological extracellular matrix of the bone. Similar to that, several studies report on the mimicry of nacre, which is composed of aragonite (calcium carbonate) crystal platelets, embedded in a complex organic matrix [90]. The combination of the calcium phosphate phase HAP and type I collagen fibrils is intended to mimic the structure and biological and mechanical properties of natural bone. Therefore, an ideal biologically inspired composite is considered to be the first choice in order to generate highly valuable biomaterials for bone substitution. There are many different strategies to synthesize collagen/HAP composites at a nanometer scale. On the one hand, they can be simple mixtures between both phases, containing HAP crystals or agglomerates in the micrometer scale or bundles of insoluble collagen fibers. Higher developed techniques apply optimized reaction conditions by controlling several parameters such as concentration, ion strength, temperature, and pH, leading to the formation of defined nanocomposite materials *in situ*. By applying several cutting-edge analysis methods, they were proven to be true artificial extracellular matrices. Some of the manufactured collagen/HAP composites were reviewed recently [1,91].

Highlighting one example out of that biomaterial group, the preparation route firstly developed by Bradt [92] and further developed by the work group of Gelinsky [50] will be described in detail. In general, the method is based on the synchronous progress of two processes: firstly, collagen fibril formation starting from tropocollagen molecules obtained from soluble type I collagen diluted in hydrochloric acid and mixed with aqueous $CaCl_2$, TRIS (tris(hydroxymethyl)aminomethane), and phosphate-buffered solution and, secondly, collagen fibril reassembly initiated by adjusting the temperature of the mixture to 37 °C. Simultaneously, the presence of calcium and phosphate ions in the mixture leads to the formation of calcium phosphate as nanoscopic HAP crystals, finally facilitating the biomimetic mineralization of the collagen. During Ostwald ripening, initially, amorphous calcium phosphate phases are formed, which are then slowly transformed into nanocrystalline HAP being the stable phase at physiological pH values. A large number of studies resulted in an optimal parameter setup for both processes to take place in a synchronized way. During that process, assembling collagen fibrils act as a template for calcium phosphate mineralization firming a homogeneous nanocomposite consisting of about 30 wt% collagen and 70 wt% HAP. The product, which can be described as mineralized collagen fibrils, can be isolated by centrifugation and used for the preparation of several types of biomaterials for testing or implantation, for example, in the form of a porous scaffold after freeze-drying [50,58,93–95]. A large number of highly qualified analysis methods confirmed the final product being similar to the natural role model from a chemical and structural point of view. The results include x-ray diffraction (XRD), Fourier transform infrared spectroscopy [92], scanning and transmission electron microscopies (SEM and TEM), and high-resolution TEM [58]. Chemical cross-linking of these composite scaffolds introduces covalent bonds between collagen molecules and enhances mechanical strength for usage as a matrix for cell cultivation. Several studies demonstrated that these scaffolds can be seeded homogeneously with mesenchymal stem cells [96], can support osteogenic differentiation toward the osteoblastic lineage [97], can be degraded by osteoclast-like cells (derived from primary human monocytes) *in vitro* [98] and *in vivo* after implantation in bone defects [99], and can be used for establishing *in vitro* models for the complex cellular processes in bone remodeling [98,100].

All composites prepared according to the biomimetic strategy share the imperfection that the complex structural principles and hierarchical organization of mammalian bone could not be recreated yet. The main consequences are much weaker mechanical properties compared to that of natural cortical or cancellous bone. This discrepancy will remain due to the lack of techniques enabling large volume biomaterial nanostructuring. Nevertheless, the state of the art enables scaffolds applicable for tissue engineering, which can be degraded and replaced by new healthy bone that exhibits mechanical strength ideally adapted to the local situation.

5.1 Fiber Spun Polymers in Polymer/Mineral Composites

In addition to the use of polymers forming fibers innately (e.g., collagen), many attempts aim on the preparation of other polymers in fibrous morphology. Recent advances in fiber spinning facilitated the introduction of this technology in biomaterial manufacturing. Starting with the wet-spinning process, electrospinning can be used for the production of thin polymer fibers down to the nanometer scale. Further processing of these fibers includes weaving, stitching, and knitting with the final three-dimensional product ranging up to several cubic centimeters in size [101]. Advantages over conventional foaming technologies include scalability, hierarchical design, anisotropic characteristics, high specific surface area, and high interconnecting porosity. These features support the adhesion of cells and the diffusion of nutrients inside the scaffold [102]. Several studies report the synthetic and natural polymers successfully applied for the tissue engineering of bone, cartilage, heart valves, bladder, and liver. However, the mechanical weakness of fiber-based constructs solely composed of organic components necessitates the introduction of mineral phases in order to achieve mechanical properties useful for structure supporting tissue substitution. The incorporation of mineral phases into fibrous polymer matrices is intended to mimic the structural organization of natural bone supplemented with some additional features. These are, for example, improved mechanical properties, favorable cell responses, and controlled degradation of the polymer caused by the neutralizing activity of the inorganic phase buffering the typical acidic degradation products of the polymers [82]. Extensive research is performed for HAP-modified polymer fibers consisting of polyhydroxybutyrate (PHB), PLA, polylactic-*co*-glycolic acid (PLGA), polycaprolactone (PCL), collagen, and gelatin, which is summarized in Table 5.

Other groups of minerals confirmed to be very useful in biomaterial applications are silica and bioactive glasses. These materials are highly biocompatible and processable via a large number of established techniques, which developed silica-based components to become an interesting alternative or supplement to calcium phosphate components in composites [110] (Table 6).

Conventional entrapment of inorganic particles in a polymer network is a strategy with limited composite characteristics. In order to achieve an

Table 5 Composites composed of fibrous polymers and calcium phosphate mineral phases

Fibrous polymer	Mineral	Reference
PCL	HAP	[103]
PHB	HAP	[104]
PLA	HAP	[105]
PLLA	Montmorillonite	[106]
Chitosan (electrospun)	HAP	[72]
Collagen (electrospun)	HAP	[107]
PLGA	HAP	[82]
PLGA, collagen	HAP	[108]
PCL, gelatin	HAP	[80]
PLA, PCL, gelatin	HAP	[109]
PCL	TCP	[87]

Table 6 Examples for polymer/mineral composite materials containing silica phases

Polymer	Mineral	Reference
PCL	Calcium silicate	[111,112]
Chitin	Nanosilica	[113]
Collagen	Bioglass	[114]
PLLA	Bioactive glass	[115]
PCL	Bioactive glass	[116,117]
pHEMA	Sol-gel silica	[118]
Collagen	Sol-gel silica	[119–121]
Collagen	Sol-gel silica, calcium phosphates	[122]
Collagen	BONITmatrix	[123]

intimate interaction between the phases of a composite, recent studies confirmed the sol-gel technique to be an appropriate method for the conjugation of inorganic materials with biological systems, owing to the compatibility of processing conditions [124]. More information on organic/inorganic hybrid nanoceramics based on sol-gel chemistry can be found in Chapter 4 of this book. Briefly, the preparation starts with molecular precursors—alkoxysilanes, whose general formula is $Si(OR)_4$ where R is an alkyl residue (C_nH_{2n+1}). The reaction is controlled by the kinetics of the hydrolytic polycondensation that occurs around the silicon atom through nucleophilic substitution [125]. In this hydrolysis process, H_2O acts as a nucleophile and displaces ROH, resulting intermediately in the soluble form of silica—orthosilicic acid ($Si(OH)_4$). Further polycondensation reactions of the silicic acid lead to siloxane bonds (Si—O—Si), which occur by heterocondensation (Si—OH + RO—Si) or homocondensation (Si—OH + HO—Si). Controlled by the pH, concentrations, solvent, catalyst, water content, temperature, and other parameters, the polymers grow to form particles or colloids [126,127]. The viscosity of the solution increases continuously up to the point where the entire solution is gelified and an amorphous three-dimensional silica network—a hydrogel—is formed [128]. One major problem when transforming silica hydrogels into dry xerogels is that the capillary forces occurring during drying often exceed the gel's strength [129]. In that case, cracking occurs and fragments or a silica powder is obtained [130].

In order to overcome this drawback, recent developments focus on the introducing polymers for reinforcement and taking advantage of the templating activity of the organic molecules. The mechanical properties of corresponding composites can be continuously tunable from ductile-tough (pure polymer) to hard-brittle (pure silica) depending on the organic/inorganic ratio [117]. Especially, collagen fibrils have been identified to be an appropriate organic matrix for the fabrication of composites with silica [122,131,132] (Figure 2). Due to the possibility of low-temperature processing *via* the sol-gel technique, the collagen's structure is preserved retaining its biological activity and additional components—of organic or inorganic nature—can be incorporated. The reactivity of both components is based on the electrostatic interaction of the negatively charged silica species and positively charged amine groups of the collagen at neutral or near-neutral pH values; thus, silicic acid preferentially polymerizes at these sites and forms nanoparticles at the surface of the collagen molecules or fibrils. Additional phases of lower reactivity, for example, calcium phosphate particles, will be physically embedded into the forming matrix.

Depending on the collagen used, indivisible composite networks exhibiting characteristic structures are formed intermediately as a composite hydrogel [121]. Further drying at high relative humidity results in shrinking and densification caused by the capillary forces. Without any heat treatment, a monolithic macroscopic nanocomposite xerogel exhibiting bone-like mechanical properties is obtained finally [122]. Conventional freeze-drying of the nanocomposite hydrogel results in highly porous scaffolds being suitable for tissue engineering applications [133].

A series of electron microscopy studies revealed the collagen morphology having distinct influence on the structure of the composite xerogels (Figure 3) [121]. The application of tropocollagen resulted in a fine-structured gel surface, similar to that of pure silica gels. Macroscopically, the tropocollagen-based xerogels were obtained as translucent flake-like fragments because capillary forces appearing during drying exceed the gel's strength. In contrast, the incorporation of collagen fibers or fibrils resulted in microporous xerogels that were obtained as homogeneous cylindrical solid monoliths. The collagen percentage showed a significant effect on the xerogel's structure and mechanical properties [132]. Pure silica is brittle (Young's modulus of about 4.2 GPa), but a distinct deformability is gained due to the presence of the organic matrix (e.g., 2 GPa for 40% collagen/60% silica). Compressive strength of the 30% collagen/70% silica composite was about 150 MPa and higher compared to that of pure silica (about 100 MPa). Strain at fracture is lowest for the pure silica xerogel due to its brittle amorphous network structure. However, the value already increases significantly when 10 wt% collagen is embedded and increases about fourfold with 30 wt % collagen. Bioactivity was slightly lowered for high collagen percentages

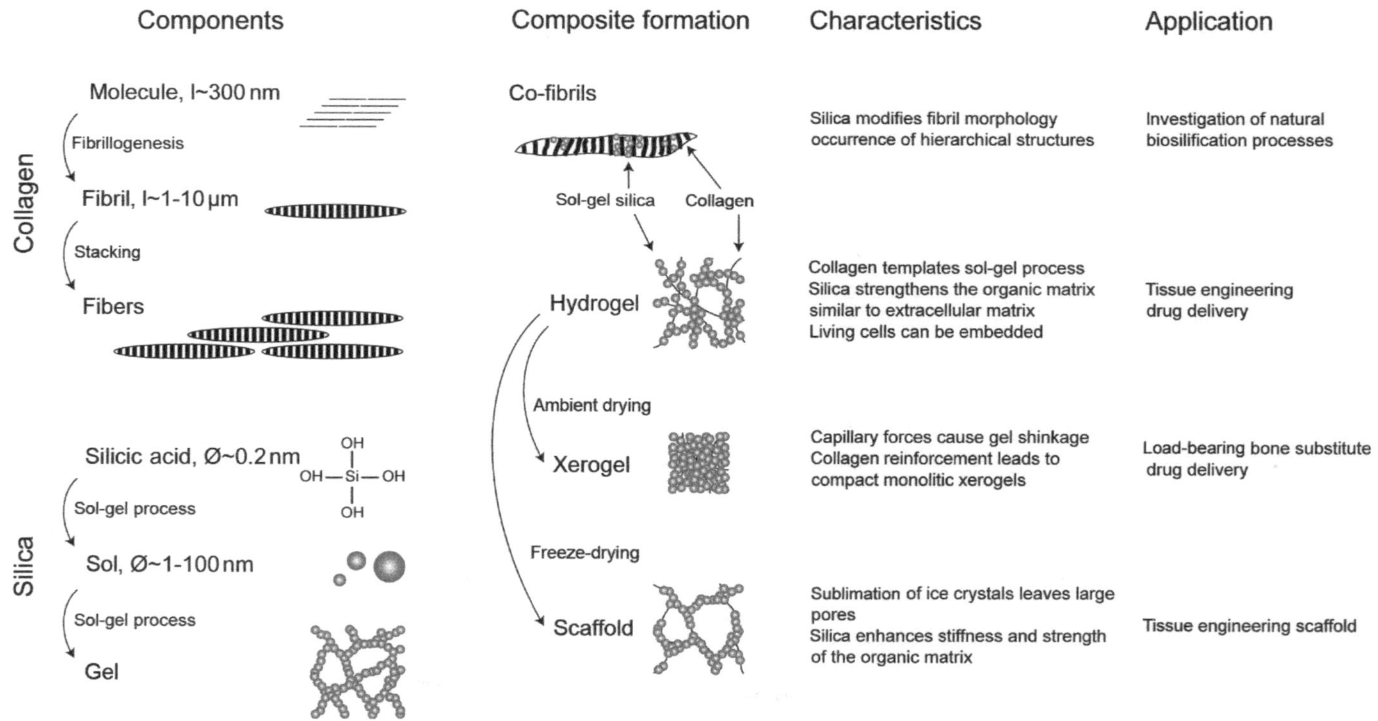

Figure 2 Schematic illustration of silica-collagen materials. Hydrogel formation starts with a sol of silica precursors (blue dots), collagen as an organic template (black lines). Conditions of composite formation determine the main characteristics and potential applications.

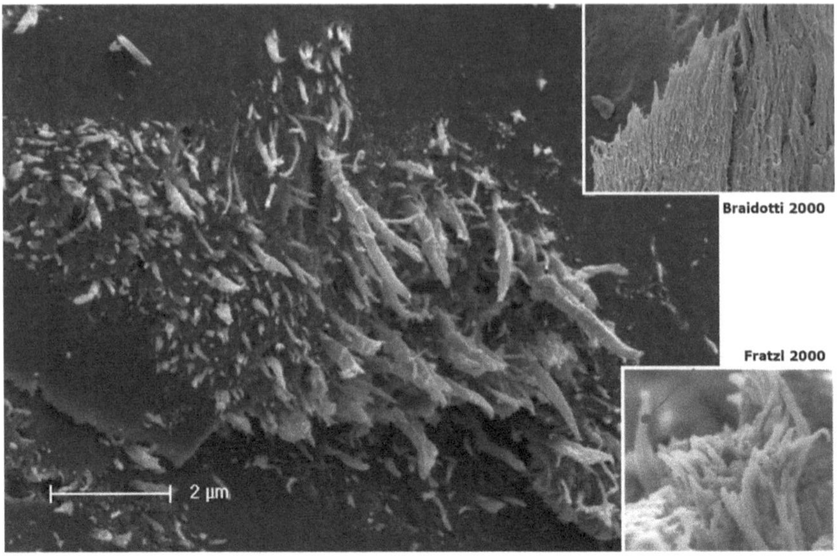

Figure 3 The SEM image of a collagen/silica xerogel fracture surface visualizes the composite character by showing collagen fibrils embedded in the silica matrix. Similarity to mammalian bone fracture sites is obvious.

reducing the amount of surface silanol groups. The addition of calcium phosphate as a third component lowered the mechanical strength but had positive effects on the bioactivity of the xerogels [122].

6. CONCLUSION

The development of composites based on polymer matrices and mineral phases as functional reinforcement is an auspicious research field that opens manifold applications, especially in the biomaterial field. Keeping in mind that nature uses composites also, often starting at the nanoscale and being hierarchically organized, forming highly specialized tissues, it is obvious that it is a promising approach to synthesize innovative biomaterials. However, not any material combination might be useful, but only those that can be adjusted to the specific needs, defined by the type of tissue to be substituted and the defect site and size. With nanocomposites, properties can be achieved that can hardly be realized with monophasic materials. This issue is the driving force for all research and development in this field, which is based on the nearly unlimited amount of combinations of polymer and mineral components, each exhibiting a unique property profile.

REFERENCES

[1] R. Murugan, S. Ramakrishna, Development o nanocomposites for bone grafting, Compos. Sci. Technol. 65 (2005) 2385–2406.
[2] G. Kickelbick, Introduction to hybrid materials, in: G. Kickelbick (Ed.), Hybrid Materials. Synthesis, Characterization, and Applications, Wiley-VCH, Weinheim, 2007, pp. 1–48.
[3] F. Watari, et al., Conversion of functions by nanosizing—from osteoconductivity to bone substitutional properties in apatite, in: M. Watanabe, O. Okuno (Eds.), Interface Oral Health Science 2007, Springer, Japan, Tokyo, 2008, pp. 139–147.
[4] C.K. Chan, et al., Biomimetic nanocomposites for bone graft applications, Nanomedicine 1 (2) (2006) 177–188.
[5] Y.Y. Liu, et al., In situ synthesis of hybrid nanocomposite with highly order arranged amorphous metallic copper nanoparticle in poly(2-hydroxyethyl methacrylate) and its potential for blood-contact uses, Acta Biomater. 4 (6) (2008) 2052–2058.
[6] D.W. Hutmacher, Scaffold design and fabrication technologies for engineering tissues—state of the art and future perspectives, J. Biomater. Sci. Polym. Ed. 12 (1) (2001) 107–124.
[7] D.W. Hutmacher, J.C. Goh, S.H. Teoh, An introduction to biodegradable materials for tissue engineering applications, Ann. Acad. Med. Singapore 30 (2) (2001) 183–191.
[8] V. Karageorgiou, D. Kaplan, Porosity of 3D biomaterial scaffolds and osteogenesis, Biomaterials 26 (27) (2005) 5474–5491.
[9] K. Rezwan, et al., Biodegradable and bioactive porous polymer/inorganic composite scaffolds for bone tissue engineering, Biomaterials 27 (18) (2006) 3413–3431.
[10] E. Polykandriotis, et al., Prevascularisation strategies in tissue engineering, Handchir. Mikrochir. Plast. Chir. 38 (4) (2006) 217–223.
[11] E.L. Chaikof, et al., Biomaterials and scaffolds in reparative medicine, Ann. N. Y. Acad. Sci. 961 (2002) 96–105.
[12] H. Ehrlich, et al., Nanostructural organization of naturally occurring composites—Part I: Silica-collagen-based biocomposites, J. Nanomater. (2008), Article ID 623838.
[13] H. Ehrlich, et al., Nanostructural organization of naturally occurring composites—Part II: Silica-chitin-based biocomposites, J. Nanomater. (2008), Article ID 670235.
[14] H. Ehrlich, et al., Mineralization of the metre-long biosilica structures of glass sponges is templated on hydroxylated collagen, Nat. Chem. 2 (12) (2010) 1084–1088.
[15] S. Weiner, H.D. Wagner, The material bone: structure-mechanical function relations, Annu. Rev. Mater. Sci. 28 (1998) 271–298.
[16] V. Ottani, et al., Hierarchical structures in fibrillar collagens, Micron 33 (7–8) (2002) 587–596.
[17] J.Y. Rho, L. Kuhn-Spearing, P. Zioupos, Mechanical properties and the hierarchical structure of bone, Med. Eng. Phys. 20 (2) (1998) 92–102.
[18] H. Gao, et al., Materials become insensitive to flaws at nanoscale: lessons from nature, Proc. Natl. Acad. Sci. U. S. A. 100 (10) (2003) 5597–5600.
[19] M.B. Bennett, et al., Mechanical properties of various mammalian tendons, J. Zool. (Lond.) 209 (1986) 537–548.
[20] T. Kokubo, H.M. Kim, M. Kawashita, Novel bioactive materials with different mechanical properties, Biomaterials 24 (13) (2003) 2161–2175.
[21] R. Murugan, S. Ramakrishna, Development of nanocomposites for bone grafting, Compos. Sci. Technol. 65 (2005) 2385–2406.
[22] K.A. Hing, Bone repair in the twenty-first century: biology, chemistry or engineering? Philos. Trans. A Math. Phys. Eng. Sci. 362 (1825) (2004) 2821–2850.
[23] K.A. Athanasiou, et al., Fundamentals of biomechanics in tissue engineering of bone, Tissue Eng. 6 (4) (2000) 361–381.

[24] M.J. Yaszemski, et al., Evolution of bone transplantation: molecular, cellular and tissue strategies to engineer human bone, Biomaterials 17 (2) (1996) 175–185.
[25] T.M. Keaveny, et al., Differences between the tensile and compressive strengths of bovine tibial trabecular bone depend on modulus, J. Biomech. 27 (9) (1994) 1137–1146.
[26] S. Mann, Biomineralization—Principles and Concepts in Bioorganic Materials Chemistry, Oxford University Press, Oxford, 2001.
[27] H.R. Ramay, M. Zhang, Biphasic calcium phosphate nanocomposite porous scaffolds for load-bearing bone tissue engineering, Biomaterials 25 (21) (2004) 5171–5180.
[28] Y. Lin, et al., Ectopic and in situ bone formation of adipose tissue-derived stromal cells in biphasic calcium phosphate nanocomposite, J. Biomed. Mater. Res. A 81 (4) (2007) 900–910.
[29] T.L. Arinzeh, et al., A comparative study of biphasic calcium phosphate ceramics for human mesenchymal stem-cell-induced bone formation, Biomaterials 26 (17) (2005) 3631–3638.
[30] E.M. Carlisle, Biochemical and morphological changes associated with long bone abnormalities in silicon deficiency, J. Nutr. 110 (5) (1980) 1046–1056.
[31] L.L. Hench, J.M. Polak, Third-generation biomedical materials, Science 295 (2002) 1014–1017.
[32] Q. Fu, et al., Preparation and characterization of a novel bioactive bone cement: glass based nanoscale hydroxyapatite bone cement, J. Mater. Sci. Mater. Med. 15 (2004) 1333–1338.
[33] G. Gupta, et al., Characterization of a novel bioactive composite using advanced X-ray computed tomography, Compos. Struct. 71 (2005) 423–428.
[34] H.H. Xu, D.T. Smith, C.G. Simon, Strong and bioactive composites containing nano-silica-fused whiskers for bone repair, Biomaterials 25 (19) (2004) 4615–4626.
[35] A. Diaz, et al., Growth of hydroxyapatite in a biocompatible mesoporous ordered silica, Acta Biomater. 2 (2) (2006) 173–179.
[36] J. Korventausta, et al., Calcium phosphate formation and ion dissolution rates in silica gel-PDLLA composites, Biomaterials 24 (28) (2003) 5173–5182.
[37] S. Radin, et al., In vivo tissue response to resorbable silica xerogels as controlled-release materials, Biomaterials 26 (9) (2005) 1043–1052.
[38] L.L. Hench, Bioceramics: from concept to clinic, J. Am. Ceram. Soc. 74 (1991) 1487–1510.
[39] N. Patel, et al., A comparative study on the in vivo behavior of hydroxyapatite and silicon substituted hydroxyapatite granules, J. Mater. Sci. Mater. Med. 13 (12) (2002) 1199–1206.
[40] L. Borum, O.C. Wilson, Surface modification of hydroxyapatite. Part II. Silica, Biomaterials 24 (21) (2003) 3681–3688.
[41] F. Zhang, et al., Preparation, mechanical properties and in vitro degradability of wollastonite/tricalcium phosphate macroporous scaffolds from nanocomposite powders, J. Mater. Sci. Mater. Med. 19 (1) (2008) 167–173.
[42] A. El-Ghannam, C.Q. Ning, J. Mehta, Cyclosilicate nanocomposite: a novel resorbable bioactive tissue engineering scaffold for BMP and bone-marrow cell delivery, J. Biomed. Mater. Res. A 71 (3) (2004) 377–390.
[43] T. Gerber, et al., Development of bioactive sol-gel material template for in vitro and in vivo synthesis of bone material, J. Sol-Gel Sci. Technol. 19 (2000) 441–445.
[44] K.O. Henkel, et al., Neuartiges Knochenaufbaumaterial auf Kalziumphosphatbasis, Mund Kiefer Gesichtschir. 8 (2004) 277–281.
[45] T. Gerber, et al., Development and in vivo test of sol-gel derived bone grafting materials, J. Sol-Gel. Sci. Technol. 26 (2003) 1173–1178.

[46] T. Traykova, et al., Silica/calcium phosphate sol-gel derived bone grafting material—from animal test to first clinical experience, Key Eng. Mater. 254–256 (2004) 679–682.
[47] S. Blindow, et al., Hydroxyapatite/SiO_2 composites via freeze casting for bone tissue engineering, Adv. Eng. Mater. 11 (11) (2009) 875–884.
[48] H. Guo, et al., Development of calcium silicate/calcium phosphate cement for bone regeneration, Biomed. Mater. 2 (3) (2007) S153–S159.
[49] J. Li, et al., Novel calcium silicate/calcium phosphate composites for potential applications as injectable bone cements, Biomed. Mater. 3 (4) (2008) 044102.
[50] M. Gelinsky, Mineralized collagen as biomaterial and matrix for bone tissue engineering, in: U. Meyer et al., (Eds.), Fundamentals of Tissue Engineering and Regenerative Medicine, Springer, Heidelberg, Berlin, 2009, pp. 485–493.
[51] D.A. Wahl, et al., Controlling the processing of collagen-hydroxyapatite scaffolds for bone tissue engineering, J. Mater. Sci. Mater. Med. 18 (2) (2007) 201–209.
[52] Y. Ohyabu, et al., A collagen sponge incorporating a hydroxyapatite/chondroitinsulfate composite as a scaffold for cartilage tissue engineering, J. Biomater. Sci. Polym. Ed. 20 (13) (2009) 1861–1874.
[53] Z. Huang, et al., A bone-like nano-hydroxyapatite/collagen loaded injectable scaffold, Biomed. Mater. 4 (5) (2009) 55005.
[54] Y. Liu, Incorporation of hydroxyapatite sol into collagen gel to regulate the contraction mediated by human bone marrow-derived stromal cells, IEEE Trans. Nanobiosci. 9 (1) (2010) 1–11.
[55] L.A. Sena, et al., Synthesis and characterization of biocomposites with different hydroxyapatite-collagen ratios, J. Mater. Sci. Mater. Med. 20 (12) (2009) 2395–2400.
[56] S. Liao, et al., Fabrication of nano-hydroxyapatite/collagen/osteonectin composites for bone graft applications, Biomed. Mater. 4 (2) (2009) 25019.
[57] A.A. Al-Munajjed, et al., Development of a biomimetic collagen-hydroxyapatite scaffold for bone tissue engineering using a SBF immersion technique, J. Biomed. Mater. Res. B Appl. Biomater. 90 (2) (2009) 584–591.
[58] M. Gelinsky, et al., Porous three-dimensional scaffolds made of mineralized collagen: preparation and properties of a biomimetic nanocomposite material for tissue engineering of bone, Chem. Eng. J. 137 (2008) 84–96.
[59] S. Liao, et al., Systematic fabrication of nano-carbonated hydroxyapatite/collagen composites for biomimetic bone grafts, Bioinspir. Biomim. 2 (3) (2007) 37–41.
[60] S. Yunoki, et al., Fabrication and mechanical and tissue ingrowth properties of unidirectionally porous hydroxyapatite/collagen composite, J. Biomed. Mater. Res. B Appl. Biomater. 80 (1) (2007) 166–173.
[61] S. Itoh, et al., Development of a novel biomaterial, hydroxyapatite/collagen (HAp/Col) composite for medical use, Biomed. Mater. Eng. 15 (1–2) (2005) 29–41.
[62] A. Tampieri, et al., Biologically inspired synthesis of bone-like composite: self-assembled collagen fibers/hydroxyapatite nanocrystals, J. Biomed. Mater. Res. A 67 (2) (2003) 618–625.
[63] X. Niu, et al., Porous nano-HA/collagen/PLLA scaffold containing chitosan microspheres for controlled delivery of synthetic peptide derived from BMP-2, J. Control. Release 134 (2) (2009) 111–117.
[64] L. Zhang, et al., Effect of chitosan as a dispersant on collagen-hydroxyapatite composite matrices, Tissue Eng. Part C Methods 16 (1) (2009) 71–79.
[65] X. Wang, et al., Synthesis and evaluation of collagen-chitosan-hydroxyapatite nanocomposites for bone grafting, J. Biomed. Mater. Res. A 89 (4) (2009) 1079–1087.
[66] Y. Wang, et al., Synthesis and characterization of collagen-chitosan-hydroxyapatite artificial bone matrix, J. Biomed. Mater. Res. A 86 (1) (2008) 244–252.

[67] J.D. Chen, Y. Wang, X. Chen, In situ fabrication of nano-hydroxyapatite in a macroporous chitosan scaffold for tissue engineering, J. Biomater. Sci. Polym. Ed. 20 (2009) 1555–1565.
[68] H. Kashiwazaki, et al., Fabrication of porous chitosan/hydroxyapatite nanocomposites: their mechanical and biological properties, Biomed. Mater. Eng. 19 (2-3) (2009) 133–140.
[69] K. Madhumathi, et al., Wet chemical synthesis of chitosan hydrogel-hydroxyapatite composite membranes for tissue engineering applications, Int. J. Biol. Macromol. 45 (1) (2009) 12–15.
[70] B.M. Chesnutt, et al., Composite chitosan/nano-hydroxyapatite scaffolds induce osteocalcin production by osteoblasts in vitro and support bone formation in vivo, Tissue Eng. Part A 15 (9) (2009) 2571–2579.
[71] W.W. Thein-Han, R.D. Misra, Biomimetic chitosan-nanohydroxyapatite composite scaffolds for bone tissue engineering, Acta Biomater. 5 (4) (2009) 1182–1197.
[72] Y. Zhang, et al., Electrospun biomimetic nanocomposite nanofibers of hydroxyapatite/chitosan for bone tissue engineering, Biomaterials 29 (32) (2008) 4314–4322.
[73] J. Liuyun, et al., Preparation and properties of a novel bone repair composite: nano-hydroxyapatite/chitosan/carboxymethyl cellulose, J. Mater. Sci. Mater. Med. 19 (3) (2008) 981–987.
[74] X. Xiao, et al., Preparation and characterization of hydroxyapatite/polycaprolactone-chitosan composites, J. Mater. Sci. Mater. Med. 20 (12) (2009) 2375–2383.
[75] X. Cai, et al., Preparation and characterization of homogeneous chitosan-polylactic acid/hydroxyapatite nanocomposite for bone tissue engineering and evaluation of its mechanical properties, Acta Biomater. 5 (7) (2009) 2693–2703.
[76] F. Despang, et al., Alginate/calcium phosphate scaffolds with oriented, tube-like pores, Mat.-wiss. u. Werkstofftech. 36 (2005) 761–767.
[77] A. Bernhardt, et al., Proliferation and osteogenic differentiation of human bone marrow stromal cells on alginate-gelatine-hydroxyapatite scaffolds with anisotropic pore structure, J. Tissue Eng. Regen. Med. 3 (1) (2009) 54–62.
[78] R. Dittrich, et al., Scaffolds for hard tissue engineering by ionotropic gelation of alginate—influence of selected preparation parameters, J. Am. Ceram. Soc. 90 (2007) 1703–1708.
[79] M. Zandi, et al., Biocompatibility evaluation of nano-rod hydroxyapatite/gelatin coated with nano-HAp as a novel scaffold using mesenchymal stem cells, J. Biomed. Mater. Res. A 92 (4) (2009) 1244–1255, http://dx.doi.org/10.1002/jbm.a.32452.
[80] J.R. Venugopal, et al., Nanobioengineered electrospun composite nanofibers and osteoblasts for bone regeneration, Artif. Organs 32 (5) (2008) 388–397.
[81] C. Spadaccio, et al., Poly-L-lactic acid/hydroxyapatite electrospun nanocomposites induce chondrogenic differentiation of human MSC, Ann. Biomed. Eng. 37 (7) (2009) 1376–1389.
[82] M.V. Jose, et al., Aligned PLGA/HA nanofibrous nanocomposite scaffolds for bone tissue engineering, Acta Biomater. 5 (1) (2009) 305–315.
[83] J. Liao, et al., Development of nanohydroxyapatite/polycarbonate composite for bone repair, J. Biomater. Appl. 24 (1) (2009) 31–45.
[84] B. Fang, et al., Proliferation and osteoblastic differentiation of human bone marrow stromal cells on hydroxyapatite/bacterial cellulose nanocomposite scaffolds, Tissue Eng. Part A 15 (5) (2009) 1091–1098.
[85] J. Huang, et al., Development of nano-sized hydroxyapatite reinforced composites for tissue engineering scaffolds, J. Mater. Sci. Mater. Med. 18 (11) (2007) 2151–2157.
[86] K.S. Katti, D.R. Katti, R. Dash, Synthesis and characterization of a novel chitosan/montmorillonite/hydroxyapatite nanocomposite for bone tissue engineering, Biomed. Mater. 3 (3) (2008) 034122.

[87] C. Erisken, D.M. Kalyon, H. Wang, Functionally graded electrospun polycaprolactone and beta-tricalcium phosphate nanocomposites for tissue engineering applications, Biomaterials 29 (30) (2008) 4065–4073.
[88] A.R. Boccaccini, J.J. Blaker, Bioactive composite materials for tissue engineering scaffolds, Expert Rev. Med. Devices 2 (2005) 303–317.
[89] S.V. Dorozhkin, Calcium orthophosphate-based biocomposites and hybrid biomaterials, J. Mater. Sci. 44 (2009) 2343–2387.
[90] S.M. Waraich, et al., Preparation of mother-of-pearl-like materials for biomedical applications by layer-by-layer deposition, Biomaterialien 10 (2009) 152.
[91] D.A. Wahl, J.T. Czernuszka, Collagen-hydroxyapatite composites for hard tissue repair, Eur. Cell Mater. 11 (2006) 43–56.
[92] J.-H. Bradt, et al., Biomimetic mineralization of collagen by combined fibril assembly and calcium phosphate formation, Chem. Mater. 11 (1999) 2694–2701.
[93] R. Burth, M. Gelinsky, W. Pompe, Collagen-hydroxyapatite tapes—a new implant material, Tech. Textiles 8 (1999) 20–21.
[94] A. Bernhardt, et al., Mineralised collagen—an artificial, extracellular bone matrix—improves osteogenic differentiation of bone marrow stromal cells, J. Mater. Sci. Mater. Med. 19 (1) (2008) 269–275.
[95] M. Gelinsky, M. Eckert, F. Despang, Biphasic, but monolithic scaffolds for the therapy of osteochondral defects, Int. J. Mater. Res. 98 (2007) 749–755.
[96] A. Lode, A. Bernhardt, M. Gelinsky, Cultivation of human bone marrow stromal cells on three-dimensional scaffolds of mineralized collagen: influence of seeding density on colonization, proliferation and osteogenic differentiation, J. Tissue Eng. Regen. Med. 2 (7) (2008) 400–407.
[97] A. Bernhardt, et al., In vitro osteogenic potential of human bone marrow stromal cells cultivated in porous scaffolds from mineralized collagen, J. Biomed. Mater. Res. A 90 (3) (2009) 852–862.
[98] H. Domaschke, et al., In vitro ossification and remodeling of mineralized collagen I scaffolds, Tissue Eng. 12 (4) (2006) 949–958.
[99] A. Yokoyama, et al., Biomimetic porous scaffolds with high elasticity made from mineralized collagen—an animal study, J. Biomed. Mater. Res. B Appl. Biomater. 75 (2) (2005) 464–472.
[100] A. Bernhardt, et al., Crosstalk of osteoblast and osteoclast precursors on mineralized collagen—towards an in vitro model for bone-remodelling, J. Biomed. Mater. Res. A 95 (3) (2010) 848–856.
[101] J.H. Jang, O. Castano, H.W. Kim, Electrospun materials as potential platforms for bone tissue engineering, Adv. Drug Deliv. Rev. 61 (12) (2009) 1065–1083.
[102] C.T. Laurencin, et al., Tissue engineering: orthopedic applications, Annu. Rev. Biomed. Eng. 1 (1999) 19–46.
[103] V. Thomas, et al., Electrospun bioactive nanocomposite scaffolds of polycaprolactone and nanohydroxyapatite for bone tissue engineering, J. Nanosci. Nanotechnol. 6 (2) (2006) 487–493.
[104] Y. Ito, et al., A composite of hydroxyapatite with electrospun biodegradable nanofibers as a tissue engineering material, J. Biosci. Bioeng. 100 (1) (2005) 43–49.
[105] H.W. Kim, H.H. Lee, J.C. Knowles, Electrospinning biomedical nanocomposite fibers of hydroxyapatite/poly(lactic acid) for bone regeneration, J. Biomed. Mater. Res. A 79 (3) (2006) 643–649.
[106] Y.H. Lee, et al., Electrospun dual-porosity structure and biodegradation morphology of Montmorillonite reinforced PLLA nanocomposite scaffolds, Biomaterials 26 (16) (2005) 3165–3172.
[107] V. Thomas, et al., Nanostructured biocomposite scaffolds based on collagen coelectrospun with nanohydroxyapatite, Biomacromolecules 8 (2) (2007) 631–637.

[108] M. Ngiam, et al., The fabrication of nano-hydroxyapatite on PLGA and PLGA/collagen nanofibrous composite scaffolds and their effects in osteoblastic behavior for bone tissue engineering, Bone 45 (1) (2009) 4–16.
[109] D. Gupta, et al., Nanostructured biocomposite substrates by electrospinning and electrospraying for the mineralization of osteoblasts, Biomaterials 30 (11) (2009) 2085–2094.
[110] S. Sakka, Sol-gel technology as representative processing for nanomaterials: case studies on the starting solution, J. Sol-Gel. Sci. Technol. 46 (2008) 241–249.
[111] J. Wei, et al., Preparation and characterization of bioactive calcium silicate and poly (epsilon-caprolactone) nanocomposite for bone tissue regeneration, J. Biomed. Mater. Res. A 90 (3) (2009) 702–712.
[112] I. Kotela, et al., Polymer nanocomposites for bone tissue substitutes, Ceram. Int. 35 (6) (2009) 2475–2480.
[113] K. Madhumathi, et al., Novel chitin/nanosilica composite scaffolds for bone tissue engineering applications, Int. J. Biol. Macromol. 45 (3) (2009) 289–292.
[114] H.W. Kim, J.H. Song, H.E. Kim, Bioactive glass nanofiber-collagen nanocomposite as a novel bone regeneration matrix, J. Biomed. Mater. Res. A 79 (3) (2006) 698–705.
[115] Z. Hong, R.L. Reis, J.F. Mano, Preparation and in vitro characterization of scaffolds of poly(L-lactic acid) containing bioactive glass ceramic nanoparticles, Acta Biomater. 4 (5) (2008) 1297–1306.
[116] A.J. Aho, et al., Injectable bioactive glass/biodegradable polymer composite for bone and cartilage reconstruction: concept and experimental outcome with thermoplastic composites of poly(epsilon-caprolactone-*co*-D,L-lactide) and bioactive glass S53P4, J. Mater. Sci. Mater. Med. 15 (10) (2004) 1165–1173.
[117] S.H. Rhee, Bone-like apatite-forming ability and mechanical properties of poly(epsilon-caprolactone)/silica hybrid as a function of poly(epsilon-caprolactone) content, Biomaterials 25 (7–8) (2004) 1167–1175.
[118] A. Costantini, et al., Bioactive poly(2-hydroxyethylmethacrylate)/silica gel hybrid nanocomposites prepared by sol-gel process, J. Biomed. Mater. Res. B Appl. Biomater. 86 (1) (2008) 98–104.
[119] Y. Ono, et al., Preparation of novel hollow fiber silica using collagen fibers as a template, Chem. Lett. 6 (1999) 475–476.
[120] S. Heinemann, et al., Biomimetically inspired hybrid materials based on silicified collagen, Int. J. Mater. Res. 98 (7) (2007) 603–608.
[121] S. Heinemann, et al., A novel biomimetic hybrid material made of silicified collagen: perspectives for bone replacement, Adv. Eng. Mater. 9 (12) (2007) 1061–1068.
[122] S. Heinemann, et al., Bioactive silica-collagen composite xerogels modified by calcium phosphate phases with adjustable mechanical properties for bone replacement, Acta Biomater. 5 (2009) 1979–1990.
[123] B.W. Thimm, et al., Biocompatibility studies of endothelial cells on a novel calcium phosphate/SiO_2-xerogel composite for bone tissue engineering, Biomed. Mater. 3 (1) (2008) 15007.
[124] G. Carturan, et al., Encapsulation of functional cells by sol-gel silica: actual progress and perspectives for cell therapy, J. Mater. Chem. 14 (2004) 2087–2098.
[125] R.J.P. Corriu, The control of nanostructured solids: a challenge for molecular chemistry, Eur. J. Inorg. Chem. 5 (2001) 1109–1121.
[126] C.C. Perry, Silicification, Rev. Mineral. Geochem. 54 (2003) 291–327.
[127] C.J. Brinker, G.W. Sherer, The Physics and Chemistry of Sol-Gel Processing, Academic Press, New York, 1990.
[128] R.K. Iler, Isolation and characterization of particle nuclei, J. Colloid Interface Sci. 75 (1) (1980) 138–148.
[129] G.W. Scherer, Theory of drying, J. Am. Ceram. Soc. 73 (1990) 3–14.

[130] J. Zhong, D.C. Greenspan, Processing and properties of sol-gel bioactive glasses, J. Biomed. Mater. Res. 53 (6) (2000) 694–701.
[131] Y. Ono, et al., Preparation of novel hollow fiber silica using collagen fibers as a template, Chem. Lett. 28 (1999) 475–476.
[132] S. Heinemann, et al., Possibilities and limitations of preparing silica/collagen/hydroxyapatite composite xerogels as load-bearing biomaterials, Compos. Sci. Technol. 71 (16) (2011) 1873–1880, http://dx.doi.org/10.1016/j.compscitech.2011.08.023.
[133] S. Heinemann, et al., Effect of silica and hydroxyapatite mineralization on the mechanical properties and the biocompatibility of nanocomposite collagen scaffolds, ACS Appl. Mater. Interfaces 3 (11) (2011) 4323–4331.

CHAPTER 2

Bio-nanoceramics and Bio-nanocomposites

Monika Šupová, Tomáš Suchý
Department of Composites and Carbon Materials, Institute of Rock Structure and Mechanics, Academy of Sciences of the Czech Republic, Prague, Czech Republic

Contents

1. BIO-NANOCERAMICS

The class of ceramics used for repairing and replacing diseased and damaged parts of the musculoskeletal systems are termed bioceramics. Bioceramics range in biocompatibility from ceramic oxides, which are inert in the body, to other extremes of resorbable materials. The most widely used bioresorbable ceramics include calcium orthophosphates, represented by hydroxyapatite (HAp). Over the past decades, many different methods have been introduced and presented for preparing pure and substituted HAp nanoparticles with precise control over microstructure, particle shape, and size [1]. The mineral phase can also be obtained from a wide range of natural sources, referred to as bioapatite (BAp), as an alternative to processes for preparing synthetic apatite. It is believed that nanosized BAp isolated or prepared from

Handbook of Nanoceramic and Nanocomposite Coatings and Materials
http://dx.doi.org/10.1016/B978-0-12-799947-0.00002-X

biogenic sources is the best material for tissue replacement and regeneration. Nanosized BAp exhibits enhanced resorbability and much higher bioactivity than micron-sized or synthetic HAp. The release of calcium ions from nanosized BAp is also similar to the behavior of the apatitic phase in native tissue. In recent decades, a number of isolation and preparation routes for producing BAp have been developed and published [2].

1.1 Bioapatites

The major components of apatite-mineralized tissue (bone, dentin, and enamel) are minerals (calcium phosphate), organics (collagen), and water. Bone and dentin consist of about 45-70 wt.% mineral and 10 wt.% water, and the rest is collagen with a small proportion of noncollagenous proteins [3]. Enamel is strikingly different in its total lack of collagen and its 96 wt.% mineral content [4].

The bone mineral is a calcium phosphate idealized as HAp ($Ca_{10}(PO_4)_6(OH)_2$). However, comprehensive studies on bone and synthetic apatites have led to the conclusion that bone mineral is not pure HAp. It is associated with minor groups and elements (e.g., CO_3^{2-}, HPO_4^{2-}, Na^+, and Mg^{2+}) and trace elements (e.g., Sr^{2+}, K^+, Cl^-, and F^-), some of them at the ppm level [3] (see Table 1). They play a vital role in the biochemical reactions

Table 1 Composition and lattice parameters of mineral phase of bone compared with stoichiometric HAp, as reported by Dorozhkin and Epple [5]

Composition		Bone	Stoichiometric HAp
Ca	(wt.%)	34.80-36.60	39.6
P		15.20-17.10	18.5
Carbonate		4.80-7.40	-
Na		0.90-1.00	-
Mg		0.60-0.72	-
Cl		0.10-0.13	-
F		0.03-0.10	-
K		0.03-0.07	-
Sr		0-0.05	-
Si	(ppm)	0-500	-
Zn		0-39	-
Cr		0-0.33	-
Co		0-0.025	-
Mn		0-0.17	-
a-axis	(Å)	9.410	9.430
c-axis		6.890	6.891

associated with bone metabolism. Several crystallochemical formulas describing the chemistry of BAp were proposed by Skinner [3] as (Ca, Na, Mg, K, Sr, Pb,. . .)$_{10}$ (PO_4, CO_3, SO_4,. . .)$_6$ (OH, F, Cl, CO_3)$_2$, whereas Cazalbou et al. [6] used $Ca_{8.3}\square_{1.7}(PO_4)_{4.3}$ (HPO_4 and CO_3)$_{1.7}$ (OH and $0.5CO_3$)$_{0.3}\square_{1.7}$ (where $\square$ is a vacancy). The chemical composition of apatite in the bone can be varied, but not with the same randomness and flexibility, because apatite has several different crystallographic sites where atomic exchanges can occur, and many different elements with different ionic charges can be accommodated or substituted in those positions.

Ionic substitution can affect the crystal structure, crystallinity, surface charge, solubility, and other vital properties, leading to major changes in biological performance of these apatites upon implantation. Apatite crystallites nucleate on and grow within the collagen network in the form of plate-shaped nanocrystals, which are 2-3 nm in thickness and tens of nanometers in length and width [7] (Figure 1). The degree of crystallinity of biological apatites is always poor when the crystallite size is of the order of 200 nm and lower. The average crystallite size of the human bone in the direction of the *c*-axis, studied by Handschin and Stern [8], was determined as 28 nm within the 0–25 age group, reaching a constant average domain size of 34 nm within the 30-80 age group. The *c*-axes of apatite crystallites lie parallel to the long axis of the microfibril [7]. Four calcium ions per unit cell, called Ca(I) (Figure 1), are in columns parallel to the *c*-axis. The other calcium ions, six per unit cell, called Ca(II), form equilateral triangles along the *c*-axis, the so-called anion channel, which is typically occupied by two monovalent anions (most commonly OH^- and F^-) per unit cell. The large size of the Ca(II) triangle allows the motion of the anion along the column axis.

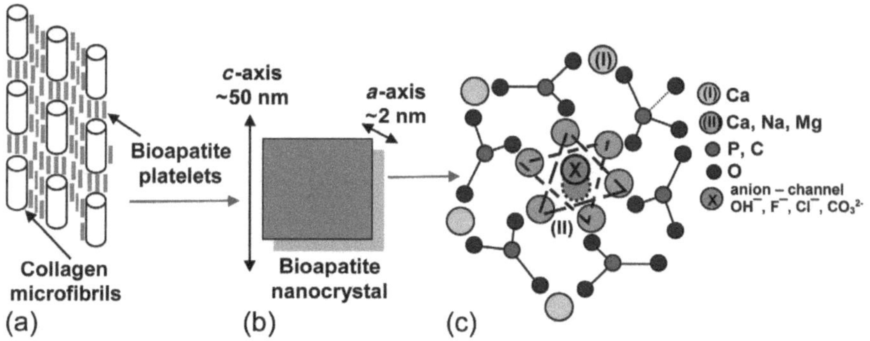

Figure 1 Arrangement of collagen microfibrils (a) with bioapatite (BAp) crystallites (b) and a view of the atomic structure of BAp—viewed down the *c*-axis (c).

The incorporation of F^- instead of OH^- results in the contraction of the unit cell along the direction of the *a*-axis. Bone apatite does not have a high concentration of OH^- groups [9], and some bone apatite may not contain any OH^- groups. Pasteris et al. [10] observed that the degree of hydroxylation of apatite covaries with its degree of atomic ordering and its crystallite size.

The major substituent in biological apatite consists of carbonates (CO_3^{2-}), which occur in bone mineral at levels typically 5-8 wt.% [11]. The carbonate ion can substitute at two sites in the apatite structure, namely, the hydroxyl position and the phosphate ion position, giving A-type and B-type carbonate apatites, respectively. B-type carbonate substitution causes changes in various physical properties, such as decreases in *a*-axial length together with increases in *c*-axial length, changes in crystallite size, and changes in the amount of crystallographic microstrain and in solubility [8,12] due to the fact that the $Ca—CO_3$ bonds are weaker than the $Ca—PO_4$ bonds. High CO_3^{2-} bone crystallites also express a platelet morphology that interfaces very effectively with collagen fibrils [11].

The deficit in the negative charge caused by the replacement of PO_4^{3-} by either CO_3^{2-} or HPO_4^{2-} can be compensated by the loss of a positive charge, for example, through the removal of Ca^{2+} from the lattice or through substitution by a monovalent ion, typically sodium for calcium [13], because the ionic radii of Ca^{2+} and Na^+ ions are quite similar (0.99 and 0.98 Å, respectively). El Feki et al. [14] found that sodium is located preferentially in the Ca (II) site (Figure 1) of the apatite structure. Another influential cationic substitute is magnesium (Mg^{2+}), which typically amounts to around 6 mol.% in the cartilage and bone tissue during the initial phases of osteogenesis, while it tends to disappear when it matures [15]. Substitution of magnesium for calcium in the HAp structure can be at most ~10 atom%. Bigi et al. [16] used Rietveld structure refinements and found that magnesium was incorporated into the HAp structure essentially in the Ca(II) site (Figure 1). They explained this preference on the basis of the different metal-oxygen interactions and the different geometries of the two distinct metal sites in the apatite structure. Partial substitution of Mg^{2+} for Ca^{2+} causes a reduction in the dimension of the *c*-axis [15–19], owing to the considerably smaller radius of Mg^{2+} (0.69 Å) than that of Ca^{2+} (0.99 Å) [17]. Magnesium inhibits apatite crystallization and destabilizes the apatitic structure. This consequently favors its thermal conversion into β-tricalcium phosphate (β-TCP) [15,18,19]. The total magnesium content in the solid phase is always significantly higher than the amount incorporated into the apatite structure.

The excess is probably located in the amorphous phase and/or on the crystallite surface [18]. However, Ren et al. [20] performed a computational study using *ab initio* generalized gradient approximation density functional theory to reveal changes in the lattice parameters. They found that the lattice constants, *a* and *c*, decreased with increasing Mg substitution and that the Ca(I) sites in the HAp lattices were energetically favored sites for Mg substitution. These discrepancies can be explained by strong dependency of the lattice parameters on the preparation method, as different methods for preparing HAp-like materials would probably result in different levels of hydration and possibly other defects in the materials. HPO_4^{2-}-containing apatites have traditionally been called Ca-deficient apatites, and they are defined as those apatitic phases with a Ca/P ratio smaller than 1.67, which is a typical value for stoichiometric HAp. Another very striking feature of biological nanocrystalline apatites is the presence of a "nonapatitic" environment [21]. It can be described as hydrated domains of the nanocrystals containing relatively mobile ions (mainly bivalent cations and anions: Ca^{2+}, HPO_4^{2-}, and CO_3^{2-}). This unstable hydrated surface layer offers a wide range of reaction possibilities, such as crystal-crystal, crystal-substrate, crystal-molecule, and crystal-cell interactions, which living organisms can benefit from [22].

1.2 Preparation of Synthetic Apatites Simulating Native BAp

BAp began to be a topic of interest several decades ago, and its attractive properties have been appreciated particularly in the past decade. Since it has better biocompatibility than HAp, many scientific studies and research projects have dealt with the preparation of "synthetic BAps," which chemically and structurally mimic and imitate real, natural BAp. The substitution can affect the biological and mechanical properties. An ionocovalent structural model of the apatite family shows a structure that can accept both cationic and anionic substituents. These substitutions induce modifications in lattice parameters and crystallinity, which substantially influence the solubility of HAp under physiological conditions, without generating significant changes in the hexagonal system of apatite. HAp particles generated from a typical dry method are usually large in size (submicron and micron) and irregular in shape. Wet methods have therefore conventionally been applied to particles having a nanosized structure with a regular morphology. The most widely used wet method for preparation is conventional chemical precipitation, and both anionic and cationic substitutions are applied.

Anionic substitutions are carried out using the carbonate, fluoride, silicate, and sulfate groups. Carbonates are among the most abundant substituents in BAp. $NaHCO_3$, $(NH_4)_2CO_3$, or a carbon dioxide-saturated solution can be used as a source for carbonates. The substitution leads to the preferential formation of B-type [14] carbonate apatites. Barralet et al. [13] indicated the presence of carbonate in the A site for low carbonate contents (<4 wt.%), while at higher contents, the carbonate was located predominantly in the B site. They also studied the effect of the reaction temperature and the concentration of bicarbonate ions on particle morphology and phase composition.

Other wet methods suitable for preparing carbonate-substituted HAp are sol-gel [23] and nanoemulsion [24] methods. Sol-gel offers the advantage of molecular-level mixing of reactants, improving the chemical homogeneity of the resulting powder. It exhibits a stoichiometric structure with a large surface area and a small cluster size (ranging from 50 nm to about 1 μm). A further notable advantage is that a several hundred degrees lower temperature is required for calcination and sintering of sol-gel HAp, leading to a decrease in degradation. An interesting modification of the sol-gel method is self-propagating combustion synthesis, which provides intimate blending among the constituents using a suitable fuel, which also acts as a complexing agent (e.g., citric acid, urea, and sucrose) in an aqueous medium and a vigorous exothermic redox reaction between the fuel and an oxidizer (e.g., nitric acid). Sasikumar et al. [25] proved that by adding citric acid to succinic acid, the autoignition temperature of succinic acid can be brought down to a lower temperature. A mixture of fuels always leads to β-TCP as a major product, while the use of a single fuel results in carbonated HAp. Emulsion processing enables the clustering to be refined and the formation of hard agglomerates to be restricted. A microemulsion is a thermodynamically stable isotopic transparent dispersion of two immiscible liquids, such as water and organics, stabilized by the presence of surfactants. Nanoemulsions, like microemulsions, are a new class of emulsions having very fine and uniform droplet sizes, typically in the range of 20-200 nm. Nanoemulsions, in contrast to microemulsions, are not thermodynamically stable, but are highly kinetically stable against sedimentation and creaming. They also offer the possibility of obtaining microemulsion-like dispersions using low surfactant concentrations. Zhou et al. [24] used the nanoemulsion technique to synthesize nanosized carbonated HAp, even without the surfactant. The nanospheres were in an amorphous state and became highly crystalline after calcination at 900 °C.

Another important anionic substituent present in the BAp structure is fluorine. Fluorine, as an important therapeutic agent in the treatment of osteoporosis, can act as a superior alternative to coating titanium hard tissue implants [26]. Fluorine-substituted HAp (FHAp) can embody the chemical composition of $Ca_{10}(PO_4)_6OH_{2-x}F_x$ (where $x=0$-2). Eslami et al. [27] proved that the thermal stability substantially is increased with increasing incorporation of fluoride into the HAp structure. The solubility can be fine-tuned, and correspondingly the biological lifetime, by varying the amount of fluoride substitutions. Cell culture results revealed that the F content affects the behaviors of cultured cells in different ways: a high F content produces a lower rate of solubility compared to HAp, which favors cell attachment. However, a reduction in Ca^{2+} release could inhibit cell proliferation. It is therefore of great value and has been widely prepared by multiple wet techniques, for example, precipitation [28], sol-gel [26], hydrolysis [29], hydrothermal preparation [30], and surfactant-assisted preparation [31]. However, most wet methods need precise controls on the processing conditions, and the composition and the properties of the final products are strongly influenced by various parameters. Wet methods are therefore not so suitable for mass production, especially on an industrial scale, but mechanochemical methods can be used for this purpose [32]. The milling procedure, applied for several hours, mechanically induced a solid-state reaction between HAp and CaF_2, and together with the hydrothermal process, this led to the formation of FHAp. Zhang et al. [32] found that the starting materials at first reacted to a poorly crystallized calcium-deficient apatite, and complete incorporation of the fluoride ions into the apatite occurred in the second stage. To investigate the substitutions of trace anions into the HAp lattice and its effect on the crystal structural and biological performance, several studies have been conducted, where ions such as Cl^- [33], SO_4^{2-} [34], and SiO_4^{4-} [35–39] were substituted into the lattice of synthetic HAp. Inorganic sulfate is the fourth most abundant anion in the human plasma and is required by cells to function properly. Alshemary et al. [34] substituted sulfate ions (SO_4^{2-}) into HAp ($Ca_{10}(PO_4)_{6-x}(SO_4)_x(OH)_{2-x}$, where $x=0.05$-0.5), via a microwave-assisted ion exchange process. The resulting needle-shaped particles were 41 nm in length, with reduced crystallinity. Kannan et al. [33] synthetized chlorapatite ($Ca_{10}(PO_4)_6(OH)_{2-x}(Cl)_x$, where $x=0.2$-2), corresponding to a weight range of 0.64-6.42 wt.%, using an aqueous precipitation method with NH_4Cl as a source for chlorine. They studied the effect of incorporated chlorine on changing the lattice parameters. Substitution of chlorine in calcium-deficient apatites tends to form biphasic mixtures of HAp and β-TCP after calcination, and the proportions in the mixtures were dependent on the deficiency of calcium in the precursors that were used. Many reports have suggested that silicon would be essential for the metabolic processes associated with growth and skeletal development and that the incorporation of silicon into a synthetic HAp structure would improve its bioactivity [35,37,38]. Si-HAp powders have been synthetized by solid-state reaction [35] and hydrothermal [38,39], precipitation [36], and sol-gel [37] methods. Si-HAp powders with the corresponding chemical formula $Ca_{10}(PO_4)_{6-x}(SiO_4)_x(OH)_{2-x}$, where $x=0.05$-0.5 usually contain up to 1.4 wt.% of Si ($x=0.5$). However, Si-HAp with content up to 1.8 wt.% have also been prepared [39]. SEM observation clearly showed that the substitution of Si in HAp inhibited the grain growth of Si-HAp [36], and XRD results confirm a reduction in crystallite size [38]. Balamurugan et al. [37] evaluated the biocompatibility effects of Si-HAp (Si contents of 0, 1, 3, and 5 mol.%) *in vitro* with human osteoblasts. A high Si content (5 mol.%) appears to promote rapid bone mineralization. However, a high Si content may result in fast dissolution of the material, owing to a decrease in crystallite size, which may not be ideal for cell attachment for prolonged time periods. An optimum level of Si appears to be 3 mol.%, which balances these effects.

Ma et al. [40] prepared selenite-substituted HAp (Se-HAp) with various Se/P ratios by a coprecipitation method, using sodium selenite as a selenium source. Selenium was incorporated into the HAp lattice by partially replacing phosphate (PO_4^{3-}) groups by selenite (SeO_3^{2-}) groups. This Se-HAp has been synthesized as a promising idea for treating bone cancers to reduce the probability of recurrence, because selenium plays an important role in protein functions and it has a significant effect on the induction of cancer cell apoptosis.

One of the most widely used cationic substitutions is the incorporation of magnesium in concentrations up to 10 mol.% in relation to Ca [41–43] by a precipitation method [41,42] or by a mechanochemical-hydrothermal route [43], usually at room temperature [20,41–43]. However, preparation temperatures of ~100 °C have also been used [16,19,20]. Landi et al. [41] had proved that ion dopings and synthesis conditions of 5.7 mol.% Mg-substituted HAp made the apatite similar to biological apatite in terms of composition, morphology, and crystallinity, and no genotoxicity, carcinogenicity, or cytotoxicity was revealed when *in vivo* tests were performed. The significance of potassium in the composition of bone lies in its tremendous influence on the biomineralization process. It has been proved that potassium can be substituted in the apatite lattice (K-HAp) without significant changes in the structural parameters [44]. A limited number of studies have been carried out on the synthesis of K-HAp. Nordsröm and Karlsson

[45] prepared it by soaking HAp in potassium carbonate and potassium chloride solutions for up to 8 weeks, and they found that after 2 weeks, one potassium ion substitutes one calcium ion. Kannan et al. [44] proved that the substitution of potassium into the apatite lattice improved its thermal stability until 1300 °C.

Recently, strontium ranelate, a newly developed drug for treating osteoporosis, has been shown to have dual effects of stimulating osteoblast differentiation and inhibiting osteoclast activity and bone resorption, which could reduce the incidence of fractures in osteoporotic patients. It has been shown that Sr can influence bone cells even if incorporated into HAp [46] and can be substituted in the crystal lattice of HAp (Sr-HAp) without any solubility limit. In addition, Sr substitution also stabilizes the HAp phase against any phase transformation and also crystal growth during high-temperature sintering. Several research works on the synthesis of Sr-HAp using a sol-gel process [47], precipitation [48], and a hydrothermal method [49] have been reported. Other methods, for example, micro-arc treatment [46] and laser ablation [50], have been used for producing Sr-HAp coatings. The studied concentrations of strontium (Sr/Sr + Ca) have varied up to 50 mol.%. Mardziah et al. [47] found that calcined powders with a doping concentration of 2 mol.% of Sr contained HAp and β-TCP as the secondary phase began to appear at a temperature of 900 °C, whereas non-Sr-doped HAp appeared at a temperature below 800 °C. It has been found that ionic extracts of Sr-HAp with 3.22 mol.% of Sr substitution showed the best proliferation, osteogenic differentiation, and angiogenic factor expression of human osteoblast-like MG-63 cells [49]. However, Chung and Long [46] proved that coatings with a concentration exceeding 38.9 at.% of Sr significantly inhibit RAW264.7 osteoclast differentiation.

Several works have dealt with multi-ionic substitutions for better imitation of the chemical composition and the structural properties of real BAps. Cationic (Na and Mg [51]; Sr, Mg, and Zn [52]) and mutual combinations of substitutions in anionic and cationic sites (e.g., Na, Mg, and F [53]; Na, Mg, and SiO_4 [54]; F and CO_3 [55]; Zn and CO_3 [56]; and F and Mg [57]) have been formed. Zinc is naturally present in the bone and stimulates both bone growth and bone mineralization [58], in addition to antimicrobial activity. Suresh Kumar et al. [56] proved the regulating effect of zinc on the apatite-forming ability of HAp when substituted along with carbonate. In addition, the presence of carbonates along with zinc in HAp improved the thermal stability compared to HAp and Zn-HAp.

A study by Gopi et al. [52] revealed that synthesized Sr/Mg/Zn-HAp not only supports the growth of apatite but also accelerates growth onto itself.

A very interesting solution for problems connected with infections after implantation may be the preparation of strontium/cerium-cosubstituted HAp (Sr/Ce-HAp) [59]. To improve the bioactivity of pure HAp, Sr^{2+} ions were substituted and further cosubstituted by various concentrations (0.05, 0.075, and 0.1 M) of Ce^{3+} to enhance the antibacterial property. A similar approach was used by Gopi et al. [60]. They synthesized antibacterial and bioactive silver/magnesium-cosubstituted HAp (Ag/Mg-HAp) with concentrations (0.5, 1.5, and 2.5 wt.%) of Ag. Magnesium compensates the cytotoxicity of Ag, as proved by *in vitro* and antibacterial activity tests.

1.3 Isolation and Preparation of BAp from Various Biogenic Sources

The utilization of waste biological materials has high potential in this respect. Waste biomaterials are accumulating, while supplies of natural raw materials are being used up. Reuse of waste biomaterials would be very advantageous. Recent studies on the production of BAp use a wide range of natural sources from various water and terrestrial environments and also from various biogenic types, for example, vegetal and animal. For simplification, the large number of methods and their modifications can be classified into three groups. This is schematically summarized in Figure 2. However, "real BAps," with a chemical composition and structure similar to that of human bone, can be acquired only by direct isolation from animal sources (Figure 2, route a). Routes b and c in Figure 2 usually lead to the acquisition of HAp, carbonated HAp, and β-TCP. The most updated review paper [2] summarizing recent and very recent work on isolating and preparing BAps from various natural sources, together with a detailed description of individual methods for acquiring BAps, was published in January 2014.

1.3.1 Isolation of BAps from Biowastes

The most widely utilized biowastes for isolating BAp are xenogenous bones, which are morphologically and structurally similar to human bone. The bone has 30-35 wt.% of organic compounds (on a dry basis). The main organic compound present in the natural bone is collagen (95 wt.%), while other organic compounds, such as chondroitin sulfate, keratin sulfate, and lipids, exist in small concentrations. The mineral/collagen ratio is dependent on the animal genus, type of bones, and age of the animal [4]. The isolation of BAp must therefore involve a defatting and deproteinating process.

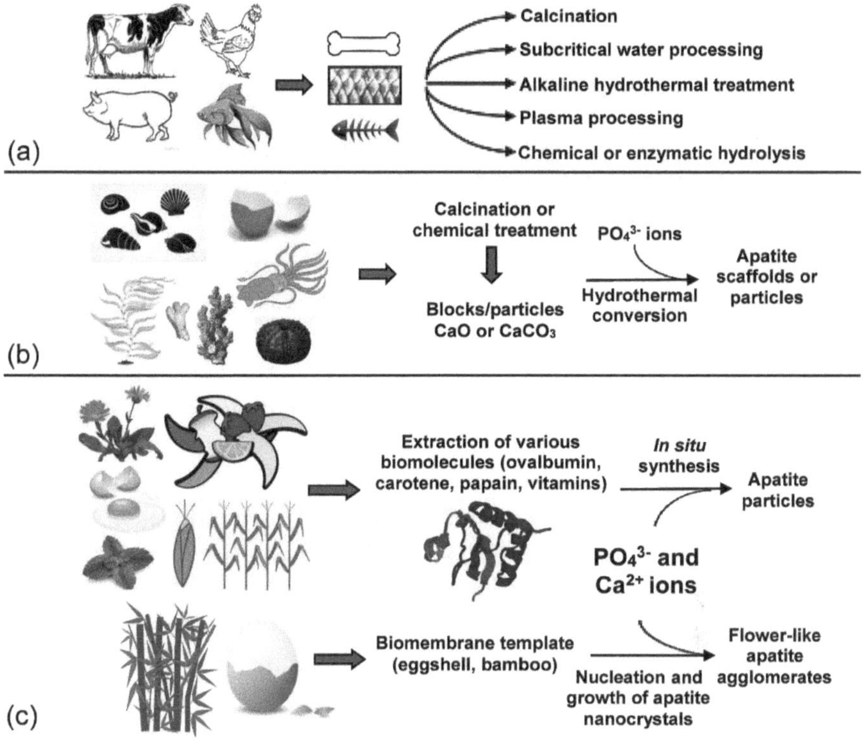

Figure 2 Acquisition of BAp via various natural sources: (a) direct isolation of BAps from biowastes of animal origin; (b) preparation from eggshells and marine creatures; (c) *in situ* synthesis using naturally derived biomolecules and biomembranes.

Calcination is a biologically safe process (no foreign chemicals are utilized), and it enables BAp to be prepared directly from biogenic materials. Calcination involves annealing for several hours (2-24) at various temperatures in the range of 600-1300 °C. An annealing process above 700 °C enhances BAp crystallinity and particle size and also enables secondary phase formation, such as β-TCP. A disadvantage of calcination is the elimination of carbonate groups and consequently the yielding of calcium oxide (CaO) [61], which can affect the biological properties of the BAp. Carbonate loss can be avoided by carrying out thermal treatments in a carbon dioxide atmosphere [62]. CaO can be eliminated by applying a repeated washing procedure [63] or by neutralizing through a reaction with diluted phosphoric acid (H_3PO_4) [64].

Another possible approach for the defatting and deproteinating processes is chemical treatment, for example, a 30% hydrogen peroxide (H_2O_2)

solution [65], warm hydrazine [15,66], or a sodium hypochlorite solution [67]. A very interesting defatting and deproteinating process that eliminates the problem of hazardous alkaline waste management was applied by Sobczak-Kupiec and Wzorek [64], using a solution of lactic acid and soft conditions: temperature of 125-135 °C and pressure of 0.26-0.30 MPa. Efficient degreasing is usually performed using ether, acetone, and/or ethanol mixtures, and deproteination is usually performed using hot alkali for several hours. This process is therefore called "alkaline hydrothermal treatment" and is very often followed by calcination to remove the rest of the organic impurities [61,68]. The most widely used alkali for deproteination is sodium hydroxide (NaOH) [61–63,68–70], in the form of a hot (70-100 °C) solution in various concentrations (1-16 wt.%) for several hours (24-48). However, Barakat et al. [69] used more drastic conditions (25 wt.% NaOH solution, 250 °C, 5 h in an autoclave). There are also other very interesting but less widely applied and less widely published methods for isolating BAp, such as extraction with the use of subcritical water [69], enzymatic hydrolysis at optimal pH and temperature [71], and isolation by a polymer-assisted thermal calcination method in the presence of various polymers [70]. Also worth mentioning is the removal of organic matter through transferred arc argon plasma processing [72]. Important problems related to these extraction processes are the general low purity of the final product and the content of organic impurities, such as peptides and proteins that can finally act as a source of immunologic responses or cytotoxicity. Many researchers have isolated BAp's from bones of various animal genera, and many of them have assumed that the final BAp's can hardly be expected to contain proteins after the use of chemicals and elevated temperatures (>400 °C). Contrary, other researchers have found protein contamination using a range of ways and methods to analyze their residues. The simplest method is visual indication according to the color of the product [64,70,72,73]. With increasing calcination temperature, the color changes from grayish to white. With infrared spectrometry, proteins can be determined through amidic bands (~1545 cm^{-1}) and C—H bands (~2800 and 2900 cm^{-1}) [69,72]. The microstructure of BAp and cross-banded collagen fibrils after calcination can be observed using scanning and transmission electron microscopies [67,70]. Proteins or peptides can be detected on the basis of the carbon content using elemental analysis [73] and on the basis of the nitrogen content using the Kjeldahl method [71]. Positivity or negativity for hydroxyproline, one of the most abundant amino acids in collagen, can be used as evidence for the presence of peptide or proteins [67].

Termine et al. [66] performed colorimetric determination with the use of Folin reagent according to Lowry et al. [74]. The results for the existence of peptide or protein are not unequivocal. The most widely mentioned temperature range for the removal of organic impurities is 500-700 °C. Protein impurity content has been determined up to 0.5 wt.% [66,71]. It is not clear what influence peptide residence can have on cell adhesion, proliferation, and viability. Their final effect can be studied and proved using *in vitro* and *in vivo* tests.

Plants provide new and very interesting sources for BAp isolation by a calcination procedure. Shaltout et al. [75] used five different plants (khat (*Catha edulis*), basil, mint, green tea, and trifolium), and their results showed that ~50 wt.% of plant matter is inorganic phase and the main inorganic constituent of khat and basil leaves is HAp.

1.3.2 Preparation of Apatites by Reacting Extracted Ca Precursor with Phosphate Ions

The approach to preparation is based on extracting Ca precursors (e.g., $CaCO_3$ and CaO), followed by reaction with phosphate ions to form apatites. Eggshell biowastes and mainly the exoskeletons of marine creatures such as corals, shells, cuttlefish, and sea urchins, which typically consist of 95-99 wt.% $CaCO_3$, can be used as Ca sources for this purpose. The first submethod in this approach is the so-called hydrothermal chemical conversion, which was first used for HAp formation directly from corals by Roy and Linnehan [76]. Complete replacement of aragonite by phosphatic material was achieved at 270 °C and 103 MPa, according to the reaction [76–78]:

$$10CaCO_3 + 6(NH_4)_2HPO_4 + 2H_2O \rightarrow$$
$$Ca_{10}(PO_4)_6(OH)_2 + 6(NH_4)_2CO_3 + 4H_2CO_3$$

The orthorhombic aragonite structure, described by a pseudohexagonal cell, closely resembles the hexagonal structure of apatite. In addition, there is a roughly two-dimensional structural relation that can be schematically represented by the following formulas (where □ is a vacancy):

$$\text{Aragonite}\ Ca_8\square_2(CO_3)_6(CO_3)_2$$

$$\text{Apatite}\ Ca_{10}(PO_4)_6(OH)_2$$

This conversion has been called a "topotactic reaction," and it was described by Eysel and Roy [79]. However, in 1998, Zaremba et al. [80] made a correction, and they showed that the HAp growth mechanism holds via a dissolution-recrystallization process. Hydrothermal chemical conversion usually occurs in the high-temperature range, but Ni and Ratner

[81] demonstrated that aragonite plates and fine powders (<100 μm) of nacreous origin are prone to transform into HAp even at room temperature. The hydrothermal process can be improved by using an excess of diammonium phosphate ($(NH_4)_2HPO_4$) at relatively low pressure (3.8 MPa) [82]. The conversion process was accelerated (from 17 days to 10 days) by using the mineralizer KH_2PO_4 [77]. Coral reefs form a delicate and important ecosystem for sustaining marine life, so less aggressive options are preferable: for example, marine algae seem to provide a new and interesting source. Felício-Fernandes et al. [83] demonstrated the possibility of producing nonstoichiometric AB-type carbonated HAp from the $CaCO_3$ of marine algae, with ammonium fluoride as another option. Cuttlefish bones were first used by Rocha et al. [78] for preparing a AB-type carbonated HAp scaffold with the preserved interconnected porous structure, featuring pore size 80-100 μm. Vecchio et al. [84] used sea urchin spines, which consist of large single crystals of Mg-rich calcite ($(Ca,Mg)CO_3$). After hydrothermal treatment, they converted to resorbable Mg-substituted tricalcium phosphate (β-TCMP) with the interconnected porous structure of the original spine. The average compressive strength of the converted spines is ~23 MPa, which is somewhat higher than the compressive strength of cancellous bone (4-12 MPa). Boonyang et al. [85] used a very interesting source of calcium—crocodile eggshells—and three different phosphate precursors for conversion.

The second preparation submethod applies calcination (~900 °C) to decompose the organic matter and convert $CaCO_3$ to CaO, which is subsequently reacted with a suitable phosphate precursor, such as H_3PO_4, $(NH_4)_2HPO_4$, and KH_2PO_4. Not only is this method very often used for processing eggshell waste, but also it has been applied to coral [86] and red algae [87]. Acid treatments (HNO_3 and HCl), to replace calcination, have been also tried, but a yellow discoloration of the final product causes complications. The reaction with a suitable phosphate precursor can proceed under various conditions, at ambient pressure and low temperature (100 °C) [87] or at very high temperature (1050 °C) [88]. The influence of temperature, pressure, time of treatment, and the concentration of $(NH_4)_2HPO_4$ on the development of the HAp structure has been studied in detail by Jokanović et al. [89].

1.3.3 *In Situ* Synthesis Using Naturally Derived Biomolecules or Biomembranes

Synthesis of nanosized particles in the presence of various biomacromolecules is based on a biomimetic approach that mimics biological mechanisms and may replicate the natural processes. Organic templates in the form of macromolecules and biomembranes can serve as crystallizers. Protein molecules have various charged groups, that is, carboxyl, amido, phosphatic, and hydroxyl, which can control apatite growth in the presence of calcium and phosphate ions. The most promising collagen is difficult to define due to its diversity. Bovine serum albumin [90] has amino acid sequencing and molecular weight comparable to that of human serum albumin. However, it is expensive and is difficult to extract, so new alternatives have been introduced. Ovalbumin [91] extracted from egg white is a protein with a very stable structure, biocompatibility, and availability and low cost, and it can be modified to obtain a variety of chemical, physical, and biological properties. Nayar and Guha [92] used biomolecules extracted from wastes such as orange peel and potato peel, eggshell, papaya leaf, and calendula flower, which contain proteins, vitamins, and glycoproteins and can exert significant control over the *in situ* synthesis of nanosized apatite particles.

Eggshell and bamboo membranes offer an environment that is similar to the biological environment, and they may control processes such as ion diffusion and nucleation. Zhang et al. studied the influences of individual properties (temperature, pH, and holding time) on the morphology of flowerlike HAp agglomerates, using eggshell [93] and bamboo [94] membranes, and proposed the crystal growth mechanism. However, bamboo membrane composed of cellulose is different from eggshell membrane, which mainly contains protein. The chemical composition (e.g., the presence of functional groups on the surfaces) and the physical properties (e.g., the shape, size, and distribution of the pores) of the membranes can generally be considered as the main factors controlling the morphologies of the final apatite products.

2. BIO-NANOCOMPOSITES

In the field of tissue engineering, a range of different multicomponent materials are currently being used as scaffolds for reconstructing soft or hard tissue defects. These engineered composite materials have to provide the functionality associated with specific living biological tissues. In other words, they should provide the appropriate biological, structural, chemical, and mechanical cues to promote normal cellular behavior and function and thus provide help for early reparation of the defected tissue. Therefore, these bio-nanocomposites have found widespread applications as biomedical materials [95]. This section will focus on composites based on various naturally derived components and on strategies for preparing them. It will summarize recent work on preparing a class of materials termed bio-nanocomposites, which can be defined as multiphase materials in which at least one of the phases has a biological origin and one of the phases has a dimension of less than 100 nm. Special attention will be paid to describing composites made of derivates of natural polymers with synthetic calcium phosphates and composites containing BAp with a polymer matrix.

2.1 Composites Containing Apatites with Natural Polymer Matrices

Among biomedically relevant polymers, we can distinguish between synthetic polymers, like PLA, PGA and their copolymers, and PCL, and polymers of biological origin, like polysaccharides (starch, alginate, chitin/chitosan, cellulose, and hyaluronic acid derivatives), and proteins (soy, collagen, fibrin, and gelatin). Natural polymers often possess highly organized structures and may contain ligands necessary for the binding of cell receptors. However, they can also contain various impurities. Synthetic polymers can be produced under controlled conditions with predictable and reproducible mechanical and physical properties, together with control over impurities [96].

Collagen is a key protein and is one of the most prevalent proteins in the connective tissue of animals. It constitutes approximately 25% of the total body protein in vertebrates. It is made up of three polypeptide strands (alpha triple left-handed helix). Fibrillar type I collagen is responsible for the tensile strength of the bone. It has excellent biocompatibility, provides easy degradation into physiological end products, and has suitable interaction with cells and other macromolecules. When the triple helix unwinds and the chains are separated, this denatured mass of tangled chains cools down, absorbs all of the surrounding water like a sponge, and forms denatured collagen—gelatin. Chitosan is a linear polysaccharide composed of randomly distributed β-(1-4)-linked D-glucosamine (deacetylated unit) and *N*-acetyl-D-glucosamine, made of deacetylated chitin. When the number of *N*-acetyl-D-glucosamine units in a chitin/chitosan mixture is higher than 50%, the biopolymer is termed chitin. 50% deacetylated chitosan has a lower inflammation reaction than the other forms when they are implanted *in vivo*. Alternative natural polymers may be starch-based polymers, hyaluronan, cellulose, or alginates. Starch is a carbohydrate consisting of a large number of glucose units joined together by glycosidic bonds. It has been demonstrated to be useful for tissue engineering of the bone, because of its interesting mechanical properties. Hyaluronan (or hyaluronic acid) is an important extracellular matrix glycosaminoglycan composed of a linear sequence of D-glucuronic acid and *N*-acetyl-D-glucosamine. Alginate is a salt of alginic acid and an anionic polysaccharide distributed widely in the cell walls of brown algae, where it forms a viscous gum by binding water. Cellulose, $(C_6H_{10}O_5)_n$, is the major component in the rigid cell walls in plants, a linear polysaccharide polymer with many glucose monosaccharide units. As an acetate ester of cellulose, cellulose acetate is a biodegradable polymer. Sodium alginate (composed of mannuronic and guluronic dimmers) has been widely used in cell encapsulation technology. The primary block, guluronic acid, contains available carboxylic acid groups that allow the alginate to be reversibly cross-linked by divalent cations, such as Ca^{2+} and Mg^{2+}, to form a relatively stable hydrogel.

Increasing interest in new materials based on blends of two or more polymers has been observed mainly during the last three decades. Blends of synthetic and natural polymers, the so-called bioartificial or biosynthetic polymeric materials, can form a new class of biomaterials with improved mechanical properties and biocompatibility compared with those of single components [97]. Polymers or polymer blends can be used in various ways, for example, sponges, nanofibers, and films for coatings [98] or surface modifications [99] aiming at activating the physicochemical and surface properties of biomaterials. These properties are absolutely crucial in enhancing the fixation of implants. These bioactive layers can also serve as a local drug delivery system for the treatment of osteomyelitis [100], with the release of antibiotics in concentrations exceeding the minimum inhibitory concentrations for the most common pathogens.

Mimicking the structure of calcified tissues, the relevant calcium orthophosphates with bioresorbable polymer hybrid biomaterials [96] provide excellent opportunities to improve conventional bone implants with suitable physical, biological, and mechanical properties and predictable degradation behavior. Calcium orthophosphates can be incorporated as either a filler or a coating (or both) and either into or onto a biodegradable polymer matrix, in the form of particles or fibers. The general behavior of these bioorganic/calcium orthophosphate biocomposites is dependent on the nature, the structure, and the concentrations of the individual components. However, other parameters, for example, the preparation conditions, can also determine the properties of the final materials.

Various preparation methods for nanoscale apatite/polymer biocomposites have been described and implemented [101], including mechanical blending, compounding, ball milling, dispersion of ceramic fillers into a polymer-solvent solution, melt extrusion of a ceramic/polymer powder mixture, freeze-drying, electrospinning, and *in situ* coprecipitation. Solvent casting of biocomposite scaffolds involves dissolving the polymer in an organic solvent, mixing it with ceramic particles, and casting the solution into a predefined 3-D mold. The solvent is subsequently allowed to evaporate. The main advantage of this fabrication technique is that it is easy to prepare and operate and there is no need for specialized equipment. Thermally induced phase separation, also known as freeze-drying, allows the interconnected porous structure to be prepared with structures of a certain pore size and with less variable porosity. When a homogeneous solution at an elevated temperature is cooled, a homogeneous multicomponent system becomes thermodynamically unstable and tends to separate into a polymer-rich phase and a polymer-lean phase. The frozen mixture is then lyophilized, using a freeze dryer in which the ice solvent evaporates. After the solvent is removed, the polymer-rich phase solidifies and a microporous structure is generated. The final structure depends on the competing phase transitions between the liquid-liquid phase separation and the solidification processes and can be optimized by the process conditions, for example, by the pH and concentration of the solution, the freezing rate, and the partial pressure [102]. The electrospinning process has attracted considerable interest as potential polymer and polymer composite fabrication technique for preparing fiber systems that can mimic the morphology of natural tissues. Natural polymeric nanofibers have been also often modified with bioactive particles to improve their mechanical properties and/or bioactivity [103], for example, collagen or gelatin or chitosan nanofibers with nano-HAp [104], silk/nano-HAp [105], PCL/nano-HAp/collagen [106], and PCL/$CaCO_3$ nanofibers [107], and their synthetic analogs PLA nanofibers with HAp or β-TCP and poly(lactide-*co*-glycolide) with diamond nanoparticles [103]. Another important approach is electrospinning of core-shell fibers, which enables the incorporation of functional biomolecules (anticancer agents, proteins, DNA, and living cells) [108]. It seems that biocomposite porous scaffolds obtained by other fabrication methods (freeze-drying, solvent casting and particulate leaching, and rapid prototyping) are more suitable than the electrospinning process for preparing scaffolds with a specific pore size and distribution for better cell seeding and proliferation and for a wide range of implant shape manipulation. *In situ* mineralization is a biomimetic process in which mineralization occurs in close association with the polymer. It is well known that natural bone consists of nanosized blade-like crystals of BAp grown in intimate contact with collagen fibers. This is the main reason for processing bone graft nanocomposites with the use of strategies found in nature, the so-called biomimetic processes [109]. The term biomimetic process can be defined as a microstructural process that mimics or inspires the biological mechanism, in part or as a whole. This biological process generates highly ordered materials with a hybrid composition, a complex texture, and ultrafine crystallites through hierarchical self-assembly. This has recently formed the new research field of biomimetic mineralization (biomineralization), a complex process that involves controlled nucleation and growth of carbonated HAp (a bone-like mineral) onto polymer scaffolds in simulated body fluids (SBFs). For example, this mechanism has been studied in detail and described by Kokubo et al. [110] using SBF simulating the chemical composition of only the inorganic part of human plasma. Calcium ions are attracted to the negative surface potential. Then, the calcium-rich amorphous calcium phosphate is deposited until the surface potential is reversed and phosphate-rich amorphous calcium phosphate is deposited. This amorphous apatitic layer finally crystallizes into apatite. Carbonated HAp was more similar to bone apatite than to synthetic apatite, as it had low crystallinity and nanoscale size, which are important factors for reabsorption and remodeling of the bone. Presently, "biomimetic" techniques include surface modification of biomaterials with bioadhesive extracellular matrix macromolecules or specific binding motifs. The role of the noncollagenous and nonfibrous proteins that establish cell-adhesive functions and calcium-binding properties and provide the necessary cell microenvironment was described, in detail, in several papers [111,112].

Polymer composite scaffolds can embody limited mechanical and swelling properties. Consequently, they have lower structural stability, which could be an important limiting factor for cell seeding, proliferation, and growth. It is a very important issue to understand the degradation process, specifically faster internal degradation and degradation-induced morphological and compositional changes. Changes in mechanical, chemical, and structural properties during degradation therefore need to be evaluated. These issues are very important from the point of view of the function of a scaffold in an organism. One of the most suitable examples of natural polymer/apatite composites is collagen matrix/HAp composite, which is, in the case of bone tissue engineering, the subject of countless investigations [107]. Further examples include PCL/β-TCP [113], HAp/poly(L-lactide)/collagen [114], or β-TCP/PCL/collagen [115] and similar analogs aiming at mimicking the structure and composition of biological materials. In the case of collagenous composites, it is important to overcome the limitations of non-cross-linked collagen (e.g., low stability, poor mechanical properties, and fast degradation in water environment). Several physical or chemical methods for collagen cross-linking are applied. These approaches can have its own limitations, for example, partial cytotoxicity of glutaraldehyde (GA) or nordihydroguaiaretic acid, at least the results of *in vitro* tests very often antagonistic [116,117]. Composites based on collagen usually contain several heterogeneous components that can potentially react with cross-linking agents. The use of a cross-linking reagent is specific for the existing composite system. Many studies have noted the remarkably low cytotoxicity and genotoxicity of genipin in comparison with GA [118]. Other cross-linking agents with good *in vitro* cytocompatibility include sugars (glucose, fructose, and sucrose) [119] and 1-ethyl-3-(3-dimethylaminopropyl)carbodiimide [120].

2.2 Composites Containing BAp with Polymer Matrices

BAps derived from natural sources are used in an attempt to overcome the limitations of synthetic apatites. Various BAps are used in medical clinical practice, predominantly for dental and orthopedic applications. A list of some commercially used BAps is presented in Table 2, together with a description [121,122].

As is shown in Table 2, commercial BAps are used mainly singly without a polymer matrix, except in some rare cases. Commercially used BAp Pep-Gen P-15® (Table 2) is, in reality, a composite containing calcined bovine bone coated with a small amount of pentadecapeptide (P-15, a part of the sequence of collagen) [121]. Other commercially used BAps Tutoplast® (Table 2) and Kiel bone are natural composites, because these are materials that still contain all or most of the organic bone matrix (20-50 wt.%). Kiel bone is a bovine bone graft material that has been treated chemically with chloroform/methanol and H_2O_2 to remove all organic components except the collagen. It is not in clinical use, probably due to unsatisfactory results of clinical tests in the 1960s [123]. In the case of noncommercial BAps, the *in vitro* [71,78] and *in vivo* [84,124] biological evaluations have also been carried out. However, they have been applied singly, not in polymers, in the form of blocks or particles. Only a few works have studied BAps from biogenic sources with polymer matrices as composites. Murugan and Ramakrishna [125] dealt with the coupling of tetracycline onto the surface of a nonmodified polymer and a polymer grafted with polyglycidylmethacrylate (PGMA) coralline BAp (CBA) obtained from *Goniopora* [126] as a drug delivery agent. The percentage loading of tetracycline was found to

Table 2 A list of some commercially used BAps, with a description

Name of BAp	Source	Description: Preparation	Description: Form
Algipore®	Algae	Hydrothermal conversion at 700 °C	Granulate with particle sizes of 0.3-2 mm and pores in the range of 5-10 μm
Bio-Oss®	Bovine	Annealing process (up to 300 °C), followed by a chemical treatment	Particle size of the granulate is 0.25-2 mm
Endobon® Cerabone®	Bovine	High-temperature sintered (>1200 °C) As a granulate	Containing HAp, highly porous blocks in the centimeter range (pore size typically 1 mm; porosity typically 50%)
PepGen P-15®	Bovine	Calcined at 1100 °C	Containing HAp coated with a pentadecapeptide, available as granulate with a particle size of 0.25-0.42 μm
Pro Osteon® 200R and 500R (formerly Interpore 200/500)	Coral	Prepared by a hydrothermal exchange	Consisting of a thin 2-10 μm layer of HAp over a layer of $CaCO_3$. Types 200R and 500R exhibit porosity 180-220 μm and 280-770 μm, respectively; the particle size of the granulate is 0.5-1 mm (200R) and 1-9 mm (500R)
Tutoplast®	Bovine	Obtained after a multistep chemical treatment	Granulate (particle size 0.25-2 mm) and as a centimeter-size block or cylinder with pore sizes of some hundred micrometers

be 41±1.46% for CBA and 74±2.24% for CBA-g-PGMA. They also proved that the CBA-g-PGMA allows more prolonged release of tetracycline than ungrafted CBA, due to coupling between the amino group of the drug and the epoxy group of the polymer chains. Sobczak-Kupiec et al. [127] used BAp isolated from pork bone. Gelatin and poly(acrylic acid) were used as the polymer matrix. Composites were obtained with the use of microwave irradiation, and they were doped with silver nanoparticles, known for a broad spectrum of antimicrobial activities.

Our research group has prepared a bioinspired bone implant (Figure 3a, unpublished results) with the desired porous structure (Figure 3b), based on a combination of natural and synthetic components. The bone implants were based on an integrated porous composite structure consisting of 38 wt.% of natural collagen as matrix (Figure 3c), 59.5 wt.% of poly(D,L-lactide) electrospun nanofibers (Figure 3d), and 5.2 wt.% of BAp nanoparticles (Figure 3e) supplemented by 0.5 wt.% of sodium hyaluronate. The nanosized mineral phase was directly isolated from bovine bone by an alkaline hydrothermal process inspired by Murugan et al. [68]. Fish collagen (isolated from the skin of *Cyprinus carpio*) was selected for the final biocomposite in order to suppress the risk of toxicity, that is, the presence of bovine prions or porcine viruses. In summary, two types of composites based on an integrated porous composite structure with two degradation periods were

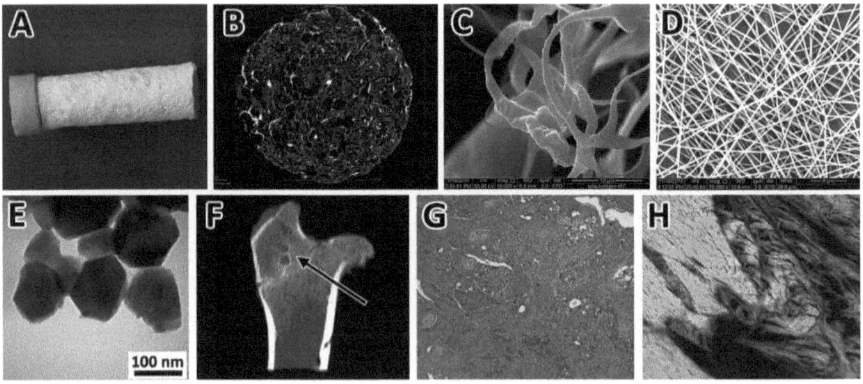

Figure 3 Prototypic biodegradable biocomposites. (A) A composite implant with a PEEK marker, (B) a micro-CT scan of the inner structure of the composite, (C) natural collagen isolated from fish skin, (D) PDLLA electrospun nanofibers, (E) BAp isolated from bovine bone, (F) a CT scan of bone defects with implanted composites at 26 weeks postsurgery. An example of hematoxylin/eosin staining of a bone defect with an implanted composite (miniature pigs, femur); (G) 52 weeks postsurgery (mag. 40×); (H) bone ingrowth and collagen formation in composite fragments (fluorescence, mag. 200×).

prepared and partially tested *in vivo*, using miniature pigs (Figure 3f). The use of pigs as an animal model is more suitable, since pig bones are closely related to human bones (in terms of bone mineral density and bone mineral concentration). On the basis of this preliminary histological evaluation, it can be stated that the composites were biocompatible and showed a sufficient rate of osseointegration (Figure 3g and h).

3. SUMMARY AND PERSPECTIVES

No single-phase biomaterial is able to provide all the essential features of bones and/or other calcified tissues. There is therefore a great need to engineer multiphase biomaterials (biocomposites) with structural, chemical, and mechanical properties and a biological performance to be able to mimic the native tissue. The proper combination of a ductile polymer matrix with a brittle, hard, and bioactive calcium orthophosphate filler offers many advantages for biomedical applications. With increasing demand for artificial replacements, there is a need for constant innovation in the field of biomaterials. The discovery that BAp ceramic and natural polymers generated partially or entirely from biogenic sources are better accepted by living organs began to make these materials more attractive. BAp ceramic generated from natural sources is available worldwide and is therefore inexpensive. Seafood wastes, slaughterhouse wastes, and eggshells have recently become a serious problem near big cities, and there is a high potential for utilizing these biowastes. New alternative natural sources, mainly of vegetal origin, provide interesting challenges for future research. As natural polymers are very expensive and sometimes exhibit low physical properties, for example, low thermal stability, there is a need to combine them with synthetic polymers to develop new materials. These newly developed materials can combine the advantages of individual polymer types. They should be biocompatible while, at the same time, possessing suitable thermal stability and mechanical properties for use in biomedical applications. In addition, polymeric blends can be strengthened by minerals to produce the appropriate size, shape, and distribution of individual crystals, similar to hard tissue. Much research work has been carried out on preparing biocomposites and the individual components of biocomposites, and many studies have been made of their physicochemical and biological properties. However, research results are in many cases very far from commercial implementation, and much work remains to be done on the long path from laboratory studies to clinical practice. This requires effective cooperation among chemists, materials scientists, biologists, bioengineers, and clinicians.

ACKNOWLEDGMENT

We are grateful for the financial support given to our work by the long-term conceptual development research organization under project no. RVO: 67985891.

REFERENCES

[1] M. Sadat-Shojai, M.T. Khorasani, E. Dinpanah-Khoshdargi, et al., Synthesis methods for nanosized hydroxyapatite in diverse structures, Acta Biomater. 9 (2013) 7591–7621.
[2] M. Šupová, Isolation and preparation of nanoscale bioapatites from natural sources: a review, J. Nanosci. Nanotechnol. 14 (1) (2014) 546–563.
[3] H.C.W. Skinner, Biominerals, Mineral. Mag. 69 (2005) 621–641.
[4] J.D. Pasteris, B. Wopenka, E. Valsami-Jones, Tooth mineralization: why apatite? Elements 4 (2008) 97–104.
[5] S.V. Dorozhkin, M. Epple, Biological and medical significance of calcium phosphates, Angew. Chem. Int. Ed. 41 (17) (2002) 3130–3146.
[6] S. Cazalbou, C. Combes, D. Eichert, et al., Adaptive physico-chemistry of bio-related calcium phosphates, J. Mater. Chem. 14 (2004) 2148–2153.
[7] A. Veis, A window on biomineralization, Science 307 (2005) 1419–1420.
[8] R.G. Handschin, W.B. Stern, X-ray diffraction studies on the lattice perfection of human bone apatite (Crista Iliaca), Bone 16 (4) (1995) 355S–363S.
[9] C. Rey, J.L. Miquel, L. Facchini, et al., Hydroxyl groups in bone mineral, Bone 16 (1995) 583–586.
[10] J.D. Pasteris, B. Wopenka, J.J. Freeman, et al., Lack of OH in nanocrystalline apatite as a function of degree of atomic order: implications for bone and biomaterials, Biomaterials 25 (2004) 229–238.
[11] R. Zapanta-LeGeros, Effect of carbonate on the lattice parameters of apatite, Nature 206 (1965) 403–404.
[12] K.D. Rogers, P. Daniels, An X-ray diffraction study of the effects of heat treatment on bone mineral microstructure, Biomaterials 23 (2002) 2577–2585.
[13] J. Barralet, S. Best, W. Bonfield, Carbonate substitution in precipitated hydroxyapatite: an investigation into the effects of reaction temperature and bicarbonate ion concentration, J. Biomed. Mater. Res. 41 (1998) 79–86.
[14] H. El Feki, J.M. Savariault, A. Ben Salah, Structure refinements by the Rietveld method of partially substituted hydroxyapatite: $Ca_9Na_{0.5}(PO_4)_{4.5}(CO_3)_{1.5}(OH)_2$, J. Alloys Compds. 287 (1999) 114–120.
[15] A. Bigi, B. Foresti, R. Gregoriani, et al., The role of magnesium on the structure of biological apatites, Calcif. Tissue Int. 50 (1992) 439–444.
[16] A. Bigi, G. Falini, E. Foresti, et al., Rietveld structure refinements of calcium hydroxylapatite containing magnesium, Acta Crystallogr. B 52 (1996) 87–92.
[17] L. Mayer, R. Schlam, J.D.B. Featberstone, Magnesium-containing carbonate apatites, J. Inorg. Biochem. 66 (1997) 1–6.
[18] E. Bertoni, A. Bigi, G. Cojazzi, et al., Nanocrystals of magnesium and fluoride substituted hydroxyapatite, J. Inorg. Biochem. 72 (1998) 29–35.
[19] D. Laurencin, N. Almora-Barrios, N. de Leeuw, et al., Magnesium incorporation into hydroxyapatite, Biomaterials 32 (2011) 1826–1837.
[20] F. Ren, Y. Leng, R. Xin, et al., Synthesis, characterization and ab initio simulation of magnesium-substituted hydroxyapatite, Acta Biomater. 6 (2010) 2787–2796.
[21] C. Rey, M. Schimizu, B. Collins, M.J. Glimcher, Resolution-enhanced FTIR study of the environment of phosphate ions in the early deposits of a solid phase of calcium

phosphate in bone and enamel and their evolution with age. I. Investigations in the v_4 PO_4 domain, Calcified Tissue Int. 46 (1990) 384–394.
[22] S.V. Dorozhkin, Nanodimensional and nanocrystalline apatites and other calcium orthophosphates in biomedical engineering, biology and medicine, Materials 2 (2009) 1975–2045.
[23] M.H. Fathi, A. Hanifi, V. Mortazavi, Preparation and bioactivity evaluation of bone-like hydroxyapatite nanopowder, J. Mater. Process. Technol. 202 (2008) 536–542.
[24] W.Y. Zhou, M. Wang, W.L. Cheung, et al., Synthesis of carbonated hydroxyapatite nanospheres through nanoemulsion, J. Mater. Sci. Mater. Med. 19 (2008) 103–110.
[25] S. Sasikumar, R. Vijayaraghavan, Solution combustion synthesis of bioceramic calcium phosphates by single and mixed fuels—a comparative study, Ceram. Int. 34 (2008) 1373–1379.
[26] C.J. Tredwin, G. Georgiou, H.W. Kim, et al., Hydroxyapatite, fluor-hydroxyapatite and fluorapatite produced via the sol-gel method: bonding to titanium and scanning electron microscopy, Dent. Mater. 29 (2013) 521–529.
[27] H. Eslami, M. Solati-Hashjin, M. Tahriri, The comparison of powder characteristics and physicochemical, mechanical and biological properties between nanostructure ceramics of hydroxyapatite and fluoridated hydroxyapatite, Mater. Sci. Eng. C 29 (2009) 1387–1398.
[28] Y. Chen, X. Miao, Thermal and chemical stability of fluorohydroxyapatite ceramics with different fluorine contents, Biomaterials 26 (2005) 1205–1210.
[29] E.Z. Kurmaev, S. Matsuya, S. Shin, et al., Observation of fluorapatite formation under hydrolysis of tetracalcium phosphate in the presence of KF by means of soft X-ray emission and adsorption spectroscopy, J. Mater. Sci. Mater. Med. 13 (2002) 33–37.
[30] L.M. Rodriguez-Lorenzo, J.N. Hart, K.A. Gross, Influence of fluorine in the synthesis of apatites. Synthesis of solid solutions of hydroxy-fluorapatite, Biomaterials 24 (2003) 3777–3785.
[31] H.G. Zhang, Q. Zhu, Surfactant-assisted preparation of fluoride-substituted hydroxyapatite nanorods, Mater. Lett. 59 (2005) 3054–3058.
[32] H.G. Zhang, Q. Zhu, Z.H. Xie, Mechanochemical-hydrothermal synthesis and characterization of fluoridated hydroxyapatite, Mater. Res. Bull. 40 (2005) 1326–1334.
[33] S. Kannan, A. Rebelo, A.F. Lemos, et al., Synthesis and mechanical behaviour of chlorapatite and chlorapatite/TCP composites, J. Eur. Ceram. Soc. 27 (2007) 2287–2294.
[34] A.Z. Alshemary, Y. Goh, M. Akram, et al., Microwave assisted synthesis of nano sized sulphate doped hydroxyapatite, Mater. Res. Bull. 48 (2013) 2106–2110.
[35] B.D. Hahn, J.M. Lee, D.S. Park, et al., Aerosol deposition of silicon-substituted hydroxyapatite coatings for biomedical applications, Thin Solid Films 518 (2010) 2194–2199.
[36] L.T. Bang, K. Ishikawa, R. Othman, Effect of silicon and heat-treatment temperature on the morphology and mechanical properties of silicon-substituted hydroxyapatite, Ceram. Int. 37 (2011) 3637–3642.
[37] A. Balamurugan, A.H.S. Rebelo, A.F. Lemos, et al., Suitability evaluation of sol-gel derived Si-substituted hydroxyapatite for dental and maxillofacial applications through *in vitro* osteoblasts response, Dent. Mater. 24 (2008) 1374–1380.
[38] A. Aminian, M. Solati-Hashjin, A. Samadikuchaksaraei, et al., Synthesis of silicon-substituted hydroxyapatite by a hydrothermal method with two different phosphorous sources, Ceram. Int. 37 (2011) 1219–1229.
[39] N. Zhang, W. Liu, H. Zhu, et al., Tailoring Si-substitution level of Si-hydroxyapatite nanowires via regulating Si-content of calcium silicates as hydrothermal precursors, Ceram. Int. 40 (2014) 11239–11243.
[40] J. Ma, Y. Wang, L. Zhou, et al., Preparation and characterization of selenite substituted hydroxyapatite, Mater. Sci. Eng. C 33 (2013) 440–445.

[41] E. Landi, G. Logroscino, L. Proietti, et al., Biomimetic Mg-substituted hydroxyapatite: from synthesis to in vivo behaviour, J. Mater. Sci. Mater. Med. 19 (2008) 239–247.
[42] L. Stipniece, K. Salma-Ancane, N. Borodajenko, et al., Characterization of Mg-substituted hydroxyapatite synthesized by wet chemical method, Ceram. Int. 40 (2014) 3261–3267.
[43] W.L. Suchanek, K. Byrappa, P. Shuk, et al., Preparation of magnesium-substituted hydroxyapatite powders by the mechanochemical-hydrothermal method, Biomaterials 25 (2004) 4647–4657.
[44] S. Kannan, J.M.G. Ventura, J.M.F. Ferreira, Synthesis and thermal stability of potassium substituted hydroxyapatites and hydroxyapatite/β-tricalciumphosphate mixtures, Ceram. Int. 33 (2007) 1489–1494.
[45] E.G. Nordsröm, K.H. Karlsson, Chemical characterization of a potassium hydroxyapatite prepared by soaking in potassium chloride and carbonate solutions, Bio-Med. Mater. Eng. 2 (1992) 185–189.
[46] C.J. Chung, H.Y. Long, Systematic strontium substitution in hydroxyapatite coatings on titanium via micro-arc treatment and their osteoblast/osteoclast responses, Acta Biomater. 7 (2011) 4081–4087.
[47] C.M. Mardziah, I. Sopyan, S. Ramesh, Strontium-doped hydroxyapatite nanopowder via sol-gel method: effect of strontium concentration and calcination temperature on phase behavior, Trends Biomater. Artif. Organs 23 (2009) 105–113.
[48] M.D. O'Donnell, Y. Fredholm, A. de Rouffignac, et al., Structural analysis of a series of strontium-substituted apatites, Acta Biomater. 4 (2008) 1455–1464.
[49] K. Lin, P. Liu, L. Wei, et al., Strontium substituted hydroxyapatite porous microspheres: surfactant-free hydrothermal synthesis, enhanced biological response and sustained drug release, Chem. Eng. J. 222 (2013) 49–59.
[50] I. Pereiro, C. Rodriguez-Valencia, C. Serra, et al., Pulsed laser deposition of strontium-substituted hydroxyapatite coatings, Appl. Surf. Sci. 258 (2012) 9192–9197.
[51] S. Kannan, J.H.G. Rocha, J.M.F. Ferreira, Synthesis and thermal stability of sodium, magnesium co-substituted hydroxyapatites, J. Mater. Chem. 16 (2006) 286–291.
[52] D. Gopi, S. Nithiya, E. Shinyjoy, et al., Spectroscopic investigation on formation and growth of mineralized nanohydroxyapatite for bone tissue engineering applications, Spectrochim. Acta A 92 (2012) 194–200.
[53] S. Kannan, J.M.F. Ferreira, Synthesis and thermal stability of hydroxyapatite-β-tricalcium phosphate composites with cosubstituted sodium, magnesium, and fluorine, Chem. Mater. 18 (1) (2006) 198–203.
[54] S. Sprio, A. Tampieri, E. Landi, et al., Physico-chemical properties and solubility behaviour of multi-substituted hydroxyapatite powders containing silicon, Mater. Sci. Eng. C 28 (2008) 179–187.
[55] F. Yao, R.Z. LeGeros, Carbonate and fluoride incorporation in synthetic apatites: comparative effect on physico-chemical properties and in vitro bioactivity in fetal bovine serum, Mater. Sci. Eng. C 30 (2010) 423–430.
[56] G. Suresh Kumar, A. Thamizhavel, Y. Yokogawa, et al., Synthesis, characterization and in vitro studies of zinc and carbonate co-substituted nano-hydroxyapatite for biomedical applications, Mater. Chem. Phys. 134 (2012) 1127–1135.
[57] M. Kheradmandfard, M.H. Fathi, Preparation and characterization of Mg-doped fluorapatite nanopowders by sol-gel method, J. Alloy Compd. 504 (2010) 141–145.
[58] Y. Sogo, T. Sakurai, K. Onuma, et al., The most appropriate (Ca+Zn)/P molar ratio to minimize the zinc content of ZnTCP/HAP ceramic used in the promotion of bone formation, J. Biomed. Mater. Res. 62 (2002) 457–463.
[59] D. Gopi, S. Ramya, D. Rajeswari, et al., Strontium, cerium co-substituted hydroxyapatite nanoparticles: synthesis, characterization, antibacterial activity towards

prokaryotic strains and in vitro studies, Colloids Surf. A: Physicochem. Eng. Asp. 451 (2014) 172–180.
[60] D. Gopi, E. Shinyjoy, L. Kavitha, Synthesis and spectral characterization of silver/magnesium co-substituted hydroxyapatite for biomedical applications, Spectrochim. Acta A 127 (2014) 286–291.
[61] K. Haberko, M.M. Bucko, J. Brzezinska-Miecznik, et al., Natural hydroxyapatite—its behaviour during heat treatment, J. Eur. Ceram. Soc. 26 (2006) 42.
[62] K. Haberko, M.M. Bućko, W. Mozgawa, et al., Behaviour of bone origin hydroxyapatite at elevated temperatures and in O_2 and CO_2 atmospheres, Ceram. Int. 35 (2009) 2537–2540.
[63] M. Lombardi, P. Palmero, K. Haberko, et al., Processing of a natural hydroxyapatite powder: from powder optimization to porous bodies development, J. Eur. Ceram. Soc. 31 (2011) 2513–2518.
[64] A. Sobczak-Kupiec, Z. Wzorek, The influence of calcination parameters on free calcium oxide content in natural hydroxyapatite, Ceram. Int. 38 (2012) 641–647.
[65] G. Penel, G. Leroy, C. Rey, et al., MicroRaman spectral study of the PO_4 and CO_3 vibrational modes in synthetic and biological apatites, Calcif. Tissue Int. 63 (1998) 475–481.
[66] J.D. Termine, E.D. Eanes, D.J. Greenfield, et al., Hydrazine-deproteinated bone mineral, Calcified Tissue Int. 12 (1973) 73–90.
[67] M. Raspanti, S. Guizzardi, V. De Pasquale, et al., Ultrastructure compact bone of heat-deproteinated, Biomaterials 15 (1994) 433–437.
[68] R. Murugan, S. Ramakrishna, Rao K. Panduranga, Nanoporous hydroxy-carbonate apatite scaffold made of natural bone, Mater. Lett. 60 (2006) 2844–2847.
[69] N.A.M. Barakat, M.S. Khil, A.M. Omran, et al., Extraction of pure natural hydroxyapatite from the bovine bones bio waste by three different methods, J. Mater. Process. Technol. 209 (2009) 3408–3415.
[70] R. Pallela, J. Venkatesan, S.K. Kim, Polymer assisted isolation of hydroxyapatite from Thunnus obesus bone, Ceram. Int. 37 (2011) 3489–3497.
[71] Y.C. Huang, P.C. Hsiao, H.J. Chai, Hydroxyapatite extracted from fish scale: effects on MG63 osteoblast-like cells, Ceram. Int. 37 (2011) 1825–1831.
[72] C.P. Yoganand, V. Selvarajan, O.M. Goudouri, et al., Preparation of bovine hydroxyapatite by transferred arc plasma, Curr. Appl. Phys. 11 (2011) 702–709.
[73] M. Ozawa, S. Suzuki, Microstructural development of natural hydroxyapatite originated from fish-bone waste through heat treatment, J. Am. Ceram. Soc. 85 (2002) 1315–1317.
[74] O.H. Lowry, N.J. Rosebrough, A.L. Farr, et al., Protein measurement with the Folin phenol reagent, J. Biol. Chem. 193 (1951) 265–275.
[75] A.A. Shaltout, M.A. Allam, M.A. Moharram, FTIR spectroscopic, thermal and XRD characterization of hydroxyapatite from new natural sources, Spectrochim. Acta A 83 (2011) 56–60.
[76] D.M. Roy, S.K. Linnehan, Hydroxyapatite formed from coral skeletal carbonate by hydrothermal exchange, Nature 247 (1974) 220–222.
[77] Y. Xu, D. Wang, L. Yang, et al., Hydrothermal conversion of coral into hydroxyapatite, Mater. Charact. 47 (2001) 83–87.
[78] J.H.G. Rocha, A.F. Lemos, S. Agathopoulos, et al., Hydrothermal growth of hydroxyapatite scaffolds from aragonitic cuttlefish bones, J. Biomed. Mater. Res. A 77 (2006) 160–168.
[79] W. Eysel, D.M. Roy, Topotactic reaction of aragonite to hydroxyapatite, Z. Kristallogr.: Cryst. Mater. 141 (1975) 11–24.
[80] C.M. Zaremba, D.E. Morse, S. Mann, et al., Aragonite-hydroxyapatite conversion in gastropod (Abalone) nacre, Chem. Mater. 10 (1998) 3813–3824.

[81] M. Ni, B.D. Ratner, Nacre surface transformation to hydroxyapatite in a phosphate buffer solution, Biomaterials 24 (2003) 4323–4331.
[82] J. Hu, J.J. Russell, B. Ben-Nissan, Production and analysis of hydroxyapatite from Australian corals via hydrothermal process, J. Mater. Sci. Lett. 20 (2001) 85–87.
[83] G. Felício-Fernandes, M.C.M. Laranjeira, Calcium phosphate biomaterials from marine algae, Quim. Nova 23 (2000) 441–446.
[84] K.S. Vecchio, X. Zhang, J.B. Massie, et al., Conversion of sea urchin spines to Mg-substituted tricalcium phosphate for bone implants, Acta Biomater. 3 (2007) 785–793.
[85] U. Boonyang, P. Chaopanich, A. Wongchaisuwat, et al., Effect of phosphate precursor on the production of hydroxyapatite from crocodile eggshells, J. Biomim. Biomater. Tissue Eng. 5 (2010) 31–37.
[86] M. Sivakumar, T.S. Sampath Kumar, K.L. Shantha, et al., Development of hydroxyapatite derived from Indian coral, Biomaterials 17 (1996) 1709–1714.
[87] P.J. Walsh, F.J. Buchanan, M. Dring, et al., Low-pressure synthesis and characterisation of hydroxyapatite derived from mineralise red algae, Chem. Eng. J. 137 (2008) 173–179.
[88] E.M. Rivera, M. Araiza, W. Brostow, et al., Synthesis of hydroxyapatite from eggshells, Mater. Lett. 41 (1999) 128–134.
[89] V. Jokanović, D. Izvonar, M.D. Dramićanin, et al., Hydrothermal synthesis and nanostructure of carbonated calcium hydroxyapatite, J. Mater. Sci.: Mater. Med. 17 (2006) 539–546.
[90] Y. Han, S. Li, X. Wang, A novel thermolysis method of colloidal protein precursors to prepare hydroxyapatite nanocrystals, Cryst. Res. Technol. 44 (2009) 336–340.
[91] H. Zhao, W. He, Y. Wang, et al., Biomineralization of large hydroxyapatite particles using ovalbumin as biosurfactant, Mater. Lett. 62 (2008) 3603–3605.
[92] S. Nayar, A. Guha, Waste utilization for the controlled synthesis of nanosized hydroxyapatite, Mater. Sci. Eng. C 29 (2009) 1326–1329.
[93] Y. Zhang, Y. Liu, X. Ji, et al., Flower-like agglomerates of hydroxyapatite crystals formed on egg-shell membrane, Colloid Surf. B: Biointerfaces 82 (2011) 490–496.
[94] Y. Zhang, Y. Liu, X. Ji, Fabrication of flower-like hydroxyapatite agglomerates with the assistant of bamboo membrane, Mater. Lett. 65 (2011) 1982–1985.
[95] C. Aim, T. Coradin, Nanocomposites from biopolymer hydrogels: blueprints for white biotechnology and green materials chemistry, J. Polym. Sci. Pol. Phys. 50 (2012) 669–680.
[96] S.V. Dorozhkin, Biocomposites and hybrid biomaterials based on calcium orthophosphates, Biomatter 1 (2011) 3–56.
[97] A. Sionkowska, Current research on the blends of natural and synthetic polymers as new biomaterials: Review, Prog. Polym. Sci. 36 (2011) 1254–1276.
[98] T. Fuchs, G. Schmidmaier, M.J. Raschke, et al., Bioactive-coated implants in trauma surgery, Eur. J. Trauma Emerg. Surg. 34 (2008) 60–68.
[99] L. Moroni, M.K. Gunnewiek, E.M. Benetti, Polymer brush coatings regulating cell behavior: passive interfaces turn into active, Acta Biomater. 10 (2014) 2367–2378.
[100] S.K. Nandi, P. Mukherjee, S. Roy, et al., Local antibiotic delivery systems for the treatment of osteomyelitis—a review, Mater. Sci. Eng. C 29 (2009) 2478–2485.
[101] F. Sun, H. Zhou, J. Lee, Various preparation methods of highly porous hydroxyapatite/polymer nanoscale biocomposites for bone regeneration, Acta Biomater. 7 (2011) 3812–3828.
[102] Q.L. Loh, C. Choong, Three-dimensional scaffolds for tissue engineering applications: role of porosity and pore size, Tissue Eng. B 19 (6) (2013) 485–502.
[103] K. Novotna, M. Zajdova, T. Suchy, et al., Polylactide nanofibers with hydroxyapatite as growth substrates for osteoblast-like cells, J. Biomed. Mater. Res. Part A 102 (11) (2014) 3918–3930.

[104] S.H. Teng, E.J. Lee, P. Wang, H.E. Kim, Collagen/hydroxyapatite composite nanofibers by electrospinning, Matter Lett. 62 (17-18) (2008) 3055–3058; C. Li, C. Vepari, H.J. Jin, et al., Electrospun silk-BMP-2 scaffolds for bone tissue engineering, Biomaterials 27 (16) (2006) 3115–3124.
[105] J. Venugopal, P. Vadgama, T.S. Kumar, et al., Biocomposite nanofibres and osteoblasts for bone tissue engineering, Nanotechnology 18 (2007) 1–8.
[106] K. Fujihara, M. Kotaki, S. Ramakrishna, Guided bone regeneration membrane made of polycaprolactone/calcium carbonate composite nano-fibers, Biomaterials 26 (19) (2005) 4139–4147.
[107] A.F. Goh, I. Shakir, R. Hussain, Electrospun fibers for tissue engineering, drug delivery, and wound dressing, J. Mater. Sci. 48 (2013) 3027–3054.
[108] A. Rogina, Electrospinning process: versatile preparation method for biodegradable and natural polymers and biocomposite systems applied in tissue engineering and drug delivery, Appl. Surf. Sci. 296 (2014) 221–230.
[109] Z. Xia, M. Wei, Biomimetic fabrication of collagen-apatite scaffolds for bone tissue regeneration, J. Biomater. Tissue Eng. 3 (2013) 369–384.
[110] Y. Abe, T. Kokubo, T. Yamamuro, Apatite coating on ceramics, metals and polymers utilizing a biological process, J. Mater. Sci. Mater. Med. 1 (1990) 233–238.
[111] M.B. Rahmany, M. Van Dyke, Biomimetic approaches to modulate cellular adhesion in biomaterials: a review, Acta Biomater. 9 (2013) 5431–5437.
[112] R. Tejero, E. Anitua, G. Orive, Toward the biomimetic implant surface: biopolymers on titanium-based implants for bone regeneration, Prog. Pol. Sci. 39 (2014) 1406–1447.
[113] A. Bianco, E. Di Federico, I. Cacciotti, Electrospun poly(ε-caprolactone)-based composites using synthesized β-tricalcium phosphate, Polym. Adv. Technol. 22 (12) (2011) 1832–1841.
[114] M.P. Prabhakaran, J. Venugopal, S. Ramakrishna, Electrospun nanostructured scaffolds for bone tissue engineering, Acta Biomater. 5 (8) (2009) 2884–2893.
[115] M. Yeo, H. Lee, G. Kim, Three-dimensional hierarchical composite scaffolds consisting of polycaprolactone, β-tricalcium phosphate, and collagen nanofibers: fabrication, physical properties, and in vitro cell activity for bone tissue regeneration, Biomacromolecules 12 (2) (2011) 502–510.
[116] T.J. Koob, T.A. Willis, D.J. Hernandez, Biocompatibility of NDGA-polymerized collagen fibers. I. Evaluation of cytotoxicity with tendon fibroblasts in vitro, J. Biomed. Mater. Res. 56 (2001) 31–39.
[117] S.C. Sahu, D.I. Ruggles, M.W. O'Donnell, Prooxidant activity and toxicity of nordihydroguaiaretic acid in clone-9 rat hepatocyte cultures, Food Chem. Toxicol. 44 (2006) 1751–1757.
[118] J.S. Yoo, Y.J. Kim, S.H. Kim, et al., Study on genipin: a new alternative natural crosslinking agent for fixing heterograft tissue, Korean J. Thorac. Cardiovasc. Surg. 44 (2011) 197–207.
[119] K. Siimon, P. Reemann, A. Põder, et al., Effect of glucose content on thermally crosslinked fibrous gelatin scaffolds for tissue engineering, Mater. Sci. Eng. C 42 (2014) 538–545.
[120] J.K. McDade, E.P. Brennan-Pierce, M.B. Ariganello, et al., Interactions of U937 macrophage-like cells with decellularized pericardial matrix materials: influence of crosslinking treatment, Acta Biomater. 9 (2013) 7191–7199.
[121] D. Tadic, M. Epple, A thorough physicochemical characterisation of 14 calcium phosphate-based bone substitution materials in comparison to natural bone, Biomaterials 25 (2004) 987–994.
[122] S.S. Jensen, M. Aaboe, E.M. Pinholt, et al., Tissue reaction and material characteristics of four bone substitutes, Int. J. Oral Maxillofac. Implants 11 (1996) 55–66.

[123] H.C. Killey, I.R.H. Kramer, H.C. Wright, The response of the rabbit to implants of processed bovine bone (Kiel bone) and the effects of varying the relationship between implant and host bone, Arch. Oral Biol. 15 (1970) 33–42.
[124] K.S. Vecchio, X. Zhang, J.B. Massie, et al., Conversion of bulk seashells to biocompatible hydroxyapatite for bone implants, Acta Biomater. 3 (2007) 910–918.
[125] R. Murugan, S. Ramakrishna, Coupling of therapeutic molecules onto surface modified coralline hydroxyapatite, Biomaterials 25 (2004) 3073–3080.
[126] R. Murugan, K. Panduranga Rao, T.S. Sampath Kumar, Microwave synthesis of bioresorbable carbonated HAP using Goniopora, Key Eng. Mater. 240–242 (2003) 51–54.
[127] A. Sobczak-Kupiec, D. Malina, M. Piatkowski, et al., Physicochemical and biological properties of hydrogel/gelatin/hydroxyapatite PAA/G/HAp/AgNPs composites modified with silver nanoparticles, J. Nanosci. Nanotechnol. 12 (12) (2012) 9302–9311.

CHAPTER 3

Organic/Inorganic Hybrid Nanoceramics Based on Sol-Gel Chemistry

Martha Geffers, Uwe Gbureck, Claus Moseke, Elke Vorndran
Department for Functional Materials in Medicine and Dentistry, University of Würzburg, Würzburg, Germany

Contents

1. INTRODUCTION

The sol-gel synthesis of advanced organic/inorganic composite materials is based on the controlled hydrolysis and polycondensation of metal alkoxides with residual organic functionalities being present in the final composite material. Nowadays, the application of the sol-gel process covers materials for photocatalysis [1], optics [2], fuel cells [3], and corrosion protective coatings [4] as well as for biomedical engineering and biomaterials science [5]. Starting materials for the sol-gel process are (liquid) metal alkoxide compounds (e.g., Me = Si, Ti, and Zr), in which the alkoxide moieties are subsequently hydrolyzed (1) after the addition of water to form a sol:

Handbook of Nanoceramic and Nanocomposite Coatings and Materials
http://dx.doi.org/10.1016/B978-0-12-799947-0.00003-1

$$Si(OR)_4 + H_2O \rightarrow HO - Si(OR)_3 + ROH \quad (1)$$

The resulting metal hydroxides are thermodynamically unstable and tend to form Me—O—Me bonds in a condensation reaction by eliminating either water (2) or alcohol (3) as by-products:

$$2HO - Si(OR)_3 \rightarrow (RO)_3Si - O - Si(OR)_3 + H_2O \quad (2)$$

$$HO - Si(OR)_3 + Si(OR)_4 \rightarrow (RO)_3Si - O - Si(OR)_3 + ROH \quad (3)$$

Further hydrolysis and condensation result in the formation of colloidal particles, which subsequently aggregate to a three-dimensional network (gelation). Quantitative reaction according to (1)–(3) will then lead to the formation of an inorganic metal oxide network. Due to the entrapped solvent, the gels are highly porous; however, drying of the gel under atmospheric conditions results in shrinkage due to capillary forces and the formation of a denser xerogel. This shrinkage can be avoided by alternative techniques like supercritical drying (e.g., exchanging the solvent by liquid carbon dioxide followed by heating above the triple point of CO_2), in order to form highly porous aerogels [6]. Depending on processing parameters, further application forms of sol-gel-derived materials can be obtained, namely, films, fibers, and (nano)particles. Major advantages of sol-gel material synthesis compared to traditional solid-state preparation routes are the high purity of the starting materials, low processing temperatures, the ability to form nanomaterials with controlled porosity in a wide range, and the possibility to introduce organic functionalities [7]. The latter ones can be manifold (e.g., amino, epoxy, carboxy, methacryl, or fluorinated carbon chains), which results in a variety of chemical and material properties of the final material (Figure 1a). This chapter aims to review recent approaches of bottom-up material fabrication by the sol-gel process, including bulk materials and functional coatings in biomedical applications:

- The synthesis of ceramic composites like mesoporous bioactive glasses (MBGs) [9] or organic/inorganic hybrid nanocomposites as dental filling materials (Ormocer®) [10]. The latter ones offer the advantage of reduced shrinkage during photopolymerization and increased wear resistance and hardness, as compared to traditional dental composites.
- The application of thin sol-gel derived coatings as coupling agents for biologically active molecules or a hydrolytically stable bonding of polymers to metals or ceramics [11].
- The use of nanoporous hybrid ceramics prepared by sol-gel chemistry for the encapsulation of enzymes and proteins as well as for controlled drug

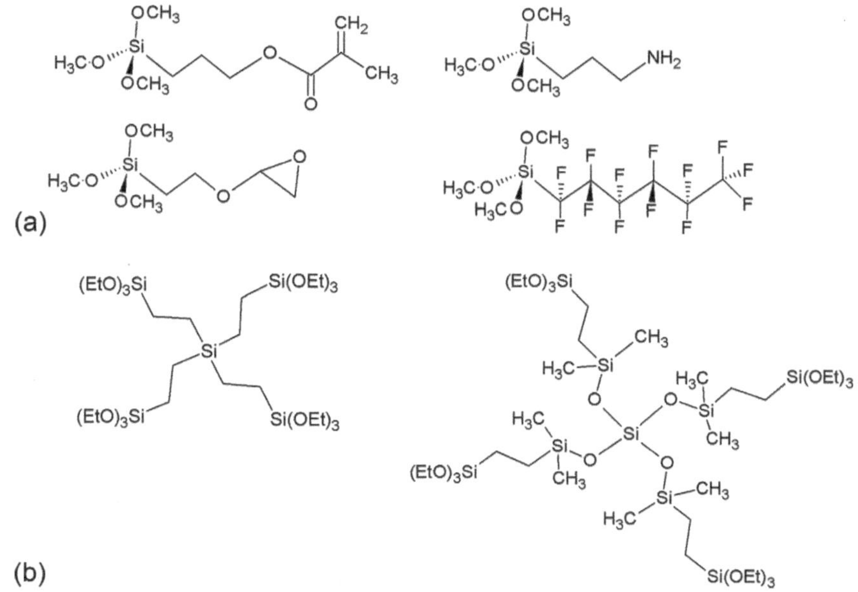

Figure 1 (a) Examples of commercially available silica precursors with organic functionalities and (b) star-gel silica precursors used in [8] for the synthesis of a bioactive sol-gel glass.

delivery purposes [12]. Encapsulation of such molecules is beneficial for protection against degradation while maintaining the activity of enzymes and proteins. At the same time, the adjustability of the nanoporosity of sol-gel hybrid nanoceramics offers the possibility to adjust the release rate of encapsulated drugs.

- The encapsulation of living microorganisms in silica-based ceramics ("biocers"), which is enabled by avoiding harsh preparation conditions like high temperature, acidic pH, and the use of organic solvents [13]. The fields of application for these biologically modified ceramic composites in biotechnology comprise the use as biosensors, the biodegradation of phenols, and bioremediation of wastewater as well as the biosynthesis of drugs.

2. PREPARATION OF BULK BIOMATERIALS BY SOL-GEL CHEMISTRY

2.1 Bone Grafts Based on Mesoporous Bioglasses

Bone is a slowly growing tissue with limited self-healing capacity. Defects above a critical size of approximately 1 cm in humans will not properly heal without further implantation of a bone graft to fill the space and to prevent the ingrowth of fibrous tissue [14]. Synthetic bone replacement materials are often based on either crystalline calcium phosphate (CaP) bioceramics [15] or CaP-containing bioactive glasses [16]. The latter have been demonstrated to form a strong bond to bone and to degrade slowly under *in vivo* conditions [17]. The mechanism of this strong bone bonding property is related to the formation of a nanocrystalline carbonated hydroxyapatite layer on the bioactive glass surface. This apatite layer has a composition similar to the mineral phase of bone and is able to interact with collagen fibrils of the host bone [9]. Bioactive glasses are traditionally processed by melting and quenching, which leads to dense monoliths or granules [9]; porosity, which is required for tissue ingrowth, has to be introduced into these glasses afterward, for example, by foam replication methods. An alternative strategy for bioactive glass synthesis is sol-gel processing using liquid silica, calcium, and phosphate precursors with typical compositions, for example, 60 mol% SiO_2, 36 mol% CaO, and 4 mol% P_2O_5 [18] or 80 mol% SiO_2, 16 mol% CaO, and 4 mol% P_2O_5. Sol-gel-derived bioactive glasses have an inherent nanoporosity due to the absence of a melting step. Although often no sodium ions are incorporated (which normally increase the solubility of traditional melt-derived bioactive glasses), the solubility of sol-gel glasses is high due to the high specific surface of 126.5-164.7 m^2/g [19]. Despite of this high specific surface area and nanoporosity, the bioactivity of sol-gel glasses was found to be much lower than that of molten bioglasses. This was demonstrated by Izquierdo-Barba et al. [20], who observed apatite precipitation from simulated body fluid on sol-gel bioglass only after 60 days compared to 3 days for conventional bioglass. An improvement of bioactivity can be achieved by synthesizing MBGs, in which a self-assembling process is induced by evaporation using a block copolymeric structure-directing template [21]. In contrast to normal sol-gel glass synthesis carried out at temperatures >100 °C in aqueous solution, MBGs were formed at room temperature in near nonaqueous conditions in the presence of block copolymers like $EO_{20}PO_{70}EO_{20}$ and $EO_{106}PO_{70}EO_{106}$ (EO, poly(ethylene oxide); PO, poly(propylene oxide)). The self-assembling process under these conditions led to the formation of well-ordered mesopores in a size range of 4.4-7.1 nm [21]. The bioactivity of such MBGs is strongly enhanced: in comparison to traditionally synthesized sol-gel bioglasses, the surface is fully covered with calcium-deficient apatite particles already after 3 days immersion in SBF.

The most common silica precursor for bioglass synthesis is tetraethoxysilane (TEOS); however, attempts have been made to improve cross-linking in the gels by using starlike silica compounds (Figure 1b) [8]. Although such star-gel bioglasses showed a slight decrease in compressive strength, lower hardness, and Young's modulus compared with conventional sol-gel glasses, their fracture toughness was increased by a factor of 3, which reflects the more flexible chemical structure of the star gels.

2.2 Organic/Inorganic Hybrid Nanocomposites for Dentistry (Ormocer®)

Photocurable resin-based composites [10] have become very popular in operative dentistry and in the fixation of ceramic inlays, root posts, and dental crowns [22], as they provide excellent esthetics and at the same time minimize the need to sacrifice sound dental hard tissue, as it is required for retaining conventional restorative materials. However, a major disadvantage of these materials is their polymerization shrinkage, which can lead to a deformation of dental hard tissue and in the worst case to a gap between filling and cavity [23]. Dental composites are usually composed of at least two phases, namely, synthetic polymers as the organic matrix and inorganic fillers as reinforcement, plus additives like initiators and activators [24]. Furthermore, as polymerization shrinkage of the restorative material cannot be completely avoided, various bonding agent systems have been developed [25]. In order to improve the adhesion between filling and tooth mineral, the latter has to be etched to increase the active surface area of enamel and/or dentin, followed by hermetic sealing of the restoration. Arising complications like postoperative sensitivity caused by excessive etching of dentine tubules and technically difficult application procedures have led to several approaches, including etching and rinsing protocols with 1, 2, or 3 steps and various chemical formulations for priming and bonding [26]. The synthesis of novel organic/inorganic hybrid bulk ceramic composites via the sol-gel process and their use as dental filling materials (Ormocer®) offer the advantages of reducing the shrinkage during photopolymerization and increasing wear resistance and hardness compared to traditional dental composites.

Ormocer® (*or*ganically *mo*dified *cer*amics) [27] have been increasingly attracting the interest of dentistry research as restorative materials and bonding agents with the most promising properties. The main advantage in comparison to classical composite filling materials arises from the formation of a strongly connected organic/inorganic dual-phase system from the sol-gel in a two-step route: The inorganic network forming in the sol-gel solution entraps the reactive organic monomers; hence, the subsequent polymerization occurs within the solidified matrix with reduced shrinkage. Furthermore, the inorganic Si—O—Si structure of the Ormocer leads to increased abrasion resistance [28], which may significantly improve the longevity of dental fillings.

The inorganic part of the two-step process of Ormocer formation occurs by controlled hydrolysis and condensation of silicon alkoxides (or other metal alkoxides), as described above (Eqs. 1–3). In contrast, the precursor of an Ormocer also contains organic segments with both functional and polymerizable groups. Figure 2a shows a cluster consisting of the inorganic siloxane network with condensed diphenylsilanediol, methacryl, and epoxy groups after polycondensation. Subsequently, the organic network is formed by photoinitiated UV polymerization of the methacryl groups and/or thermally induced ring-opening polymerization [29] of the epoxy groups, finally resulting in the cross-linked hybrid network shown in Figure 2b.

Ongoing clinical studies with Ormocer as restorative materials have already supported their suitability for durable treatment of cavities in molars and premolars [30]. Esthetical issues are given by eventual discoloration of the filling but can be overcome by treatment with whitening gels [31] of the Ormocer with greater resistance to adverse microleakage and hardness effects in comparison to conventional composites [32]. Besides *in vivo* studies, also, extensive preclinical investigations have been conducted, which—despite the disadvantage of being less close to real-life conditions—provide prompt availability of reproducible and comparable experimental data. Here, Ormocer-based restoration systems were proved to provide superior marginal quality and resistance to thermomechanical loading in comparison to mild-etching all-in-one adhesives and hence outweighed the simplified

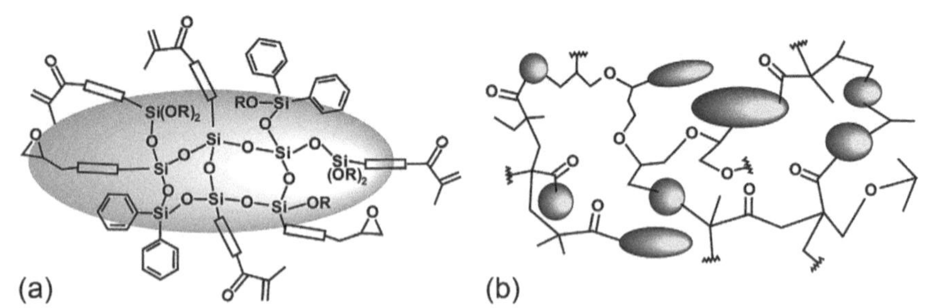

Figure 2 Example of a four-component Ormocer-forming sol-gel system (a) reactive organically modified clusters after polycondensation (R=H, CH_3, and CH_2CH_3; rectangles=ether chains) and (b) cross-linked inorganic/organic polymer network after thermal polymerization of epoxy groups.

applicability of the latter ones [25]. Furthermore, the utilization of Ormocer in both restorative materials and bonding agents turned out to be slightly preferable due to the combination of chemical affinities of the bonding agent to the calcium of the tooth as well as to the filling material. It can be concluded that also in the future, Ormocer will play a significant role in the further improvement of restorative materials and bonding agents.

3. METAL-ORGANIC COUPLING AGENTS

3.1 Covalent Bonding of Bioactives to Implant Surfaces

In many studies, the sol-gel process is used to achieve a covalent binding of bioactive molecules to inorganic (metallic and ceramic) surfaces by using bifunctional silanes, predominantly with amino groups. The principle of the surface modification is demonstrated in Figure 3a for 3-(trimethoxysilyl)-propylamine (APMS) as commonly used spacer. At first, the silane is bound to the inorganic surface (after prehydrolyzing with small amounts of water), for example, by dip coating. Afterward, bioactives are coupled

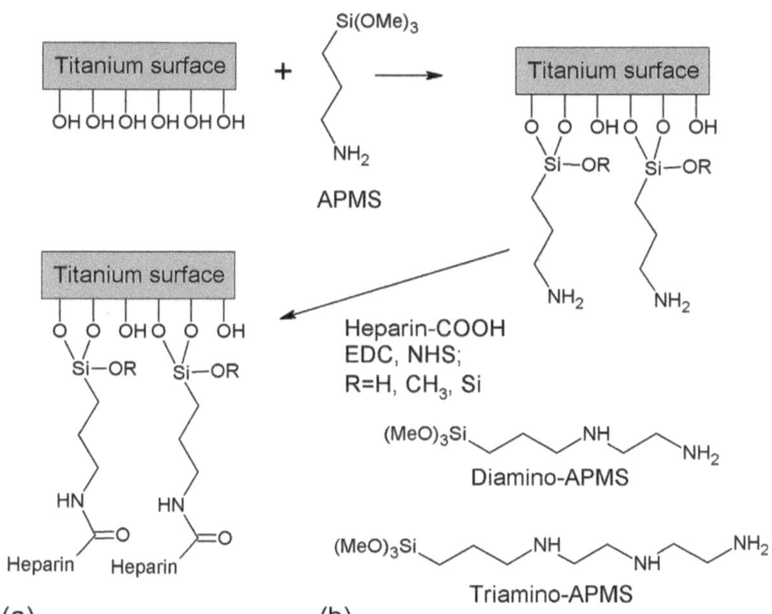

Figure 3 (a) Reaction scheme of the immobilization of heparin on titania surfaces by dimensions of APMS coupling agent and (b) further spacer molecules used as coupling agents for heparin [33].

to the free amino groups followed by the activation of carboxylic groups (of the bioactive) with carbodiimides (e.g., *N*-(3-dimethylaminopropyl)-*N'*-ethylcarbodiimide hydrochloride (EDC)) and *N*-hydroxysuccinimide (NHS) [33]. This processing regime has been used to covalently bond collagen type I to stainless steel [34], BMP-2 to titanium [35] or bioactive glass [36], and fibronectin, BMP-2, and BMP-9 peptide motifs to glass [37].

Alternative coupling agents (Figure 3b) with two or three amino moieties, such as *N*-[3-(trimethoxysilyl)propyl]ethylenediamine (diamino-APMS) and *N*1-[3-(trimethoxysilyl)-propyl]diethylenetriamine (triamino-APMS), were used by Tebbe et al. [33] due to their longer spacer chains, which lead to higher mobility of the immobilized bioactive and hence improved biological potency of the drug. It is noteworthy that only the primary amino groups of the spacer molecules are reactive, whereas the secondary amino groups (R—NH—R′) are not able to form amide bonds with carboxylic groups. The successful modification of TiO_2 surfaces with coupling agent and heparin was demonstrated by measuring zeta-potential changes in aqueous solution, since both the spacer and the drug led to a change of the potential due to charged amino and sulfate groups.

Aminosilanes (e.g., APMS), which are commonly used for surface modification, contain three reactive alkoxide groups for bonding to the surface. However, also, cross-linking between the silanes may occur, which in turn leads to nonuniform coating structures. To overcome this problem, monofunctional aminosilanes (3-aminopropyldimethylethoxysilane) have been applied, which resulted in thinner and more uniform coatings compared with trifunctional silanes. Despite of only one binding group per molecule, a high stability of the coating in acidic environment (up to 17 days) was observed [38].

The abovementioned reaction with aminosilanes requires carboxylic acid groups within the bioactive; if those are not available, epoxy-functionalized silanes can be used instead. This was demonstrated by Mohorčič et al. [39], who functionalized an inert glass substrate with a silane coating with epoxide rings, to which an antimicrobial peptide (polymyxin B) was coupled by means of a catalyst (WCl_6 and $H_3P(W_3O_{10})_4$). The successful peptide immobilization on the surface was proved by FTIR, XPS, and AFM analyses. Antimicrobial properties of the coated surface against *Escherichia coli* were investigated, showing a bacterial reduction by five orders of magnitude within 24 h. A direct antimicrobial property of the silane coating could be achieved by using quaternary ammonium salts as functionalities, for example, 3-(trimethoxysilyl)-propyldimethyloctadecyl ammonium chloride [40]. Such modified polymeric surfaces showed a reduction of bacterial activity from 90% to 0% for gram-positive staphylococci and from 90% to 25% for gram-negative *E. coli* and *Pseudomonas aeruginosa.*

3.2 Coupling Agents for Dental Applications

Thin sol-gel derived coatings are commonly applied in dentistry as coupling agents for hydrolytically stable bonding of polymers to either metals or ceramics [11]. Without such a coupling agent, the adhesion between the materials is only based on short-range physical forces (e.g., van der Waals); the aggressive environment in the oral cavity will lead to the diffusion of water molecules into the interface of polymer and metal/ceramic and hence to a loss of adhesion. Similarly, changes of temperature due to cold and hot food consumption cause mechanical stress in the interface due to the different thermal expansion coefficients of the materials. The use of thin coupling agent layers (e.g., methacryloxypropyltrimethoxysilane) is dealing with these problems by creating a direct covalent bond between the materials (Figure 4). The alkoxide moieties of the coupling agent are (after hydrolysis) able to bind to hydroxyl groups at the inorganic surface, while at the same time, the organic functionality can copolymerize with monomers from an applied resin [41]. Since these coupling agent layers are typically 20-50 nm thick and have flexural properties, they also reduce the effect of thermally induced stresses as described above [42]. Indeed, many studies could prove the positive effect of such interface engineering by silane coupling agents on the long-term stability of metal/polymer [43] or ceramic/polymer [44,45] joints. Apart from the abovementioned compound, also other functional silanes like *N*-(3-(trimethoxysilyl)-propyl)ethylenediamine [46], (3-(triethoxysilyl)propyl)urea [46], vinyltriisopropoxysilane [43], tris(3-trimethoxysilylpropyl)isocyanurate [43], 3-isocyanatopropyltriethoxysilane [47], and allyltrimethoxysilane [41] can be applied. This enables shear strengths of metal polymer joints of up to 12-20 MPa after aging by thermocycling (5000-6000 cycles between 5 and 55 °C), whereas the addition of a cross-linking silane (1,2-bis-(triethoxysilyl)ethane) to the abovementioned

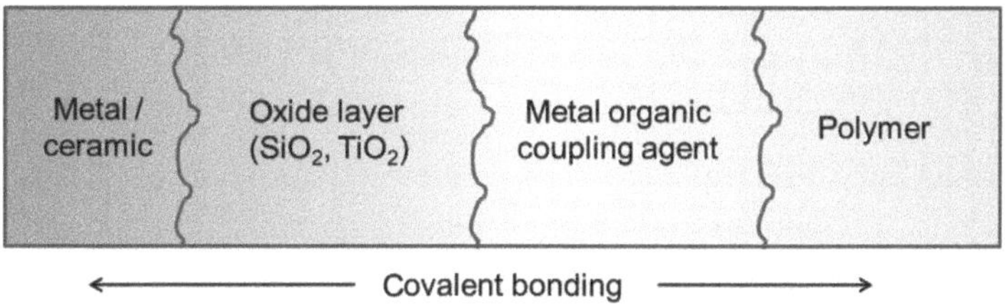

Figure 4 Principle of silane coupling agents used in dentistry to covalently bind inorganic (metals and ceramics) and organic (polymers) materials.

functional silanes seems to be favorable to achieve a highly hydrolytically stable bonding [46].

Since the composition of dental alloys varies over a broad range and particularly noble metal alloys do not provide enough surface OH groups, it is necessary to apply a primary coating to the metal surface to ensure sufficient adhesiveness of the silane coupling agents. In many cases, this is done by using silicate layers, which can be deposited by either sol-gel process [48], flame pyrolysis [49], or tribochemical surface modification [50]. The latter uses the high kinetic energy transfer of the particles to a surface during sandblasting to apply thin coatings of silica [50] or titania [51]. Dental ceramics may be also pretreated by the same methods as described above; in addition, etching of the ceramic surface with hydrofluoric acid can be used to create an activated surface with OH groups [45].

Surface silanization is also useful to improve the bonding ability of polymethylmethacrylate (PMMA) bone cements to metal implants in orthopedic applications [52]. For the long-term stability of cemented hip and knee prostheses, the interface between implant and bone cement is considered to be highly susceptible to mechanical and/or hydrolytic failure [53]. An improvement of the hydrolysis resistance of the titanium/bone cement interface was obtained by using a tribochemical TiO_2 modification of the metal surface in combination with either applying a 2% solution of methacryloxypropyltrimethoxysilane to the titanium surface or adding 5 and 20 wt.% coupling agent to the PMMA cement. The latter had the advantage of an easy clinical applicability of the procedure, as the processing and implantation techniques of the cement material were maintained. In addition to an improvement of the hydrolysis resistance of the implant/cement interface, the modification of PMMA with methacryloxypropyltrimethoxysilane can also improve the adhesion of the cement to bone. A simultaneous modification of PMMA cement with silane (20 wt.% in the MMA monomer) and $CaCl_2$ (20 wt.% in the cement powder) was demonstrated to enhance the osteoconductivity of the material due to an accelerated apatite formation of the cement surface [54].

4. NANOPOROUS SOL-GEL HYBRID CERAMICS FOR DRUG AND PROTEIN DELIVERY

Nanoporous hybrid ceramics are of high interest in biotechnological and medical applications as delivery and carrier systems for small and large biomolecules, as they provide controlled surface chemistry as well as an intrinsic porosity on a nanometer scale, which results in high surface areas [55] (Table 1). Without any

Table 1 Drug delivery systems based on sol-gel-derived materials

Material	Precursor	Shape	Drug
Silica gel [56]	TEOS	Monolith	Vancomycin
Silica gel [57]	TEOS	Mesoporous tubular nanocarriers	Sodium ampicillin, cytochrome *c*
Silica gel [58]	TEOS	Microspheres	Vancomycin, bupivacaine
Calcium/silica and silica/polycaprolactone hybride [59]	TMOS	Monolith	Sodium ampicillin
Silica gel [60]	TEOS	Microparticles	Gentamicin
Silica-coated calcium pectinate [61]	TEOS	Beads	Theophylline
Zinc-silica [62]	TEOS	Microparticles	Insulin
Silica gel [63]	TEOS	Monolith	Doxorubicin hydrochloride
Hybrid silica gel [64]	γ-Methacryloxypropyl trimethoxysilane (γ-MPS)	Bulk, films	Heparin
Silica gel [65]	TEOS	Microspheres	Dexmedetomidine HCl

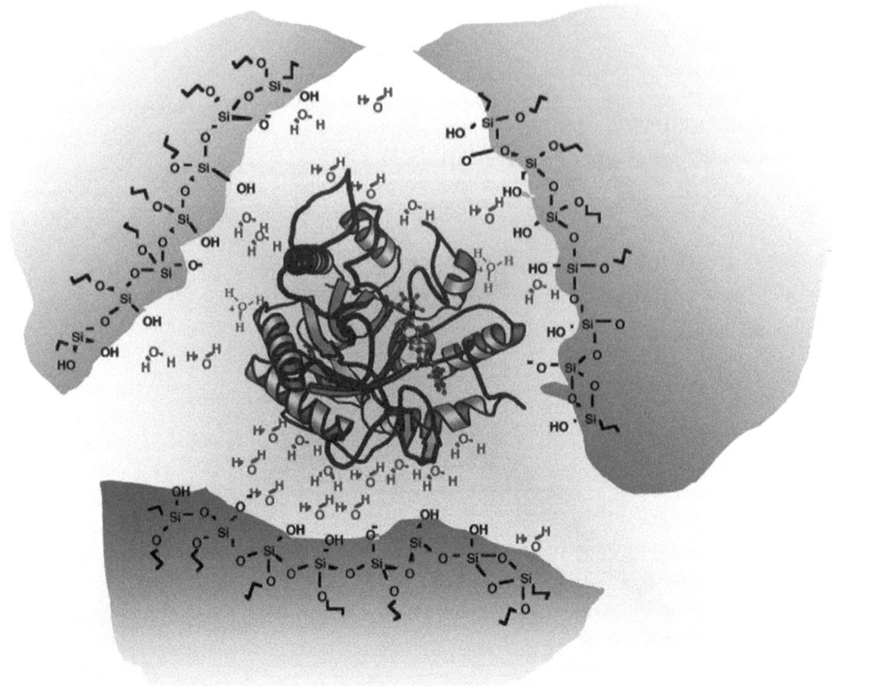

Figure 5 Schematic view of an entrapped enzyme with a few water molecules inside the pore of a silica gel network. *(Copyright 2005, Journal of the American Chemical Society, reprinted with kind permission).*

immobilization, large biomolecules often show a reduced half-life, because they are very sensitive to environmental parameters such as pH, temperature, and ionic strength. An extension of their half-life in comparison to solution environment can be achieved by the encapsulation of enzymes in silica gel or silica gel-based hybrid materials with organic polymers [66,67]. Silica gel forms a protection cage around the enzymes and prevents them from aggregation or unfolding (Figure 5). In comparison to organic polymers, silica gel offers the advantages of improved mechanical strength, chemical stability, biological inertness, low-temperature synthesis, and no swelling in aqueous solutions [68].

The crucial factor for a successful encapsulation of biological molecules is the maintenance of the biological activity during the sol-gel process. The synthesis reactions (especially the condensation) are too harsh for the most proteins and enzymes. For example, the alcohol generated during the hydrolysis of tetraethyl orthosilicate (TEOS) or tetramethyl orthosilicate (TMOS) precursors can initiate the denaturation process. To overcome this problem, the alcohol could be removed by evaporation prior to the addition of the biomolecules [69]. Ellerby et al. described a method that became the basic standard procedure for the encapsulation of sensitive proteins. No extra alcohol is included, and after the condensation reaction, a buffer is added to increase the pH prior to encapsulation [70]. Bhatia et al. developed another new synthesis route without any alcohol on the base of sodium silicate. They successfully encapsulated enzymes such as glucose-6-phosphatase and horseradish peroxidase, which are both of high interest in biotechnological applications [71].

For the application as delivery or carrier system, the geometric character of the silica sol-gel has to be adapted to the biological agent. Density, pore size distribution, and nanostructure as release-determining parameters can be controlled by the variation of the reaction kinetics [72], namely, by changes in pH, water and alcohol content, and catalyst selection. Radin et al. showed for dextran (10-70 kDa) as a model drug that the release rate depends on the molecular weight of the delivered molecule [73]. Another factor for controlling drug release was found to be the degree of cross-linking of organic moieties. Tebbe et al. [64] studied the release of heparin (average molecular weight between 13 and 18 kDa) from an organically modified porous silica matrix (poly(methacryloxypropyl)-poly(silsesquioxane), P-MA-PS). The matrices were obtained from methacryloxypropyltrimethoxysilane (MAS) via the sol-gel process under acidic conditions, followed by photochemical polymerization and cross-linking of the organic matrix. Modulation of the polymerization degree of the organic matrix in the range of 0-71% allowed for adjusting the release kinetics of heparin according to therapeutic requirements. It was demonstrated that higher drug loads and decreased polymerization degree resulted in a faster release profile of heparin, which followed a square root of time kinetics according to the Higuchi model. The hydrolytic degradation of hybrid xerogel was found to follow a zero-order kinetics, whereas the heparin concentration did not show an influence on the degradation rate of the matrix. Since the released heparin retained its biological activity, the P-MA-PS matrices are of clinical interest, for example, as a coating for drug-eluting coronary stents.

For the encapsulation of enzymes as biological catalysts in silica-based sol-gel ceramics, the maintenance of their conformation and hence their biological activity is a prerequisite during the immobilization, since even small changes in their tertiary or quaternary structure can result in a dramatic loss of activity. In general, three types of immobilization approaches exist: (I) covalent binding of the enzymes with the silica material by surface functionalization, (II) physical adsorption via hydrogen bonding and electrostatic interactions, and (III) physical entrapment [74,75]. In comparison to physical adsorption, covalent binding of the enzyme to silica gel brings the advantage of a reduced washout due to the highly stable conjunction [74]. However, the covalent binding could significantly drop down the enzyme activity due to a possible conformational change of the enzyme. In contrast, it is postulated that a multiple-point covalent bonding on the carrier matrix could stabilize the native protein conformation. Particularly for the use of enzymes in nonaqueous environment, the lifetime could be significantly extended in comparison to the free enzyme, whereby the enzyme is protected against denaturation by the forced structure [75].

The adsorption and biological activity of immobilized enzymes are strongly influenced by the pH and electrostatic interactions. The isoelectric point of silica gel is in the range of pH 3 [76]. Therefore, in most applications, the silica surface is negatively charged and electrostatic interactions between silica gel surface and the immobilized protein strongly depend on the isoelectric point of the protein [68]. On one hand, an opposite charge could be used as driving force for adsorption [77]; on the other hand, it could become a crucial factor for protein unfolding and reduced catalytic activity [78]. The latter can be avoided by building an enzyme-polyelectrolyte complex, whereby the critical charged sites are protected [79]. In relation to the pH in these nanoscaled pores, it is important to consider that the pH could be lowered by the overlapping electrical double layer, induced by the negatively charged silica gel surface and hence the excess of H^+ [71].

5. ENCAPSULATION OF MICROORGANISMS AND CELLS IN SILICA-BASED CERAMICS

5.1 General Aspects of Cell Encapsulation in Inorganic Matrices

Lithobiontic microorganisms, including cyanobacteria, fungi, algae, yeast, diatoms, and lichens, are naturally occurring archetypes of biohybrids, practicing biomineralization as product of their synergetic life-form with stones or as protection against harsh environmental conditions or predators [80,81]. A technical implementation of biomineralization is given by sol-gel-processed silica matrices that offer optimal properties for application as cell hosts in various fields in biotechnology or medicine. Prominent characteristics of these materials are biological and chemical inertness, thermal and mechanical stability, resistance against fouling, and negligible swelling in organic or inorganic solvents, which prevents leaching of the cells. A special feature of silica is given by its transparency, which makes it feasible for the immobilization of photoactive microorganisms like algae or cyanobacteria [82]. Moreover, the possibility of the precise tuning of physical properties, particularly the porosity, by synthesis parameters, prevents cell leaching and provides control over inevitably required diffusion processes like the exchange of metabolites. However, the sol-gel process has to be optimized, as mentioned before, to provide cytocompatible conditions particularly with regard to pH, ionic strength, temperature, the absence of cytotoxic substances like alcohol, gentle solidification, and drying of the gel to prevent cell lysis. Pioneer work of encapsulation of microorganisms within alkoxide-based sol-gel silica was performed by Carturan et al. [83], who could show that the immobilization of *Saccharomyces cerevisiae*, an acid- and alcohol-tolerating yeast, in silica films resulted in a higher catalytic activity compared to free yeast cells. The dramatically improved efficiency of the production process of immobilized cells originated from the protection against environmental stress or other harmful conditions and the restriction of cell division or proliferation as energy-consuming and energy-competitive process to metabolic activity [82]. Immobilized yeast cells showed a 10 times higher activity than free cells [84], while for the encapsulation of plant cells, a 100-fold alkaloid production could be proved [85].

In comparison to relatively robust microorganisms, the immobilization of plant or mammalian cells is a much more challenging task, since these cells are bigger, more fragile, and more sensitive to variations in their optimal environment conditions. These challenges of the encapsulation of sensitive microorganisms were met by developing alternative, cell-friendly processing routes. First approaches were made by the minimization of the contact time between the sol and the biocomponents by adding the microorganisms or cells at a late state of the sol-to-gel transformation, when hydrolysis and condensation of the alkoxides were completed [70]. Using this method, the contact time can be limited to the gelation time, which ranges from 30 s to 5 min. Peterson et al. practiced this sol-gel route for the encapsulation of pancreatic islets without significantly reducing cell activity and viability [86]. Another possibility to avoid ethanol in the sol-gel process as cosolvent or by-product is addressed in the use of aqueous precursors like tetrakis(2-hydroxyethyl) orthosilicate (THEOS) [87]. The condensation of THEOS results in a release of ethylene glycol, which builds a protective fluidic layer in the interface between the cell and the silica network. Coiffier et al. investigated the viability of *E. coli* after immobilization in sol-gel-derived silica matrices using two different routes involving alkoxide or aqueous silica gels [87]. The resulting activities evidenced that the aqueous precursors were less harmful than the alkoxide-based ones. Chen et al. immobilized bacteria (*Methylomonas* species, strain GYJ3) within sol-gel silica matrices derived from orthosilicate precursors or processed via an alcohol-free route with sodium silicate. Cells entrapped within aqueous silica gels showed a 1.5-fold higher activity and a better resistance against basic pH or increasing temperatures, as compared to free cells. Furthermore, the immobilized cells could be repeatedly used for more than 25 times without significant loss of activity [88].

The cell viability can also be successfully preserved by the addition of osmoprotective agents, for example, glycerol, poly(vinyl alcohol), and glycine betaine, to the sol-gel/biocomponent mixture [89]. These agents act as a physical barrier between the microorganisms and the silica and stabilize the cell membrane and avoid cell lysis induced by stressing factors like matrix shrinkage [90]. Altering the synthesis procedure is a further approach of maintaining cell viability and activity. One alternative route is given by a two-step strategy, where cells are first preencapsulated in biodegradable polymers (e.g., alginate) and then coated with sol-gel silica to form a heterogeneous biological-inorganic network [91].

Other procedures involve the removal of alcoholic by-products released during the sol-gel transformation. This principle was tracked by Carturan and coworkers, who developed the so-called Biosil technology [92]. In this method, vaporized alkoxide precursors were used to immobilize living microorganisms previously deposited on scaffolding materials like silk, polyester, glass fibers, and sponges [93] within a sol-gel silica film. The solidification of the silica film takes place directly on the cell surface by the reaction of the gaseous precursors with water adsorbed on the cell surface. Toxic alcohol generated as sol-gel reaction by-product will be eliminated simultaneously to the silica gel deposition via gas flux. A precise control of membrane thickness is given by the exposure period of the cells to the gas, while membrane porosity and stability can be adjusted by the choice of the precursor. This process is suitable for yeast and bacteria as well as for large cells, like mammalian (e.g., hepatocytes [93,94] or pancreatic islets [92]) or plant cells, since the size of the cells is not constrained by the pore size of the sol-gel silica network [92].

As a further alternative and promising method for the immobilization of microorganisms in silica matrices, Soltmann et al. developed the freeze-gelation technique of biologically modified nanosols [95]. This method offers beneficial aspects like low-cost and nearly zero shrinkage of the ceramic parts, which allows the fabrication of extremely stable and complexly shaped silica matrices with tunable porosities. Furthermore, mild preservation of microorganisms by freeze-drying results in long-term stability and allows the storage under dry conditions. The four essential steps of this procedure start with the preparation of a colloidal dispersion (nanosol) of a ceramic powder (e.g., alumina or mullite), water, and additives like glycerol, polyacrylic acid, and polyvinyl alcohol. This aqueous nanosol is mixed with the biocomponent followed by casting of the slurry in a mold. The solidification is attained by freezing at −40 to −70 °C, causing the irreversible transformation of the colloidal dispersion by the sol-gel transition, which is induced by phase separation and packing of the nanoparticles due to ice crystal formation. After demolding of the biologically modified ceramic (biocer) structure, drying will be performed by lyophilization or evaporation of water [96]. Microstructure and porosity can be tailored by the freezing and drying conditions, the water content of the slurry, and the incorporation of leachable pore-forming agents (e.g., salts and saccharides [97]) or water-soluble polymers like poly(ethylene glycol) [98]. This method has been used for the encapsulation of different microorganisms like *Bacillus sphaericus* [95], *Rhodococcus ruber* [99], and *S. cerevisiae* [95].

Following these cytocompatible processing routes, a variety of microorganisms and cells, such as yeasts [83,97,100], bacteria [87,95,96], algae [101], cyanobacteria [82,102], and plant or mammalian [92] cells, have been immobilized in silica networks. The various application fields comprise biotechnology, bioremediation, and medicine, particularly biosensors, biosorbents, bioreactors, or biocatalysts, used in particular for wastewater recycling, production of pharmaceuticals or biofuel, cell transplantation, artificial organs, or immunoassays [103] (Table 2).

5.2 Medical Application of Biohybrids

Sol-gel encapsulation of microorganisms or whole cells into biocompatible silica becomes increasingly important for applications as implantable therapeutic systems. These biohybrids can be used as delivery systems for bioactive molecules [114] or for the degradation of harmful substances [112] with the general aim to treat chronic or degenerative diseases like diabetes, epilepsy, Alzheimer's disease, Huntington's disease, Parkinson's disease, amyotrophic lateral sclerosis, hepatic failure, hemophilia, and infertility. One outstanding characteristic of the silica network is the precise control over porosity and pore size to create a selective barrier for the

Table 2 Application field of sol-gel silica-encapsulated microorganisms

Application field	Cell/microorganism	Application
Bioreactor	Cyanobacteria	CO_2 remediation
	Synechococcus [104]	Biofuel synthesis
	Synechococcus species [102]	
	Chlorella vulgaris [105]	
	Haematococcus pluvialis [101]	Astaxanthin production
	Klebsormidium flaccidum [106]	Gold nanoparticle synthesis
	Plant cells	Alkaloid production
	Catharanthus roseus [85]	
Bioremediation	Bacteria	Heavy metal binding
	Bacillus sphaericus [107]	
	Algae	
	Chlorella vulgaris [108]	
	Yeast cells [109] (not specified)	Phenol biodegradation
	Bacteria	
	Rhodococcus ruber [99]	
	Fungi	Glycol biodegradation
	Aspergillus spores [110]	
Artificial pancreas	Islets of Langerhans [92]	Insulin production
Artificial liver	Hepatocytes [93]	Bilirubin conjugation, ammonia removal, urea production, and diazepam metabolism
Immunoassay	*Leishmania guyanensis* [111]	Leishmaniasis diagnosis
Artificial kidney	Bacteria	Urea and uric acid degradation
	Escherichia coli [112] (genetically modified)	
Drug delivery system	Mammalian cells	Hormone delivery (estradiol and progesterone)
	Ovarian follicles [113]	

immunologic protection of entrapped bioagents while allowing free nutrient and metabolic exchange [115]. Current research efforts concern the immobilization of the pancreatic islets of Langerhans for the use as bioartificial pancreas. Boninsegna et al. and Sakai et al. demonstrated that encapsulated islets were still active and allowed the glucose stimulation and insulin release through the silica membrane [91,92]. In a subsequent *in vivo* study in diabetic rats, these encapsulated islets were proved to be a suitable diabetes treatment by means of successful restoration of glycemia levels within 24 h and up to over 100 days, without signs of immunologic response or inflammatory reactions [91]. Other researchers follow the idea of entrapped cells for the locally restricted production and secretion of antibiotics [116], hormones [113], or chemotherapeutics [117], in order to minimize undesirable side effects in the treatment of reproductive or endocrine disorders like Turner's syndrome or hypothyreosis, infections, and cancer.

Cell encapsulation of bacteria or whole cells is also of high interest for biosensoring applications. Immunoassays are a special kind of biosensors, based on specific antigen-antibody recognition, and used for the detection of specific diseases. For this purpose, Copello et al. immobilized the pathogenic protozoon *Leishmania guyanensis* in thin sol-gel silica films on microplates [111]. A specific antibody-antigen reaction between the protozoon and the antibodies in the patient's serum can be detected via a conjugated enzyme color reaction indicating positive sera and leishmaniasis infection. The sol-gel parasite hybrid material maintained its properties over months of storage at 4 °C, which makes this method suitable for the direct detection of pathogens with minimal equipment and under restricted storage possibilities.

5.3 Biotechnological Application of Biohybrids

Silica-immobilized microorganisms meet the biotechnological requirements of long-term stability, high activity, and ease of biomass production. Furthermore, the immobilization of the biomass will simplify the technical process by the ease of biocatalyst/metabolite separation, which increases the efficiency of the metabolic process by the reusability of the bioreactor. Moreover, several studies demonstrated the increasing metabolic activity of immobilized cells compared to free cells. According to the majority of scientific opinions, cell membrane lysis is responsible for this phenomenon. Stress actors during encapsulation of cells damage the cell membrane. This results in an enhanced diffusibility of nutrients and metabolists, which is equivalent to a faster enzymatic activity [118].

Especially, the photosynthetic activity of plant cells [13], cyanobacteria [102], or microalgae [101] is of high interest for biotechnological applications, including the synthesis of pharmaceutics, pesticides, flavorings and fragrances, recombinant proteins, nanoparticles, or biofuels. Fiedler et al. created a continuously working bioreactor by the immobilization of the microalgae *Haematococcus pluvialis* within sol-gel silica layers. These algae produce astaxanthin, a carotene dye used as supplement in fish food [101]. In this study, the microalgae maintained their viability over 40 days, whereas the metabolic activity could be optimized by the addition of stressors like iron ions during the dye production. The ability of recultivation of algae within the bioreactor offers a more efficient process, as compared to the commercial production of astaxanthin, where the extraction process is associated with the complete biomass destruction. In addition, the bioreactors can be combined with microorganisms with high sensitivity and high selectivity to detect changes in their environment. This can be used to stage the fitness of microorganisms like bacteria or immobilized cells in bioreactors [90] or to detect water pollution with drugs, toxins, or chemicals [89,119].

5.4 Environmental Applications of Biohybrids

Several microorganisms like yeasts [120], cyanobacteria [121], and bacteria [107] provide selective and extremely retentive binding capacity or degradation potential for specific pollutants and are therefore being utilized as fundamental compounds in biological wastewater treatment. The immobilization of these scouring organisms in physically and chemically stable sol-gel-based filters facilitates an easy separation of the sorbing phase and the purified filtrate as well as an easy removal of adsorbed compounds. The resulting reusability of the filter systems is an essential factor for economical bioremediation processes. In general, the purification by biological species is based on different mechanisms, namely, bioaccumulation, biotransformation, biomineralization, and biosorption, without the premise of viable cells for each case. While accumulation and degradation are intracellular metabolic processes, biosorption for the uptake of heavy metal ions [107] or dyes [122] is a passive binding process on the cell surface, which is possible even for dead microorganisms. In the case of biodegradation, the transformation of the pollutant, for example, phenol [99], hormones [123], or pesticide [124], into nontoxic components is a metabolic and hence an active process. An example of bioremediation with immobilized bacteria is given by a study from Alvarez et al. [125]. The authors could show that immobilized bacteria reduced reactive ions like Cr(VI) to Cr(III). Silica encapsulation served as protector against toxic chrome ions, since the porosity of the silica network limited ion diffusion. The result was a higher viability of immobilized cells compared to free cells. With regard to CO_2/biofuel conversion of photoactive microorganisms immobilized in host materials, biotechnologically used bioreactors contribute also to economical rehabilitation and protection [103].

REFERENCES

[1] T. Fröschl, U. Hörmann, P. Kubiak, et al., High surface area crystalline titanium dioxide: potential and limits in electrochemical energy storage and catalysis, Chem. Soc. Rev. 41 (15) (2012) 5313.
[2] P.W. de Oliveira, C. Becker-Willinger, M.H. Jilavi, Sol-gel derived nanocomposites for optical applications, Adv. Eng. Mater. 12 (5) (2010) 349–361.
[3] C. Laberty-Robert, K. Vallé, F. Pereira, C. Sanchez, Design and properties of functional hybrid organic-inorganic membranes for fuel cells, Chem. Soc. Rev. 40 (2) (2011) 961.
[4] I. Jerman, A. Šurca Vuk, M. Koželj, F. Švegl, B. Orel, Influence of amino functionalised POSS additive on the corrosion properties of (3-glycidoxypropyl) trimethoxysilane coatings on AA 2024 alloy, Prog. Org. Coat. 72 (3) (2011) 334–342.
[5] R. Mishra, B. Basu, A. Kumar, Physical and cytocompatibility properties of bioactive glass-polyvinyl alcohol-sodium alginate biocomposite foams prepared via sol-gel processing for trabecular bone regeneration, J. Mater. Sci. Mater. Med. 20 (12) (2009) 2493–2500.
[6] N. Hüsing, U. Schubert, Aerogels—airy materials: chemistry, structure, and properties, Angew. Chem. Int. Ed. 37 (1-2) (1998) 22–45.
[7] U. Schubert, N. Huesing, A. Lorenz, Hybrid inorganic-organic materials by sol-gel processing of organofunctional metal alkoxides, Chem. Mater. 7 (11) (1995) 2010–2027.
[8] M. Manzano, D. Arcos, M. Rodríguez Delgado, E. Ruiz, F.J. Gil, M. Vallet-Regí, Bioactive star gels, Chem. Mater. 18 (24) (2006) 5696–5703.
[9] J.R. Jones, Review of bioactive glass: from Hench to hybrids, Acta Biomater. 9 (1) (2013) 4457–4486.
[10] N.B. Cramer, J.W. Stansbury, C.N. Bowman, Recent advances and developments in composite dental restorative materials, J. Dent. Res. 90 (4) (2011) 402–416.
[11] C.Y.K. Lung, J.P. Matinlinna, Aspects of silane coupling agents and surface conditioning in dentistry: an overview, Dent. Mater. 28 (5) (2012) 467–477.
[12] E.T. Hwang, M.B. Gu, Enzyme stabilization by nano/microsized hybrid materials, Eng. Life Sci. 13 (1) (2013) 49–61.
[13] C.F. Meunier, P. Dandoy, B.-L. Su, Encapsulation of cells within silica matrixes: towards a new advance in the conception of living hybrid materials, J. Colloid Interface Sci. 342 (2) (2010) 211–224.
[14] M. Aaboe, E.M. Pinholt, E. Hjorting-Hansen, Healing of experimentally created defects: a review, Br. J. Oral Maxillofac. Surg. 33 (5) (1995) 312–318.
[15] S.V. Dorozhkin, Calcium orthophosphate-based bioceramics, Materials 6 (9) (2013) 3840–3942.
[16] G. Kaur, O.P. Pandey, K. Singh, D. Homa, B. Scott, G. Pickrell, A review of bioactive glasses: their structure, properties, fabrication and apatite formation, J. Biomed. Mater. Res. A 102 (1) (2014) 254–274.
[17] L.L. Hench, R.J. Splinter, W.C. Allen, T.K. Greenlee, Bonding mechanisms at the interface of ceramic prosthetic materials, J. Biomed. Mater. Res. 5 (6) (1971) 117–141.
[18] R. Li, A.E. Clark, L.L. Hench, An investigation of bioactive glass powders by sol-gel processing, J. Appl. Biomater. 2 (4) (1991) 231–239.
[19] P. Sepulveda, J.R. Jones, L.L. Hench, Characterization of melt-derived 45S5 and sol-gel-derived 58S bioactive glasses, J. Biomed. Mater. Res. 58 (6) (2001) 734–740.
[20] I. Izquierdo-Barba, L. Ruiz-González, J.C. Doadrio, J.M. González-Calbet, M. Vallet-Regí, Tissue regeneration: a new property of mesoporous materials, Solid State Sci. 7 (8) (2005) 983–989.

[21] X. Yan, C. Yu, X. Zhou, J. Tang, D. Zhao, Highly ordered mesoporous bioactive glasses with superior in vitro bone-forming bioactivities, Angew. Chem. Int. Ed. 43 (44) (2004) 5980–5984.
[22] R.S. Schwartz, J.W. Robbins, Post placement and restoration of endodontically treated teeth: a literature review, J. Endod. 30 (5) (2004) 289–301.
[23] M. Irie, K. Suzuki, D.C. Watts, Marginal gap formation of light-activated restorative materials: effects of immediate setting shrinkage and bond strength, Dent. Mater. 18 (3) (2002) 203–210.
[24] M.-H. Chen, Update on dental nanocomposites, J. Dent. Res. 89 (6) (2010) 549–560.
[25] R. Frankenberger, F.R. Tay, Self-etch vs etch-and-rinse adhesives: effect of thermo-mechanical fatigue loading on marginal quality of bonded resin composite restorations, Dent. Mater. 21 (5) (2005) 397–412.
[26] S.R. Armstrong, M.A. Vargas, Q. Fang, J.E. Laffoon, Microtensile bond strength of a total-etch 3-step, total-etch 2-step, self-etch 2-step, and a self-etch 1-step dentin bonding system through 15-month water storage, J. Adhes. Dent. 5 (1) (2003) 47–56.
[27] H. Schmidt, New type of non-crystalline solids between inorganic and organic materials, J. Non-Cryst. Solids 73 (1-3) (1985) 681–691.
[28] K.-H. Haas, S. Amberg-Schwab, K. Rose, Functionalized coating materials based on inorganic-organic polymers, Thin Solid Films 351 (1) (1999) 198–203.
[29] H. Wolter, W. Storch, A new silane precursor with reduced polymerization shrinkage, J. Sol-Gel Sci. Technol. 2 (1-3) (1994) 93–96.
[30] A.C. Hennig, E.B. Helbig, E. Haufe, G. Richter, H.W. Klimm, Restoration of Class V cavities with the Ormocer-based filling system Admira, Schweiz Monatsschrift Für Zahnmed Rev Mens Suisse Odonto-Stomatol Riv Mens Svizzera Odontol E Stomatol SSO 114 (2) (2004) 104–114.
[31] G. Pruthi, V. Jain, H.C. Kandpal, V.P. Mathur, N. Shah, Effect of bleaching on color change and surface topography of composite restorations, Int. J. Dent. 2010 (2010) 695748.
[32] N. Ayad, A. Bedewi, S. Hanafy, S. Saka, Effect of bleaching on microleakage, surface hardness, surface roughness, and color change of an Ormocer and a conventional hybrid resin composite, Internet J. Dent. Sci. 6 (2) (2009).
[33] D. Tebbe, R. Thull, U. Gbureck, Influence of spacer length on heparin coupling efficiency and fibrinogen adsorption of modified titanium surfaces, Biomed. Eng. OnLine 6 (1) (2007) 31.
[34] R. Müller, J. Abke, E. Schnell, et al., Surface engineering of stainless steel materials by covalent collagen immobilization to improve implant biocompatibility, Biomaterials 26 (34) (2005) 6962–6972.
[35] M. Chatzinikolaidou, T.K. Lichtinger, R.T. Müller, H.P. Jennissen, Peri-implant reactivity and osteoinductive potential of immobilized rhBMP-2 on titanium carriers, Acta Biomater. 6 (11) (2010) 4405–4421.
[36] K. Schickle, K. Zurlinden, C. Bergmann, et al., Synthesis of novel tricalcium phosphate-bioactive glass composite and functionalization with rhBMP-2, J. Mater. Sci. Mater. Med. 22 (4) (2011) 763–771.
[37] C. Przybylowski, T. Quinn, A. Callahan, et al., MC3T3 preosteoblast differentiation on bone morphogenetic protein-2 peptide ormosils, J. Mater. Chem. 22 (21) (2012) 10672.
[38] E.C. Pegg, G.S. Walker, C.A. Scotchford, D. Farrar, D. Grant, Mono-functional aminosilanes as primers for peptide functionalization, J. Biomed. Mater. Res. A 90A (4) (2009) 947–958.
[39] M. Mohorčič, I. Jerman, M. Zorko, et al., Surface with antimicrobial activity obtained through silane coating with covalently bound polymyxin B, J. Mater. Sci. Mater. Med. 21 (10) (2010) 2775–2782.

[40] B. Gottenbos, H.C. van der Mei, F. Klatter, P. Nieuwenhuis, H.J. Busscher, In vitro and in vivo antimicrobial activity of covalently coupled quaternary ammonium silane coatings on silicone rubber, Biomaterials 23 (6) (2002) 1417–1423.
[41] J.P. Matinlinna, K. Laajalehto, T. Laiho, I. Kangasniemi, L.V.J. Lassila, P.K. Vallittu, Surface analysis of Co-Cr-Mo alloy and Ti substrates silanized with trialkoxysilanes and silane mixtures, Surf. Interface Anal. 36 (3) (2004) 246–253.
[42] J.M. Antonucci, S.H. Dickens, B.O. Fowler, H.H.K. Xu, W.G. McDonough, Chemistry of silanes: interfaces in dental polymers and composites, J. Res. Natl. Inst. Stand. Technol. 110 (5) (2005) 541–558.
[43] J.P. Matinlinna, M. Özcan, L.V.J. Lassila, P.K. Vallittu, The effect of a 3-methacryloxypropyltrimethoxysilane and vinyltriisopropoxysilane blend and tris(3-trimethoxysilylpropyl)isocyanurate on the shear bond strength of composite resin to titanium metal, Dent. Mater. 20 (9) (2004) 804–813.
[44] T. Hooshmand, J.P. Matinlinna, A. Keshvad, S. Eskandarion, F. Zamani, Bond strength of a dental leucite-based glass ceramic to a resin cement using different silane coupling agents, J. Mech. Behav. Biomed. Mater. 17 (2013) 327–332.
[45] T. Hooshmand, R. van Noort, A. Keshvad, Bond durability of the resin-bonded and silane treated ceramic surface, Dent. Mater. 18 (2) (2002) 179–188.
[46] J.P. Matinlinna, L.V.J. Lassila, P.K. Vallittu, The effect of three silane coupling agents and their blends with a cross-linker silane on bonding a bis-GMA resin to silicatized titanium (a novel silane system), J. Dent. 34 (10) (2006) 740–746.
[47] J.P. Matinlinna, L.V.J. Lassila, I. Kangasniemi, P.K. Vallittu, Isocyanato- and methacryloxysilanes promote bis-GMA adhesion to titanium, J. Dent. Res. 84 (4) (2005) 360–364.
[48] O. Hansson, L.-E. Moberg, Evaluation of three silicoating methods for resin-bonded prostheses, Eur. J. Oral Sci. 101 (4) (1993) 243–251.
[49] E. Jakob, R. Marx, Silicoater method for bonded bridgework, Dtsch. Zahnärztl. Z. 43 (4) (1988) 461–464.
[50] R. Guggenberger, Rocatec system—adhesion by tribochemical coating, Dtsch. Zahnarztl. Z. 44 (11) (1989) 874–876.
[51] U. Gbureck, A. Masten, J. Probst, R. Thull, Tribochemical structuring and coating of implant metal surfaces with titanium oxide and hydroxyapatite layers, Mater. Sci. Eng. C-Biomim. Supramol. Syst. 23 (3) (2003) 461–465.
[52] U. Gbureck, S. Grubel, R. Thull, J.E. Barralet, Modified PMMA cements for a hydrolysis resistant metal-polymer interface in orthopaedic applications, Acta Biomater. 1 (6) (2005) 671–676.
[53] W.J. Maloney, T. Schmalzried, W.H. Harris, Analysis of long-term cemented total hip arthroplasty retrievals, Clin. Orthop. 405 (2002) 70–78.
[54] T. Miyazaki, C. Ohtsuki, M. Kyomoto, M. Tanihara, A. Mori, K. Kuramotoz, Bioactive PMMA bone cement prepared by modification with methacryloxypropyltrimethoxysilane and calcium chloride, J. Biomed. Mater. Res. A 67A (4) (2003) 1417–1423.
[55] A.E. Garcia-Bennett, Synthesis, toxicology and potential of ordered mesoporous materials in nanomedicine, Nanomedicine 6 (5) (2011) 867–877.
[56] S. Radin, P. Ducheyne, T. Kamplain, B.H. Tan, Silica sol-gel for the controlled release of antibiotics. I. Synthesis, characterization, and in vitro release, J. Biomed. Mater. Res. 57 (2) (2001) 313–320.
[57] R.K. Singh, T.-H. Kim, J.-J. Kim, E.-J. Lee, J.C. Knowles, H.-W. Kim, Mesoporous silica tubular nanocarriers for the delivery of therapeutic molecules, RSC Adv. 3 (23) (2013) 8692–8704.
[58] S. Radin, T. Chen, P. Ducheyne, The controlled release of drugs from emulsified, sol gel processed silica microspheres, Biomaterials 30 (5) (2009) 850–858.

[59] F. De Gaetano, L. Ambrosio, M.G. Raucci, A. Marotta, M. Catauro, Sol-gel processing of drug delivery materials and release kinetics, J. Mater. Sci.: Mater. Med. 16 (3) (2005) 261–265.

[60] P. Munusamy, M.N. Seleem, H. Alqublan, R. Tyler, N. Sriranganathan, G. Pickrell, Targeted drug delivery using silica xerogel systems to treat diseases due to intracellular pathogens, Mater. Sci. Eng. C: Mater. Biol. Appl. 29 (8) (2009) 2313–2318.

[61] A. Assifaoui, F. Bouyer, O. Chambin, P. Cayot, Silica-coated calcium pectinate beads for colonic drug delivery, Acta Biomater. 9 (4) (2013) 6218–6225.

[62] E. Vanea, C. Moraru, A. Vulpoi, S. Cavalu, V. Simon, Freeze-dried and spray-dried zinc-containing silica microparticles entrapping insulin, J. Biomater. Appl. 28 (8) (2014) 1190–1199.

[63] M. Prokopowicz, Bioactive silica-based nanomaterials for doxorubicin delivery: evaluation of structural properties associated with release rate, Mater. Sci. Eng. C: Mater. Biol. Appl. 33 (7) (2013) 3942–3950.

[64] D. Tebbe, R. Thull, U. Gbureck, Correlation between heparin release and polymerization degree of organically modified silica xerogels from 3-methacryloxypropylpolysilsesquioxane, Acta Biomater. 3 (6) (2007) 829–837.

[65] P. Kortesuo, M. Ahola, M. Kangas, I. Kangasniemi, A. Yli-Urpo, J. Kiesvaara, In vitro evaluation of sol-gel processed spray dried silica gel microspheres as carrier in controlled drug delivery, Int. J. Pharm. 200 (2) (2000) 223–229.

[66] B. Lillis, C. Grogan, H. Berney, W.A. Lane, Investigation into immobilisation of lactate oxidase to improve stability, Sens. Actuators B-Chem. 68 (1-3) (2000) 109–114.

[67] J.M. Harris, G.P. Lopez, W.M. Reichert, Silica-dispersed glucose oxidase for glucose sensing: in vitro testing in serum and blood and the effect of condensation pH, Sens. Actuators B-Chem. 174 (2012) 373–379.

[68] J. Livage, T. Coradin, C. Roux, Encapsulation of biomolecules in silica gels, J. Phys. Condens. Matter 13 (33) (2001) R673–R691.

[69] M.L. Ferrer, F. del Monte, D. Levy, A novel and simple alcohol-free sol-gel route for encapsulation of labile proteins, Chem. Mater. 14 (9) (2002) 3619.

[70] L. Ellerby, C. Nishida, F. Nishida, et al., Encapsulation of proteins in transparent porous silicate-glasses prepared by the sol-gel method, Science 255 (5048) (1992) 1113–1115.

[71] R.B. Bhatia, C.J. Brinker, A.K. Gupta, A.K. Singh, Aqueous sol-gel process for protein encapsulation, Chem. Mater. 12 (8) (2000) 2434–2441.

[72] C.J. Brinker, G.W. Scherer, Sol-Gel Science: The Physics and Chemistry of Sol-Gel Processing, Gulf Professional Publishing, Houston, TX, 1990.

[73] S. Radin, S. Bhattacharyya, P. Ducheyne, Nanostructural control of the release of macromolecules from silica sol-gels, Acta Biomater. 9 (8) (2013) 7987–7995.

[74] H.K. Karagulyan, V.K. Gasparyan, S.R. Decker, Immobilization of fungal beta-glucosidase on silica gel and kaolin carriers, Appl. Biochem. Biotechnol. 146 (1-3) (2008) 39–47.

[75] P. Wang, S. Dai, S.D. Waezsada, A.Y. Tsao, B.H. Davison, Enzyme stabilization by covalent binding in nanoporous sol-gel glass for nonaqueous biocatalysis, Biotechnol. Bioeng. 74 (3) (2001) 249–255.

[76] T. Yanagisawa, T. Shimizu, K. Kuroda, C. Kato, The preparation of alkyltrimethylammonium-kanemite complexes and their conversion to microporous materials, Bull. Chem. Soc. Jpn. 63 (4) (1990) 988–992.

[77] M. Malmsten, N. Burns, A. Veide, Electrostatic and hydrophobic effects of oligopeptide insertions on protein adsorption, J. Colloid Interface Sci. 204 (1) (1998) 104–111.

[78] K.F. Tipton, H.B.F. Dixon, Effects of pH on enzymes, in: D.L. Purich (Ed.), Methods in Enzymology, Academic Press, New York, 1979, pp. 183–234.

[79] Q. Chen, G.L. Kenausis, A. Heller, Stability of oxidases immobilized in silica gels, J. Am. Chem. Soc. 120 (19) (1998) 4582–4585.
[80] S.E. Favero-Longo, C. Gazzano, M. Girlanda, et al., Physical and chemical deterioration of silicate and carbonate rocks by meristematic microcolonial fungi and endolithic lichens (Chaetothyriomycetidae), Geomicrobiol. J. 28 (8) (2011) 732–744.
[81] N. Cutler, H. Viles, Eukaryotic microorganisms and stone biodeterioration, Geomicrobiol. J. 27 (6-7) (2010) 630–646.
[82] D.J. Dickson, R.L. Ely, Silica sol-gel encapsulation of cyanobacteria: lessons for academic and applied research, Appl. Microbiol. Biotechnol. 97 (5) (2013) 1809–1819.
[83] G. Carturan, R. Campostrini, S. Dire, V. Scardi, E. Dealteriis, Inorganic gels for immobilization of biocatalysts—inclusion of invertase-active whole cells of yeast (*Saccharomyces cerevisiae*) into thin-layers of SiO_2 gel deposited on glass sheets, J. Mol. Catal. 57 (1) (1989) L13–L16.
[84] L. Inama, S. Dire, G. Carturan, A. Cavazza, Entrapment of viable microorganisms by SiO_2 sol-gel layers on glass surfaces—trapping, catalytic performance and immobilization durability, J. Biotechnol. 30 (2) (1993) 197–210.
[85] G. Carturan, R.D. Monte, G. Pressi, S. Secondin, P. Verza, Production of valuable drugs from plant cells immobilized by hybrid sol-gel SiO_2, J. Sol-Gel Sci. Technol. 13 (1-3) (1998) 273–276.
[86] K.P. Peterson, C.M. Peterson, E.J.A. Pope, Silica sol-gel encapsulation of pancreatic islets, Proc. Soc. Exp. Biol. Med. 218 (4) (1998) 365–369.
[87] A. Coiffier, T. Coradin, C. Roux, O.M.M. Bouvet, J. Livage, Sol-gel encapsulation of bacteria: a comparison between alkoxide and aqueous routes, J. Mater. Chem. 11 (8) (2001) 2039–2044.
[88] J.B. Chen, Y. Xu, J.Y. Xin, S.B. Li, C.G. Xia, J.R. Cui, Efficient immobilization of whole cells of *Methylomonas* sp strain GYJ3 by sol-gel entrapment, J. Mol. Catal. B-Enzym. 30 (3-4) (2004) 167–172.
[89] J.C. Harper, D.M. Lopez, E.C. Larkin, et al., Encapsulation of *S. cerevisiae* in poly(glycerol) silicate derived matrices: effect of matrix additives and cell metabolic phase on long-term viability and rate of gene expression, Chem. Mater. 23 (10) (2011) 2555–2564.
[90] N. Nassif, O. Bouvet, M.N. Rager, C. Roux, T. Coradin, J. Livage, Living bacteria in silica gels, Nat. Mater. 1 (1) (2002) 42–44.
[91] S. Sakai, T. Ono, H. Ijima, K. Kawakami, In vitro and in vivo evaluation of alginate/sol-gel synthesized aminopropyl-silicate/alginate membrane for bioartificial pancreas, Biomaterials 23 (21) (2002) 4177–4183.
[92] S. Boninsegna, P. Bosetti, G. Carturan, G. Dellagiacoma, R.D. Monte, M. Rossi, Encapsulation of individual pancreatic islets by sol-gel SiO_2: a novel procedure for perspective cellular grafts, J. Biotechnol. 100 (3) (2003) 277–286.
[93] M. Muraca, M.T. Vilei, G.E. Zanusso, et al., SiO_2 entrapment of animal cells: liver-specific metabolic activities in silica-overlaid hepatocytes, Artif. Organs 26 (8) (2002) 664–669.
[94] M. Muraca, M.T. Vilei, E. Zanusso, et al., Encapsulation of hepatocytes by SiO_2, Transplant. Proc. 32 (8) (2000) 2713–2714.
[95] U. Soltmann, H. Bottcher, D. Koch, G. Grathwohl, Freeze gelation: a new option for the production of biological ceramic composites (biocers), Mater. Lett. 57 (19) (2003) 2861–2865.
[96] A. Pannier, Ć. Oehm, A.R. Fischer, P. Werner, U. Soltmann, H. Boettcher, Biodegradation of fuel oxygenates by sol-gel immobilized bacteria *Aquincola tertiaricarbonis* L108, Enzyme Microb. Technol. 47 (6) (2010) 291–296.
[97] U. Kunzelmann, H. Bottcher, Biosensor properties of glucose oxidase immobilized within SiO_2 gels, Sens. Actuators B-Chem. 39 (1–3) (1997) 222–228.

[98] J.F.T. Conroy, M.E. Power, J. Martin, et al., Cells in sol-gels I: a cytocompatible route for the production of macroporous silica gels, J. Sol-Gel Sci. Technol. 18 (3) (2000) 269–283.
[99] A. Pannier, M. Mkandawire, U. Soltmann, W. Pompe, H. Boettcher, Biological activity and mechanical stability of sol-gel-based biofilters using the freeze-gelation technique for immobilization of *Rhodococcus ruber*, Appl. Microbiol. Biotechnol. 93 (4) (2012) 1755–1767.
[100] E. Callone, R. Campostrini, G. Carturan, A. Cavazza, R. Guzzon, Immobilization of yeast and bacteria cells in alginate microbeads coated with silica membranes: procedures, physico-chemical features and bioactivity, J. Mater. Chem. 18 (40) (2008) 4839–4848.
[101] D. Fiedler, U. Hager, H. Franke, U. Soltmann, H. Boettcher, Algae biocers: astaxanthin formation in sol-gel immobilised living microalgae, J. Mater. Chem. 17 (3) (2007) 261–266.
[102] A. Leonard, J.C. Rooke, C.F. Meunier, H. Sarmento, J.-P. Descy, B.-L. Su, Cyanobacteria immobilised in porous silica gels: exploring biocompatible synthesis routes for the development of photobioreactors, Energy Environ. Sci. 3 (3) (2010) 370–377.
[103] A. Leonard, P. Dandoy, E. Danloy, et al., Whole-cell based hybrid materials for green energy production, environmental remediation and smart cell-therapy, Chem. Soc. Rev. 40 (2) (2011) 860–885.
[104] J.C. Rooke, A. Leonard, H. Sarmento, J.-P. Descy, B.-L. Su, Photosynthesis within porous silica gel: viability and activity of encapsulated cyanobacteria, J. Mater. Chem. 18 (24) (2008) 2833–2841.
[105] J.C. Rooke, A. Leonard, H. Sarmento, C.F. Meunier, J.-P. Descy, B.-L. Su, Novel photosynthetic CO_2 bioconvertor based on green algae entrapped in low-sodium silica gels, J. Mater. Chem. 21 (4) (2011) 951–959.
[106] C. Sicard, R. Brayner, J. Margueritat, et al., Nano-gold biosynthesis by silica-encapsulated micro-algae: a "living" bio-hybrid material, J. Mater. Chem. 20 (42) (2010) 9342–9347.
[107] J. Raff, U. Soltmann, S. Matys, et al., Bacteria-based bioceramics for bioremediation of uranium mining waste waters, in: Abstracts of the General Meeting of the American Society for Microbiology, 2003, p. Q-310.
[108] P.C. Stark, G.D. Rayson, Comparisons of metal-ion binding to immobilized biogenic materials in a flowing system, Adv. Environ. Res. 4 (2) (2000) 113–122.
[109] T. Branyik, G. Kuncova, J. Paca, The use of silica gel prepared by sol-gel method and polyurethane foam as microbial carriers in the continuous degradation of phenol, Appl. Microbiol. Biotechnol. 54 (2) (2000) 168–172.
[110] D. Fiedler, A. Thron, U. Soltmann, H. Bottcher, New packing materials for bioreactors based on coated and fiber-reinforced biocers, Chem. Mater. 16 (16) (2004) 3040–3044.
[111] G.J. Copello, M.C. De Marzi, M.F. Desimone, E.L. Malchiodi, L.E. Diaz, Antibody detection employing sol-gel immobilized parasites, J. Immunol. Methods 335 (1-2) (2008) 65–70.
[112] J.A. O'Loughlin, J.M. Bruder, M.J. Lysaght, In vivo and in vitro degradation of urea and uric acid by encapsulated genetically modified microorganisms, Tissue Eng. 10 (9-10) (2004) 1446–1455.
[113] P. Nicolas Catalano, N. Soledad Bourguignon, G. Solange Alvarez, et al., Sol-gel immobilized ovarian follicles: collaboration between two different cell types in hormone production and secretion, J. Mater. Chem. 22 (23) (2012) 11681–11687.
[114] S. Sakai, T. Ono, H. Ijima, K. Kawakami, Proliferation and insulin secretion function of mouse insulinoma cells encapsulated in alginate/sol-gel synthesized aminopropyl-silicate/alginate microcapsule, J. Sol-Gel Sci. Technol. 28 (2) (2003) 267–272.

[115] P.J. Morris, Immunoprotection of therapeutic cell transplants by encapsulation, Trends Biotechnol. 14 (5) (1996) 163–167.

[116] J.D. Marwick, P.C. Wright, J.G. Burgess, Bioprocess intensification for production of novel marine bacterial antibiotics through bioreactor operation and design, Mar. Biotechnol. 1 (5) (1999) 495–507.

[117] T. Joki, M. Machluf, A. Atala, et al., Continuous release of endostatin from microencapsulated engineered cells for tumor therapy, Nat. Biotechnol. 19 (1) (2001) 35–39.

[118] R. Yongsunthon, S.K. Lower, Force spectroscopy of bonds that form between a *Staphylococcus* bacterium and silica or polystyrene substrates, J. Electron Spectrosc. Relat. Phenom. 150 (2-3) (2006) 228–234.

[119] E. Pena-Vazquez, E. Maneiro, C. Perez-Conde, M. Cruz Moreno-Bondi, E. Costas, Microalgae fiber optic biosensors for herbicide monitoring using sol-gel technology, Biosens. Bioelectron. 24 (12) (2009) 3538–3543.

[120] J. Szilva, G. Kuncova, M. Patzak, P. Dostalek, The application of a sol-gel technique to preparation of a heavy metal biosorbent from yeast cells, J. Sol-Gel Sci. Technol. 13 (1-3) (1998) 289–294.

[121] S. Ramachandran, T. Coradin, P.K. Jain, S.K. Verma, *Nostoc calcicola* immobilized in silica-coated calcium alginate and silica gel for applications in heavy metal biosorption, Silicon 1 (4) (2009) 215–223.

[122] M. Perullini, M. Jobbagy, N. Mouso, F. Forchiassin, S.A. Bilmes, Silica-alginate-fungi biocomposites for remediation of polluted water, J. Mater. Chem. 20 (31) (2010) 6479–6483.

[123] L. Lloret, G. Eibes, G. Feijoo, M.T. Moreira, J.M. Lema, F. Hollmann, Immobilization of laccase by encapsulation in a sol-gel matrix and its characterization and use for the removal of estrogens, Biotechnol. Prog. 27 (6) (2011) 1570–1579.

[124] C.G. Kauffmann, R.T. Mandelbaum, Entrapment of atrazine-degrading enzymes in sol-gel glass, J. Biotechnol. 51 (3) (1996) 219–225.

[125] G. Solange Alvarez, M. Lucia Foglia, D. Edhit Camporotondi, M. Victoria Tuttolomondo, M. Federico Desimone, D.L. Eduardo, A functional material that combines the Cr(VI) reduction activity of *Burkholderia* sp. with the adsorbent capacity of sol-gel materials, J. Mater. Chem. 21 (17) (2011) 6359–6364.

CHAPTER 4

A Novel Approach for Facile Synthesis of Biocompatible PVA-Coated PLA Nanofibers as Composite Membrane Scaffolds for Enhanced Osteoblast Proliferation

Abdalla Abdal-hay[1,2,3], Abdel Salam Hamdy Makhlouf[4], Pablo Vanegas[1]

[1]Department of Computer Science, Faculty of Engineering, Universidad de Cuenca, Cuenca, Ecuador
[2]Department of Engineering Materials and Mechanical Design, Faculty of Engineering, South Valley of University, Qena, Egypt
[3]Department of Bionano System Engineering, College of Engineering, Chonbuk National University, Jeonju, Republic of Korea
[4]Department of Manufacturing Engineering, College of Engineering and Computer Science, University of Texas Pan-American, Edinburg, Texas, USA

Contents

Handbook of Nanoceramic and Nanocomposite Coatings and Materials
http://dx.doi.org/10.1016/B978-0-12-799947-0.00004-3

1. INTRODUCTION

The uses of biodegradable polymers as effective bone substitutes have attracted a great interest in biomedical fields during the last decade. Polylactide (PLA) is one of the most widely used biodegradable polymers for bone tissue engineering (TE) as it degrades to nontoxic lactic acid, which is naturally present in the human body, and FDA-approved [1,2]. However, some of the major limitations in the use of PLA for scaffolds are its hydrophobic character, resulting in suboptimal adhesion, spreading and growth of cells on the surface of the material [1,3], and decreased hydraulic permeability through the scaffold. The hydrophobic nature of PLA can become a serious problem in a predominantly hydrophilic bioenvironment. Further, it is considered as a too brittle material, which makes it nonoptimal for multibiomedical applications [4,5].

Considerable efforts have been made to address the aforementioned limitations of PLA aliphatic polyester polymer [2,6,7]. Some of these attempts were carried out by blending or alloying PLA with other polymers. The combination of the two polymers proved to be a viable scheme to generate new materials of desirable properties. Different types of organic compounds, such as polyethylene oxide, polypropylene oxide, poly(vinyl alcohol) (PVA), 4,4-methylene diphenyl diisocyanate, and ε-caprolactone [2,6–11], have been combined with PLA to improve its physiochemical, biological, and mechanical properties for multidesirable applications.

The previous studies have simply used a premixing solution and melt-blended the PLA with organic compounds. For example, Martin and Avérous [12] have melt-blended PLA with thermoplastic starch (TPS). However, microscopic observations revealed nonuniformly dispersed PLA inclusions in the TPS matrix. Pitarresi et al. [13] produced fiber scaffolds using electrospinning technique, employing PHEA-g-PLA copolymer as a starting material. However, these scaffolds didn't provide significant biocompatible properties. Meneghello et al. [14] found that a simple blending of PLGA with PVA shows detrimental effect on mechanical properties, with a loss of 31% in Young's modulus and more than 60% in tensile strength compared to pure PLGA membranes. Elakkiya et al. [15] synthesized PLA/PVA with a conventional premixing solution, and finally, electrospinning technique was used to fabricate fiber scaffolds. As expected, PVA showed higher water uptake and thus higher degradation rate than PLA polymer [16]. However, mixing both phases of PLA and PVA at the macromolecular level is difficult because PVA is more hydrophilic than PLA.

Consequently, conventional blending or alloying of PLA-based materials with PVA causes severe degradation of the fibrous scaffold, which can compromise scaffold strength and affect the biodegradation rate before the healing process occurs completely [15]. Accordingly, the conventional blending of PLA with other polymers seems an improper approach as it results in a product lacking many functions, including physicochemical properties and the mechanical stability due to the dissimilar macromolecular properties of the same organic compounds. The above copolymerization or modifications methods of PLA involved tedious and multistep processes and showed several disadvantages such as lack of ductility and low chemical interfacial bonding between the two polymers. Hence, continuous development of the existing scaffolds using promising materials should be continued in order to provide efficient templates for bone tissue growth and repair.

Surface modification of biomaterials is a popular method to improve device multifunctionality and mechanical properties [17]. Physiochemical modifications of PLA membrane after electrospinning can provide the membrane with enhanced properties and suitable functionality for promoting cell attachment and proliferation for bone defects applications [18]. However, in some cases, surface modifications can be time-consuming and costly compared to the conventional mats [18,19]. Under controlled processing conditions, PLA may remain mainly amorphous because of its slow crystallization kinetic [12]. Ribeiro et al. [20] reported that poly(L-lactide) electrospun fibers are amorphous but contain numerous crystal nuclei that could grow rapidly when the sample is heated up to 140 °C. Accordingly, we expect that the degree of crystallinity of the fibers can be tailored and controlled by heat treatment. In this regard, novel substrates composed of an electrospun PLA nanofiber mat coated with PVA can be successfully endowed by a simple hydrothermal approach (Scheme 1) with desired surface properties.

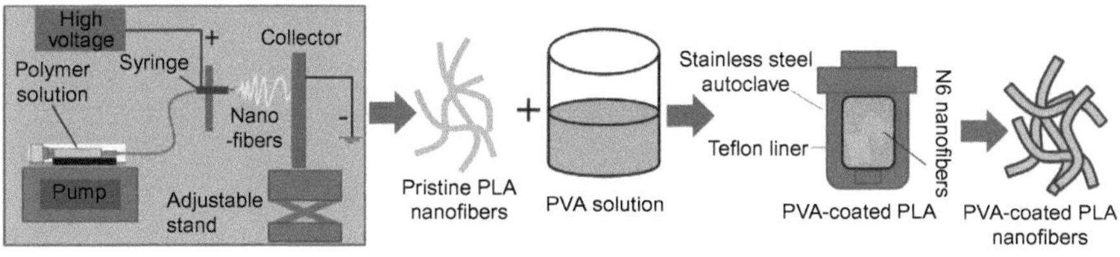

Scheme 1 Schematic diagram of the PLA electrospun nanofibers coated with PVA during hydrothermal treatment.

In the present study, we selected PVA due to its water solubility as a synthetic polymer, ease of preparation, biodegradability, and excellent physical and mechanical properties, in addition to its versatility in biomaterial applications [16,21]. PVA can interact with PLA by forming strong hydrogen bond with the oxygen atom of the ester group to improve the mechanical and structural stabilities [3]. In addition, PVA can significantly improve the mechanical properties of some organic compounds compared with pristine one. PVA, in general, has a high water content and tissue-like elasticity. The abundant hydroxyl groups on PVA can be readily modified to attach growth factors, adhesion proteins, or other molecules of biological importance [14].

The formation of PVA thin layer on PLA single nanofibers and formation of interconnected (bonded) or joined fibers could provide new insights toward better fabrication of durable high-performance bone tissue scaffolds. An improvement in the crystalline structure of the PLA coated with PVA associated with an increase of the reaction temperature can occur during hydrothermal treatment. We found out that the hydrothermal process induced the crystalline conformation due to enhanced chain mobility and significantly improved the ductility of PLA. Therefore, maximizing the hydrogen bonding between PLA as a substrate and PVA deposited as thin layer was found to provide superior mechanical properties with much higher elongation compared to PLA alone. From the viewpoint of biomedical applications, the composites do not release the acidic by-products in a burst-like situation after degradation. This helps in maintaining a pH level, which helps in cell attachment and proliferation. We suggest that degradable PVA-coated PLA nanofiber mats with increased hydrophilicity can impart desirable mechanical properties and support ECM deposition.

2. EXPERIMENTAL

2.1 Materials

The PLA type used in this study was Ingeo Biopolymer 2003D, a NatureWorks LLC (USA) product supplied by Green Chemical Co., Ltd (Korea). PLA was derived from an annually renewable resource and was specifically designed for use in food packing and biomedical applications (approved by the US Food and Drug Administration). The PLA sheets were transparent and amorphous, with a glass transition temperature of 53 °C [1]. The molecular weight of PVA ranged from 146,000 to 186,000 g mol^{-1}; 99% mol hydrolyzed was supplied by Sigma-Aldrich. Dichloromethane (DCM) (Junsei Chemical Co., Ltd., Japan) was used as a solvent. All of the chemicals used in the experiments were reagent grade.

2.2 PVA-Coated Electrospun PLA Scaffolds (Composite Mats)

The electrospinning setup used in the present work was described in detail in a previous publication [22]. The solution for electrospinning was prepared by dissolving PLA in a DCM at a concentration of 10 wt%. PLA solution was continuously stirred to obtain a transparent viscous solution appropriate for electrospinning. The solution was loaded into a 5 ml plastic syringe with a 22 gauge needle attached and dispensed using a syringe pump. The working distance between the tip of the needle and the collector was 20 cm and the applied voltage was 15 kV. The collected mat was placed in a vacuum oven overnight at 40 °C to remove any potential residual solvents.

PVA-coated PLA membrane mat was prepared following these procedures: (1) PLA as-electrospun mat was cut into rectangular specimens (30×20 mm^2). (2) These specimens were immersed into freshly prepared 1 and 2 wt% PVA aqueous solution. (3) 40 ml of PVA solution that contains PLA membrane specimens was transferred to a Teflon-lined autoclave container and heat-treated at 120 and 130 °C for 30 min (see Scheme 1). Previous studies reported that both the two polymers have good thermal resistivity up to these temperatures [21,23]. (4) After the reaction process, the PVA-coated PLA mats were gently rinsed in distilled water. (5) The resultant coated samples were left to dry at room temperature for 4 days until steady weight and then introduced into a vacuum oven (10 mbar) at 35 °C for 72 h. (6) The amount of PVA adhering to the surface was evaluated from the dry weight of the substrates.

2.3 Characterization

The surface properties of the fabricated mats were characterized by a field emission scanning electron microscope (FESEM; Hitachi S-7400, Hitachi Co., Tokyo, Japan) and X-ray diffraction analysis (XRD) (GA-XRD; Philips X'Pert, Holland). To measure the fiber diameters, the FESEM images were processed and analyzed with ImageJ (National Institutes of Health, Bethesda, MD, USA). Fourier transform infrared spectrometry (FT-IR) analysis was used to identify the phase structure of pristine and treated mats. The FT-IR spectra in transmission mode were collected using an ABB Bomem MB100 spectrometer (Bomem, Canada) with an FT-IR 4000 in the range of 4000-400 cm^{-1}. Differential scanning calorimetry (DSC; DSC Q20 V24.4 Build 116 instrument) was used to evaluate the thermal properties of the pristine PLA and PVA-PLA composite membranes. Two series of parameters (underlying rate, modulation, and period) were used to study the glass transitions, melting temperatures, and relaxation enthalpies of the samples. For all experiments, the purge flow gas was nitrogen (50 ml/min) to prevent the nanocomposite membranes from oxidization. The samples were first heated to 200 °C at a heating rate of 10 °C min^{-1} in order to eliminate any thermal history and were then isothermally conditioned at 200 °C for 1 min, quickly cooled to 0 °C, isothermally conditioned at 0 °C for 1 min, and then heated again to 200 °C at the same rate. The glass transition temperature (T_g), the crystallization temperature (T_c), the melting temperature (T_m), and the degree of crystallinity (X_c) were determined from the second heating scans [12]. The thermal history of the samples was discarded by means of the first heating scans. The T_m and T_c were taken at the beak value of the respective of the endotherms and exotherms, and the T_g at the midpoint of heat capacity changes. The degree of crystallinity (X_c) of PLA in the composites is calculated by the following equation:

$$X_c = \frac{\Delta H_f}{W \times \Delta H_{f^0}} \tag{1}$$

where ΔH_f is the apparent melt enthalpy of the composite mats, W is the weight fraction of PLA in the composite, and ΔH_f is the melt enthalpy of in its completely crystalline state.

Flat mat was used to evaluate the hydrophilicity of PLA and PVA-PLA mats, using the water contact angle (WCA) method. Drops of deionized water (3 μl) were deposited onto the top shiny side of 50 × 30 mm^2 mats positioned on the stage of a bench-type contact angle goniometer (GBX; Digidrop, France), ensuring that the membrane mat was completely flat. Measurements were repeated at different positions on the flat sheet. The measurement was taken after 1 s, five specimens for each sample were selected, five different locations on each specimen were recorded, and STD (±) was calculated.

2.4 Mechanical Characterization of Fabricated Scaffolds

An Instron universal testing machine (LLOYD Instruments, LR5K Plus, UK) was used to study the stress-strain behavior pristine and composite samples. Samples were trimmed into a "dog bone" (see inset of Figure 7) profile with offset ends to reduce grip effects according to the procedures of ASTM D-638 via die cutting from the as-obtained mats for mechanical tests. Testing was conducted with the tissue grips moving at a rate of 5 mm min^{-1} and the load was applied until the specimen shows a complete failure. Polymer mat layer thickness was assessed by ultrasonic measurement (coating thickness gauge OMEGA Instrument, OM179-745) with a precision of

1 μm (thicknesses of the fiber mat were listed in Table 2), and each sample's width and thickness were measured before testing. The tensile modulus was calculated as the slope of the initial linear portion of the stress-strain curve. Five membrane mat samples of each approach were subjected to this test. All testing was performed at room temperature. The data presented are expressed as the mean ± standard deviation.

2.5 Cytocompatibility Test (Cell Attachment and Proliferation)

For evaluation of cell attachment and growth, MC3T-E1 preosteoblast cells were cultured in Dulbecco's modified Eagle's medium (Gibco Co, USA), supplemented with 10% fetal bovine serum in a humidified incubator with 5% CO_2 and 95% relative humidity at 37 °C. The extraction media was at 100% concentration after 1-day incubation in a humidified atmosphere with 5% CO_2 at 37 °C. The cell cultures for composite (coated) and pristine samples were performed by using a direct contact method. Prior to cell culture, the samples were sterilized by ethylene oxide gas. Cells were incubated in 24-well flat-bottomed cell culture plates with 5×10^4 cells/ml medium per well and incubated for 24 h to allow cell attachment on the samples surface. The substrates were washed twice with Phosphate Buffered Saline (PBS) after an interval of culturing. Cell morphology was examined by a scanning electronic microscope (SEM; JEOL JSM 820, USA).

MTT assay was carried out according to the previous publication [17], in brief; MC3T3-E1 cells were seeded to the electrospun scaffold surface at a density of 5×10^4 cells/ml and were grown in the α-MEM medium. The culture medium was incubated at 37 °C in a humidified 5% CO_2 incubator for different days (1, 3, or 7 days) and then subsequently removed and washed twice with sterile PBS. A 100 μl of MTT solution (0.5 mg/ml) was then added to each well. After 3 h incubation at 37 °C, 100 μl of dimethyl sulfoxide was added to dissolve the formazan crystals. The optical density of the formazan solution was detected by a 24-well plate reader at 540 nm using ELISA (Benchmark Microplate Reader, BIO-RAD, CA, USA). All results are reported as mean standard deviation.

3. RESULTS AND DISCUSSION

3.1 Effect of the Reaction Temperature (Hydrothermal Process)

It is stated that electrospinning provides a simpler, versatile, and cost-effective method to fabricate scaffolds with an interconnected pore structure (based on solvent evaporation rate) and fiber-free beads with diameters down to the submicron or nanometer range [24]. Regarding bone tissue application, electrospun scaffolds have been shown to support the attachment and proliferation of ECM [25]. In the present study, the surface topography of pristine PLA electrospun membrane mats before (noted as m_1) and after (noted as m_{1T}) hydrothermal treatment at 130 °C for 30 min was investigated using high-resolution SEM as shown in Figure 1a-d. The m_1 mat shows a random structure and smooth morphology with moderate uniform diameter along their lengths as well as non-interconnected (linear) fiber structure (Figure 1a and b). FESEM analysis of a pristine mat nanofiber verified the nanometer scale of the diameter of tissue scaffold to be 780 ± 212 nm.

For the sample m_{1T}, a considerable number of the fibers had broken down spontaneously with partially adhering one to another (dashed circular area inset of panel c in Figure 1). In addition, the nanofiber matrix of PLA changed into chunks consisting of short fiber fragments with porous structure on their surface (Figure 1c and d). These morphological changes in electrospun PLA matrices during in hydrothermal treatment can be attributed to the simple hydrolysis of ester backbone of aliphatic polyester under aqueous conditions (the mechanism of hydrolysis of ester was discussed in details

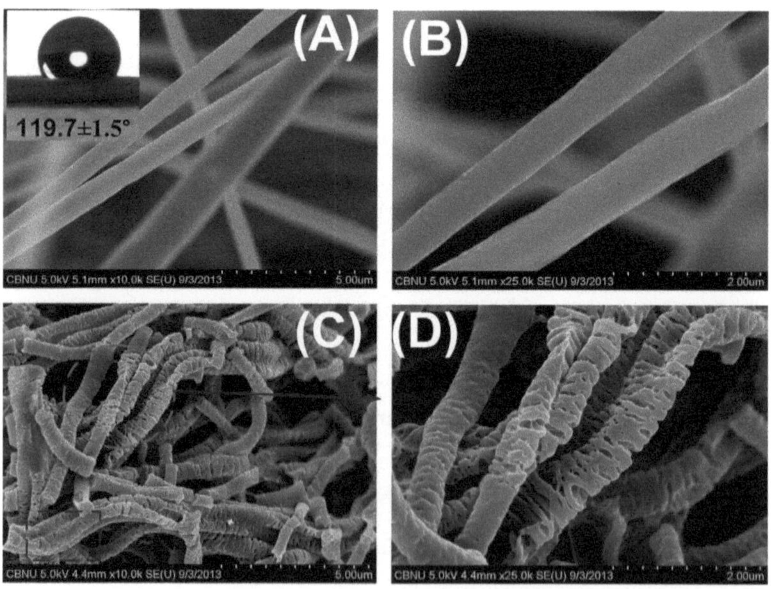

Figure 1 Scanning electron microscopy images of (a, b) as-electrospun PLA fiber mat; (c, d) treated fiber mats at 130 °C for 30 min. Inset of panel (a) shows the respective water contact angle (WCA) images.

elsewhere) [26]. In addition, this behavior can be attributed to the fiber contraction induced by crystallization and the release of internal residual stresses generated during processing. These observations are in good agreement with the data reported by Ribeiro et al. [20]. Our results confirmed the creation of porous structure on the fiber surface during the hydrothermal process, which agrees with the previous work done by other research groups [26,27] who observed this kind of porous morphology on electrospun fiber surfaces after immersion into aqueous solution.

Conversely, PVA-coated PLA fiber mat after hydrothermal process at 120 °C (noted as m_2) and 130 °C (m_3) for 30 min showed cylindrical morphology with few defects along the fiber axis as observed in FESEM morphologies (Figure 2a-d). No porous structure was observed on the fiber surface indicating that PVA plays a significant role to protect PLA fibers from the breakdown and fragment formation. The deposition of PVA onto the PLA provides interconnected fibers (point-bonded morphology) and fiber junctions as shown in m_2 mat (dashed areas in Figure 2a), which are absent in the pristine one (m_1). Further, it is clear that PVA completely covered

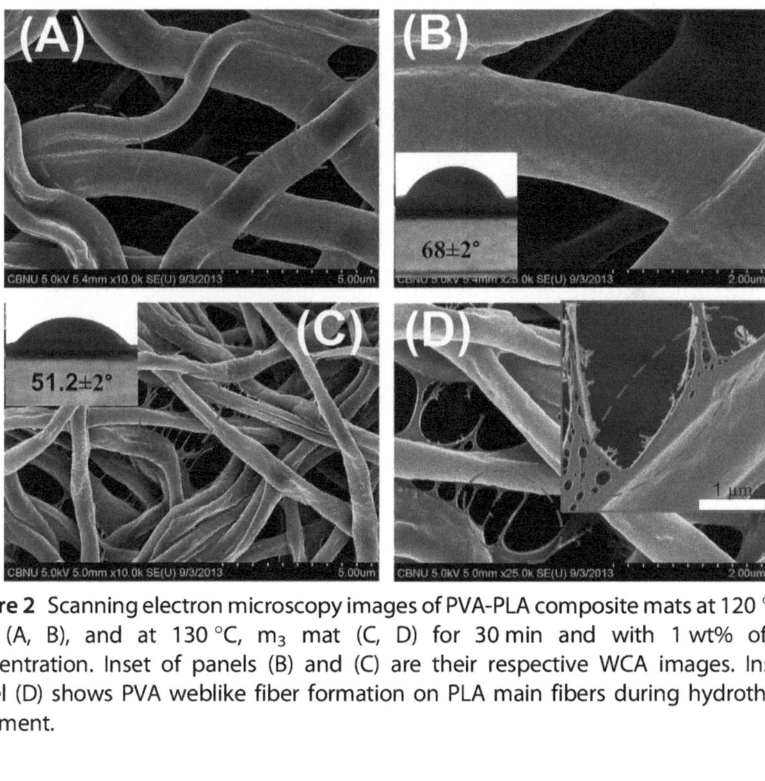

Figure 2 Scanning electron microscopy images of PVA-PLA composite mats at 120 °C, m_2 mat (A, B), and at 130 °C, m_3 mat (C, D) for 30 min and with 1 wt% of PVA concentration. Inset of panels (B) and (C) are their respective WCA images. Inset of panel (D) shows PVA weblike fiber formation on PLA main fibers during hydrothermal treatment.

PLA fibers. Figure 2 shows a well-defined layer deposited on the main nanofibers. High magnification panels reveal that PVA layer grew on the main nanofibers, which can be noticed also from the inset of panel d (the marked areas). Deposition of PVA layer was also evident from the apparent increase in fiber diameter of PLA electrospun fiber. The fiber diameters of the m_2 and m_3 composite mats were ranged from 770 to 2478 and 420 to 933 nm, respectively, as shown in Figure 2a-d. Accordingly, the reaction temperature has a significant effect on the coating formation properties. It is noteworthy mentioning that higher reaction temperature creates thinner layer. The mats coated at higher reaction temperature showed significant decrease in the coating thickness of 0.62 and 0.48 mg PVA/1.0 cm^2 of PLA mat for m_2 and m_3 composite mats, respectively. These data demonstrate a decrease in diameter size of the composite scaffold corresponding to an increase in reaction temperature. It is believed that with increasing temperature, the solution viscosity decreases resulting in higher solution flow onto the PLA single fibers, causing lower PVA thickness layer than that obtained at 120 °C. At 130 °C, highly branched bridge weblike PVA fibers were identified to form a kind of network connection between the coated fibers (dashed areas in Figure 2c and d). However, this behavior has not been detected at 120 °C (Figure 2a and b). It seems that the weblike fibers act as joints between the main fibers. It's likely that lower viscosity due to increasing the reaction temperature might cause higher polymer molecule mobility, which can explain the formation of branched PVA bridge fibers. PVA surfaces at high magnification (25,000 ×) appeared to be slightly rough. However, the reasons for this surface unevenness have not been investigated in the present work. We cannot judge whether it is an artifact or another behavior linked to the increase in reaction temperature.

3.2 Effect of the PVA Solution Concentration

In the present study, one of the major determinants of coated fiber morphology was the solution concentration. Figure 3 shows the effect of increasing the PVA solution concentration (2 wt%) at 130 °C for 30 min on the morphology of the as-obtained composite mat. From this figure, it is observed that both the PLA individual fibers and porous fibers were fully covered with PVA (Figure 3a and b), indicating higher solution viscosity that resulted in a reduction of the solution mobility and increase of the thickness of the polymer layer formed on PLA mats. Therefore, we suggest that increasing the PVA solution concentration (>1 wt%) affects negatively the surface

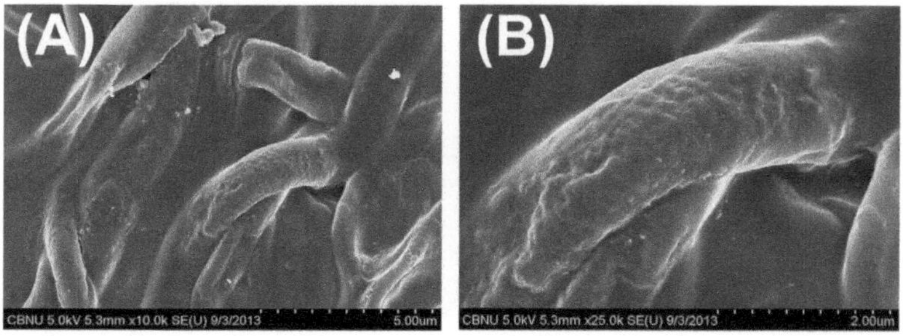

Figure 3 (a, b) Scanning electron microscopy images of PVA-PLA composite mats at 130 °C for 30 min and with 2 wt% of PVA concentration.

morphology of the nanofiber mats. As mentioned above, one of the main requirements in bone regeneration is to have scaffold materials with a 3-D interconnected pore structure organization, which has a prominent influence on cell intrusion, proliferation, and function and also on nutrient delivery, fluid absorption, and oxygen supply in bone TE [28,29].

3.3 Physicochemical Changes in Fibrous Mat During Composite Formation

To evaluate the physicochemical changes in fibrous mat during composite formation, we performed spectroscopic analysis using XRD, FT-IR, and DSC. The crystalline properties of the electrospun fibers are also of prime concern when considering the materials for biomedical application [30]. Thus, the chain conformation of biocompatible polymers, which are generally the materials of choice for the scaffold applications, directly reflects their physical, mechanical, and biological functions [31].

XRD experiments were performed on samples of as-received, pristine (m_1), hydrothermally treated (m_{1T}), and composite mats (m_2 and m_3). As-received PLA shows semicrystalline with broad and low diffraction intensity, as shown in Figure 4. This figure shows a negative change in crystal structure induced by the electrospinning process, where fully amorphous PLA structure was observed, which is in good agreement with the work ofDeitzel et al. [32]. On the other hand, XRD diffractograms of the treated (m_{1T} mat) samples showed several sharp and well-defined peaks compared with as-received and m_1 mat induced, indicating that the crystalline orders in the treated nanofiber mat significantly induced the chain conformation in the presence of hydrothermal treatment. In general, all the treated samples induced that the strongest diffraction peak at $2\theta = 16.9°$, corresponding to the (200, 110) reflection of the α-form crystals and other diffraction peaks

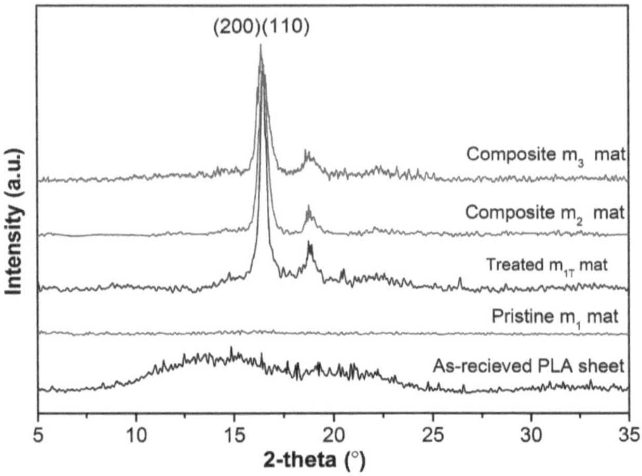

Figure 4 XRD profiles of the as-obtained samples.

with positions located at $2\theta = 19.0°$ (Figure 4), which can be ascribed to the crystal structure of PLA [15,33]. Upon PVA coatings onto PLA, the peak positions in each diffraction pattern are essentially identical, indicating that there is no significant change in crystal structure induced by the PVA coatings. Furthermore, the two diffractograms recorded for the two PVA-PLA (m_2 and m_3) composite mats were found to be similar, with same peaks and almost the same intensity. XRD results suggest that the hydrothermal process has no negative degradation effect on the PLA chain structure but it induces crystalline conformation on PLA. This finding is in good agreement with the work of Liu et al. [34] who reported that the crystallinity of as-electrospun polymer nanofibers was less and the reflection peak was broader than for the annealed fiber and solution-cast film samples. On the other hand, Elakkiya et al. [15] found that the incorporation of PVA using a conventional blending process into PLA fibers induced a negative crystalline effect on PLA, which leads to the amorphous nature of the blend, indicating the deposition of thin layer of PVA onto PLA fibers.

To verify our results, FT-IR spectra were used to study the effect of hydrothermal treatment in the presence of PVA solution on the molecular conformation of the polymer as shown in Figure 5 (and, in more details, Figure 6a and b that highlight the differences in the intensities and wave vibrations). The comparison of FT-IR spectra of as-fabricated mats before and after treatment clearly shows that sufficient amount of PVA was deposited on the surface of PLA fibers as the intensity of different bands of PVA-PLA mats was sufficiently decreased and became narrower. It was reported

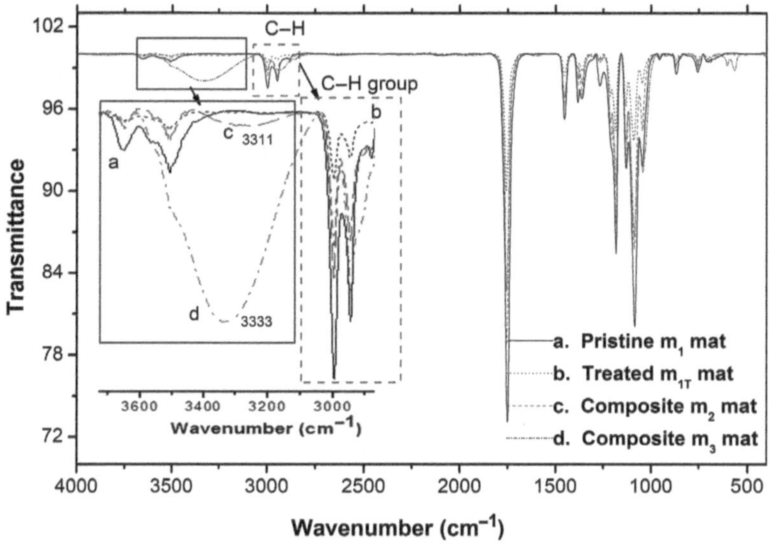

Figure 5 FT-IR spectra of the as-obtained samples. Inset shows the respective peak position at rectangular selected areas.

that the bands of the carbonyl group of PLA and the bands of the hydroxyl group of PVA occur at 1758 and 3333 cm^{-1}, respectively [16]. Accordingly, our results could clearly identify both bands in hydrothermal treated PVA-coated mats (blue selected area, inset of Figure 6b); the 3333 cm^{-1} peak of pristine PLA was not detected (inset of Figure 6a), which can be a sign of depression of this peak on PLA by deposition of PVA. Further details regarding the deposition of PVA molecules and their structure changes have been discussed by Asran et al. [16].

In particular, it is worth noting that the 921 cm^{-1} absorption band, which is characteristic of the α-crystal in PLA [20,23], is absent in the electrospun pristine mat (m_1) and becomes more intense in a composite mats (Figure 6b) indicating that the crystallinity of the polymer increases in the presence of PVA solution. As shown in Figure 6a, the increase in the crystallization degree is accompanied by a slight shift to higher frequencies and significant narrows and change in the shape of the absorption band between 830 and 890 cm^{-1} [35] for the treated mats, particularly composite mats, as expected for chain conformation [20]. Also, Figure 6b illustrates the changes in the region of the C=O stretching band (1700-1800 cm^{-1}), where a shift to higher frequency corresponds to the increasing crystallization. These results confirm that the presence of PVA solution during thermal treatment stimulates the crystal growth and nucleation of polymer.

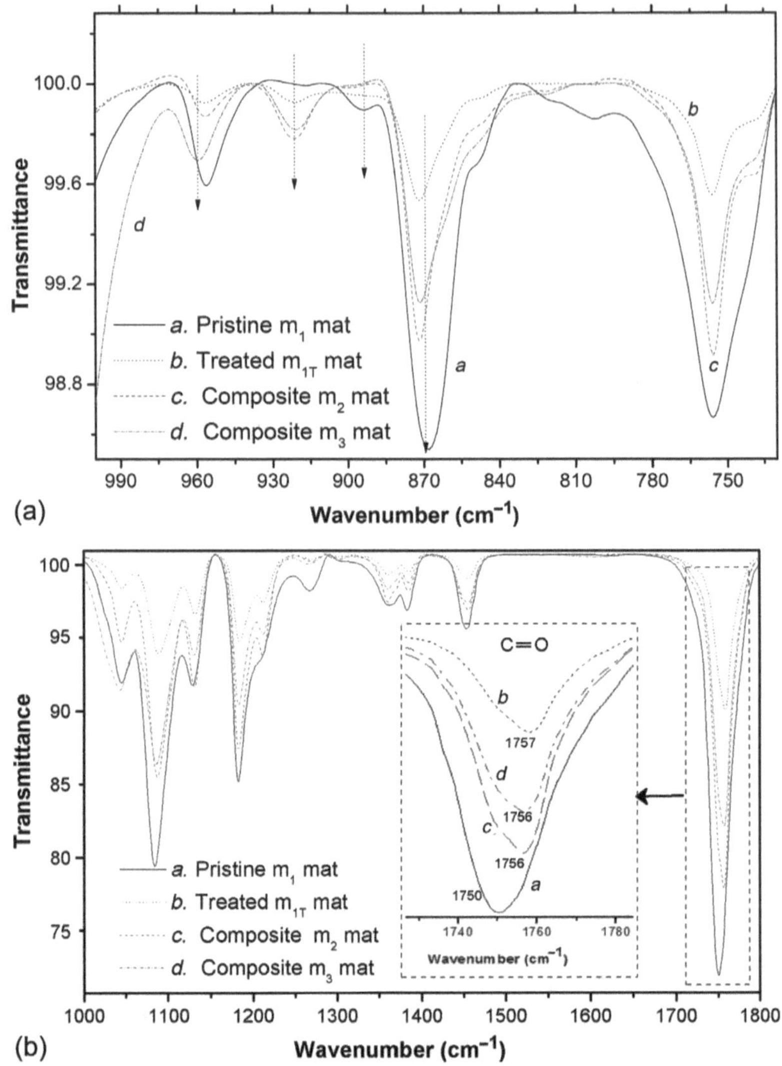

Figure 6 (a, b) Illustrate higher magnifications of FT-IR spectra that presented in Figure 5 of the as-obtained samples. Inset of (b) shows the respective peak position at rectangular selected areas.

In the FT-IR spectra of polymer composites, the functional groups involved in strong intermolecular hydrogen bonding often exhibit obvious shifts in their vibration frequencies as shown in Figure 6a and b. Taking the treated pristine PLA mat (m_{1T}) as a base, the FT-IR spectra of composites showed that the characteristic peaks of carbonyl group and hydroxyl groups were shifted toward higher frequencies due to the intermolecular hydrogen bonding interaction between the hydroxyl groups of PVA and carbonyl group of PLA (inset of Figure 5). Likewise, the crystalline-related peaks at 1417 and 1326 cm^{-1} in pure PVA disappeared in the bands of composite fibers, which can be also attributed to intermolecular interaction between the polymers at molecular level as reported elsewhere [16]. It was interestingly noticed that there was no such intensity peaks of these bands in m_{1T} mat (which was similar to m_1 mat). Furthermore, the band intensity in PVA-PLA composite mats was monotonically decreased with increasing PVA loading supporting the formation of outer PVA layer on the surface of PLA fibers. Increasing the reaction temperature (m_3 composite mat) enhances the extent of shift of the carbonyl and hydroxyl group indicating the formation of stronger hydrogen bonding and suggesting that such favorable interactions between the two polymers can lead to a miscible system [16,36]. Additionally, it can also be identified that the intensity of the C—H stretching groups definitely varied (inset of Figure 5) [37]. For the PVA-PLA composite mats, the peak intensity of around 2943 cm^{-1} gradually increases with an increasing of reaction temperature, but the peak intensity of around 2995 cm^{-1} gradually decreases. Indeed, the presence of PVA solution at high hydrothermal temperature can assure strong intermolecular hydrogen bonding interactions [16,37].

Chain crystalline conformation of PLA coated with PVA during hydrothermal treatment was investigated by DSC thermal analysis. The hydrothermal scans were helpful in understanding the role of deposition process on the change in crystallinity. The heating scans with PVA coatings and the results obtained from these scans are summarized in Table 1. It is known that the variations in values of T_g, T_c, T_m, and X_c obtained from

Table 1 Thermal properties of the as-prepared scaffold mats

Sample	**Glass transition temperature, T_g (°C)**	**Crystallization temperature, T_c (°C)**	**Melting peak, T_m (°C)**	**Melting peak, T_m (°C)-PVA**	**Crystallinity, X_c (%)**
Pristine m_1 mat	42.3	-	155.95	-	28.14
Composite m_2 mat	46.14	114.11	155.25	218	47.2
Composite m_3 mat	45.92	117.65	155.35	220	40.15

the DSC are usually assigned to the interactions between components and crystallinity properties. The DSC results showed that pure PLA doesn't show a clear melting endothermal peak and the melting peak becomes clear and sharper after PVA loading and can be attributed to the increase in the degree of perfection of the crystallites (curves are not shown here). However, the T_m values of the composite mats showed no significant difference (Table 1). On the other hand, the T_m of PVA could clearly be detected at round of 218 and 220 °C for both m_2 and m_3 composite mats, respectively, indicating that the PVA chains are highly confined during the hydrothermal treatment and further verify the deposition of PVA onto the PLA fibers [37].

The glass transition temperature (T_g) of the membrane samples is estimated from their DSC heating run shown in Table 1. It is well known that the T_g of a polymer compound is one of the most important criteria for the study of miscibility [3]. We found that the T_g value of PLA increases after the treatment process and can be attributed to decreasing of the amorphous phase content. These observations provide another evidence that PLA and PVA can provide a good miscible system (good compatibility) [16]. Tsuji and Muramatsu [38] reported that PLA and PVA showed phase separation in their blend films after solvent evaporation. Other studies confirmed that PLA is partially miscible with PVA [32]. The definition of miscibility and partial miscibility blends has been well discussed [39]. T_g for these composite membranes increased relatively from 42 °C of pristine m_1 mat to around 46 and 45.92 °C for m_2 and m_3 composite mats, respectively. Compared to pristine PLA (m_1 mat), the addition of PVA to PLA in the presence of heat energy was found to induce the crystallization (T_c) of composite samples due to enhanced chain mobility [12], as shown in Table 1. To further verify these results, Equation (1) was used to determine the percent crystallinity (X_c) using the ratio of the measured enthalpy of melting PLA composites (93 J/g) and the enthalpy of melting for fully crystallized pure PLA sample [40]. The crystallinity increased dramatically from 28.14% for pristine m_1 mat to 47.2% and 40.15% for the m_2 and m_3 composite mats, respectively (Table 1), indicating that the overall crystallinity of the treated mats was improved. This can be attributed to the decrease in the content of the amorphous domains due to crystalline growth [1]. This numerous growth (also known as nucleation) can be attributed to nonequilibrium chain conformations imposed by the electrospinning process that can be frozen upon the evaporation of the solvent [20]. The retardation process is due to the rapid solidification of stretched chains at high elongational rates during the last stages of electrospinning process, which significantly hinders the formation of crystals as noted by Zong et al. [41]. The crystallization behavior of PLA has been discussed elsewhere and results showed that the crystallinity can be obtained at more than 100 °C [42]. The degree of crystallinity increases when the treatment temperature increases over T_g and below T_m. Under standard processing conditions, PLA may remain mainly amorphous because of its low crystallization kinetics [43]. It is known that the higher the crystallinity, the higher the energy necessary to move the chains in the amorphous phase. Therefore, the temperature required for transition from a glassy state to a rubbery state will be higher. This indicates that intermolecular chemical interactions between the PLA fiber mat and PVA molecules result in an improvement in mechanical properties [44], which will be discussed in the next section. These results are in qualitative agreement with the results presented previously [1,32] in which relatively low crystallinity was observed for polymer fibers electrospun from the solution state.

3.4 Mechanical Properties

From the standpoint of human bone tissue bioengineering, the mechanical properties of the scaffold must be sufficient and not collapse during handling and during the patient's normal activities [28]. Further, these bone graft substitute materials should maintain adequate mechanical strength over critical phases of the tissue-healing process [29]. To expand the clinical applicability of biodegradable polymers, materials with better mechanical properties should be engineered. To investigate the mechanical properties of the pristine m_1 (m_{1T} is breakdown fibers) and m_2 and m_3 composite mats, load displacement curves were obtained from an average of five dog bone samples cut from each fiber mat to ensure there was no heterogeneity in thickness within the nonwoven mat. Tensile strengths and percent of elongation (ductility) were calculated and are summarized in Table 2. The load displacement curves (Figure 7) show that the tensile strength of the composite mats was

Table 2 The mechanical properties of the as-prepared scaffold mats (standard division are given as ±)

Sample	The thickness of fiber mat (μm)	Tensile strength (MPa)	Elongation at break (%)
Pristine m_1 mat	57 ± 5.7	8.1 ± 1.5	78 ± 9
Composite m_2 mat	72 ± 12.5	10.5 ± 0.8	207 ± 8
Composite m_3 mat	65 ± 9.5	11.49 ± 1.8	187 ± 5

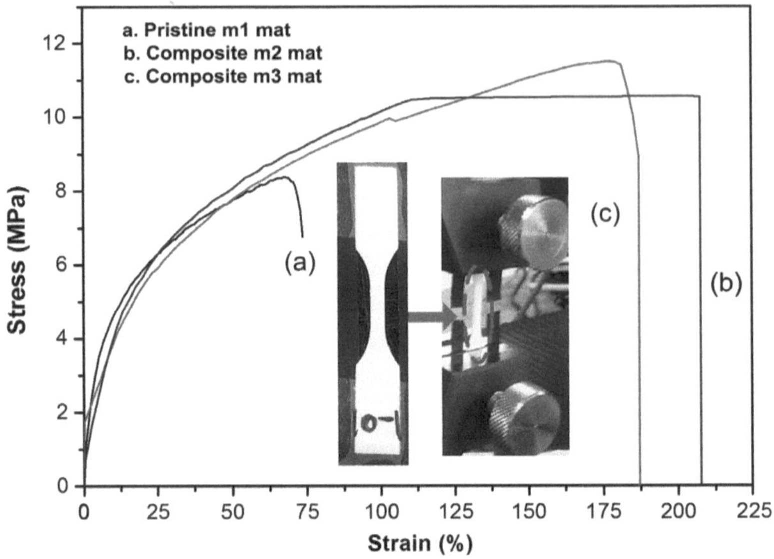

Figure 7 Stress-strain curves of pristine and composites fiber mats under tensile loading. Inset illustrates dog bone samples and their fixation at machine grip.

higher than that of the pristine m_1 mat, which can be attributed to the increased hydrogen bonding interactions. Further, the highest tensile strength was observed for the m_3 (11.49 MPa), followed by m_2 (10.5 MPa) composite mats. The linearity of displacement curves suggests that there is defined elastic region for these mats. Both composite fabrics induced excellent elongations compared to pristine one (around three times). This indicates that the PVA coating has a strong affinity on elongation improvements due to the more elastic characteristic of m_1 mat [21]. Ma and coworkers found that the addition of PVA to the blend films improves the extensibility of these composite blend films as a potential bone TE matrix [45]. The m_2 scaffold exhibited higher elongation (207%) before break than the m_3 (187%) scaffold. The possible reason is that m_2 has higher PVA coating layer thickness than that of m_3, resulting in increasing the stretching of the fiber composite mat. Our results are in good agreement with the previous studies by Yang et al. on PVA/gelatin composite polymers [46]. It was also reported that the heat treatment further improves the mechanical properties of the membranes [47]. Results showed that the higher the amount of PVA, the higher the tensile strength and the elongation at break for composite polymers, because of the good deformability and flexibility of PVA.

The improved elongation behavior of the m_2 compared with m_3 can be attributed to increased tear resistance of the branched PVA fibers (weblike fibers and parallel PLA fibers bounded to each other by PVA fibers of different diameters) compared with linear (nonbranched) membrane at high elongation. The interconnected network structure of the coated fibers at 130 °C hindered further elongation, which resulted in higher tensile strength and lower elongation prior to break. It was already reported that the mechanical properties of nonwoven mats increase with increasing the fiber junctions [41,48]. Thus, it is possible that the enhanced mechanical properties for the PVA-coated scaffolds are also due to the fiber junction rather than the crystallinity improvement of the treated scaffold. The fiber diameter of the m_3 (Figure 2c and d) mat was much thinner than the m_2 mat (Figure 2a and b). These results describe the fact that the mechanical behavior of nonwoven mats depends mainly on the chemical composition, treatment condition, morphology, and bonding structure of fibers. Hence, the relative improvement in the mechanical properties of composite mats may be explained in terms of the increased affinity between the two macromolecules. The miscibility resulted in difficult slippage of chains under loading because of more entanglements and strong physical interactions among the chains of composite polymers, such as hydrogen bond. Therefore, the addition of PVA in the presence of heat energy was quite helpful to improve the mechanical properties of PLA nanofiber scaffolds, overcome the brittleness nature of PLA and, turn it to ductile biomaterial.

3.5 Cell Attachment and Proliferation

Cell attachment and proliferation are crucial for a scaffold to support and guide tissue regeneration [29]. The design and fabrication of scaffolds are essential tasks for a functional vital engineered tissue. A provisional scaffold serves as a temporal cell adhesion and proliferation bed until natural ECM is generated by inoculated cells to form a tissue morphology resembling that of a native tissue. Biodegradable hydrophobic polyesters, generally, lack cell affinity for TE. It was reported that the incorporation of a proper hydrophilic polymer segment can increase the cell affinity [14]. The cell attachment and proliferation of membrane mat are highly affected by hydrophilicity of the membrane surface [29]. Therefore, we have measured WCA of flat membrane mats (see the inset of FESEM images Figures 1 and 2). The pristine treated m_{1T} mat cannot be used because of its short (breakdown) fibers. The WCA for m_1 membrane mat at 1 s was 119.7° (inset of Figure 1a), which is similar to the author's previous report [1] on pristine PLA mat, which indicates that the membrane surface is hydrophobic in nature. However, m_2 and m_3 membrane mats seem to be hydrophilic $68.2 \pm 2°$ and

51.2±2° (inset of Figure 2b and c). Therefore, PVA-PLA composite mats could be highly effective for cell response. These data confirm that 130 °C is the optimal treatment temperature of PVA to be coated onto PLA in order to obtain the highest improvement in hydrophilicity with minor loss of ductility compared to the composite mat obtained at 120 °C. It was also observed that pure PLA flat membrane mats float when immersed into water, showing their hydrophobic characteristic, whereas composite mats sink, qualitatively indicating that they are more hydrophilic. The WCA results indicate that PVA molecules covered the surface of the PLA fiber mat adequately. Hydrophilicity is very important when considering permeation of nutrients across the membrane and cell attachment [1,14,49]. Hydrophobic materials are also poorly wetted by cell culture medium, resulting in limited cell attachment to the scaffold and poor transfer of nutrients and waste products across the membrane. Investigation of the cell proliferation and attachment is of prime concern in TE to evaluate the cells biological response of different biomaterials *in vitro* [17]. To verify our insights, the number of the osteoblast cells initially seeded on each substrate as a percentage of the number of cells on the different membrane scaffolds cultured for 1, 3, or 7 days and then the densities (numbers) of cells were determined by MTT assay. Figure 8 shows that the proliferation of the cells varied with scaffold surface properties and chemical composition started on

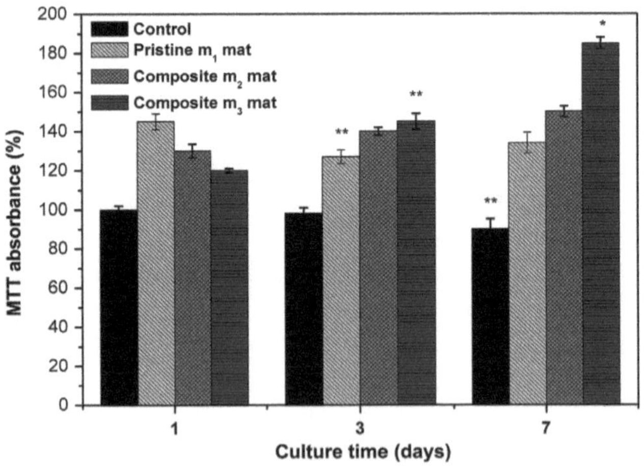

Figure 8 MTT cytotoxicity test on pristine and different composite PLA mats after 1, 3, and 7 days culture. The viability of control cells was set at 100%, and the viability relative to the control was expressed. The data are reported as the mean ± standard deviation ($n=4$). Asterisks indicate significance level of proliferation as compared to control obtained by *t*-test. (*) represents $p \leq 0.05$ and (**) represents $p \leq 0.001$.

the first day of culture. At day 1 of culture, the cell survival on the pristine PLA mat is higher than that on the composite mats. Conversely, cellular growth on the pristine PLA fiber samples showed negative progression after 3 and 7 days of culture. Comparable growth rates were observed on PVA-PLA membrane from day 3 to day 7. The proliferation of cells on the m_2 and m_3 composite scaffolds is higher and grew more continuously than the pristine one, particularly for the branched-like fibrous scaffold (m_3 mat). These results indicate that the thin PVA layer on the fiber surface and also formation of bridges across the pores of the membrane scaffolds might have accelerated the growth and proliferation of osteoblast cells (Figure 8). Our previous results proved that hydrophilic polymers have a positive effect on MC3T3-E1 osteoblast cells [17]. We assume that PVA underwent crystallization during heating [50] and water molecules may not be able to penetrate inside the fibers but were drawn into the fiber network because of the hydrophilicity of the materials. It must be noted, however, that the incorporation of the PVA molecules plays a significant role in the outcome of the cell response. Another key reason is that the enzymatic degradation rate of PLA decreases with increasing its crystallinity. This is due to higher enzymatic degradation rate of the PLA chains in the amorphous region than that in the crystalline region [38]. This is a major concern in orthopedic applications where implants with considerable degradation rate control would be required, which may result in release of degradation products with high local acid concentrations. Accordingly, we expect that the treatment process on PLA and producing composite samples from substrates could contribute to decelerating the rapid acidic phase release from the PLA molecules in the surrounding physiological environment, with a subsequent adjustment of pH *in vitro*, thereby improving the cell proliferation and ingrowth. A similar decrease in the pH of a PLA membrane would probably cause inflammatory tissue reactions *in vivo* [51]. The rapid release of acidic monomers to create acidosis would stimulate an inflammatory response [52]. Leiggener et al. [53] reported through the *in vivo* study that the higher degree of crystallinity results in a higher chemical strength and loading capacity, which promises advantages for long-term bone implantation. Results showed that the addition of PVA has influenced increased growth rates compared to pristine PLA.

Figure 9 shows the morphologies of cells cultured for 3 days on the prepared scaffolds. In general, the results showed that cell lines appeared to adhere well and exhibited a normal morphology on both surfaces. This might be due to the large surface area available for cell attachment. Notwithstanding, PVA coating on PLA fibers promoted much higher degrees of cell

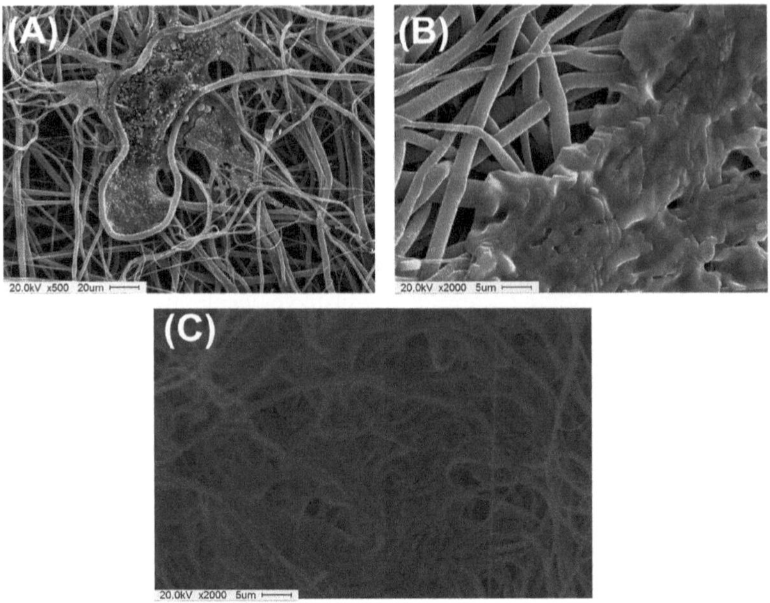

Figure 9 Bioscanning electron micrographs of MC3T3-E1 osteoblast-like cells cultured on the surfaces of (a) pristine PLA mat and (b) m_2 and (c) m_3 composite mats for 3 days.

crowding and exhibited a distribution of higher uniformity (Figure 9b and c) than that of pristine sample (Figure 9a). The composite nanofibrous scaffold surface appeared to be sufficiently covered with cell-secreted ECM and particularly the cells almost covered a majority of the m_3 membrane surface during the third day (Figure 9c), indicating confluence growth of osteoblast cell [17]. These observations indicated a better integration of cells with composite nanofiber scaffolds. This is probably because these scaffolds have good wettability and branched-like (weblike) fibrous structure compared to the pristine mat. We hypothesize that degradable PVA networks with increased hydrophobicity would support extracellular matrix deposition. Accordingly, the PVA component would enable adequate elasticity, and the abundant hydroxyl groups on PVA could be used to incorporate molecules of biological significance [54,55]. Asran et al. [16] studied the effects of blended substrates containing different PVA fraction molecules on the HaCaT epidermal cells behavior. Their blended fiber results showed a negative effect on the HaCaT cells, which can be assumed to be due to the differences of type of cells, culture condition, samples preparation conditions, chemical composition, and morphologies of the prepared scaffolds. Our results proved that the PVA-PLA composite scaffolds had not only an excellent biocompatibility *in vitro* but also a significant ability to enhance cell adhesion and proliferation. The increase in both cell proliferation and attachments on the composite fibers, particularly branched-like fiber mats, can be attributed to (a) increased protein absorption ability of the composite mat due to the good surface wettability, as confirmed by WCA; (b) the addition of PVA molecules into PLA fibers in the presence of hydrothermal treatment, inducing chain conformation on the composite, resulting in counteraction of the acidic degradation of the biodegradable PLA polymer [1]; (c) good biocompatibility of PVA, although at high MW (146,000-186,000 g mol^{-1}), this polymer is not degradable physiologically [47] and may be applicable if the implant is semipermanent (low-MW PVA $<$ 10,000 g mol^{-1}) and non-acidic bioresorbable polymers are being considered for further studies on biomaterials that will have efficient biodegradation *in vitro* and *in vivo*; (d) rougher surface due to the deposition of PVA layer on the fibers, which can provide extra sites for cell binding.

It can be concluded that surface modifications using chemical and topographical characteristics on PLA electrospun fibers can improve the cellular adhesion and proliferation. Therefore, PVA-PLA composite membrane scaffolds can be used for guided bone regeneration and can provide a beneficial frame for bone TE. It is worth mentioning that long-term *in vitro* and *in vivo* studies or degradation behavior studies are crucial to confirm the expected advantages of PVA-PLA composite membrane scaffolds.

4. CONCLUSION

This study reports a novel approach for the synthesis of composite mats by means of PLA electrospun mat followed by PVA deposition using hydrothermal treatment. The synthesized composites showed marked improvements in morphology and mechanical and physicochemical properties compared to pristine PLA electrospun mat. The PVA was found to be evenly coated throughout the individual nanofibers. The deposited layer thickness of PVA onto PLA fibers was controlled by the hydrothermal reaction temperature. It was found that the deposition of PVA via the hydrothermal process increased the hydrogen interaction bonding and induced crystalline conformation of PLA. The exponential decrease in WCA from 119.7° to 51.2° indicates that PVA dramatically improves wettability of PLA fiber mat, resulting in well-adhered and proliferated MC3T3-E1 osteoblast cell lines. Therefore, PVA-coated tridimensional biomaterials are more advantageous for human tissue bioengineering applications than their uncoated counterparts. The as-synthesized PVA-PLA composite nanofiber mats seem to be useful as provisional functional scaffolds in cardiovascular muscular and bone TE. However, further research, currently under course in our laboratories, will reveal whether the favorable features exhibited by such composite biomaterials *in vitro* hold their promise also *in vivo*.

ACKNOWLEDGMENTS

This chapter has been financially supported by the Ministry of Education, Science Technology (MEST) and National Research Foundation of Korea (NRF) through the Human Resource Training Project for Regional Innovation.

REFERENCES

[1] A. Abdal-hay, F.A. Sheikh, J.K. Lim, Air jet spinning of hydroxyapatite/poly(lactic acid) hybrid nanocomposite membrane mats for bone tissue engineering, Colloids Surf. B: Biointerfaces 102 (2013) 635–643.
[2] T. Takami, Y. Murakami, Development of PEG-PLA/PLGA microparticles for pulmonary drug delivery prepared by a novel emulsification technique assisted with amphiphilic block copolymers, Colloids Surf. B: Biointerfaces 87 (2011) 433–438.
[3] J.-T. Yeh, M.-C. Yang, C.-J. Wu, X. Wu, C.-S. Wu, Study on the crystallization kinetic and characterization of poly (lactic acid) and poly (vinyl alcohol) blends, Polym. Plast. Technol. Eng. 47 (2008) 1289–1296.
[4] B.-K. Chen, C.-C. Shih, A.F. Chen, Ductile PLA nanocomposites with improved thermal stability, Compos. A: Appl. Sci. Manuf. 43 (2012) 2289–2295.
[5] B.-K. Chen, C.-H. Shen, S.-C. Chen, A.F. Chen, Ductile PLA modified with methacryloyloxyalkyl isocyanate improves mechanical properties, Polymer 51 (2010) 4667–4672.
[6] D. Cohn, A. Hotovely Salomon, Designing biodegradable multiblock PCL/PLA thermoplastic elastomers, Biomaterials 26 (2005) 2297–2305.
[7] T. Takayama, M. Todo, Improvement of impact fracture properties of PLA/PCL polymer blend due to LTI addition, J. Mater. Sci. 41 (2006) 4989–4992.
[8] D.W. Hutmacher, T. Schantz, I. Zein, K.W. Ng, S.H. Teoh, K.C. Tan, Mechanical properties and cell cultural response of polycaprolactone scaffolds designed and fabricated via fused deposition modeling, J. Biomed. Mater. Res. 55 (2001) 203–216.
[9] C. Liu, C. Gong, Y. Pan, Y. Zhang, J. Wang, M. Huang, Y. Wang, K. Wang, M. Gou, M. Tu, Synthesis and characterization of a thermosensitive hydrogel based on biodegradable amphiphilic PCL-pluronic (L35)-PCL block copolymers, Colloids Surf. A Physicochem. Eng. Asp. 302 (2007) 430–438.
[10] Z. Li, X. Zhou, C. Pei, Synthesis and characterization of MPS-g-PLA copolymer and its application in surface modification of bacterial cellulose, Int. J. Polym. Anal. Charact. 15 (2010) 199–209.
[11] A. Bhatia, R. Gupta, S. Bhattacharya, H. Choi, Compatibility of biodegradable poly (lactic acid)(PLA) and poly (butylene succinate)(PBS) blends for packaging application, Korea-Aust. Rheol. J. 19 (2007) 125–131.
[12] O. Martin, L. Avérous, Poly(lactic acid): plasticization and properties of biodegradable multiphase systems, Polymer 42 (2001) 6209–6219.
[13] G. Pitarresi, C. Fiorica, F.S. Palumbo, F. Calascibetta, G. Giammona, Polyaspartamide-polylactide electrospun scaffolds for potential topical release of Ibuprofen, J. Biomed. Mater. Res. Part A 100A (2012) 1565–1572.

[14] G. Meneghello, D.J. Parker, B.J. Ainsworth, S.P. Perera, J.B. Chaudhuri, M.J. Ellis, P. A. De Bank, Fabrication and characterization of poly(lactic-*co*-glycolic acid)/polyvinyl alcohol blended hollow fibre membranes for tissue engineering applications, J. Membr. Sci. 344 (2009) 55–61.
[15] T. Elakkiya, R. Sheeja, K. Ramadhar, T.S. Natarajan, Biocompatibility studies of electrospun nanofibrous membrane of PLLA-PVA blend, J. Appl. Polym. Sci. 128 (5) (2012) 2840–2846.
[16] A. Asran, K. Razghandi, N. Aggarwal, G.H. Michler, T. Groth, Nanofibers from blends of polyvinyl alcohol and polyhydroxy butyrate as potential scaffold material for tissue engineering of skin, Biomacromolecules 11 (2010) 3413–3421.
[17] A. Abdal-hay, L.D. Tijing, J.K. Lim, Characterization of the surface biocompatibility of an electrospun nylon 6/CaP nanofiber scaffold using osteoblasts, Chem. Eng. J. 215–216 (2013) 57–64.
[18] A. Martins, E.D. Pinho, S. Faria, I. Pashkuleva, A.P. Marques, R.L. Reis, N.M. Neves, Surface modification of electrospun polycaprolactone nanofiber meshes by plasma treatment to enhance biological performance, Small 5 (2009) 1195–1206.
[19] J.-P. Chen, C.-H. Su, Surface modification of electrospun PLLA nanofibers by plasma treatment and cationized gelatin immobilization for cartilage tissue engineering, Acta Biomater. 7 (2011) 234–243.
[20] C. Ribeiro, V. Sencadas, C.M. Costa, J.L.G. Ribelles, S. Lanceros-Méndez, Tailoring the morphology and crystallinity of poly(L-lactide acid) electrospun membranes, Sci. Technol. Adv. Mater. 12 (2011) 015001.
[21] A. Abdal-hay, C.I. Kim, J.K. Lim, An in situ hydrothermal fabrication process of poly (vinyl alcohol)/apatite-like nanocomposites with improved thermal and mechanical properties, Ceram. Int. 40 (3) (2014) 4995–5000.
[22] A. Abdal-hay, N.A.M. Barakat, J.K. Lim, Influence of electrospinning and dip-coating techniques on the degradation and cytocompatibility of Mg-based alloy, Colloids Surf. A Physicochem. Eng. Asp. 420 (2013) 37–45.
[23] S. Kang, S.L. Hsu, H.D. Stidham, P.B. Smith, M.A. Leugers, X. Yang, A spectroscopic analysis of poly(lactic acid) structure, Macromolecules 34 (2001) 4542–4548.
[24] D. Li, Y. Xia, Electrospinning of nanofibers: reinventing the wheel? Adv. Mater. 16 (2004) 1151–1170.
[25] R. Fang, E. Zhang, L. Xu, S. Wei, Electrospun PCL/PLA/HA based nanofibers as scaffold for osteoblast-like cells, J. Nanosci. Nanotechnol. 10 (2010) 7747–7751.
[26] E.A. Schmitt, D. Flanagan, R.J. Linhardt, Importance of distinct water environments in the hydrolysis of poly (DL-lactide-*co*-glycolide), Macromolecules 27 (1994) 743–748.
[27] Y. You, B.M. Min, S.J. Lee, T.S. Lee, W.H. Park, In vitro degradation behavior of electrospun polyglycolide, polylactide, and poly (lactide-*co*-glycolide), J. Appl. Polym. Sci. 95 (2005) 193–200.
[28] K. Rezwan, Q.Z. Chen, J.J. Blaker, A.R. Boccaccini, Biodegradable and bioactive porous polymer/inorganic composite scaffolds for bone tissue engineering, Biomaterials 27 (2006) 3413–3431.
[29] A. Abdal-hay, H.R. Pant, J.K. Lim, Super-hydrophilic electrospun nylon-6/hydroxyapatite membrane for bone tissue engineering, Eur. Polym. J. 49 (2013) 1314–1321.
[30] A. Abdal-hay, J. Lim, M. Shamshi Hassan, J.K. Lim, Ultrathin conformal coating of apatite nanostructures onto electrospun nylon 6 nanofibers: mimicking the extracellular matrix, Chem. Eng. J. 228 (2013) 708–716.
[31] J.S. Stephens, D.B. Chase, J.F. Rabolt, Effect of the electrospinning process on polymer crystallization chain conformation in nylon-6 and nylon-12, Macromolecules 37 (2004) 877–881.

[32] J.M. Deitzel, J. Kleinmeyer, D. Harris, N.C. Beck Tan, The effect of processing variables on the morphology of electrospun nanofibers and textiles, Polymer 42 (2001) 261–272.
[33] J.W. Park, S.S. Im, Phase behavior and morphology in blends of poly (L-lactic acid) and poly (butylene succinate), J. Appl. Polym. Sci. 86 (2002) 647–655.
[34] Y. Liu, L. Cui, F. Guan, Y. Gao, N.E. Hedin, L. Zhu, H. Fong, Crystalline morphology and polymorphic phase transitions in electrospun nylon 6 nanofibers, Macromolecules 40 (2007) 6283–6290.
[35] G. Kister, G. Cassanas, M. Vert, Effects of morphology, conformation and configuration on the IR and Raman spectra of various poly(lactic acid)s, Polymer 39 (1998) 267–273.
[36] T. Kasuga, Y. Ota, M. Nogami, Y. Abe, Preparation and mechanical properties of polylactic acid composites containing hydroxyapatite fibers, Biomaterials 22 (2000) 19–23.
[37] X. Shuai, Y. He, N. Asakawa, Y. Inoue, Miscibility and phase structure of binary blends of poly(L-lactide) and poly(vinyl alcohol), J. Appl. Polym. Sci. 81 (2001) 762–772.
[38] H. Tsuji, H. Muramatsu, Blends of aliphatic polyesters. IV. Morphology, swelling behavior, and surface and bulk properties of blends from hydrophobic poly(L-lactide) and hydrophilic poly(vinyl alcohol), J. Appl. Polym. Sci. 81 (2001) 2151–2160.
[39] A.M. Gajria, V. Davé, R.A. Gross, S.P. McCarthy, Miscibility and biodegradability of blends of poly(lactic acid) and poly(vinyl acetate), Polymer 37 (1996) 437–444.
[40] X. Cao, A. Mohamed, S.H. Gordon, J.L. Willett, D.J. Sessa, DSC study of biodegradable poly(lactic acid) and poly(hydroxy ester ether) blends, Thermochim. Acta 406 (2003) 115–127.
[41] X. Zong, K. Kim, D. Fang, S. Ran, B.S. Hsiao, B. Chu, Structure and process relationship of electrospun bioabsorbable nanofiber membranes, Polymer 43 (2002) 4403–4412.
[42] X. Liu, M. Dever, N. Fair, R. Benson, Thermal and mechanical properties of poly (lactic acid) and poly (ethylene/butylene succinate) blends, J. Environ. Polym. Degrad. 5 (1997) 225–235.
[43] F. Hernández Sánchez, J. Molina Mateo, F.J. Romero Colomer, M. Salmerón Sánchez, J.L. Gómez Ribelles, J.F. Mano, Influence of low-temperature nucleation on the crystallization process of poly(L-lactide), Biomacromolecules 6 (2005) 3283–3290.
[44] A. Baji, Y.-W. Mai, S.-C. Wong, M. Abtahi, X. Du, Mechanical behavior of self-assembled carbon nanotube reinforced nylon 6,6 fibers, Compos. Sci. Technol. 70 (2010) 1401–1409.
[45] G. Ma, D. Yang, D. Su, X. Mu, J.F. Kennedy, J. Nie, Preparation and properties of water-soluble chitosan and polyvinyl alcohol blend films as potential bone tissue engineering matrix, Polym. Adv. Technol. 21 (2010) 189–195.
[46] D. Yang, Y. Li, J. Nie, Preparation of gelatin/PVA nanofibers and their potential application in controlled release of drugs, Carbohydr. Polym. 69 (2007) 538–543.
[47] R.A. Franco, A. Sadiasa, B.-T. Lee, Utilization of PVPA and its effect on the material properties and biocompatibility of PVA electrospun membrane, Polym. Adv. Technol. 25 (2014) 55–65.
[48] Y. Jung, H. Kim, D. Lee, S. Park, M. Khil, Characterization of PVOH nonwoven mats prepared from surfactant-polymer system via electrospinning, Macromol. Res. 13 (2005) 385–390.
[49] S.H. Oh, S.G. Kang, E.S. Kim, S.H. Cho, J.H. Lee, Fabrication and characterization of hydrophilic poly(lactic-*co*-glycolic acid)/poly(vinyl alcohol) blend cell scaffolds by melt-molding particulate-leaching method, Biomaterials 24 (2003) 4011–4021.
[50] L.A. Vol'f, Y.K. Kirilenko, Z.A. Urban, E.Y. Danilova, O.V. Bukalova, G. I. Kudryavtsev, Imparting thermal stability to fibres based on polyvinyl alcohol, Fibre Chem. 1 (1970) 258–261.

[51] C. Martin, H. Winet, J.Y. Bao, Acidity near eroding polylactide-polyglycolide in vitro and in vivo in rabbit tibial bone chambers, Biomaterials 17 (1996) 2373–2380.
[52] H. Winet, J.Y. Bao, Comparative bone healing near eroding polylactide-polyglycolide implants of differing crystallinity in rabbit tibial bone chambers, J. Biomater. Sci. Polym. Ed. 8 (1997) 517–532.
[53] C.S. Leiggener, R. Curtis, A.A. Müller, D. Pfluger, S. Gogolewski, B.A. Rahn, Influence of copolymer composition of polylactide implants on cranial bone regeneration, Biomaterials 27 (2006) 202–207.
[54] R. Hou, G. Zhang, G. Du, D. Zhan, Y. Cong, Y. Cheng, J. Fu, Magnetic nanohydroxyapatite/PVA composite hydrogels for promoted osteoblast adhesion and proliferation, Colloids Surf. B: Biointerfaces 103 (2013) 318–325.
[55] M. Kaiwang, W. Xiang, D. Yingping, C. Shaoxi, Primary study on the assay of superficial hydroxyl group content upon the solid polyvinyl alcohol, Colloids Surf. B: Biointerfaces 50 (2006) 72–75.

CHAPTER 5

Green Synthesis for Advanced Materials of Graphene Oxide (GO) with ZnO for Enhanced Photocatalytic Activity at Room Temperature

Kaviyarasu Kasinathan
Department of Physics, Sri Sankara Arts & Science College, Kanchipuram, Tamil Nadu, India

Chapter Contents

1. INTRODUCTION

Carbon nanostructures such as carbon nanotubes, fullerene, and graphene with prominent electrical and structural characteristics have drawn much attention, because they can contribute in the development of composites with improved catalytic performance [1–9]. As one atom thick sheet of sp^2-bonded carbon atoms in a hexagonal 2-D lattice, it has quickly appeared to be high promising nanomaterials with unique properties to open up a new research area for materials science and condensed matter physics [10–14] and a wide range of technological applications [15–17]. Therefore, nowadays, among known strategies for producing graphene-based materials, the utilization of graphene oxide (GO) is the most versatile and scalable method [18–21]. GO is usually used as the starting material for the generation of graphene-based materials. In this regard, graphene-containing composite materials have been attracting much attention. Fabrication of such composites requires not

Handbook of Nanoceramic and Nanocomposite Coatings and Materials
http://dx.doi.org/10.1016/B978-0-12-799947-0.00005-5

only the high-quality production of graphene nanosheets (GNSs) but also their effective incorporation in various and desirable matrices. But, in a suspension, the reduced GNS tends to aggregate because of their strong *van der Waals* interactions [22,23]. To prevent the aggregation of exfoliated GNSs, both electrostatic stabilization and chemical functionalization [24] methods have been effectively utilized. Concerning this, the composition of metal [25] or metal oxide (MO) [26] nanoparticles with GNSs have been reported. These kinds of graphene-based composites provide some advantages. For example, the reduction of the GO nanosheet incorporated in such compositions can be controlled through a photocatalytic reaction [27]. Moreover, the graphene reduced by the photocatalytic reaction can act as a sensitizer for improving the photocatalytic performance of the MO:GO (graphene oxides). Recently, Xu et al. [28] showed that Au, Pt, and Pd nanoparticles absorbed on GO nanosheets play a pivotal role in the catalytic reduction of GO with ethylene glycol, leading to the formation of graphene-metal particle composites. Williams et al. [29] also showed the photocatalytic reduction of GO nanosheets by TiO_2 and ZnO nanoparticles in ethanol suspension for a specific irradiation time. Since the controllable chemical reduction of GO nanosheets to graphene ones is a key point, for example, in future electronic applications, the chemical and electrical properties of the GO nanosheet reduced by a photocatalytic reduction process should be also investigated in some more detail.

2. EXPERIMENTAL

In this experimental part, all the reagents are analytic grade and used without further purification. In a typical synthesis, 5 g of graphite was added into 150 ml of H_2SO_4, followed by the addition of 3 g of zinc oxide, $NaNO_3$, $KMnO_4$, and H_2SO_4, and they were continuously stirred vigorously. Then, 5 ml of H_2O_2 was dissolved with the resulting solution to remove the sulfate and chloride ions. The as-obtained precipitate was dissolved with 5 g of sodium borohydride, which acts as a surfactant. After ultrasonication for 2 h, the suspension was transferred into a 250 ml Teflon-lined stainless steel autoclave. The autoclave was maintained at a temperature of 250 °C for a day. After cooling to room temperature (RT), the black product was washed with distilled water several times and then dried under vacuum at 75 °C for 24 h. Then, the black dispersion was rinsed three times with deionized water by centrifugation. Finally, nearly 1 g of potassium hydroxide was added into the suspension and stirred for 3 h. The reduced GO was dispersed in 100 ml of *N*-methylpyrrolidone by ultrasonication to prevent reduced GO nanosheet aggregation. The as-obtained precipitate was collected and was used with deionized water and anhydrous ethanol several times. The final product was dried under vacuum at 90 °C for 7 h.

The X-ray powder diffraction (XRD) of the product was measured using a D/max 2500 V (Rigaku, Japan) and X-ray powder diffractometer at a scanning rate 2θ ranging from 10° to 70° using Cu K_α radiation ($\lambda = 0.15418$ nm). Raman spectroscopy was performed at RT using a Raman microprobe (HR-800, Jobin Yvon) with 532 nm Nd:YAG excitation source. The XPS (X-ray photoelectron spectroscopy) spectrum was recorded on an ESCALAB 250 photoelectron spectrometer (Thermo VG Scientific, USA) with Al K_α (1486.6 eV) as the X-ray source. The nanoscale structures were observed using high-resolution transmission electron microscopy (HRTEM, JEOL JEM-2100F) with an accelerating voltage of 200 kV. The elemental composition was determined using the energy-dispersive X-ray spectroscopy (EDS) (IE300X) analysis that was performed at several points. Current-voltage curves of the GO:ZnO nanosheet were obtained by using a Keithley 485 autoranging picoammeter. Photoluminescence (PL) measurements of the as-synthesized product were carried out by an F-4500 KIMON fluorescence spectrophotometer at RT with a Xe lamp as the excited light source; the excitation wavelength was 340 nm. Photocatalytic activities of GO:ZnO nanosheet samples were evaluated by the photocatalytic degradation of MB under UV lamp (Philips 18 W) as a light source. The sample was prepared by dispersing a small amount of dry powder in ethanol or water.

3. RESULTS AND DISCUSSION

The GO:ZnO nanocomposite crystalline study was carried out by using XRD diffractogram of a GO nanosheet as shown in Figure 1. The diffraction peaks centered at (012), (104), (110), (113), (202), (018), and (116) are due to the planes associated with the hexagonal carbon structures concordant to the pure GO [30]. Additionally, the highest intensity peak (012) when compared with other peaks ascertains the plane of the hexagonal carbon phase. The raw graphite shows a very strong (002) peak at $2\theta = 26°$ [31]. A peak at $2\theta = 30°$ may be due to larger spacing. It is mainly attributed to the presence of oxygenated functional groups of GO. The result concludes that there is a π-π stacking between GO in solution, which induces the formation of nanosheets.

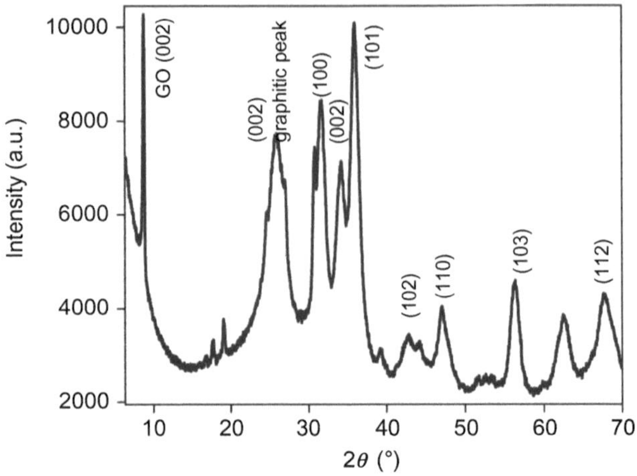

Figure 1 XRD pattern of GO:ZnO nanosheets.

Raman spectroscopy is a suitable technique to examine the ordered crystal structures of carbonaceous materials and the single-, bi-, and multilayer characteristics of GO layers. Figure 2 shows the Raman spectrum of the as-prepared GO:ZnO composites. The usual features of carbon materials in Raman spectra are the G line (~1570 cm^{-1}), which is usually assigned to the E_{2g} phonon of C sp^2 atoms, and the D line (~1350 cm^{-1}) as a breathing

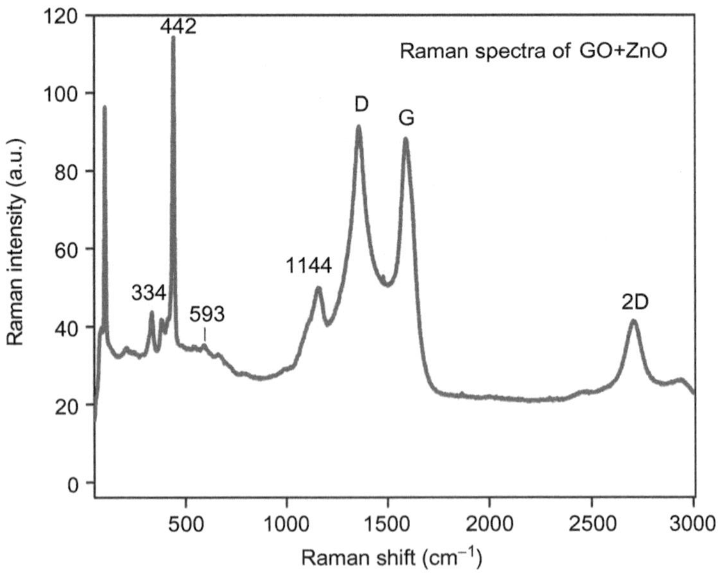

Figure 2 Raman spectra of GO:ZnO nanosheets.

mode of k-point phonons of A_{1g} symmetry [32], which is assigned to local defects and disorders especially at the edge of GO [33]. The Raman spectra in Figure 2 display the G line at 1584 cm^{-1} and the D band at 1350 cm^{-1}. It was previously shown that the peak position of the G band of the single-layer GNS (~1585 cm^{-1}) shifts about 6 cm^{-1} into lower wave numbers after stacking more graphene layers (for two to six layers, G band shifted to 1578 cm^{-1}) [34]. Hence, the value of ~1584 cm^{-1} for the G band of our sample can be primarily attributed to the single-layer GNSs. Moreover, it has been recently reported that the shape and position of the overtone of the D line (2-D band at ~2700 cm^{-1}) are a significant fingerprint that can be related to the formation and the layer numbers of GNSs [35]. The 2-D peak position of the single-layer GNS is observed at 2678 cm^{-1}, while the 2-D band of multilayer (two to four layers) GNS shifts to higher frequencies by 19 cm^{-1} [36]. In this section, we observed in Figure 2, the peak position of the observed 2-D band of the GO:ZnO nanosheet with a symmetrical shape was located at 2680 cm^{-1}, indicating the formation of single-layer GO nanosheets, which was also consistent; the substantial formation of single-layer GNSs was also observed by HRTEM analysis.

The high-resolution XPS peaks of the Zn (2P) core level of the ZnO powder and the GO:ZnO nanosheet have been shown in Figure 3. Two peaks in the Zn (2P) binding energy region were observed [37]. The peak located at a binding energy of 1023 eV corresponds to the Zn ($2P_{3/2}$) and

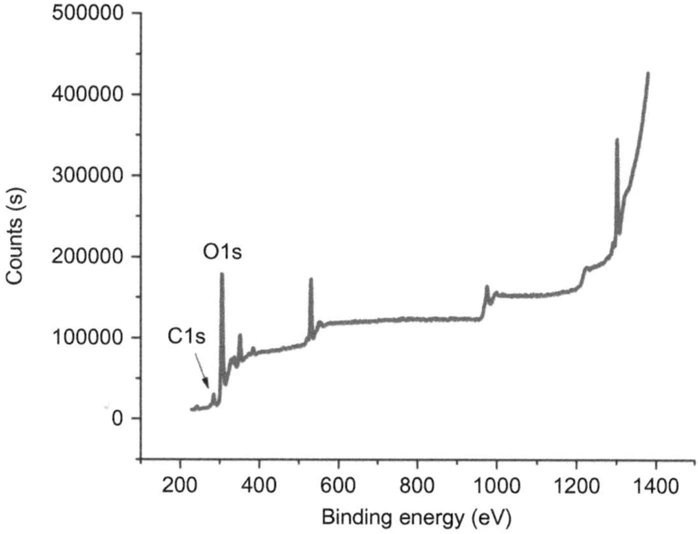

Figure 3 XPS images of GO:ZnO nanosheets.

another one located at 1064 eV is attributed to the Zn ($2P_{1/2}$). It is seen that the peak positions of the Zn (2P) core level of ZnO-hybridized GO were identical to those of ZnO powder [38]. This means that in our experimental conditions, no chemical Zn—C bond was formed between the GO nanosheet and the ZnO nanoparticles, as similarly reported for ZnO:CNTs (carbon nanotubes) by other research groups [39]. Based on the chemical SF-modified peak area ratios of the Zn ($2P_{3/2}$), C (1s), and O (1s) XPS core levels, the elemental compositions of the synthesized GO:ZnO nanosheets were obtained, 3.7 and 4.3 for Zn:C and O:C atomic ratios, respectively.

Surface micro-nanostructural HRTEM images of GNSs are shown in Figure 4. A typical HRTEM image of the GNS is shown in Figure 4a and b. The GNS are entangled with each other and scrolled, similar to crumpled paper [40]. Figure 4c shows the cross-sectional HRTEM images of the GNS. The stacking sheet structure and individual monolayers of the GNS could be clearly observed. It could be found that the thickness of the GNS is about 1-3 nm, which corresponds to approximately three to nine layers of stacked individual monatomic GNSs [41,42]. Figure 4d shows an HRTEM image of the GNS, which displays a typical crinkly and rippled structure of the GNS.

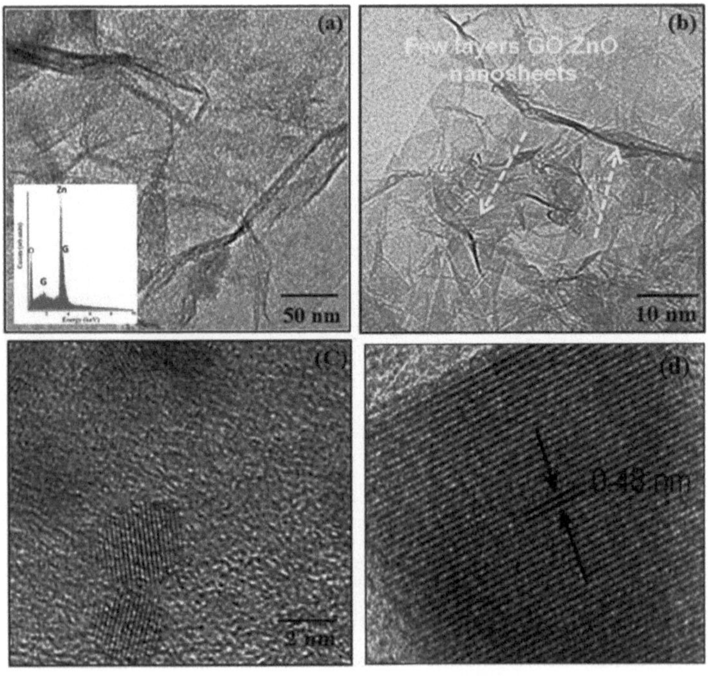

Figure 4 (a-d) TEM and EDS images of GO:ZnO nanosheets.

The size of the GNS is around 5 nm as shown by the HRTEM, which is consistent with the value calculated by XRD. It is evidently clear from HRTEM image that large GO nanosheets were observed to be more intense and the central portion of the image resembling like graphene plates [43]. They are transparent and entangled with each other. HRTEM also illustrates the amorphous nature of GO nanosheets. Evidently, the nanosheet and the surface are clean and amorphous without any dislocation [44].

In Figure 5, the linear behavior of the *C-V* curves indicates the metallic property of the nanosheets and formation of ohmic contact between the nanosheet and the electrodes. Based on the *CV* diagrams, nanosheet resistance of the GO:ZnO nanosheets was also estimated. The resistance value of the as-prepared GO:ZnO nanosheets was exhibited so high that it was nearly out of range of our measurement instruments ($>10^{10}$ ohm/sq). However, it was easily found that by increasing the irradiation time from 0.5 to 2 h, the resistance value decreased from 4.2×10^8 to 1.7×10^7 ohm/sq, indicating chemical reduction of the GO nanosheets in the photocatalytic process [45]. Comparing the resistance value of the GO:ZnO nanocrystals photocatalytically reduced for 2 h (1.7×10^7 ohm/sq) with the resistance value reported for GO reduced by hydrazine ($\sim 1 \times 10^8$ ohm/sq) and GO reduced by hydrazine and thermal reduction at 400 °C ($\sim 1 \times 10^5$ ohm/sq) [46] shows that the photocatalytic process was effective in the chemical reduction of the GO nanosheets in the GO:ZnO nanocomposites.

The PL optical spectrum studies of GO:ZnO obtained at an excitation wavelength of 340 nm are shown in Figure 6. The PL of the as-prepared GO

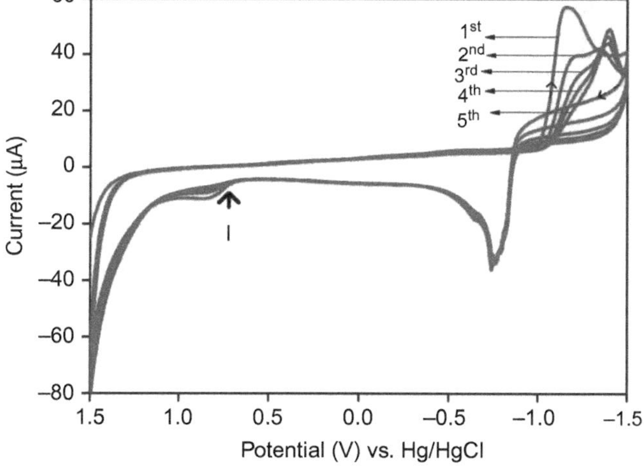

Figure 5 Electrochemical behavior of the GO:ZnO nanosheets.

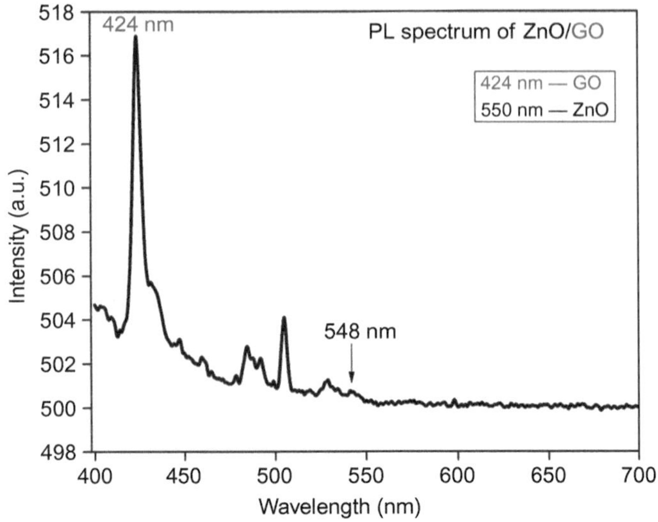

Figure 6 Comparison of the PL spectra of GO:ZnO nanosheets.

shows an emission peak in the visible spectrum at 443 nm similar to that reported by Mahato et al. [47] for liquid suspension of GO, the blueshifts of PL spectra by 0.15 eV with resulting emission peak at 424 nm. The PL in such carbon systems is a consequence of geminate recombination of localized electron-hole pairs in sp^2 clusters, which essentially behave as the luminescent centers. Since the band gap depends on the size, shape, and fraction of sp^2 domains, tunable PL could be achieved by controlling the sp^2 sites. In other words, the PL intensity increases with an increase in sp^2 content in the disordered carbon systems [48]. This undoubtedly shows the role played by lithium base in the reduction of GO. Luminescence of GO has also been reported in red and near-infrared regions [49,50], which can result from the presence of multilayered and aggregated nanosheets [51]. Centrifugation is therefore done at an early stage as illustrated in Section 2 to remove these particles. Blue PL is also considered to be an indicative of the quality of graphene suspensions. The PL spectrum of bare ZnO (black curve) clearly shows a broad green emission peak at 550 nm, which is generally believed to arise from the transition in defect states, particularly oxygen vacancies [52]. Interestingly, PL from GO:ZnO is not only blueshifted as compared with that of bare ZnO but also significantly quenched. This suggests that an additional pathway for diminishing the charge carriers dominates because of the interactions between the excited ZnO and the GO nanosheets. In other words, the quenching behavior could be explained by interfacial charge transfer from ZnO to graphene [53]. Figure 6, also shows the effect of increasing the concentration of GO to the initial solution. As the concentration of GO is increased, the extent of PL quenching increases; an observation similar to that was reported by Williams and Kamat [54]. However, they didn't observe any shift in PL measurements since their preparation involves anchoring of as-prepared ZnO particles over GO. In their case, the electrons from the n-type ZnO were primarily used up in reduction of GO upon irradiation of UV light. However, we propose another mechanism for the observed large blueshift of 0.15 eV in PL. In our case, due to simultaneous in-pot reduction of both ZnO and GO, the path of photocatalytic reduction of GO by electron transfer from ZnO is eliminated. Since it is well known that ZnO intrinsically behaves as an n-type semiconductor due to oxygen vacancies and zinc interstitials [55] and graphene behaves like a p-type semiconductor in ambient atmosphere, hence, n-p heterostructure or a depletion layer is formed at the GO:ZnO interface. A blueshift of 0.15 eV can be envisaged as a significant step towards band gap tuning of graphene-based heterostructures for future application involving optoelectronics and imaging. If the interaction between the excited ZnO and GO is indeed the root cause for the quenching of PL emission, then it should be possible to time resolve the process from the PL emission decay. Hence, PL was performed on the ethanolic solutions of intrinsic GO and GO/ZnO composites to obtain their exciton decay lifetimes of GO:ZnO dramatically decreases by many times as compared with intrinsic GO. The excitation and emission wavelengths of 340 and 424 nm were used for the experiment to record the lifetimes. The obtained decay patterns are multiexponential in nature and can be fit to a biexponential expression in order to derive decay time constants, which could be further used to calculate the average lifetimes of the PL. The inset shows the decrease in average lifetime as a function of various GO:ZnO concentrations. As prepared, the GO is insulating in nature because sp^2 clusters are being interrupted by oxygen functional groups, which in turn decrease the creation of electron-hole pairs upon excitation. The analysis of the average lifetimes conclude that the highly influential static interaction prevails between ZnO and GO nanosheets, leading to increased number of localized states allowing radiative recombination to occur. Further, the exciton lifetime was found to be weakly dependent on the duration of stirring or the RT. In any case, the reduction process is also expected to produce defects such as vacancies in the GO:ZnO composites. An increased number of defects and larger delocalization lengths in the reduced GO result in higher probability of nonradiative recombination and therefore faster decay of PL study observed. The nonradiative recombination can also occur by tunneling or hopping the mechanisms of electron-hole pairs into defect sites due to the probable presence of the deep traps.

Photocatalytic reduction of Cr(VI) by GO:ZnO was performed under UV irradiation as shown in Figure 7. The normalized temporal concentration changes (C/C_O) of Cr(VI) during the photocatalytic process are proportional to the normalized maximum absorbance (A/A_O), which can be derived from the change in the Cr(VI) absorption profile at a given time interval [56]. It is observed that Cr(VI) cannot be reduced under the dark condition even in the presence of photocatalytic GO:ZnO nanocomposites that exhibit better photocatalytic performance than pure ZnO. It is known that during photocatalysis, the light absorption and the charge transportation and separation are crucial factors [57]. The enhancement of the photocatalytic performance should be ascribed to the increase of the light absorption intensity and range and the reduction of electron-hole pair recombination in ZnO with the introduction of GO in the composite, which can be confirmed from absorption and PL measurements [58]. However, when the GO content is further increased above its optimum value, the photocatalytic performance deteriorates [59]. This is ascribed to the following reasons: (i) GO may absorb some UV light, and thus, there exists a light-harvesting competition between ZnO and GO with the increase of GO content, which leads to the decrease of the photocatalytic performance, and (ii) the excessive GO can act as a kind of recombination center instead of providing an electron pathway and promote the recombination of electron-hole pairs in GO nanosheets [60].

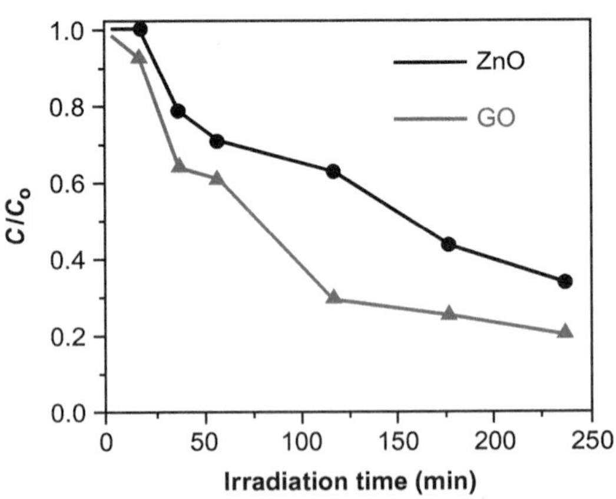

Figure 7 Photocatalytic studies of GO:ZnO nanosheets.

4. CONCLUSIONS

Raman spectroscopy confirmed the formation of single-layer GO nanosheets. XPS analysis showed the compositional and the oxidization states of these materials suspended in ethanol. Variations of the optical absorption of GO:ZnO graphene oxide suspension and decrease in the resistive value of the nanosheets also confirmed the chemical reduction by the photocatalytic process. The reduced GO nanosheets hybridized by the ZnO nanoparticles were soluble in ethanol, as a polar solvent. The photocatalytic reduction mechanism was described based on the electron transfer from the photoexcited ZnO nanoparticles into the GO nanosheets and the scavenging of the holes by the ethanol solution. The synthesized GO:ZnO not only can be applied in tuning the electrical conductivity of the GO layers but also can be used as a photoactive graphene semiconductor composite in various future photocatalytic applications.

REFERENCES

[1] H. Fu, T. Zhu, S. Zhu, Photo corrosion Inhibition and Enhancement of Photo catalytic Activity for ZnO via Hybridization with C_{60}, Environ. Sci. Technol. 42 (2008) 8064.
[2] G. Williams, B. Seger, P.V. Kamat, TiO_2-graphene nanocomposites UV-assisted photocatalytic reduction of graphene oxide, ACS Nano 2 (2008) 1487.
[3] Y. Xie, S.H. Heo, S. Hwa, G. Ali, S. Cho, Synthesis and Photocatalytic Activity of Anatase TiO2 Nanoparticles-coated Carbon Nanotubes, Nanoscale Res. Lett. 5 (2009) 603.
[4] O. Akhavan, M. Abdolahad, Y. Abdi, S. Mohajerzadeh, Synthesis of titania/carbon nanotube heterojunction arrays for photoinactivation of E. coli in visible light irradiation, Carbon 47 (2009) 3280.
[5] G. An, W. Ma, Z. Sun, Z. Liu, B. Han, S. Miao, Z. Miao, Preparation of titania/carbon nanotube composites using supercritical ethanol and their photocatalytic activity for phenol degradation under visible light irradiation, Carbon 45 (2007) 1795.
[6] H. Yu, X. Quan, S. Chen, Y. Zhang, Carbon nanotubes supported by titanium dioxide nanoparticlesas recyclable and green catalyst for mild synthesisof dihydropyrimidinones/thiones, J. Photochem. Photobiol. A Chem. 200 (2008) 301.
[7] O. Akhavan, E. Ghaderi, Photocatalytic Reduction of Graphene Oxide Nanosheets on TiO2 Thin Film for Photoinactivation of Bacteria in Solar Light Irradiation, J. Phys. Chem. C 113 (2009) 20214.
[8] K.S. Novoselov, A.K. Geim, S.V. Morozov, D. Jiang, Y. Zhang, S.V. Dubonos, I. Grigorieva, A.A. Firsov, Electric field effect in atomically thin carbon films, Science 306 (2004) 666.
[9] M.I. Katsnelson, K.S. Novoselov, Graphene: new bridge between condensed matter physics and quantum electrodynamics, Solid State Commun. 143 (2007) 3.
[10] J.C. Meyer, A.K. Geim, M.I. Katsnelson, K.S. Novoselov, T.J. Booth, S. Roth, The structure of suspended graphene sheets, Nature 446 (2007) 446.
[11] Y.B. Zhang, Y.W. Tan, H.L. Stormer, P. Kim, Experimental observation of the quantum Hall effect and Berry's phase in grapheme, Nature 438 (2005) 201.

[12] R. Arsat, M. Breedon, M. Shafei, P.G. Spizziri, Graphene-like nano-sheets for surface acoustic wave gas sensor applications, Chem. Phys. Lett. 467 (2009) 344.
[13] S. Gilje, S. Han, M. Wang, K.L. Wang, R.B. Kaner, A chemical route to graphene for device applications, Nano Lett. 7 (2007) 3394.
[14] X. Liang, Z. Fu, S. Chou, Fabricated via Transfer-Printing in Device Active-Areas on Large Wafer, Nano Lett. 7 (2007) 3840.
[15] C. Gomez, R.T. Weitz, A.M. Bittner, M. Burghard, Elastic Properties of Chemically Derived Single Graphene Sheets, Nano Lett. 7 (2007) 3499.
[16] B. Standley, W. Bao, H. Zhang, C.N. Lau, M. Bockrath, Graphene-based atomic-scale switches, Nano Lett. 8 (2008) 3345.
[17] S. Stankovich, D. Dikin, K. Dommett, K.M. Kohlhaas, E.J. Zimney, E.A. Stach, Graphene-based composite materials, Nature 442 (2006) 282.
[18] W.S. Hummers, R.E. Offeman, Functionalized Graphene and Graphene Oxide: Materials Synthesis and Electronic Applications, J. Am. Chem. Soc. 80 (1958) 1339.
[19] G. Huey-Shya, A. Rohana, A.F. Muhammad, Air Pollution Control over Nano-catalysts, Turk. J. Chem. 35 (2011) 375.
[20] N.A. Kotov, Carbon sheet solutions, Nature 442 (2006) 255.
[21] J.T. Robinson, F.K. Perkins, E.S. Snow, P.E. Sheehan, Reduced graphene oxide molecular sensors, Nano Lett. 8 (2008) 3137.
[22] D. Wang, D. Choi, J. Yang, Z. Nei, R. Kou, Self-assembled TiO2-graphene hybrid nanostructures for enhanced Li-ion insertion, ACS Nano 3 (2009) 907.
[23] A.P. Yu, P. Ramesh, M.E. Itkis, E. Bekyarova, Graphite Nanoplatelet-Epoxy Composite Thermal Interface Materials, J. Phys. Chem. C 111 (2007) 7565.
[24] X. Wang, L.J. Zhi, N. Tsao, Z. Tomovic, Transparent carbon films as electrodes in organic solar cells, Angew. Chem. Int. Ed. 7 (2008) 2990.
[25] S. Watcharotone, A. Dikin, R. Piner, I. Jung, H.B. Dommet, Graphene-Silica Composite Thin Films as Transparent Conductors, Nano Lett. 7 (2007) 1888.
[26] S. Niyogi, E. Bekyarova, S. Itkis, M.A. Hamon, Solution Properties of Graphite and Graphene, J. Am. Chem. Soc. 128 (2006) 7720.
[27] H.C. Schniepp, J.L. Li, M.J. McAllister, H. Sai, M. Herrera, D.H. Adamson, Functionalized single graphene sheets derived from splitting graphite oxide, J. Phys. Chem. B 110 (2006) 8535.
[28] L.X. Li, X.R. Wang, L. Zhang, S.W. Lee, H.K. Dai, Chemical Approaches toward Graphene-Based Nanomaterials and their Applications in Energy-Related, Science 319 (2008) 1229.
[29] C. William, X. Wang, J. Zhu, Graphene/Semiconductor Nanocomposites: Preparation and Application for Photocatalytic Hydrogen Evolution, J. Phys. Chem. C 112 (2008) 19841.
[30] H. Zeng, Y. Cao, S. Xie, J. Yang, X. Wang, L. Sun, Large cluster and hollow microfibers by multicomponent self-assembling of citrate stabilized gold nanoparticles with temperature-responsive amphiphilic dendrimers, Nanoscale Res. Lett. 8 (2013) 133.
[31] J. Wua, X. Shen, L. Jiang, K. Wang, K. Chen, Preparation and Characterization of Reduced Graphene Oxide Sheets via Water-Based Exfoliation and Reduction Methods, Appl. Surf. Sci. 256 (2010) 2826.
[32] G. Wang, B. Wang, J. Park, X. Shen, J. Yao, Synthesis and characterisation of hydrophilic and organophilic graphene nanosheets, Carbon 47 (2009) 68.
[33] A.C. Ferrari, J. Robertson, Interpretation of Raman spectra of disordered and amorphous carbon, Phys. Rev. B 61 (2000) 14095.
[34] S. Dato, V. Radmilovic, Z. Lee, P. Phillips, M. Frenkach, Substrate-free gas-phase synthesis of graphene sheets, Nano Lett. 8 (2008) 2012.
[35] A.C. Ferrari, J.C. Meyer, V. Scardaci, C. Casiraghi, M. Lazzeri, F. Mauri, Raman Spectrum of Graphene and Graphene Layers, Phys. Rev. Lett. 97 (2006) 187401.
[36] D. Graf, F. Molitor, K. Ensslin, C. Stampfer, C. Hierold, Spatially resolved Raman spectroscopy of single- and few-layer grapheme, Nano Lett. 7 (2007) 238.

[37] F. Tuinstra, J.L. Koenig, Raman Spectrum of Graphite, J. Chem. Phys. 53 (1970) 1126.
[38] H. Kim, H.L. Park, H. Jeon, J.H. Chang, J.K. Chang, In-situ observation and characterization of structural evolution in a phase-change memory device by TEM-STM, Ceram. Int. 35 (2009) 131.
[39] I. Sameera, R. Bhatia, V. Prasad, Preparation, characterization and electrical conductivity studies of MWCNT/ZnO nanoparticles hybrid, Physica B 405 (2010) 17094.
[40] T.C. Chiang, F. Seitz, Chemical analysis of graphene oxide films after heat and chemical treatments by X-ray photoelectron and Micro-Raman spectroscopy, Ann. Phys. 10 (2001) 61.
[41] P. Guo, H. Song, X. Chen, Electrochemical performance of graphene nanosheets as anode material for lithium-ion batteries, Electrochem. Commun. 11 (2009) 1320.
[42] C. Wang, D. Li, C.O. Too, C.C. Wallace, Comparative studies on electrochemical activity of graphene nanosheets and carbon nanotubes, Chem. Mater. 21 (2009) 2604.
[43] D. Pan, S.S. Wang, B. Zhao, M.W. Wu, Z. Jiao, Graphene and Graphene Oxide: Synthesis, Properties, and Applications, Chem. Mater. 21 (2009) 3136.
[44] S.M. Paek, E. Yoo, I. Honma, Graphene-based Composites for Electrochemical Energy Storage, Nano Lett. 9 (2009) 72.
[45] P. Poizot, S. Laruelle, S. Grugeon, L. Dupont, J.M. Tarascon, Nano-sized transition-metal oxides as negative-electrode materials for lithium-ion batteries, Nature 407 (2000) 496.
[46] S. Yumitori, Role of Peroxide Ions in Formation of Graphene Nanosheets by Electrochemical Exfoliation of Graphite, J. Mater. Sci. 35 (2000) 139.
[47] T.H. Mahato, G.K. Prasad, B. Singh, J. Acharya, A.R. Srivastava, R. Vijayaragjavan, Nanocrystalline zinc oxide for the decontamination of sarin, J. Hazard. Mater. 165 (2009) 928.
[48] G. Eda, Y. Lin, C. Mattevi, H. Yamaguchi, I. Chen, Blue Photoluminescence from Chemically Derived Graphene Oxide, Adv. Mater. 22 (2010) 505.
[49] J.D. Fowler, A.J. Allen, R.B. Kaner, N.B. Weiller, Practical chemical sensors from chemically derived grapheme, ACS Nano 3 (2009) 301.
[50] J. Robertson, Recombination and photoluminescence mechanism in hydrogenated amorphous carbon, Phys. Rev. B 53 (1996) 16302.
[51] Z. Liu, J.T. Robinson, X. Sun, H. Dai, PEGylated Nano-Graphene Oxide for Delivery of Water Insoluble Cancer Drugs, J. Am. Chem. Soc. 130 (2008) 1087.
[52] P. Luo, M.P. Vora, E.J. Johnson, J.M. Kikkawa, Tuning electronic and magnetic properties of partially hydrogenated graphene by biaxial tensile strain: a computational study, Appl. Phys. Lett. 94 (2009) 111901.
[53] J.Q. Hu, Q. Lu, X.M. Meng, C.S. Lee, Thermal Reduction Route to the Fabrication of Coaxial Zn/ZnO nanocables and ZnO nanotubes, Chem. Mater. 15 (2003) 305.
[54] G. Williams, P.V. Kamat, Graphene-semiconductor nanocomposites: excited-state interactions between ZnO nanoparticles and graphene oxide, Langmuir 25 (2009) 13869.
[55] D.C. Look, J.W. Hemsky, J.R. Sizelove, Residual Native Shallow Donor in ZnO, Phys. Rev. Lett. 82 (1999) 2552.
[56] M. Qazi, T. Vogt, G. Koley, Trace gas detection using nanostructured graphite layers, Appl. Phys. Lett. 91 (2007) 233101.
[57] H. Zhang, X.J. Lu, Y.M. Li, Y. Wang, TiO_2 Nanocrystals Grown on Graphene as Advanced Photocatalytic Hybrid Materials, ACS Nano 4 (2010) 380.
[58] T.G. Xu, L.W. Zhang, X.H. Cheng, Y. Zhu, Enhanced photocatalytic properties of graphene oxide/ZnO nanohybrid by Mg dopants, Appl. Catal. B Environ. 101 (2011) 382.
[59] N.L. Yang, J. Zhai, D. Wang, Y.S. Chen, Microwave-assisted synthesis of CdS–reduced graphene oxide composites for photocatalytic reduction of Cr(VI), ACS Nano 2 (2010) 887.
[60] G. Zhu, T. Xu, L.K. Pan, Q.F. Zhao, Z. Sun, Microwave-assisted synthesis of ZnO–graphene composite for photocatalytic reduction of Cr(VI), J. Electro Analyt. Chem. 650 (2011) 248.

CHAPTER 6

Polyolefin/Graphene Nanocomposite Materials

Muthukumaraswamy Rangaraj Vengatesan, Vikas Mittal
Department of Chemical Engineering, The Petroleum Institute, Abu Dhabi, United Arab Emirates

Chapter Contents

1. INTRODUCTION

Polymer nanocomposites, in which at least one dimension of the reinforcement phase is less than 100 nm, have resulted in a tremendous research effort in order to generate functional materials by synergistically combining the properties of constituent materials [1–7]. Graphene, which consists of atomic thick sheets of covalently sp^2-bonded carbon atoms in a hexagonal arrangement, has received attention of researchers as reinforcing filler for the generation of polymer nanocomposites [8–11]. Its choice as a filler is due to its excellent electrical and mechanical properties, which are significantly better than for other inorganic filler materials. A single defect-free graphene layer has Young's modulus of 1.0 TPa, intrinsic strength of 42 N/m, thermal conductivity of 4840-5300 W/(m K), electron mobility exceeding 25,000 cm^2/V s, excellent gas impermeability, and specific

Handbook of Nanoceramic and Nanocomposite Coatings and Materials
http://dx.doi.org/10.1016/B978-0-12-799947-0.00006-7

surface area of 2630 m^2/g [9,12–18]. All these properties make this material even superior to carbon nanotubes for use in polymer nanocomposites. A number of studies on polymer nanocomposites based on graphene have been published in a short span of time since its development [9,19–29]. Additionally, uses of graphene-based materials in application like carbon films, sensors, electrodes, and solar cells have been proposed underlying the vast potential of these materials [30–34].

The synthesis of graphene (G) platelets has been reviewed extensively in the literature [8,19,20]. Though bottom-up approaches of graphene synthesis generate defect-free platelets of superior properties, these methods are not suitable for bulk production of graphene. Various bottom-up approaches of graphene synthesis include arc discharge, chemical vapor deposition, unzipping carbon nanotubes, and epitaxial growth [35–43]. In the top-down processes, graphite oxide (GO) is produced first by treatment of graphite with strong mineral acids and oxidizing agents [44]. Oxidation of graphite to form GO was reported initially by Brodie in 1859 using potassium chlorate and nitric acid oxidants [45]. The process leads to some non-GO products, which need to be carefully washed [27,46,47]. The structure of GO has generated many theoretical and experimental research efforts [44,48–57] and is believed to contain hydroxyl and epoxy groups on the basal plane, whereas carboxylic groups are present around the periphery of sheets [44]. Both chemical reduction and thermal exfoliation methods have been suggested to exfoliate the GO to form graphene.

The macroscopic properties of polymer nanocomposites are mainly dependent on the thermodynamic factors such as interfacial compatibility of polymer and filler phases and polarity match between the filler surface and the polymer chains. In addition, nanoscale dispersion and distribution of the filler also depends on the size, shape, dispersion techniques and equipment, time of mixing and applied shear, etc. The full advantage of nanofillers can be only achieved by considering these factors, which could lead to uniform transfer of superior properties of nanofillers to host polymer matrix [58–63]. Significant research effort has focused on attaining full potency of nanofillers by using different mixing techniques, compatibilizer (functional polymers), and coupling agents, modifying polymer backbone or filler surface, etc. Modification of filler surface represents a very versatile approach to achieve compatibilization. Nanotubes have been studied extensively for covalent and noncovalent modifications of the surface [64–73]. Similarly, various physical and chemical modes of surface modification of aluminosilicates have also been reported [74–84]. In the case of graphene, the presence of less number of functional groups (such as carboxyl, epoxide, and hydroxyl) on the surface of pristine graphene leads to lower compatibility with polar polymer matrices. Similarly, the dispersion of polar graphene oxide in nonpolar polymers is not optimal owing to the absence of positive interactions between them. Thus, similar surface modification processes for graphene platelets are also needed to achieve effective dispersion in the polymer matrices. The knowledge gained specially for the nanotube surface modification has immense relevance for graphene platelets' modification due to the similar chemical structure of these materials. The use of amphiphilic compatibilizers having polar and nonpolar groups has also been reported extensively for filler systems like aluminosilicates and nanotubes to improve filler dispersion [85–89].

2. PREPARATION METHODS FOR POLYOLEFIN/GRAPHENE NANOCOMPOSITES

The miscibility of graphene layers and the polyolefin matrix mainly depends on hydrophobic interaction, reactive blending of functionalized graphene oxide (fGO) with compatibilizer, catalyst support GO in *in situ* polymerization, etc. The following methods are widely used for the incorporation of layered graphene into the polyolefin matrix.

2.1 *In Situ* Method for the Preparation of Polyolefin/Graphene Nanocomposites

In this method, polyolefin/graphene nanocomposites are prepared by using the propylene or ethylene monomer and different types of catalysis such as metallocene complex rac-Me_2Si (Ind) $2ZrCl_2$ catalyst and methyl aluminoxane as cocatalyst [90,91], Ziegler-Natta catalyst [92], and Ni(II) catalyst precursor [93]. This technique results in a stronger interfacial interaction and better miscibility of graphene layer within the polyolefin matrix. The composites prepared from this method show better mechanical and electrical properties compared to other solution and melt mixing method.

2.2 Solution Blending

This is a facile method for the bulk-scale fabrication of the composites. In this method, polyolefin dissolved in a proper solvent and thermally reduced graphene oxide (TRG) or functionalized reduced graphene oxide (FRGO) is added to the polyolefin solution in different weight ratios. The mixtures are subjected to ultrasound followed by shearing. After a complete blending of graphene layer into polyolefin matrix, subsequently, the solvent has been removed from the mixture and finally gets an exact shape of the composite using specific mold. A huge number of polyolefin/graphene nanocomposites were prepared from this method using polyolefin matrices with FRGO or TRG [29,94–98].

2.3 Melt Mixing Method

In this method, polyolefin matrices are mixed with FRGO or TRG using extrusion, internal mixing, and two-roll mill at elevated temperature. This method is cost-effective, eco-friendly, and industry valuable for bulk scale compared with other methods [29,98].

3. PROPERTIES OF POLYOLEFIN/GRAPHENE NANOCOMPOSITES

As mentioned above, graphene sheets incorporated polyolefin composites are prepared through melt mixing, solution blending, and *in situ* polymerization approaches. Small amount of graphene sheets influences a significant improvement on thermal, mechanical, and electrical properties of the resulted polyolefin nanocomposites. The properties of the resulting composites depend on the graphene sheets and the polymerization techniques. The chemical functionalization on the graphene sheets also influences the properties of the resulting polyolefin composites.

3.1 Mechanical Properties of Polyolefin/Graphene Nanocomposites

Recently, [97] reported different chemical routes to functionalize graphene with short-chain polyethylene (PE), which influences the mechanical properties of graphene-based high-density polyethylene (HDPE) nanocomposites. The authors observed that the PE grafting on the graphene through thiol-yne click chemistry showed a significant effect on Young's modulus and elongation at break of the resultant composites [97]. Docoslis et al. [98] studied the dispersion and properties of graphene/PE composite using maleated linear low-density polyethylene (LLDPE) derivatives and TRG through a noncovalent compatibilization approach. It was observed that the pyridine-functionalized PE matrix showed a better dispersion with TRG due to the π-π interactions and significantly improved Young's moduli without reduction of tensile stress and elongation at the break [98]. Macosko et al. [29] reported the graphene/PE nanocomposites with the effect of PE functionalization and blending methods. It was suggested that the TRG is more compatible with the amine-functionalized PE and the composites fabricated from solution blending method resulted in good mechanical properties compared to melt mixing method [29]. Galland et al. prepared graphene/PE composites through *in situ* polymerization method [99]. It was observed that the storage modulus of the composites increased with increase in weight percentage of graphene nanosheets (GNS) and a slight decrease in tensile strength. Lee et al. [100] developed low-density polyethylene (LDPE)/graphene nanocomposites *via* solution blending method. It was concluded that the storage modulus of the composites increases with the increase in graphene content [100]. Wang et al. prepared LDPE/graphene nanocomposites using vinyl-functionalized graphene sheets and LDPE through solution blending method [101]. It was inferred that the tensile strength and Young's modulus of the composites increased up to 27% and 98.2% compared to pristine LDPE. Mittal and Matsko studied the low-temperature brittleness behavior of PE/chlorinated PE blends with graphene oxide nanofiller [28]. The authors concluded that the GO did not reduce the flexibility of the composites. Figure 1 demonstrates the fracture analysis performed on the polymer as well as nanocomposite samples at two temperatures: −140 and −195 °C. The PE and PE + CPE25 and PE + CPE35 blends (with 5% CPE content) were flexible at −140 °C, whereas only PE + CPE35 sample was still flexible at −195 °C. The other two samples had a brittle failure at −195 °C. It confirmed the findings from the AFM analysis about the flexibility of CPE35 containing samples at −180 °C and indicated that the samples were still flexible at −195 °C. Similar to the PE + CPE35 blend, the nanocomposite PE + G + CPE35 was also flexible at −195 °C indicating that addition of graphene to this system did not lead to reduction in flexibility. The other

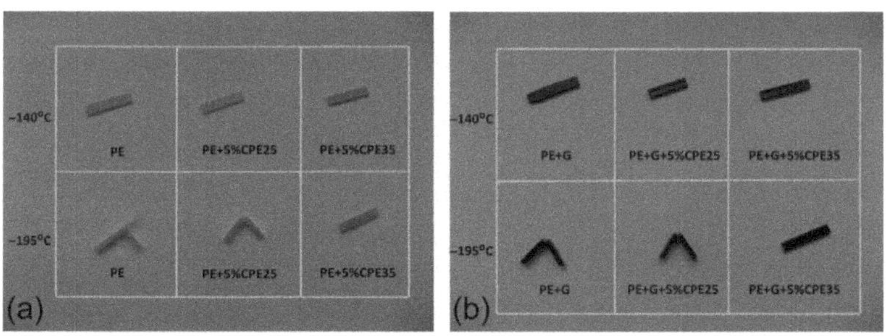

Figure 1 Fracture analysis of samples: (a) PE and PE+CPE blends, (b) PE+G+CPE nanocomposites. *(Reproduced from Ref. [28] with permission from Springer).*

two composites PE + G and PE + G + CPE25 were flexible only near the glass transition temperature of PE and failed in a brittle manner at −195 °C.

In another study, Wang et al. prepared graphene/LDPE composites using functionalized reduced graphene oxide (fGO) with different chain lengths of alkyl amine and LDPE via solution blending [102]. Dodecyl amine-modified reduced GO used composites possessing higher mechanical properties compared with other composites. Jin et al. also studied the reinforcing effect of alkylated graphene oxide (AGO) on a nonpolar polypropylene (PP) matrix compared with alkylated carbon nanotubes [103]. It was found that the addition of 0.1 wt% of AGO on PP increased Young's modulus by more than 70%. It was further revealed that the reinforcing effect of AGO was superior to ACNT in relation to Young's modulus enhancement of PP composites.

In a related study, Hu et al. reported covalently fGO/PP nanocomposites using melt-blending method [104]. It was observed that the addition of 0.5wt % of fGO increased only the storage modules of PP up to 15%, but it decreased the elongation at break and there was no change in the value of tensile strength of PP composites with increase in fGO content. Shanmugaraj et al. prepared long-chain AGO materials using different chain lengths of alkylated amines and incorporated into PP matrix [105]. 1 wt% of octadecylamine-modified GO-incorporated PP composites showed an improved Young's modulus up to 47% and tensile strength up to 29.4% compared to other alkylated amine-modified GO and pristine GO-incorporated PP composites. In another study, Qaiss et al. developed graphene/PP nanocomposites via melt mixing method [106]. By increasing the content of GNS into the PP matrix, the tensile strength and Young's modulus increased but the elongation at break decreased. Fu et al. fabricated graphene-coated PP latex and incorporated into PP matrix through melt-blending method [107]. The authors concluded that the addition of 0.42 vol% of graphene increased the tensile strength and modulus by 75% and 74%, respectively (Figure 2). Cao et al. also prepared polypropylene-grafted-graphene oxide (PP-g-GO) and incorporated into polypropylene/polyphenylene oxide (PP/PPO) blends [96]. The increase in weight percentage of PP-g-GO increased the elongation at break and tensile strength of PP/PPO blend.

In a recent study, HDPE/graphene nanocomposites were generated using different amounts of chlorinated PE compatibilizer [27]. Chlorinated PEs with 25% chlorine content (named as CPE25) and 35% chorine content (named as CPE35) were used. The nanocomposites were prepared by either solution mixing to form CPE + G masterbatch or melt mixing of the

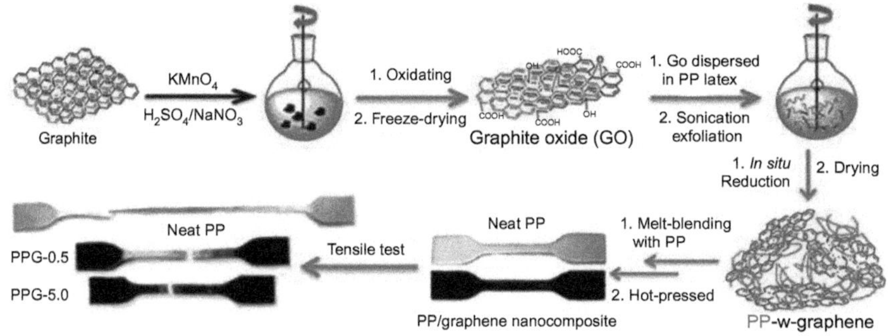

Figure 2 Schematic representation of process flow of polypropylene (PP)/graphene nanocomposites fabrication and the tensile tests. *(Reproduced from Ref. [107] with permission).*

masterbatch with PE. PE + G composites were also generated by direct melt mixing. Morphology of the masterbatches as well as nanocomposites is shown in Figure 3 (for 5% compatibilizer content). The CPE35 masterbatch (Figure 3a) was observed to have better graphene dispersion as compared to CPE25 containing masterbatch (Figure 3b). Though complete nanoscale delamination of the graphene platelets was not observed in the composites, however, the composites with CPE35 compatibilizer also had much better filler dispersion as compared to corresponding CPE25 nanocomposites. Graphene stacks of varying thicknesses (single layers to multiple layers) could be observed for CPE35-containing composites as shown in Figure 3c and d, whereas the stack thickness was much higher for CPE25 nanocomposite as shown in Figure 3e. As the compatibilizers differ in the extent of chlorination, therefore, resulting morphology can be related to the interaction of polar chlorine atoms with the graphene surface. Higher extent of chlorination in the matrix caused higher magnitude of interfacial interactions between the polymer and the filler surface resulting in increased extent of filler delamination. Thus, increased chlorination content not only decreased the polymer crystallinity but also increased the susceptibility of filler platelets to delaminate in the polymer matrix. Table 1 also reports the tensile properties of PE and nanocomposites. The tensile modulus for pure polymer was observed to be 1063 MPa, which was enhanced to 1150 MPa in PE + G nanocomposite without compatibilizer. The addition of semicrystalline compatibilizer CPE25 increased the modulus gradually until 5 wt%, after which a reduction in the modulus was recorded. Similar observations have also been made earlier, where the modulus was described to be affected by a balance between increase in modulus owing to enhanced filler dispersion by

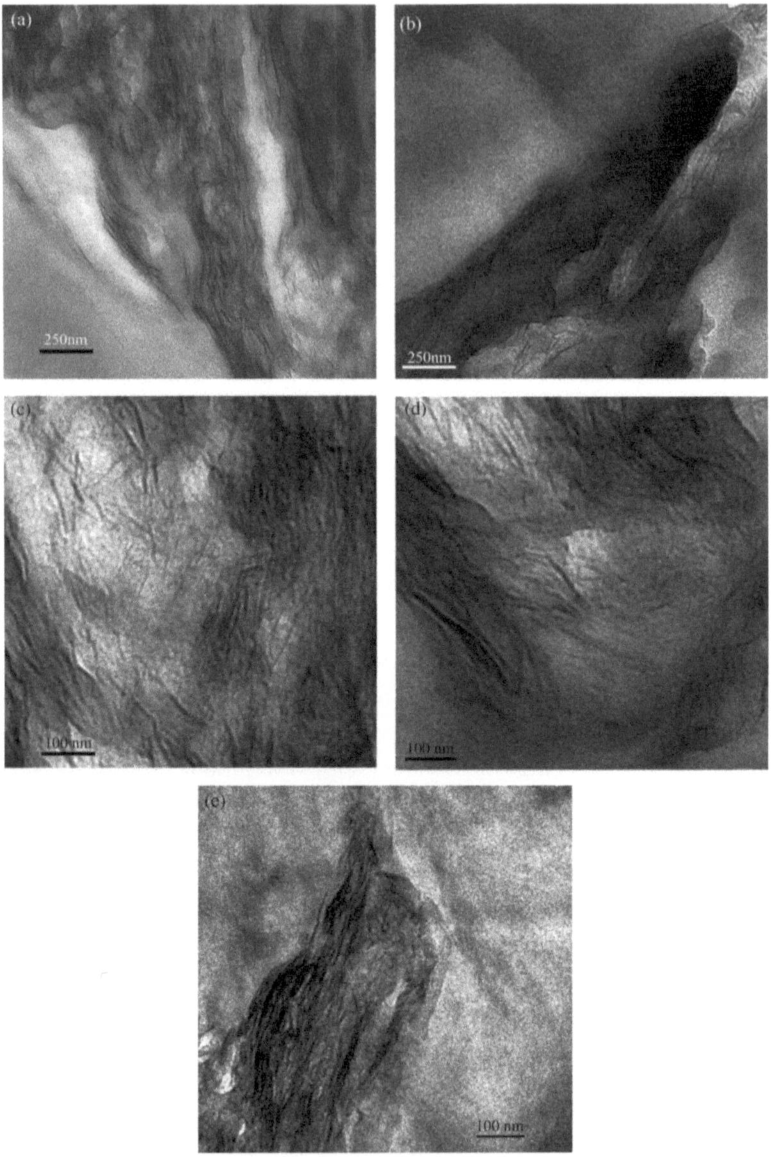

Figure 3 TEM micrographs of (a) CPE35 + G masterbatch, (b) CPE25 + G masterbatch, (c) and (d) PE + 5%CPE35 + G, and (e) PE + 5%CPE25 + G nanocomposites. The black lines are the intersection of graphene platelets. *(Reproduced from Ref. [27] with permission from Wiley).*

Table 1 Tensile properties of the pure polymer and polymer-graphene nanocomposites

Sr. no.	Polymer nanocomposite	Young's modulus[a] (MPa)	Peak stress[b] (MPa)	Peak strain[c] (%)
1	PE	1063	62	8.4
2	PE + G	1150	58	7.8
3	PE + 1%CPE25 + G	974	56	8.5
4	PE + 2%CPE25 + G	1181	69	7.8
5	PE + 5%CPE25 + G	1228	75	7.6
6	PE + 10%CPE25 + G	1160	65	7.8
7	PE + 5%CPE35 + G	905	48	8.7
8	PE + 10%CPE35 + G	1105	68	8.0

[a]Relative probable error 2%.
[b]Relative probable error 2%.
[c]Relative probable error 5%.
Reproduced from Ref. [27] with permission from Wiley.

the compatibilizer and a simultaneous decrease in modulus due to plasticization of the matrix caused by it. Until 5 wt% content of compatibilizer in the composite, the delamination affect dominated resulting in the increment of 16% in the modulus as compared to pure PE. On further increasing the compatibilizer content, the plasticization effect dominated the performance resulting in the decrease in tensile modulus. The increased extent of interfacial interactions (hence filler delamination) also resulted in different behaviors of CPE25 and CPE35 compatibilizers, as CPE35-containing composites exhibited an increase in modulus even at 10% compatibilizer content. However, the modulus of CPE35-containing composites was lower than the corresponding CPE25 composites owing to amorphous nature of CPE35.

It should also be noted that though the increments in the modulus are not tremendous, however, these enhancements are still significant owing to a very low amount of graphene used to achieve them. The importance of masterbatch approach was also confirmed by comparing the tensile modulus value of the PE + G + 5%CPE25 with the similar composite generate by only melt mixing. A value of 969 MPa was obtained for such melt-mixed composite, which was much lower than 1228 MPa for the composite generated with masterbatch approach. The peak stress also showed similar behavior as tensile modulus. Addition of graphene to PE without compatibilizer resulted in a slight decrease of peak stress due to restriction in segmental mobility via mechanical interlocking with graphene tactoids. On the other hand, addition of CPE25 gradually enhanced the strength until 5 wt% content resulting in an increment of 21% as compared to pure polymer. The CPE35-containing composites showed an increase of 10% in strength at a compatibilizer content of 10 wt%. The composites still remained rigid as the peak strain was not significantly affected on the addition on compatibilizers. A comparison of the mechanical properties of the generated PE + G nanocomposites with other systems like PE-clay nanocomposites also revealed their high potential. For example, PE-clay nanocomposites without compatibilizer showed an increase of 4% in Young's modulus at 2 wt% filler content [34], which was much lower than the graphene nanocomposites even with 0.5 wt% filler content.

3.2 Thermal and Flame-Retardant Behaviors of Polyolefin/Graphene Composites

As mentioned above, Cao et al. studied the flame-retardant and thermal stability of PP/PPO blends incorporated with PP-g-GO [96]. The authors observed that the PP-g-GO acts as multifunctional filler for PP/PPO blends and it enhances the thermal stability and flame-retardant behavior of PP/PPO blends. Similarly, Ma et al. successfully grafted the PP on the surface of TRG covalently and studied the effect of compatibility and thermal stability of isotactic polypropylene (iPP) matrix with the incorporation of PP-g-TRG [108]. It was observed that the PP-g-TRG acts as a heterogeneous nucleation agent and shows a significant improvement in thermal stability of iPP. In another study, Li et al. studied the isothermal and nonisothermal behavior of iPP with the incorporation of graphene oxide nanosheets (GONs) [109]. They observed that the GONs acts as α-nucleating agent and it significantly elevated the crystallization temperature of iPP (Figure 4).

Haung et al. studied the combination effect of carbon nanotube with graphene on intumescent flame-retardant (IFR) behavior of PP nanocomposites [110]. The IFR material was prepared using melamine polyphosphate and pentaerythritol in 80:20 wt% ratio, and the IFR-PP/graphene nanocomposites are prepared using 18 wt% of IFR blend material, 1 wt% of reduced graphene oxide (rGO), and 1 wt% of CNT with PP through melt-blending method. The composites showed the limiting oxygen index value of 31.4% and UL-94 VO grade and also exhibited 40 s delayed time of ignition compared to neat PP. Similarly, Yu et al. prepared a flame-retardant functionalized GO (FRs-FGO) using amine functional GO and phosphoramide oligomer [111]. FRs-FGO was incorporated into PP and simultaneously compatibilized with PP-grafted maleic anhydride (PP-g-MA). With increase in content of FRs-FGO into PP matrix, crystallization temperature increased and peak heat release rate and total heat release decreased

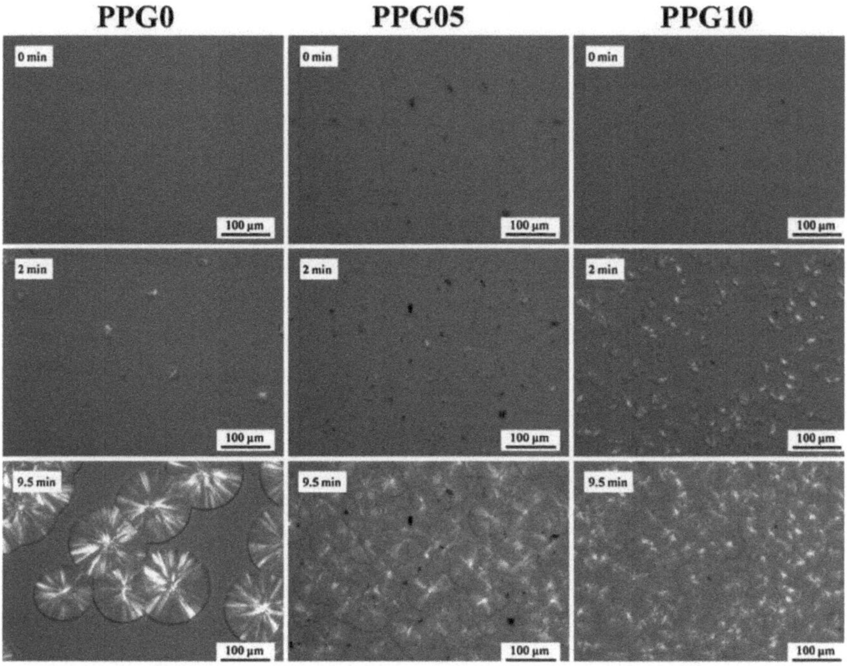

Figure 4 Polarized optical micrographs of PPG0, PPG05, and PPG10 during isothermally crystallizing at 130 °C. *(Reproduced from Ref. [109] with permission).*

along with decrease in the fire growth rate index of the resultant composites. The authors concluded that the incorporation of FRs-FGO into PP matrix compatibilized with PP-g-MA and improves its fire safety. Hu et al. prepared hyperbranched FRs-FGO using *N*-aminoethylpiperazine, hydroxyethyl acrylate, methyldichlorophosphate, and GO via *in situ* polymerization technique [112]. FGO was incorporated into cross-linked PE matrix and its flammability and thermal stability were studied. It was observed that the incorporation of FGO into cross-linked PE endowed a highly flame-retardant and thermally stable PE composite. In a similar work, Han et al. studied the flame retardancy behavior of PE/alumina trihydrate (PE/ATH) composites with the effect of incorporation of graphene nanoplatelets (GNPs) [113]. The composites were prepared through melt-blending method. The incorporation of 0.2 wt% of GNPs into the PE/ATH composite decreased the heat release peak by 18%. Inuwa et al. studied the flammability and thermal stability of polyethylene terephthalate/polypropylene (PET/PP) blends with the influence of exfoliated graphite nanoplatelets (GNPs) [114]. It was observed that the thermal conductivity of the nanocomposites increased about 80%, but the degree of crystallinity of the composites was decreased with increase in GNPs content. With increasing amount of GNPs into PET/PP blends, the molar loss heat and delayed time of ignition of the nanocomposites increased. Ran et al. developed HDPE/brominated polystyrene/GNPs composites *via* melt blending and studied its thermal stability and flammability with effect of the Friedel-Crafts reaction [115]. Anhydrous aluminum trichloride ($AlCl_3$) was used as a Lewis catalyst and it was added to initiate the Friedel-Crafts reaction for the effective dispersion of GNPs into the polymer blends. The results revealed that the GNPs were partly unfolded and the domains became smaller in the presence of $AlCl_3$. The incorporation of $AlCl_3$ into polymer-graphene nanocomposites reduced the flammability and slowed down the heat release rate.

3.3 Electrical Properties of Polyolefin/Graphene Composites

Ramasamy et al. developed PP/graphene composites that could be converted into metamaterial (negative dielectric) at desired frequencies [116]. The authors observed that by changing the concentration of graphene, the negative dielectric material turned into positive dielectric material. Less than 5% of graphene in PP composites reduced the electrical conductivity and also showed low dielectric loss at high frequencies near the percolation threshold. It was concluded that the low concentration of graphene in PP nanocomposites can enable the composites to behave as a high dielectric material. Shevchenko et al. prepared iPP-graphene nanocomposites using liquid propylene, GNPs, and metallocene catalyst through ultrasonication followed by *in situ* polymerization techniques [91]. The authors suggested that the ultrasonication reduces the aspect ratio of GNP and provides a more uniform distribution of GNP in PP matrix. Composites with sonicated GNP showed a percolation threshold of 2-3 vol%, whereas the composites with pristine GNP showed a percolation threshold of 0.25 vol%. Zhao et al. prepared i-PP-graphene composites using graphene-supported Ziegler-Natta catalyst and liquid propylene through *in situ* slurry polymerization method [117]. The authors investigated the morphological behavior and conductivity of the composites with the effect of graphene concentration and annealing time. It was reported that the low content of graphene cannot form a macroscopic network and the electrical conductivity can increase a several orders of magnitude only at long period of annealing. Similarly, Wang et al. prepared reduced graphene oxide/polypropylene (rGO/PP) composites with an ultralow percolation threshold as low as 0.033 vol% [118].

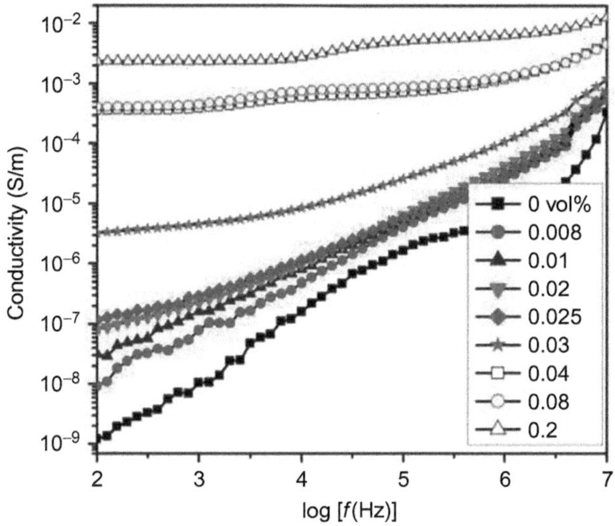

Figure 5 Conductivity of the rGO/PP composites with a different volume fraction of rGO as a function of frequency at room temperature. *(Reproduced from Ref. [118] with permission).*

Increasing content of rGO in PP composites resulted in an insulator-to-conductor percolation transition (Figure 5).

In another study, Carotenuto et al. fabricated an electrically conductive PE-supported graphene film based on simple mechanical approach [119]. The film showed a semiconducting nature at lower temperature. It was observed that the film exhibited good conductivity with increasing graphene content but it lowered the optical transparency. Wang et al. developed electrically conductive ultrahigh-molecular-weight polyethylene (UHMWPE)/graphene nanocomposites with segregated structure [120]. The authors found that the UHMWPE powder could generate a static electricity after high-speed mechanical friction and adsorb fluffy graphene on the surface, and the composite exhibited high electrical conductivity with percolation threshold of 0.1 vol%. Du et al. prepared multiwalled carbon nanotube (MWCNT)/HDPE composites and GNS/HDPE nanocomposites with segregated network structure and compared its electrical properties [121]. MWCNT/HDPE composites showed higher conductivity than that of GNS/HDPE composites at the same filler content. This was due to different conductivity mechanisms between the two composites, while the MWCNT/HDPE composites possessed three-dimensional conductive system, whereas the GNS/HDPE composites showed a two-dimensional conductive system.

3.4 Other Properties

In a recent study [122], ethylene-*co*-acrylic acid (EAA) compatibilizer with 9% acrylic acid content in its structure was used to compatibilize PE/graphene nanocomposites. In addition, ionomer ethylene-*co*-methacrylic acid (with zinc ion) (EMAZ) was also used as a compatibilizer to study the effect of ionic species in generating ionic interactions with the filler surface. In this copolymer, methacrylic acid content was 10% and a part of the methacrylic acid along the copolymer backbone was neutralized with zinc ions. Though the melting point of EMAZ was lower than EAA and overall much lower than the pure PE, the density of the PE and compatibilizers was similar. EMAZ had significantly low MFI as compared to EAA indicating that the molecular weight of EMAZ was higher than EAA. Mapping of the components in the nanocomposites is desired in order to confirm the translation of interfacial interactions into the distribution of components. Energy-dispersive X-ray analysis would not be suitable for such mapping purposes due to common elements present in all filler, compatibilizer, and PE phases. Detection of nanometer thin filler platelets may also be challenging for this method. Raman mapping has also been reported to be a functional method for such mapping, but the high-energy beam used for scanning damages the polymer samples significantly, thus requiring the use of methods involving lower-energy beams. One such option is hyperspectral imaging and mapping in visible-near-infrared range in which polymers can sustain the lower-energy beams for longer periods. This method also allows the detection of component distribution at nanoscale. For the samples in this study, all individual components (graphene, PE, EAA, and EMAZ), binary components (PE-G, PE-EAA, PE-EMAZ, EAA-G, and EMAZ-G), and nanocomposites were analyzed in order to generate spectral libraries of each component of the nanocomposites. Once each sample had a data set, analysis was performed to create a spectral profile specific to each component. This was performed by systematically comparing the spectral library of each component to the other components until only unique spectra were left. Due to the varying extents of spectral mixing present in the samples, traditional mapping was unsuccessful. Pure constructs did not map well or give a clear picture of how each compatibilizer facilitated the dispersal of graphene platelets throughout PE. To better illustrate the location of graphene and compatibilizer, the peak location classifier tool was used. The compiled spectral libraries were examined for consistencies in the strongest spectral response. The consistent characteristic peak for PE was determined to be

650 nm, graphene 700 nm, EAA 625 nm, and EMAZ 675 nm. The peak location classifier routine was run using the reference peak locations. A classification file was created, which generated a pseudo-color overlay. Each pixel in the overlay showed the location of the characteristic peaks determined from the reference libraries. By aligning the overlay images, the location of the graphene and compatibilizer in the PE matrix could be mapped. Figure 6 shows the mapping of distribution of graphene and EAA at two different locations in PE-EAA-G nanocomposite. Pixels in red represented the presence of graphene peak, whereas pixels in green represented the EAA peak in the nanocomposite. Similarly, Figure 7 also shows the mapping of

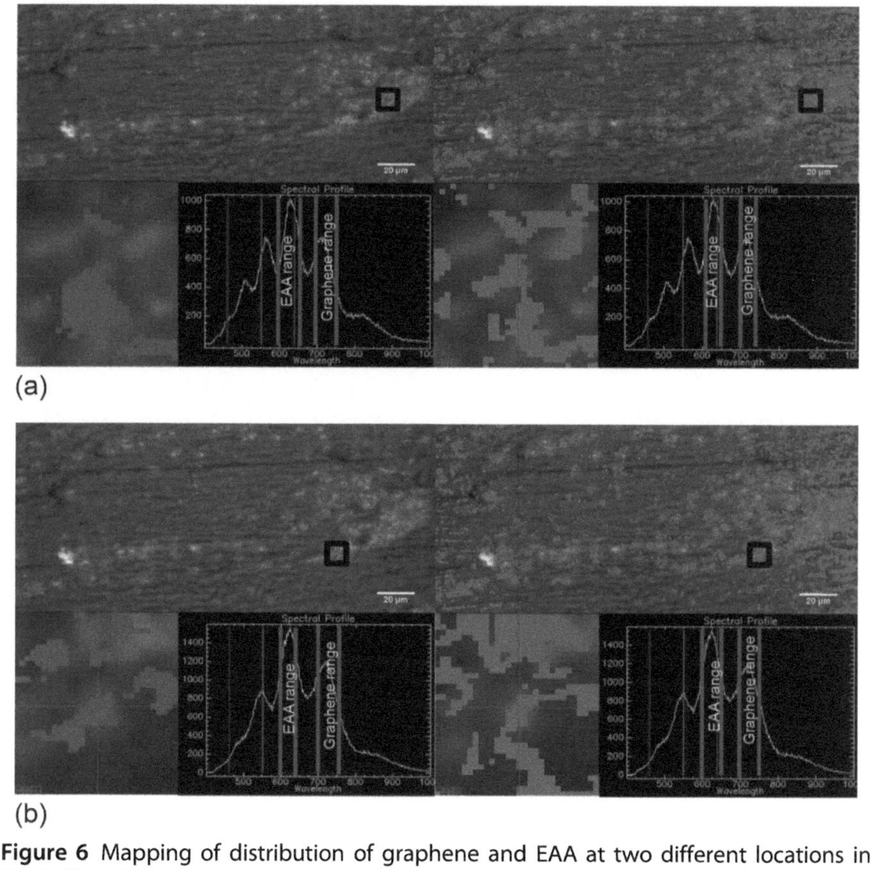

Figure 6 Mapping of distribution of graphene and EAA at two different locations in PE-5%EAA-G nanocomposite. Pixels in red represent the presence of graphene peak, whereas pixels in green represent the presence of EAA peak in the nanocomposite. *(Reproduced from Ref. [122] with permission from Springer).*

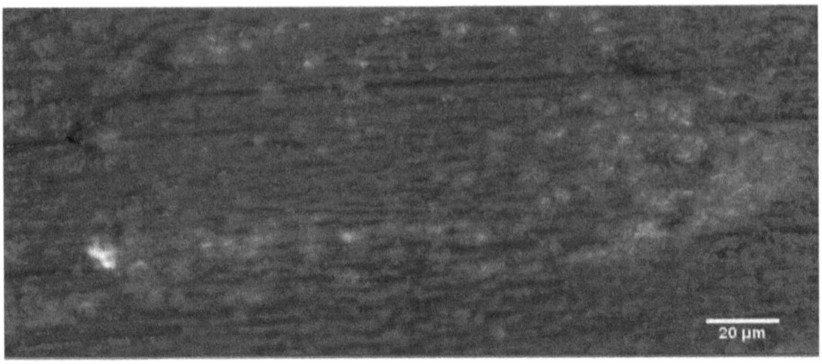

Figure 7 Mapping of spectral mixing of the graphene and EAA peaks in PE-5%EAA-G nanocomposite. Pixels in red indicate the presence of graphene peak, pixels in green indicate the presence of the EAA peak, whereas pixels in yellow indicate the presence of both peaks. *(Reproduced from Ref. [122] with permission from Springer).*

spectral mixing of the graphene and EAA peaks in PE-EAA-G nanocomposite, where pixels in red indicated the presence of graphene peak, pixels in green the EAA peak, and pixels in yellow both peaks. EAA in the PE-EAA-G nanocomposite sample was observed to bond/mix well with the graphene causing spectral shifting and mixing. Several areas were observed to contain characteristic peaks of both EAA and graphene in the nanocomposite.

In one of the recent studies on the degradation of polyolefin-based graphene nanocomposites, the effect of amount of metal ion-based pro-oxidant and different fillers on the mechanical and thermal performance as well as photodegradation of thick PE moldings was investigated. Fillers included 0-D and 2-D synthetic nanomaterials like silica, aluminosilicate (without surface modification), and thermally reduced graphene. The composites were generated by melt blending and the weathering properties were studied in conjunction with mechanical, thermal, rheological, and morphological properties. The presence of additive resulted in slight increment in the modulus of PE, though a reduction in tensile strength and enhancement in elongation at break were observed with increasing additive content due to a small extent of matrix plasticization. It should also be noted that the observed effects resulted from a combination of factors like increase in crystallinity due to the nucleating effect of the metal particles in the additive and plasticization. Addition of fillers resulted in enhancement of tensile modulus, which was further improved on increasing the extent of additive. Thus, combination of both increment in crystallinity and stress transfer to filler resulted in enhanced modulus in the composites. The graphene composite with 2.5% additive content exhibited nearly 200% increase in the modulus as compared to pure PE. Similar to rheological performance, silica composites exhibited least increase in the modulus among the composites. The tensile strength was observed to slightly decrease in the composites due to strain hardening as well as entrapment of polymer chains in the filler interlayers. The elongation at break for graphene nanocomposites was decreased significantly and the composites became quite brittle. In contrast, silica and silicate nanocomposites had still fairly high flexibility, which was enhanced on increasing the additive content in the composite. As oxo-degradation causes the formation of a carbonyl group at every scission location in the polymer chains, the onset and level of carbonyl group development in the specimen are more accurate measures of induced degradation by the metal ion pro-oxidant. Figure 8 demonstrates the degree of embrittlement of the samples with 2.5% additive calculated from the carbonyl index measurements as a function of exposure time adjusted to real environment in a constant temperature of 30 °C in sunlight. PE sample containing only 2.5% additive exhibited the fastest rate of degradation reaching 100% embrittlement after ~795 accelerated aging hours (equivalent to around 8.7 months at 30 °C). The two nanocomposites with 5% silica and 5% silicate performed similarly to each other, with the silica containing composite reaching 100% embrittlement after ~1443 accelerated aging hours (equivalent to around 15.8 months at 30 °C) and the silicate composite after ~1279 accelerated aging hours (equivalent to around 14 months at 30 °C). The graphene-based composite did not reach 100% embrittlement during the test period, but was still ~48.2% embrittled at the end of the test.

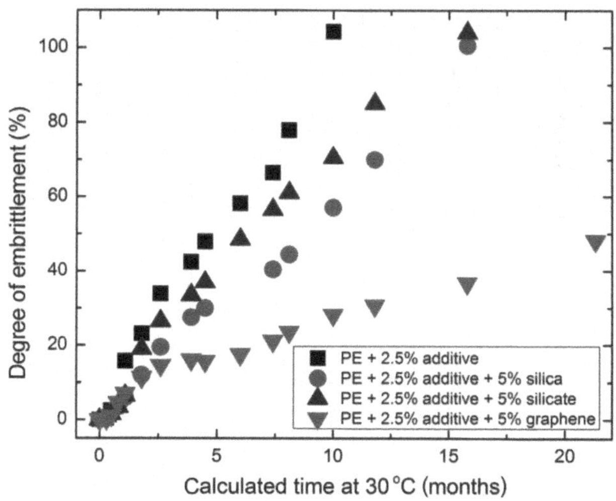

Figure 8 Degree of embrittlement of the materials during photodegradation testing plotted as a function of time of exposure.

3.5 Other Related Studies

In other related studies, Kim et al. [29] reported the effect of PE functionalization and blending methods on the properties of PE/graphene nanocomposites. TRG was blended with LLDPE and its amine-, nitrile-, and isocyanate-functionalized analogs using both melt and solution blending. Graphene platelets were well exfoliated in functionalized LLDPE matrices, whereas phase separation was observed for the unmodified LLDPE. Solvent-based blending was also observed to be more effective in dispersing the graphene platelets as compared to melt compounding. Jiang and Drzal [123] also reported that the thermal stability and thermal conductivity of PE/exfoliated GNP-based nanocomposites enhanced with increasing graphene concentration.

Kuila et al. [100] synthesized polymer nanocomposites by solution mixing of LLDPE with dodecylamine-modified graphene (DA-G). The storage modulus of the nanocomposite with 8 wt% DA-G was 293 MPa, which was 118% higher than that of neat polymer. However, only a marginal change in both β-relaxation (T_g) and α-relaxation for both pure polymer and the nanocomposites was observed. Achaby and Qaiss [124] compared the performance of HDPE/GNS and HDPE/MWCNT nanocomposites prepared using the melt mixing. At a filler content of 3 wt%, the tensile strength of the graphene nanocomposite was 77% higher than for the pure polymer, whereas the corresponding nanotubes nanocomposite resulted in an increase of 58%. Similarly, the modulus of graphene and nanotubes reinforced nanocomposites was enhanced by 87% and 57%, respectively, as compared to pure polymer.

Electrical percolation of 0.1-0.3 vol% of GNPs was reported by Kalaitzidou et al. [125] for PP nanocomposites. The platelets were derived from thermal expansion of acid-intercalated graphite. Milani et al. [90] also investigated the synthesis and properties of isotactic polypropylene/graphene nanosheet nanocomposites. *In situ* metallocene polymerization was used to generate the nanocomposites. GNS were uniformly dispersed in the polymer matrix and significantly improved the crystallization and the degradation temperatures. Similarly, Young's modulus and storage modulus were observed to enhance with nanosheet content. The modulus was observed to enhance from 1280 MPa for neat polymer to 1920 MPa in the nanocomposite with the 17.4 wt% GNS content. Electrical conductivity in the composite was observed to increase by a factor of 10^8 as compared to neat polymer. Song et al. [107] reported an eco-friendly strategy for fabricating the exfoliated polyolefin nanocomposites via first coating graphene using PP latex followed by melt-blending the coated graphene with PP matrix. At a filler volume fraction of 0.42%, a 75% increase in yield strength and 74% increase in Young's modulus as compared to pure PP were achieved. The glass transition temperature in the nanocomposite with 0.042 vol% was observed to be higher by 2.5 K as compared to pure PP. The thermal stability of PP was also observed to significantly improve on addition of graphene, and initial degradation temperature is enhanced by 26 °C at 0.42 vol% graphene content. Covalently functionalized GNS with dodecyl (12C) trimethoxysilane and incorporated into PP matrix by solution mixing method were reported by Yang et al. [95]. At 3 wt% content of the filler, the peak degradation temperature in the nanocomposite was enhanced by 16.7 °C. Upon incorporation of only 0.5 wt% of filler in the PP matrix, tensile strength, Young's modulus, and elongation at break in the nanocomposite increased by 19.7%, 16.5%, and 26.8%, respectively.

4. CONCLUSIONS

Graphene has become a material of choice for generating 3-D nanomaterials using polymer matrices. Since its discovery, the synthetic methods to generate the graphene and graphene oxide platelets have also seen significant modifications. Numerous surface modification techniques to generate compatibility with the polymers have evolved. The following general conclusions regarding the polymer-graphene nanocomposites can be made:

A. The functionality and capability to enhance properties of graphene have been confirmed to be superior to other fillers like layered silicates and nanotubes.

B. Graphene provides possibilities to tune the surface in order to achieve nanoscale dispersion of platelets in the polymer matrices in order to achieve high aspect ratio of platelets resulting in superior properties.

C. Significant enhancements in electrical, mechanical, thermal, and rheological properties of the polyolefins, especially PE and PP, have been reported at lower-filler loadings. This helps to retain transparency as well as lower density of these polymers.

In spite of the high degree of polymer reinforcement, some challenges do persist, which need further research attention. The cost of graphene generation and allied modification process causes an increase in the price. Synthesis in large amounts and reproducing graphene with same characteristics is also challenging. Toxicology aspects of graphene-related materials have also been not fully established.

REFERENCES

[1] M. Alexandre, Ph. Dubois, Polymer-layered silicate nanocomposites: preparation, properties and uses of a new class of materials, Mater. Sci. Eng. R 28 (2000) 1.
[2] A. Okada, A. Usuki, T. Kurauchi, O. Kamigito, Polymer- clay hybrids, ACS Symp. Ser. 585 (1995) 55.
[3] E.P. Giannelis, Polymer layered silicate nanocomposites, Adv. Mater. 8 (1996) 29.
[4] P.C. LeBaron, Z. Wang, T.J. Pinnavaia, Polymer-layered silicate nanocomposites: an overview, Appl. Clay Sci. 15 (1999) 11.
[5] Y. Brechet, J.Y. Cavaille, E. Chabert, L. Chazeau, R. Dendievel, L. Flandin, C. Gauthier, Polymer based nanocomposites: effect of filler-filler and filler-matrix interactions, Adv. Eng. Mater. 3 (2001) 571.
[6] S. Pavlidoua, C.D. Papaspyrides, A review on polymer–layered silicate nanocomposites, Prog. Polym. Sci. 33 (2008) 1119.
[7] J.E. Mark, Ceramic-reinforced polymers and polymer-modified ceramics, Polym. Eng. Sci. 36 (1996) 2905.
[8] M. Song, V. Mittal (Ed.), Polymer graphene nanocomposites, royal society of chemistry, Cambridge, UK, 2012 (Chapter 1).
[9] P. Mukhopadhyay, R.K. Gupta, Trends and frontiers in graphene-based polymer nanocomposites, Plast. Eng. 67 (2011) 32.
[10] A.K. Geim, K.S. Novoselov, The rise of graphene, Nat. Mater. 6 (2007) 183.
[11] O.C. Compton, S.B.T. Nguyen, Graphene oxide, highly reduced graphene oxide, and graphene: versatile building blocks for carbon-based materials, Small 6 (2010) 711.
[12] C. Lee, X. Wei, J.W. Kysar, J. Hone, Measurement of the elastic properties and intrinsic strength of monolayer graphene, Science 321 (2008) 385.
[13] H. Zhao, K. Min, N.R. Aluru, Size and chirality dependent elastic properties of graphene nanoribbons under uniaxial tension, Nano Lett. 9 (2009) 3012.
[14] G.V. Lier, C.V. Alsenoy, V.V. Doren, P. Greelings, Ab initio study of the elastic properties of single-walled carbon nanotubes and graphene, Chem. Phys. Lett. 326 (2000) 181.
[15] D.H. Robertson, D.W. Brenner, J.W. Mintmire, Energetics of nanoscale graphitic tubules, Phys. Rev. B 45 (1992) 12592.
[16] M.E. Itkis, F. Borondics, A. Yu, R.C. Haddon, thermal conductivity measurements of semitransparent single-walled carbon nanotube films by a bolometric technique, Nano Lett. 7 (2007) 900.
[17] S. Park, S. Rouff, Chemical methods for the production of graphenes, Nat. Nanotechnol. 4 (2009) 217.
[18] D. Li, M.B. Muller, S. Gilje, R.B. Kaner, G.G. Wallac, Processable aqueous dispersions of graphene nanosheets, Nat. Nanotechnol. 3 (2007) 101.
[19] H. Kim, A.A. Abdala, C.W. Macosko, Graphene/Polymer nanocomposites, Macromolecules 43 (2010) 6515.
[20] T. Kuilla, S. Bhadra, D. Yao, N.H. Kim, S. Bose, J.H. Lee, Recent advances in graphene based polymer composites, Prog. Polym. Sci. 35 (2010) 1350.
[21] D. Cai, M. Song, Recent advance in functionalized graphene/polymer nanocomposites, J. Mater. Chem. 20 (2010) 7906.
[22] M.A. Rafiee, J. Rafiee, Z. Wang, H. Song, Z.Z. Yu, N. Koratkar, Enhanced mechanical properties of nanocomposites at low graphene content, ACS Nano 3 (2009) 3884.
[23] P. Steurer, R. Wissert, R. Thomann, R. Muelhaupt, Functionalized graphenes and thermoplastic nanocomposites based upon expanded graphite oxide, Macromol. Rapid Commun. 30 (2009) 316.
[24] H. Kim, Y. Miura, C.W. Macosko, Graphene/Polyurethane nanocomposites for improved gas barrier and electrical conductivity, Chem. Mater. 22 (2010) 3441.

[25] D.A. Nguyen, Y.R. Lee, A.V. Raghu, H.M. Jeong, C.M. Shin, B.K. Kim, Morphological and physical properties of a thermoplastic polyurethane reinforced with functionalized graphene sheet, Polym. Int. 58 (2009) 412.
[26] M. Fang, K. Wang, H. Lu, Y. Yang, S. Nutt, Covalent polymer functionalization of graphene nanosheets and mechanical properties of composites, J. Mater. Chem. 19 (2009) 7098.
[27] A.U. Chaudhry, V. Mittal, High-density polyethylene nanocomposites using masterbatches of chlorinated polyethylene/graphene oxide, Polym. Eng. Sci. 53 (2013) 78.
[28] V. Mittal, N. Matsko, PE-CPE blends and their graphene oxide nanocomposites with reduced low temperature brittleness, Colloid Polym. Sci. 291 (2013) 1949.
[29] H. Kim, S. Kobayashi, M.A. AbdurRahim, M.J. Zhang, A. Khusainova, M.A. Hillmyer, A.A. Abdala, C.W. Macosko, Graphene/polyethylene nanocomposites: effect of polyethylene functionalization and blending methods, Polymer 52 (2011) 1837.
[30] M.J. Allen, V.C. Tung, R.B. Kaner, Honeycomb Carbon: A review of graphene, Chem. Rev. 110 (2010) 132.
[31] Y. Matsuo, K. Hatase, Y. Sugie, Selective intercalation of aromatic molecules into alkyltrimethylammonium ion-intercalated graphite oxide, Chem. Lett. 28 (1999) 1109.
[32] Y.B. Zhang, Y.W. Tan, H.L. Stormer, P. Kim, Experimental observation of the quantum hall effect and berry's phase in graphene, Nature 438 (2005) 201.
[33] H. Becerril, J. Mao, Z. Liu, M. Stoltenberg, Z. Bao, Y. Chen, Evaluation of solution-processed reduced graphene oxide films as transparent conductors, ACS Nano 2 (2008) 463.
[34] X. Wang, L. Zhi, K. Mullen, Transparent, Conductive graphene electrodes for dye-sensitized solar cells, Nano Lett. 8 (2008) 323.
[35] X. Wang, H. You, F. Liu, M. Li, L. Wan, S. Li, Q. Li, Y. Xu, R. Tian, Z. Yu, D. Xiang, J. Cheng, Large-scale synthesis of few-layered graphene using CVD, Chem. Vapor Depos. 15 (2009) 53.
[36] E. Dervishi, Z. Li, F. Watanabe, A. Biswas, Y. Xu, R.B. Alexandru, V. Saini, S.B. Alexandru, Large-scale graphene production by RF-cCVD method, Chem. Commun. 4061 (2009).
[37] S. Karmakar, N.V. Kulkarni, A.B. Nawale, N.P. Lalla, R. Mishra, V.G. Sathe, S. V. Bhoraskar, A.K. Das, A novel approach towards selective bulk synthesis of few-layer graphenes in an electric arc, J. Phys. D. Appl. Phys. 42 (2009) 115201.
[38] Z.H. Ni, W. Chen, X.F. Fan, J.L. Kuo, T. Yu, A.T.S. Wee, Z.X. Shen, Raman spectroscopy of epitaxial graphene on a SiC substrate, Phys. Rev. B 77 (2008) 115416.
[39] W.A. de Heer, C. Berger, X. Wu, P.N. First, E.H. Conrad, X. Li, T. Li, M. Sprinkle, J. Hass, M.L. Sadowski, M. Potemski, G. Martinez, Epitaxial graphene, Solid State Commun. 143 (2007) 92.
[40] A. Hirsch, Unzipping Carbon Nanotubes: A peeling method for the formation of graphene nanoribbons, Angew. Chem. Int. Ed. 48 (2009) 6594.
[41] L. Zhi, K. Muellen, A bottom-up approach from molecular nanographenes to unconventional carbon materials, J. Mater. Chem. 18 (2008) 1472.
[42] L. Jiao, L. Zhang, X. Wang, G. Diankov, H. Dai, Narrow graphene nanoribbons from carbon nanotubes, Nature 458 (2009) 877.
[43] T. Seyller, A. Bostwick, K.V. Emtsev, K. Horn, L. Ley, J.L. McChesney, T. Ohta, J.D. Riley, E. Rotenberg, F. Speck, Epitaxial graphene: a new material, Phys. Status Solidi B 245 (2008) 1436.
[44] J.R. Potts, D.R. Dreyer, C.W. Bielawski, R.S. Ruoff, Graphene-based polymer nanocomposites, Polymer 52 (2011) 5.
[45] B.C. Brodie, On the atomic weight of graphite, Philos. Trans. R. Soc. London 149 (1859) 249.

[46] M.J. McAllister, J.L. Li, D.H. Adamson, H.C. Schniepp, A.A. Abdala, J. Liu, M. Herrera-Alonso, D.L. Milius, R. Car, R.K. Prud'homme, I.A. Aksay, Single sheet functionalized graphene by oxidation and thermal expansion of graphite, Chem. Mater. 19 (2007) 4396.
[47] W.S. Hummers, R.E. Offeman, Preparation of graphitic oxide, J. Am. Chem. Soc. 80 (1958) 1339.
[48] D.W. Boukhvalov, M.I. Katsnelson, Modeling of graphite oxide, J. Am. Chem. Soc. 130 (2008) 10697.
[49] W.H. Zhang, V. Carravetta, Z.Y. Li, Y. Luo, J.L. Yang, Oxidation states of graphene: Insights from computational spectroscopy, J. Chem. Phys. 131 (2009) 244505.
[50] H.-K. Jeong, Y.P. Lee, R.J.W.E. Lahaye, M.-H. Park, K.H. An, I.J. Kim, C.-W. Yang, C.Y. Park, R.S. Ruoff, Y.H. Lee, Evidence of graphitic ab stacking order of graphite oxides, J. Am. Chem. Soc. 130 (2008) 1362.
[51] T. Szabo, O. Berkesi, P. Forgo, K. Josepovits, Y. Sanakis, D. Petridis, I. Dekany, Evolution of surface functional groups in a series of progressively oxidized graphite oxides, Chem. Mater. 18 (2006) 2740.
[52] W. Cai, R.D. Piner, F.J. Stadermann, S. Park, M.A. Shaibat, Y. Ishii, D. Yang, A. Velamakanni, S.J. An, M. Stoller, J. An, D. Chen, R.S. Ruoff, Synthesis and solid-state NMR structural characterization of 13C-labeled graphite oxide, Science 321 (2008) 1815.
[53] W. Scholz, H.P. Boehm, Z. Anorg, Untersuchungen am graphitoxid. VI. Betrachtungen zur Struktur des Graphitoxids, Allg. Chem. 369 (1969) 327.
[54] H. He, J. Klinowski, M. Forster, A. Lerf, A new structural model for graphite oxide, Chem. Phys. Lett. 287 (1998) 53.
[55] T. Nakajima, A. Mabuchi, R. Hagiwara, A new structure model of graphite oxide, Carbon 26 (1988) 357.
[56] K.A. Mkhoyan, A.W. Contryman, J. Silcox, D.A. Stewart, G. Eda, C. Mattevi, S. Miller, M. Chhowalla, Atomic and electronic structure of graphene-oxide, Nano Lett. 9 (2009) 1058.
[57] A. Clauss, R. Plass, H.P. Boehm, U.Z. Hofmann, Untersuchungen zur Struktur des Graphitoxyds, Anorg. Allg. Chem. 291 (1957) 205.
[58] H. Liu, L.C. Brinson, Reinforcing efficiency of nanoparticles: a simple comparison for polymer nanocomposites, Compos. Sci. Technol. 68 (2008) 1502.
[59] B. Reddy, Advances in diverse industrial applications of nanocomposites, InTech, Rijeka, 2011.
[60] L. Hui, R.C. Smith, X. Wang, J.K. Nelson, L.S. Schadler, Conference on electrical insulation and dielectric phenomena, (2008) p. 317.
[61] I. Manas-Zloczower, Analysis of mixing in polymer processing equipment, Rheol. Bull. 66 (1997) 5.
[62] S.H. Lee, E.N.R. Cho, S.H. Jeon, J.R. Youn, Rheological and electrical properties of polypropylene composites containing functionalized multi-walled carbon nanotubes and compatibilizers, Carbon 45 (2007) 2810.
[63] M.A. Serageldin, H. Wang, Effect of operating parameters on time to decomposition of high density polyethylene and chlorinated polyethylenes, Thermochim. Acta 117 (1987) 157.
[64] J. Zhu, M. Yudasaka, M. Zhang, S. Ijima, Dispersing Carbon Nanotubes in Water: a Noncovalent and Nonorganic Way, J. Phys. Chem. B 108 (2004) 11317.
[65] F. Balvlavoine, P. Schultz, C. Richard, V. Mallouh, T.W. Ebbeson, C. Mioskowski, Helical crystallization of proteins on carbon nanotubes: a first step towards the development of new biosensors, Angew. Chem. Int. Ed. Engl. 38 (1999) 1912.
[66] M. Shim, N.W.S. Kam, R.J. Chen, Y. Li, H. Dai, Functionalization of carbon nanotubes for biocompatibility and biomolecular recognition, Nano Lett. 2 (2002) 285.

[67] J.M. Harris, S. Zalipsky, Poly(Ethylene Glycol): Chemistry and biological application, american chemical society, Washington, DC, 1997.
[68] M. Szycher, Biocompatible polymers, metals and composites, Technomic, Lancaster, PA, 1983.
[69] E. Ostuni, R.G. Chapman, R.E. Holmlin, S. Takayama, G.M. Whitesides, A survey of structure–property relationships of surfaces that resist the adsorption of protein, Langmuir 17 (2001) 5605.
[70] M.J. O'Connell, P. Boul, L.M. Ericson, C. Huffman, Y. Wang, E. Haroz, C. Kuper, J. Tour, K.D. Ausman, R.E. Smalley, Reversible water-solubilization of single-walled carbon nanotubes by polymer wrapping, Chem. Phys. Lett. 342 (2001) 265.
[71] Q. Fu, C. Lu, J. Liu, Selective coating of single wall carbon nanotubes with thin SiO2 layer, Nano Lett. 2 (2002) 329.
[72] F.J. Gomez, R.J. Chen, D. Wang, R.M. Waymouth, H. Dai, Ring opening metathesis polymerization on non-covalently functionalized single-walled carbon nanotubes, Chem. Commun. 190 (2003).
[73] Y.-T. Liu, W. Zhao, Z.-Y. Huang, Y.-F. Gao, X.-M. Xie, X.-H. Wang, X.-Y. Ye, Noncovalent surface modification of carbon nanotubes for solubility in organic solvents, Carbon 44 (2006) 1613.
[74] V. Mittal, Polymer chains grafted "to" and "from" layered silicate clay platelets, J. Colloid Interface Sci. 314 (2007) 141.
[75] V. Mittal, Esterification reactions on the surface of layered silicate clay platelets, J. Colloid Interface Sci. 315 (2007) 135.
[76] V. Mittal, V. Herle, Physical adsorption of organic molecules on the surface of layered silicate clay platelets: A thermogravimetric study, J. Colloid Interface Sci. 327 (2008) 295.
[77] X. Fu, S. Qutubuddin, Polymer–clay nanocomposites: exfoliation of organophilic montmorillonite nanolayers in polystyrene, Polymer 42 (2001) 807.
[78] P. Uthirakumar, C.J. Kim, K.S. Nahm, Y.B. Hahn, Y.S. Lee, Preparation and characterization of new difunctional cationic radical initiator-montmorillonite hybrids, Colloids Surf. A: Physicochem. Eng. Aspects 247 (2004) 69.
[79] L.P. Meier, R.A. Shelden, W.R. Caseri, U.W. Suter, Polymerization of styrene with initiator ionically bound to high surface area mica: grafting via an unexpected mechanism, Macromolecules 27 (1994) 1637.
[80] X. Fan, C. Xia, R.C. Advincula, Grafting of polymers from clay nanoparticles via in situ free radical surface-initiated polymerization: monocationic versus bicationic initiators, Langmuir 19 (2003) 4381.
[81] X. Fan, C. Xia, R.C. Advincula, Intercalation of polymerization initiators into montmorillonite nanoparticle platelets: free radical vs. anionic initiator clays, Colloids Surf. A 219 (2003) 75.
[82] U. Velten, R.A. Shelden, W.R. Caseri, U.W. Suter, Y. Li, Polymerization of styrene with peroxide initiator ionically bound to high surface area mica, Macromolecules 32 (1999) 3590.
[83] S. Su, C.A. Wilkie, Exfoliated poly(methyl methacrylate) and polystyrene nanocomposites occur when the clay cation contains a vinyl monomer, J. Polym. Sci. Part A: Polym. Chem. 41 (2003) 1124.
[84] O. Prucker, J. Ruhe, Mechanism of radical chain polymerizations initiated by azo compounds covalently bound to the surface of spherical particles, Macromolecules 31 (1998) 602.
[85] M. Kato, A. Usuki, A. Okada, Cure processing modeling and cure cycle simulation of epoxy-terminated poly(phenylene ether ketone). IV. Cure cycle simulation, J. Appl. Polym. Sci. 66 (1997) 1781.
[86] M. Kawasumi, N. Hasegawa, M. Kato, A. Usuki, A. Okada, Preparation and mechanical properties of polypropylene–clay hybrids, Macromolecules 30 (1997) 6333.

[87] N. Hasegawa, M. Kawasumi, M. Kato, A. Usuki, A. Okada, Preparation and mechanical properties of polypropylene-clay hybrids using a maleic anhydride-modified polypropylene oligomer, J. Appl. Polym. Sci. 67 (1998) 87.
[88] E. Manias, A. Touny, L. Wu, K. Strawhecker, B. Lu, T.C. Chung, Polypropylene/montmorillonite nanocomposites. Review of the Synthetic Routes and Materials Properties, Chem. Mater. 13 (2001) 3516.
[89] P. Reichert, H. Nitz, S. Klinke, R. Brandsch, R. Thomann, R. Muelhaupt, Poly(propylene)/organoclay nanocomposite formation: Influence of compatibilizer functionality and organoclay modification, Macromol. Mater. Eng. 275 (2000) 8.
[90] M.A. Milani, D. Gonzalez, R. Quijada, N.R.S. Basso, M.L. Cerrada, D.S. Azambuja, G.B. Galland, Polypropylene/graphene nanosheet nanocomposites by in situ polymerization: Synthesis, characterization and fundamental properties, Compos. Sci. Technol. 84 (2013) 1.
[91] V.G. Shevchenko, S.V. Polschikov, P.M. Nedorezova, A.N. Klyamkina, A. N. Shchegolikhin, A.M. Aladyshev, V.E. Muradyan, In situ polymerized poly(propylene)/graphene nanoplatelets nanocomposites: dielectric and microwave properties, Polymer 53 (2012) 5330.
[92] Y. Huang, Y. Qin, Y. Zhou, H. Niu, Z.Z. Yu, J.Y. Dong, Polypropylene/graphene oxide nanocomposites prepared by in situ ziegler–natta polymerization, Chem. Mater. 22 (2010) 4096.
[93] A. Tchernook, M. Krumova, F.J. Töolle, R. Müulhaupt, S. Mecking, Composites from aqueous polyethylene nanocrystal/graphene dispersions, Macromolecules 47 (2014) 3017.
[94] J. Fan, R. Huang, S. Ye, T. Li, J. Feng, The probable influence of in situ thermal reduction of graphene oxides on the crystallization behavior of isotactic polypropylene, Polymer 55 (2014) 4341.
[95] X. Yang, X. Wang, J. Yang, J. Li, L. Wan, Functionalization of graphene using trimethoxysilanes and its reinforcement on polypropylene nanocomposites, Chem. Phys. Lett. 570 (2013) 125.
[96] Y. Cao, J. Feng, P. Wu, Polypropylene-grafted graphene oxide sheets as multifunctional compatibilizers for polyolefin-based polymer blends, J. Mater. Chem. 22 (2012) 14997.
[97] M. Castelaín, G. Martínez, C. Marco, G. Ellis, H.J. Salavagione, Effect of click-chemistry approaches for graphene modification on the electrical, thermal, and mechanical properties of polyethylene/graphene nanocomposites, Macromolecules 46 (2013) 8980.
[98] A.A. Vasileiou, M. Kontopoulou, A. Docoslis, A noncovalent compatibilization approach to improve the filler dispersion and properties of polyethylene/graphene composites, ACS Appl. Mater. Interfaces 6 (2014) 1916.
[99] F.C. Fim, N.R.S. Basso, A.P. Graebin, D.S. Azambuja, G.B. Galland, Thermal, electrical, and mechanical properties of polyethylene–graphene nanocomposites obtained by in situ polymerization, J. Appl. Polym. Sci. 128 (2013) 2630.
[100] T. Kuila, S. Bose, A.K. Mishra, P. Khanra, N.H. Kim, J.H. Lee, Effect of functionalized graphene on the physical properties of linear low density polyethylene nanocomposites, Polym. Test. 31 (2012) 31.
[101] J. Wang, C. Xu, H. Hu, L. Wan, R. Chen, H. Zheng, F. Liu, M. Zhang, X. Shang, X. Wang, Synthesis, mechanical, and barrier properties of LDPE/graphene nanocomposites using vinyl triethoxysilane as a coupling agent, J. Nanopart. Res. 13 (2011) 869.
[102] X. Yang, T. Mei, J. Yang, C. Zhang, M. Lv, X. Wang, Synthesis and characterization of alkylamine-functionalized graphene for polyolefin-based nanocomposites, Appl. Surf. Sci. 305 (2014) 725.
[103] Y.S. Yun, Y.H. Bae, D.H. Kim, J.Y. Lee, I.J. Chin, H.J. Jin, Reinforcing effects of adding alkylated graphene oxide to polypropylene, Carbon 49 (2011) 3553.

[104] B. Yuan, C. Bao, L. Song, N. Hong, K.M. Liew, Y. Hu, Preparation of functionalized graphene oxide/polypropylene nanocomposite with significantly improved thermal stability and studies on the crystallization behavior and mechanical properties, Chem. Eng. J. 237 (2014) 411.
[105] S.H. Ryu, A.M. Shanmugharaj, Influence of long-chain alkylamine-modified graphene oxide on the crystallization, mechanical and electrical properties of isotactic polypropylene nanocomposites, Chem. Eng. J. 244 (2014) 552.
[106] M.E. Achaby, F.E. Arrakhiz, S. Vaudreuil, A.K. Qaiss, M. Bousmina, O.F. Fehri, Mechanical, thermal, and rheological properties of graphene-based polypropylene nanocomposites prepared by melt mixing, Polym. Compos. 33 (2012) 733.
[107] P. Song, Z. Cao, Y. Cai, L. Zhao, Z. Fang, S. Fu, Fabrication of exfoliated graphene-based polypropylene nanocomposites with enhanced mechanical and thermal properties, Polymer 52 (2011) 4001.
[108] M.C. Hsiao, S.H. Liao, Y.F. Lin, C.A. Wang, N.W. Pu, H.M. Tsai, C.C.M. Ma, Preparation and characterization of polypropylene-graft-thermally reduced graphite oxide with an improved compatibility with polypropylene-based nanocomposite, Nanoscale 3 (2011) 1516.
[109] J.Z. Xu, Y.Y. Liang, H.D. Huang, G.J. Zhong, J. Lei, C. Chen, Z.M. Li, Isothermal and nonisothermal crystallization of isotactic polypropylene/graphene oxide nanosheet nanocomposites, J. Polym. Res. 19 (2012) 9975.
[110] G. Huang, S. Wang, P. Song, C. Wu, S. Chen, X. Wang, Combination effect of carbon nanotubes with graphene on intumescent flame-retardant polypropylene nanocomposites, Compos. Part A 59 (2014) 18.
[111] B. Yu, X. Wang, X. Qian, W. Xing, H. Yang, L. Ma, Y. Lin, S. Jiang, L. Song, Y. Hu, S. Lo, Functionalized graphene oxide/phosphoramide oligomer hybrids flame retardant prepared via in situ polymerization for improving the fire safety of polypropylene, RSC Adv. 4 (2014) 31782.
[112] W. Hu, J. Zhan, N. Hong, B. Wang, L. Song, A.A. Stec, T.R. Hull, J. Wang, Y. Hu, Effect of functionalized graphene oxide with hyper-branched flame retardant on flammability and thermal stability of cross-linked polyethylene, Ind. Eng. Chem. Res. 53 (2014) 3073.
[113] Z. Han, Y. Wang, W. Dong, P. Wang, Enhanced fire retardancy of polyethylene/alumina trihydrate composites by graphene nanoplatelets, Mater. Lett. 128 (2014) 275.
[114] I.M. Inuwa, A. Hassan, D.Y. Wang, S.A. Samsudin, M.K. Mohamad Haafiz, S. L. Wong, M. Jawaid, Influence of exfoliated graphite nanoplatelets on the flammability and thermal properties of polyethylene terephthalate/polypropylene nanocomposites, Polym. Degrad. Stab. 110 (2014) 137.
[115] S. Ran, Z. Guo, L. Han, Z. Fang, Effect of Friedel–Crafts reaction on the thermal stability and flammability of high-density polyethylene/brominated polystyrene/graphene nanoplatelet composites, Polym. Int. 63 (2014) 1835.
[116] R.P. Ramasamy, K. Yang, M.H. Rafailovich, Polypropylene–graphene – a nanocomposite that can be converted into a meta-material at desired frequencies, RSC Adv. 4 (2014) 44888.
[117] S. Zhao, F. Chen, C. Zhao, Y. Huang, J.Y. Dong, C.C. Han, Interpenetrating network formation in isotactic polypropylene/graphene composites, Polymer 54 (2013) 3680.
[118] D. Wang, X. Zhang, J.W. Zha, J. Zhao, Z.M. Dang, G.H. Hu, Dielectric properties of reduced graphene oxide/polypropylene composites with ultralow percolation threshold, Polymer 54 (2013) 1916.
[119] C. Carotenuto, S.D. Nicola, G. Ausanio, D. Massarotti, L. Nicolais, G.P. Pepe, Synthesis and characterization of electrically conductive polyethylene-supported graphene films, Nanoscale Res. Lett. 9 (2014) 475.

[120] B. Wang, H. Li, L. Li, P. Chen, Z. Wang, Q. Gu, Electrostatic adsorption method for preparing electrically conducting ultrahigh molecular weight polyethylene/graphene nanosheets composites with a segregated network, Compos. Sci. Technol. 89 (2013) 180.
[121] J. Du, L. Zhao, Y. Zeng, L. Zhang, F. Li, P. Liu, C. Liu, Comparison of electrical properties between multi-walled carbon nanotube and graphene nanosheet/high density polyethylene composites with a segregated network structure, Carbon 49 (2011) 1094.
[122] V. Mittal, L. Krauss, Compatibilized polyethylene—thermally reduced graphene nanocomposites: Interfacial interactions and hyperspectral mapping for component distribution, Colloid Polym. Sci. 292 (2014) 2509.
[123] X. Jiang, L.T. Drzal, Multifunctional high-density polyethylene nanocomposites produced by incorporation of exfoliated graphene nanoplatelets 2: crystallization, thermal and electrical properties, Polym. Compos. 33 (2012) 636.
[124] M. El Achaby, A. Qaiss, Processing and properties of polyethylene reinforced by graphene nanosheets and carbon nanotubes, Mater. Design 44 (2013) 81.
[125] K. Kalaitzidou, H. Fukushima, P. Askeland, L.T. Drzal, The nucleating effect of exfoliated graphite nanoplatelets and their influence on the crystal structure and electrical conductivity of polypropylene nanocomposites, J. Mater. Sci. 43 (2008) 2895.

CHAPTER 7

UV-Curable Nanocomposite Coatings and Materials

Saeed Bastani[1,2], Majid Mohseni[1]
[1]Surface Coatings and Corrosion Department, Institute for Color Science and Technology, Tehran, Iran
[2]Center of Excellence for Color Science and Technology, Tehran, Iran

Chapter Contents

Handbook of Nanoceramic and Nanocomposite Coatings and Materials
http://dx.doi.org/10.1016/B978-0-12-799947-0.00007-9

1. THE CHEMISTRY OF UV CURING

1.1 UV-Curing Process: Theoretical Aspects

The technology of radiation-curable coatings has recently undergone significant progress and has attracted significant attention in scientific and industrial fields [1].

Radiation-curable coatings are in the forms of water-based, liquid [2], and powder coatings [3] and typically formulated using polyester, polyether, and epoxy, as well as urethane acrylate oligomers [4,5], along with active diluents or acrylic monomers and photoinitiators. Higher curing rate, lower requirement for energy of curing, high solid content, lower viscosity, lower initial cost, higher cross-linking degree, efficient chemical, and mechanical properties are considered as the advantages of these systems [6]. Additionally, radiation-curing coatings are more economical, particularly in continuous product lines, which have enabled them to grow fast in a range of different fields such as computer components, coatings, printing inks, composite materials, and dental restorative materials [1].

1.1.1 Linear and Cross-Linking Systems

Radiation-curable systems can be divided into two types based on average functionality, linear and cross-linking systems [7]. The existence of at least one multifunctional component is necessary for the formation of a three-dimensional network, but it is not sufficient enough [7]. General steps of reaction are initiation, propagation, chain transfer, and termination. Chain transfer and termination reactions are conducted using different mechanisms. In linear systems, by starting polymerization reactions, termination reactions also begin immediately. In free radical polymerization, dead polymer chains are formed with two mechanisms of chain transfer (to monomer, polymer, solvent, initiator, or additive) and bimolecular termination [7]. Autoacceleration-autodeceleration polymerization behavior can be observed in UV-curing systems. Proceeding the conversion of system leads to the decrease of the mobility of polymer radicals and thereby the rate of termination reaction. On the other hand, the rate of polymerization increases (autoacceleration) [7].

In higher conversions, the effect of cross-linking becomes dominant on the chain progress reaction and the polymerization rate decreases (autodeceleration) [8]. Autoacceleration-autodeceleration behavior can be assessed by conversion ratio at the maximum rate of polymerization. At higher conversions, the gel effect vanishes and the rate of polymerization reaches its maximum value (R_p^{max}). Vitrification effect leads to decrease the polymerization rate until reaching the final conversion [7].

The use of multifunctional monomers causes the network to form more complex. Regarding the kinetics of network formation, cross-linking systems are affected by factors not existing in the linear systems. Along with the progress of the reaction through the addition of the next monomer or adjacent chains, radicals attack the intermolecular double bonds. This is called cyclization (intermolecular cross-linking), which leads to the formation of a three-dimensional network. Cyclization occurs when a polymer has flexible chains to form a ring [7]. A hard or too long bridge between two unsaturated zones prevents intermolecular cross-linking. These reactions lead to the formation of compact structures at the start of polymerization, called microgels. Macrogelation occurs through the chemical reactions of microgel particles together. Termination process is controlled by translational diffusion (diffusion of the mass center of a polymer coil) and segmental diffusion (rearrangement of the two ends of macroradical next to each other for a more effective reaction) [7].

The following provides further explanation concerning the mechanisms of termination reactions in linear and cross-linked systems.

1.1.2 Photoinitiation Process

UV-curing process is started by photoinitiators. Photoinitiators are classified according to the type of polymerization reaction taking place, such as free radical, cationic and anionic. Free radical initiators are placed in two groups of α-cleavage (type 1) and H abstraction (type 2). Most initiators of type 1 are substituted carbonyl and aromatic compounds. Benzoin and its derivatives, benzyl ketals, acetophenones, aminoalkyl phenones, *O*-acyl-α-oximinoketones, α-hydroxyalkyl ketones, and acylphosphine oxide fall into this category. Free radical creation of benzoin by UV absorption and α-cleavage mechanisms is shown in Figure 1.

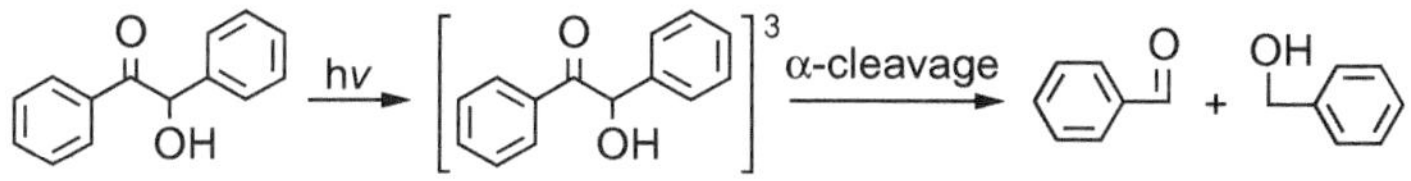

Figure 1 α-Cleavage mechanism for benzoin [9].

Unlike the type 1 initiators, the reactions of type 2 initiators are bimolecular; thus, their curing rate will be lower compared to the type 1 initiators [9]. Aromatic ketones such as benzophenone, xanthone, and thioxanthone, as well as benzyl, quinones, ketone aromatics, phenylglyoxalates, 3-ketocoumarin, and camphorquinone, are in this category. This type of initiators creates a ketyl radical and another radical from the hydrogen donor (a radical precursor with a weak covalent bond) through the absorption of rays [9]. Polymerization of vinyl monomers usually begins with the radicals, which are obtained from a hydrogen donor. The overall reactions of these types of initiators are shown, as an example of a benzophenone, in Figure 2 [9].

Selection of an initiator agent (hydrogen donor) that reacts rapidly with excited initiators is of great importance. Amines, ethers, and alcohols can be used as hydrogen donors. It is worth noting that tertiary amines are more active initiation agents compared to alcohols or ethers [9]. Each photoinitiator has advantages and disadvantages, and the selection of photoinitiators highly depends on the requirement of particular application. The advantages and disadvantages of some of photoinitiators are listed in Table 1.

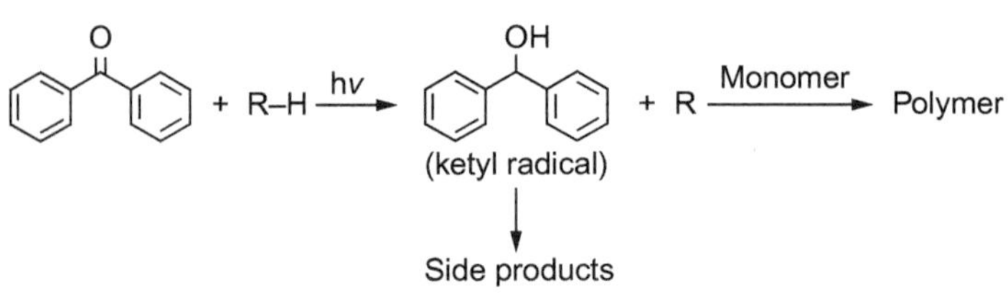

Figure 2 Mechanism for type 2 initiation with a benzophenone and electron donor [9].

Table 1 Advantages and disadvantages of some usual of photoinitiators [9]

Photoinitiator	Advantage	Disadvantage
Benzoin	High quantum efficiency, high reactivity, suitable for styrene monomers	Low thermal stability
Benzyl monoketals	Good thermal stability, little yellowing effect, suitable for clear coats	-
Aminoalkyl phenones	Long wavelength absorption, used for formulations containing pigments	-
Acylphosphine oxide	Thermal stability, high chemical reactivity	-
Benzodioxinones	Increases shelf life	-
Thioxanthones	High efficiency	-

1.1.3 The Mechanisms of Termination

By cross-linking and the formation of entanglement in network structure, transmission diffusion gets significantly low, but polymer radicals can still get close to each other by segmental diffusion or through the progress of reaction. When reaction proceeds and cross-linking density gets higher (by about 15% conversion, after the effect of gelation), free radicals are entangled in such a way that their mass centers are immobile and the only intrusion possible for radical chains is through the progress of polymerization. In this stage, coefficient of termination reactions is proportional to the coefficient of propagation [7].

An important feature in the formation of networks is radical trapping. Since the mobility of radicals is extremely low in cross-linked systems, they tend to exhibit radical trapping. Polymer chains surround radicals, and as the space for chain mobility is severely limited, radicals are trapped and are not available for continuing radical reactions [9]. Accordingly, polymer radicals are divided into three categories: (1) free radicals, which are not connected to the network; (2) radicals that are poorly connected to one part of the network and, although they are relatively limited, still have movement; and (3) trapped radicals that are surrounded by dead polymer chains and can no longer react [9].

Consequently, there are two types of termination reactions:

The reaction of polymer radicals (bimolecular termination) [7]:

$$P_i^{\cdot} + P_j^{\cdot} \xrightarrow{k_t^b} \text{Polymer} \tag{1}$$

with a bimolecular termination rate coefficient of k_t^b.

First-order reaction in which one polymer radical is involved (monomolecular termination) [7]:

$$P_i^{\cdot} \xrightarrow{k_t^m} \left(P_i^{\cdot}\right)_{\text{trapped}} \tag{2}$$

with a monomolecular termination rate coefficient of k_t^m.

The type of termination reaction in cross-linked polymerization depends on the degree of double-bond conversion and reaction conditions involved, such as a monomer structure (cross-linking density) and temperature [7]. It should be noted that bimolecular termination is irreversible, but radicals removed by monomolecular processes can again become active through factors such as increasing the temperature and mobility of chains [7].

2. UV-CURABLE NANOCOMPOSITE SYSTEMS

From a general point of view, composite materials can be prepared in micro- [10] and nanoscale [11,12]. Nanocomposites are a new class of composite materials. The size of their dispersed phase particles is in the range of 1-100 nm, and their properties are significantly improved compared to traditional composites. Nanoparticles can be in different morphologies including spherical (nanotitan [13], nanosilica [14], etc.), rodlike (carbon nanotube, rodlike nanosilicate [15], etc.), and platelet (graphene [16], exfoliated nanoclay [17], etc.). Nanocomposite radiation-curable systems have the advantage ofboth radiation-curing systems and nanocomposites. As a result, they have been considered extensively in industries of coatings, inks, adhesives, and printing. Most of the works done on radiation-curing nanocomposites have been based on nanosilica [12,18] and nanoclay. Preparing coatings with high scratch and mar resistance are examples of these materials [19].

Nanocomposite systems can be divided into two categories based on their method of preparation: the incorporation of nanoparticles and the *in situ* method [20]. More details about these methods are provided in the following sections.

2.1 Incorporation of Nanoparticles

In this method of nanocomposite preparation, nanoparticles are prepared or treated separately and then mixed with the organic phase. The high-efficiency homogenization methods such as high-pressure homogenization, ultrasonic homogenization, and a combination of these methods by simpler mixing methods can be utilized in this method of preparation. By using the dispersion of nanoparticles, more condensed inorganic structures with a higher fractal structure can be dispersed within the organic phase, and the control of particle morphology is less complicated [20]. Uniform distribution of the particles is desired in order to achieve the high transparency, but the primary particles size is difficult to obtain with this method. Radiation-curable monomers and resins are sensitive to high temperatures and high shear rates since their structures could be decomposed. Broken molecules and structures may remain in the system, soften the polymer network, and reduce cross-linking density, which causes a drop in the final properties of the nanocomposites. To circumvent this problem, first, the inorganic particles can be dispersed in a suitable solvent; then, the obtained dispersion can be introduced into the organic phase; and the solvent can be removed from the system. Solvent remaining in the system may cause problems for the radiation-curing process and mechanical properties of the final composite. The remaining solvents, along with the high cost of these methods and the time-consuming processes, impose restrictions for these procedures. For better dispersion of mineral nanoparticles in the organic phase and the enhanced miscibility of both organic and inorganic phases, coupling agents can be used. This approach can be applied in different forms. Inorganic particles can be treated with coupling agents before the incorporation process, or in another method, coupling agent is added into the organic phase and improves the miscibility of organic and inorganic phases and facilitates mixing processes.

2.1.1 Colloidal Stability

Uniform distribution of nanoparticles has a strong influence on the optical and mechanical properties of nanocomposites [20]. The density of dispersed phase is usually much larger than monomers and organic phase components; thus, in organic/inorganic systems, there is tendency to phase separation and the loss of colloidal stability. In many applications and especially in the case of printing inks, radiation-curable formulations have low viscosity, which makes it more difficult to provide stability.

To stabilize the radiation-curable dispersions and increase the miscibility of organic and inorganic phases, surface preparation of nanoparticles can be used. The sol-gel methods are among the usual methods for conducting the surface treatment. Results of AFM tests from cured film and SALS from a radiation-curable formulation have shown that using coupling agents causes the uniform and rapid dispersion of nanoparticles at the nanoscale.

Another method for enhancing the stability of dispersions is the modification of nanoparticles by surface active agent materials and polyelectrolytes [21]. Nanoparticles can be modified before adding them to the organic phase in an aqueous environment, followed by drying and milling of the modified particles.

In another method, a dendritic polymer can be used for preparing nanocomposites in order to achieve the desired morphology. For example, for the distribution of clay nanoparticles and to reach an exfoliate state, dendritic polymers with appropriate functionality can be used [22]. Dendritic polymers can enter the empty space between nanoclay planes, known as galleries, and, by increasing their distance, increase their stability and facilitate their uniform distribution. Dispersion stability can be assessed by measuring the surface potential of particles, measuring zeta potential, or direct monitoring of particles by AFM. If these methods are not feasible, transparency or haze degree can be examined and its variations alongside time can be studied.

2.1.2 Coupling Agents

Precursors are the materials that initiate the sol-gel processes and are commonly sharing two features. First, they can dissolve in a reaction medium, and second, the chemical reactivity of their alkoxy groups is sufficient for participating in the gel formation reactions [23]. Some salts, oxides, hydroxides, complexes, alkoxides, acylates, and amines can be used as precursors if they are soluble in a suitable solvent. Precursors that are often used in radiation-curing systems are selected based on the type of oligomers used in formulation and typically include vinyltrimethoxysilane (VTMS) and 3-methacryloxypropyltrimethoxysilane (MPS) [24]. In order to increase functionality of the sol-gel process and the fractal dimension of the inorganic phase, these inorganic precursors are usually mixed with tetraethoxysilane (TEOS) [12,25]. In such cases, TEOS is referred to as a network-forming agent and MPS or VTMS is referred to as the network-modifying agent. Generally, sol-gel processes involve the hydrolysis of these precursors and their self-condensation reactions. Hydrolyzed precursors can connect to the UV-curing organic phase by reacting with acrylic, methacrylate, or vinyl sites and the formation of a hybrid nanocomposite. These hybrid materials contain organic and inorganic phases that are connected by strong covalent bonds. As previously mentioned, the organic parts have the task of adhesion and flexibility, while the inorganic parts improve the mechanical properties.

2.2 *In Situ* Method

To prepare nanocomposites via the *in situ* method, an inorganic constituent or precursor, along with solvent and a calculated amount of water, is mixed into the organic phase to *in situ* formation of inorganic phases through the sol-gel process. Nanocomposites prepared by this method are stable and uniform dispersions of inorganic phases into an organic phase can be achieved. In contrast to the method of incorporation of particles, inorganic phases have lower density and fractal dimension and the control of its particle size and morphology is limited [20].

2.2.1 Sol-Gel Process

In recent years, using the sol-gel technique for the preparation of hybrid organic/inorganic material has become much more widespread. In the sol-gel process, silanol groups are condensed to form a siloxane network, and ultimately, an inorganic amorphous network is created from the solution phase. During these reactions, a transition state from colloidal (liquid) to an irreversible gel (solid) is seen, thus labeling this type of reaction as sol-gel. The advantage of using the sol-gel method is obtaining pure, uniform inorganic particles at low temperature, which can be directly used in polymer latex and radiation-curing systems [20,26]. The sol-gel process can be catalyzed by acting as either an acidic or a basic catalyst. The type of catalyst, concentration of water, medium, and temperature all affect the structure and morphology of the formed silica [23,27].

3. CHARACTERIZATION

3.1 Structural Characterization

As previously stated, one of the most important factors in nanocomposite properties is the heterogeneity. The control of heterogeneity, particle size, and morphology by providing appropriate interaction with the continuous phase is of great importance. For the nanocomposites, two or more techniques are often used, and by correlating the obtained results, the structure of a nanocomposite can be characterized. In the following, the different methods for characterization of the structure of nanocomposites will be discussed.

3.1.1 Nuclear Magnetic Resonance Spectroscopy

Nuclear magnetic resonance (NMR) spectroscopy is used to study the structure of nanocomposites by examining the bonds of target atom with adjacent carbon or other similar atoms. For example, the covalent bonds of silicon atoms with organic phase can be observed; additionally, the degree of conversion of silica domain can be approximated [12]. The NMR spectroscopy can also be used for investigating inorganic domain variations during radiation and the curing process by performing the NMR spectroscopy before and after the curing process in liquid state and solid state, respectively [28,29]. Table 2 shows the different displacements of silicon atoms in nanocomposites.

As shown in Figure 3, T^i are varieties of silicon atoms that are attached to an organic group and Q^i shows varieties of silicon atoms that are not bonded to organic groups; in both of them, *i* specifies the number of —O—Si groups attached to the considered silicon atom.

3.1.2 FTIR Spectroscopy

FTIR spectroscopy is often used for preliminary identification of the inorganic phase. Usually, nanocomposites are spectrum in the wave number

Table 2 Chemical displacements of different types of silicon atoms in the spectra of ^{29}Si-NMR [28,29]
Chemical shift (ppm)

Structure	T^1	T^2	T^3	Q^1	Q^2	Q^3	Q^4
	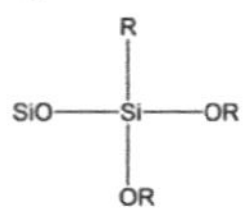	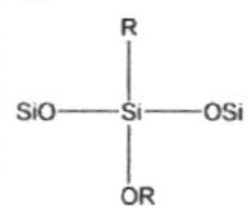	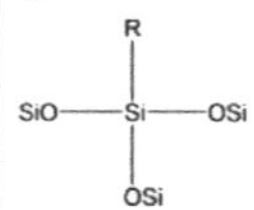	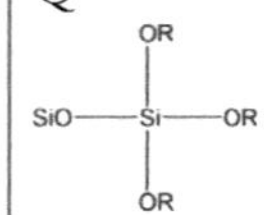	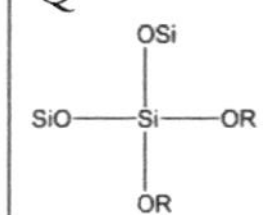	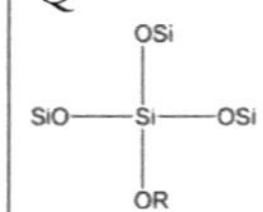	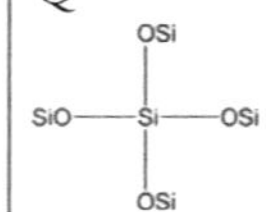

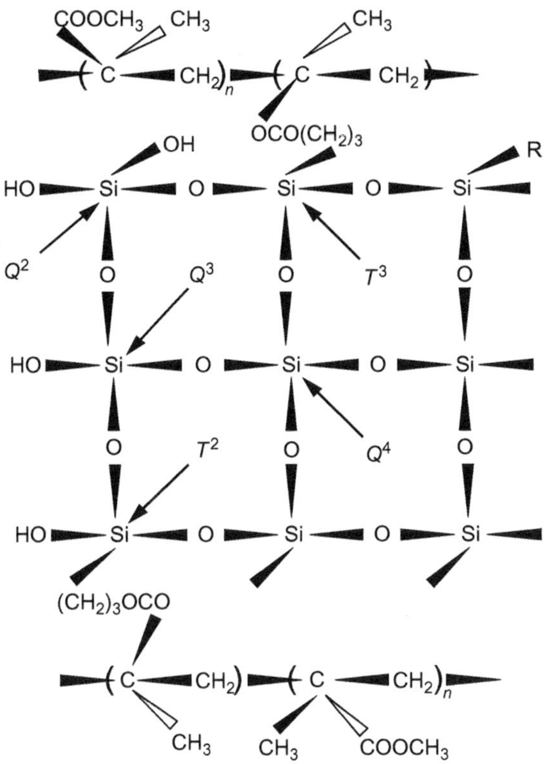

Figure 3 Theoretical structure of a hybrid material and different types of silicon atoms [28].

range of 400-4000 cm^{-1}. The peaks of CH_2 groups, carbonyl groups, and ND dual bonds are observed in the range of 2900, 1720, and 1600 cm^{-1}, respectively. The type and degree of network formed based on Si—O—Si by sol-gel method are important in analyzing the spectrum of a material. For siloxane bonds, there are mainly three absorption bands in the FTIR spectrum, which are attributed to different modes of vibration the of Si—O—Si bond. The first state is the rocking motion of an oxygen atom bridged between two silicon atoms, which occurs perpendicular to the Si—O—Si plane at around 460 cm^{-1}. The second state occurring at 800 cm^{-1} is the symmetrical stretching of an oxygen atom in the direction of bisecting the angle of Si—O—Si bridge. This peak has low intensity, so that it often cannot be observed in the spectrum. The third state has more intensity and, as such, has attracted the most attention; it is attributed to the vertical movements of oxygen atoms parallel to the imaginary line that connects the silicon atoms. FTIR is used to obtain the conversion ratio of double bonds; these are provided in detail in the next section.

3.1.3 Electron Microscopy

Electron microscopes can be used for studying dispersed phase morphology in the nanocomposite. Furthermore, the distribution and orientation of the particles and their relationships with the dispersion methods can be investigated by the electron microscopes. Transmission electron microscopy (TEM) and scanning electron microscopy (SEM) fall into this class of microscopies. For example, Figure 4 illustrates the TEM of a UV-curable epoxy acrylate nanocomposite films containing 3 wt% of modified montmorillonites. The dark regions are the clay layers dispersed in the polymer matrix. Higher magnification insets show that the clay sheets are oriented in layers.

3.1.4 SAXS and Other X-ray Techniques

Materials prepared by the sol-gel method have a fractal dimension that can change, depending on the method used for employing a precursor.

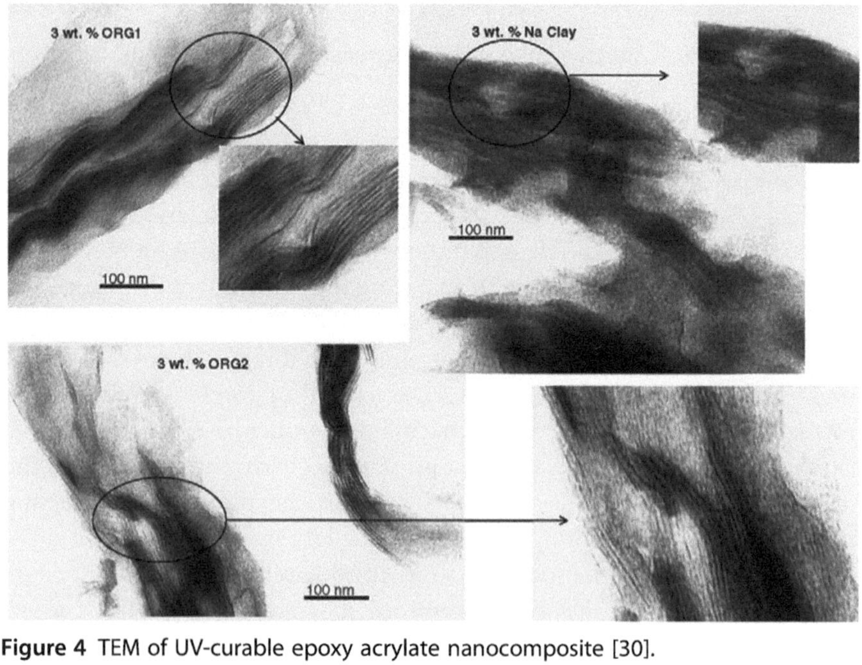

Figure 4 TEM of UV-curable epoxy acrylate nanocomposite [30].

To evaluate this fractal dimension and the morphological behavior of nanocomposites prepared by the sol-gel process, SAXS can be used. For example, it has been shown that the use of optimum ratio of precursors in a mixture can lead to enhance the fractal percentage of the inorganic phase and, as a result, the mechanical properties of prepared nanocomposites [19].

Other X-ray-based tests can be used for identifying the structure of nanocomposites. For example, for many years, nanocomposites had been divided into two categories, intercalated and exfoliated, based on X-ray diffraction (XRD). Although this is a dated classification, XRD and the displacement of scattering angles are still widely used to detect the inorganic structures into an organic phase.

3.2 Characterization of UV Cross-Linking Process

To investigate the kinetics of cross-linking, the initial required curves are conversion percentage in terms of photopolymerization time. These curves can be achieved through two methods of real-time FIIR spectroscopy [31–33] or photo-differential scanning calorimetry [12,18,34,35]. The details for using these tools are explained below.

3.2.1 Real-Time FTIR Spectroscopy

To calculate the percentage of the final conversion of radiation-curable coatings using real-time FTIR, its spectra are taken from curing material during UV radiation. The obtained spectra become normal relative to the length of peaks that remain unchanged during the curing process (e.g., the length of the peak of the carbonyl bond of 1700 cm^{-1}), and the conversion can be obtained from the peak area of a carbon-carbon double bond in the range of 810 cm^{-1} or 1600 cm^{-1}, by Equation (3) [36–56]:

$$\text{Conversion}\% = \frac{(A_{810})_0 - (A_{810})_t}{(A_{810})_0} \times 100 \tag{3}$$

where $(A_{810})_0$ represents the peak area of carbon=carbon double bond prior to irradiation and $(A_{810})_t$ is area of this peak at moment t of radiation.

3.2.2 Photo-DSC Methods

Calculating the conversion using the calorimetry method can be done by applying the assumption that the ratio of heat released at any point in the polymerization process to the total heat of double bonds in the system is proportional to the conversion of double bonds. To calculate the conversion (P)

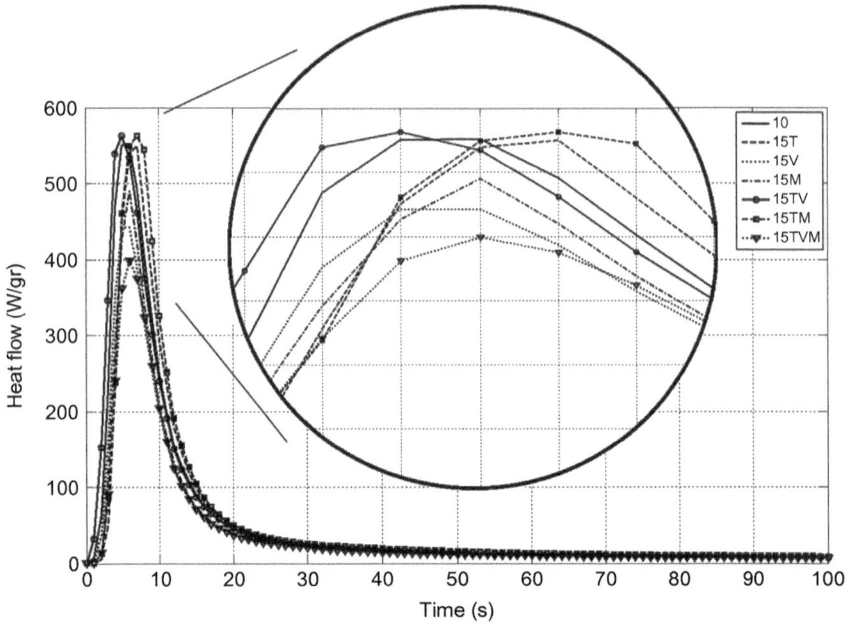

Figure 5 Usual exotherms for the UV-curable nanocomposite coatings [12].

from primary data (heat flow vs. time, Figure 5), the following formula was applied [10,57–60]:

$$P\% = \frac{100}{\Delta H_{\mathrm{T}}} \int_0^t \frac{\mathrm{d}H}{\mathrm{d}t} \mathrm{d}t \tag{4}$$

where ΔH_{T} stands for the total enthalpy and H represents the enthalpy at any stage of polymerization. The final conversion can be calculated by substituting t with the final time. Different methods have been proposed for calculating ΔH_{T}. There are some reports on calculating the total enthalpy by integrating from $t=0$ to the time of reaching equilibrium (approximately 20 min [61]) [57]. Other reports showed that ΔH_{T} can be calculated by measuring the heat generated by the all double bonds in heat-initiated polymerization reactions. A more recent technique is the theoretical calculation of the total heat of reaction according to the number of all double bonds and the energy released from each of the reactions of bonds [12,18,62]:

$$\Delta H_{\mathrm{T}}^{\mathrm{theory}} = \frac{\Delta H_0^{\mathrm{theory}}(\text{acrylate, methacrylate, or vinyl})}{\mathrm{Mw}^{\mathrm{theory}}} \times \text{functionality} \tag{5}$$

To calculate the conversion rate of radiation curing, the following equation can be used [35,60,63,64]:

$$\text{Rate of Conversion} = \frac{\mathrm{d}P}{\mathrm{d}t} = \frac{\mathrm{d}H/\mathrm{d}t}{\Delta H_{\mathrm{T}}} \quad (6)$$

3.2.3 UV-Curing Kinetics

In cross-linking systems, the rate-determining reactions, especially in the early stages, are complex, and assessing real coefficients of polymerization rate in these systems is very difficult. There are some reports concerning the investigation of reaction rate coefficients [65]. However, the calculated values are not real of reaction rate coefficients; in fact, they are a combinative value of the rate coefficient of propagation and termination processes. However, when these rate coefficients are measured in different conversion degrees, they produce an overall measure of the reaction rate and the cross-linking density. Assessing the rate of cross-linking systems is highly dependent on the kinetic model that explains polymerization behavior.

Various models are used to describe polymerization reactions; these are briefly described below.

3.2.3.1 Partly Integrated Models

This model, first used in 1979 by Tryson and Shultz [61], was later widely implemented. In this simple model, the rate coefficients of propagation and termination reactions are calculated assuming that the only termination reaction is bimolecular. The method follows that polymerization starts with radiation at a specific and known time, and at a specified conversion, the radiation cuts off (dark polymerization). The reaction in the dark conditions can be monitored by differential scanning calorimetry techniques at isotherm mode and real-time infrared spectroscopy. The propagation and termination reactions are shown as simple equations below, respectively [61]:

$$R_{\mathrm{p}} = -\frac{\mathrm{d}[\mathrm{M}]}{\mathrm{d}t} = k_{\mathrm{p}}[\mathrm{M}][\mathrm{P}^{\cdot}] \quad (7)$$

$$R_{\mathrm{t}}^{\mathrm{b}} = -\frac{\mathrm{d}[\mathrm{P}^{\cdot}]}{\mathrm{d}t} = 2k_{\mathrm{t}}^{\mathrm{b}}[\mathrm{P}^{\cdot}]^2 \quad (8)$$

where [M] stands for the concentration of double bonds, $[\mathrm{P}^{\cdot}]$ is radical concentration, R_{p} is the rate of propagation reaction, and $R_{\mathrm{t}}^{\mathrm{b}}$ is the rate of bimolecular termination reactions. The value of $[\mathrm{P}^{\cdot}]$ can be determined by integration from the rate of bimolecular termination reactions equation in the dark. By placing this value in the propagation reaction equation, the equation of the partly integrated model can be obtained as shown below [61]:

$$\frac{[\mathrm{M}]_\mathrm{t}}{(R_\mathrm{p})_\mathrm{t}} = \frac{2k_\mathrm{t}^\mathrm{b}}{k_\mathrm{p}} t + \frac{[\mathrm{M}]_0}{(R_\mathrm{p})_0} \tag{9}$$

where index 0 is related to the polymerization parameters at the beginning of dark polymerization and index **t** is related to parameters at time **t** after the light cutoff. The term $2k_\mathrm{t}/k_\mathrm{p}$ can be calculated from the slop of $[\mathrm{M}]_\mathrm{t}/(R_\mathrm{p})_\mathrm{t}$ against time. In some systems, at the start of dark polymerization, deviation from a linear state is viewed as being related to the response time of DSC and the thermal conductivity of the system. This deviation may be due to the lack of the model in describing termination mechanisms or a change in rate coefficients in dark conditions. Since the steady-state approximation for the radical concentration in the early stages (autoacceleration) of polymerization is not true [66], it is assumed that the concentration of radicals is fixed during measurement intervals and the rate of polymerization is expressed by the following equation [61]:

$$R_\mathrm{p}^\mathrm{b} = \frac{k_\mathrm{p}}{\left(k_\mathrm{t}^\mathrm{b}\right)^{0.5}} [\mathrm{M}] \varphi^{0.5} \tag{10}$$

where $\varphi = I_\mathrm{a}\Phi$ where I_a is the absorption intensity Einstein length and Φ is quantum yield of radiation. By calculating the value of $k_\mathrm{t}^\mathrm{b}/k_\mathrm{p}$ and that of $k_\mathrm{p}/(k_\mathrm{t}^\mathrm{b})^{0.5}$ from the previous equation, the values of k_t and k_p can be calculated.

3.2.3.2 Fully Integrated Models

The models reviewed in the previous section are very popular. In fact in most likely state, both monomolecular and bimolecular mechanisms occur in parallel during termination. Therefore, three different termination mechanisms can occur: (1) monomolecular, where all radicals are trapped; (2) bimolecular, in which the reaction is between two macroradicals; and (3) mixture, in which the monomolecular and bimolecular processes occur in parallel [7]:

$$R_\mathrm{t}^\mathrm{mix} = -\frac{\mathrm{d}[\mathrm{P}^\cdot]}{\mathrm{d}t} = 2k_\mathrm{t}^\mathrm{b}[\mathrm{P}^\cdot]^2 + k_\mathrm{t}^\mathrm{m}[\mathrm{P}^\cdot] \tag{11}$$

Taking the procedure previously described and combining it with polymerization rate, three models of termination can be identified that describe the dark polymerization [7,18]:

Monomolecular termination model (model 1):

$$-\ln(1-p_{\mathrm{d}})=\frac{k_{\mathrm{p}}}{k_{\mathrm{t}}^{\mathrm{m}}}[\mathrm{P}^{\cdot}]_0\left(1-\exp\left(-k_{\mathrm{t}}^{\mathrm{m}}t\right)\right) \tag{12}$$

Bimolecular termination model (model 2):

$$-\ln(1-p_{\mathrm{d}})=\frac{1}{2}\frac{k_{\mathrm{p}}}{k_{\mathrm{t}}^{\mathrm{b}}}\ln\left(1+2[\mathrm{P}^{\cdot}]_0k_{\mathrm{t}}^{\mathrm{b}}t\right) \tag{13}$$

Mixed termination model (model 3):

$$-\ln(1-p_{\mathrm{d}})=\frac{1}{2}\frac{k_{\mathrm{p}}}{k_{\mathrm{t}}^{\mathrm{b}}}\ln\left(1+2[\mathrm{P}^{\cdot}]_0\frac{k_{\mathrm{t}}^{\mathrm{b}}}{k_{\mathrm{t}}^{\mathrm{m}}}\left(1-\exp\left(-k_{\mathrm{t}}^{\mathrm{m}}t\right)\right)\right) \tag{14}$$

where $[\mathrm{P}^{\cdot}]_0$ is macroradical concentration at the beginning of the dark polymerization, t is the time that has passed after the start of the dark polymerization, and p_{d} is the degree of conversion of double bonds during the dark polymerization.

Recent models have been indicated as nonlinear, and as there is no assumption of linearity, they have greater accuracy. These models only require information about the conversion percentage of double bonds in terms of the dark period as a function of time.

Fully integrated models were first presented in 1991 by Timpe and Strehmel and have been further developed by others.

The steady-state equation for evaluating rate coefficients from model 3 is provided below [7]:

$$R_{\mathrm{p}}^{\mathrm{mix}}=\frac{k_{\mathrm{p}}}{4k_{\mathrm{t}}^{\mathrm{b}}}[\mathrm{M}]\left\{\left[\left(k_{\mathrm{t}}^{\mathrm{m}}\right)^2+16\varphi k_{\mathrm{t}}^{\mathrm{b}}\right]^{0.5}-k_{\mathrm{t}}^{\mathrm{m}}\right\} \tag{15}$$

3.2.3.3 Autocatalytic Model

The autocatalytic effect usually occurs when the formed intermediate species causes the rate to increase. For example, the epoxy-amine curing process accelerates with the formed hydroxyl groups in the reaction. Processes of this class are often described by the Kamal model. The general model shows that when the initial rate of reaction is not negligible, the equation is as follows [7,67]:

$$\frac{\mathrm{d}p}{\mathrm{d}t}=(k_1+k_2p^m)(1-p)^n \tag{16}$$

where k_1 and k_2 are the Arrhenius rate constants, n is a component of reaction degree, and m is the autoacceleration component.

In this equation, p is the conversion, k_1 is the rate constant for the noncatalyzed process, and k_2p^m indicates the autoacceleration nature of the process. The sum of n and m indicates the general degree of reaction.

3.3 Physical and Mechanical Properties of Radiation-Curable Nanocomposite Coatings

3.3.1 Optical Properties

The optical properties of nanocomposites are important in some special applications such as optical fiber coatings and lens covers. The weathering resistance of nanocomposites is largely dependent on its optical properties, particularly where it concerns the absorption of ultraviolet. Optical properties heavily depend on particle size, shape, crystallinity, and structure. In transparent systems, to achieve suitable transparency, the dispersion of particles should be appropriately and effectively [19].

For studying the transparency, UV-visible spectroscopy can be performed in transmission and absorption. In radiation-curing systems with high thickness, the scattering of radiation by the dispersed phase causes problems for curing samples and in the kinetic investigation of photopolymerization. This problem is on peak on opaque systems, and in radiation-curing systems containing titanium oxide pigments, absorption and scattering of light cause a reduction in conversion. This problem can be solved by increasing the photoinitiator content and using initiators with high efficiency.

3.3.2 Surface Properties

The surface properties of nanocomposites have been widely investigated. The surface properties of nanocomposites can be examined in terms of topology and chemistry. Surface topology can be investigated by methods such as AFM [12] and SEM. The presence of functional groups that are effective in terms of chemistry can be assessed by ATR, SEM-EDX, and XPS. One of the main characteristics of surface properties is wettability. Surface wetting is effective in many instances such as flowability of fluids on surface, dispersion of particles in a medium, and self-cleaning properties of surface. Using the properties of hydrophobic, hydrophilic, and wettability properties, coatings can be prepared for specific applications such as antifog and self-cleaning.

Wetting can be examined and quantified by various methods, including the static contact angle of liquids with certain surface tension, the slip angle, and hysteresis of the contact angle. Some models have specifically been provided for these purposes, including the Young model, Wenzel model, and Cassie-Baxter model. Photocuring reactions are very sensitive to temperature, light intensity, photoinitiator, and the percentage and type of functional groups of monomers and oligomers. Because of these various controlling factors in radiation-curing nanocomposite systems, specific surface patterns can be created and the surface properties can also be controlled.

3.3.3 DMTA Analysis

In this method, changes in mechanical properties can be measured according to oscillatory tension within a certain frequency range at different temperatures. In this method, oscillatory stress is exerted on the sample and its dynamic modulus. The mechanical module of its loss as a function of temperature or frequency is measured at a specific temperature. For this purpose, examples of nanocomposite films can be prepared and used for analysis.

When an inorganic material is introduced into an organic system, to achieve the best mechanical properties, the boundary layer between the polymer and the mineral levels should have two mutual features:

1. For optimal tension transfer between the high modulus mineral phase and resin with low modulus, the border area between these two phases must be a module between the modulus of organic phase and that of inorganic phase.
2. To achieve a material with high toughness and resistance against tension between the two phases, a flexible boundary zone is needed for easily transferring of localized tensions.

3.3.4 Scratch Resistance

Scratch is one of the most serious types of mechanical damage imposed on materials and coatings and gives rise to problems related to the type and amount of scratch. Scratch resistance in transparent coatings is of great importance. Scratching occurs with two general mechanisms of plastic and failure. Failure scratches have sharp edges and irregular shape; thus, they can scatter light. In contrast, plastic scratches are smoother and cause less light scattering.

The first approach to increasing scratch resistance in radiation-curing coatings is to optimize the cross-linking density of the continuous phase with an appropriate formulation; the second is to use nanofillers and prepare nanocomposites. In nanocomposite systems, inorganic parts increase scratch resistance and organic parts provide resistance against weathering. Modulus inorganic materials have higher hardness compared to organic materials. Addition of nanoparticles with a good dispersion increases the hardness of coatings and prevents the penetration of sharps objects into the nanocomposite that can create a scratch. Many reports have documented evidences detailing the preparation of radiation-curing nanocomposites that are resistant against scratching using the nanoparticles Al_2O_3, SiO_2, ZrO_2, and TiO_2 [68,69]. It was demonstrated that the most influential parameters in scratch resistance of coatings would be elastic modulus and hardness. Coatings with higher values for these two parameters can resist much more comfortably against plastic deformation.

Nanoscale scratch testing is very useful to measure the scratch resistance of a coating and to understand the relationship between deformation, removal mechanisms, and scratch morphology. The AFM pictures of the scratched films of a UV-curable nanocomposite film are depicted in Figure 6. As can be seen, a straight (i.e., rectangular) and completely clear scratch was obtained for the sample without nanoparticles. On the other hand, the profile of the scratch was not straight, and it looked like a zebra profile for the sample with nanoparticles. It seems that the presence of nanoparticles prevented the straight movement of the probe and a zebra profile was achieved. The results show that the scratch recovery was much higher for the sample with nanoparticles.

Figure 7 shows the lateral force time plots. It appears that the presence of hard inorganic particles in polymer matrix divided the single lateral force into different force components. Therefore, the required force to create scratch in the UV-cured films that included nanoparticles was higher than those without nanoalumina.

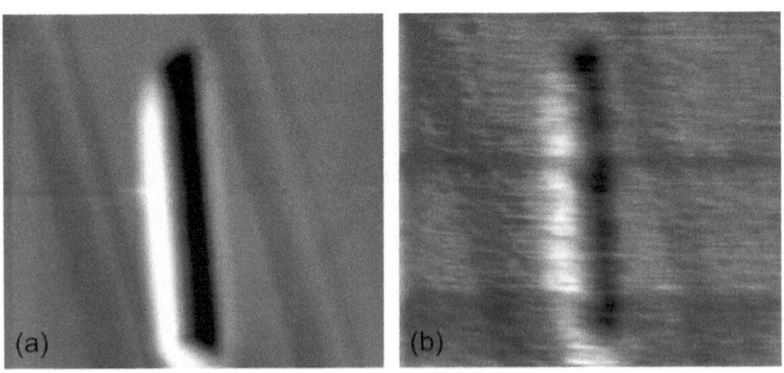

Figure 6 Two-dimensional AFM of nanocomposite films without nanoalumina (a) and with nanoalumina (b) [69].

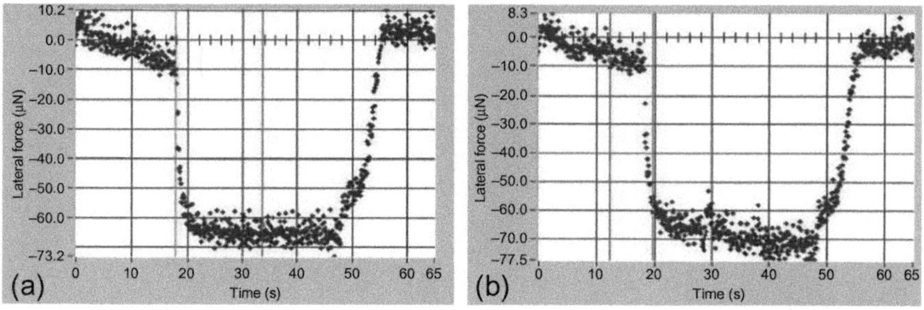

Figure 7 Lateral force vs. time without nanoalumina (a) and with nanoalumina (b) [69].

4. NEW APPROACHES AND NOVEL FINDINGS

4.1 Sandwich Structures

In the last few decades, sandwich composites have been the subject of many studies as new materials because of their peculiar properties. Sandwich composites are a specialized form of laminated composites in which a thick, soft, and lightweight core is sandwiched between two thin and stiff fiber-reinforced plastic skins. The use of light core reduces the weight so this weight reduction results in a number of benefits, including increased range, higher payloads, and decreased fuel consumption.

The challenge of making a structure as light as possible, and without sacrificing strength, is crucial in some applications such as aircraft design. The use of sandwich composites is becoming popular in structural design because of their ability to substantially decrease weight while maintaining mechanical performance. Inevitably, the requirement leads to the need to stabilize thin surfaces to withstand tensile and compressive loads and the combination of the two, in tension, torsion, and bending [70].

Sandwich structure origins from nature, and hence, the concept is older than mankind itself. The branches of the elder tree are a good example for a foam core sandwich structure. The bones in the skeletons of animals and humans are sandwich structures with foamlike core materials as well. The microstructure of these nanocomposites was characterized by wide-angle XRD, TEM, and SEM.

Sandwich structures are mostly used in applications in which mechanical performance as well as lightness is important. An epoxy-clay nanocomposite with low degree of gas permeability was made based on sandwich structure for the use in aircraft. A transparent *cis*-1,4-polybutadiene (BR)/clay nanocomposite was prepared via blending method, and the morphology of

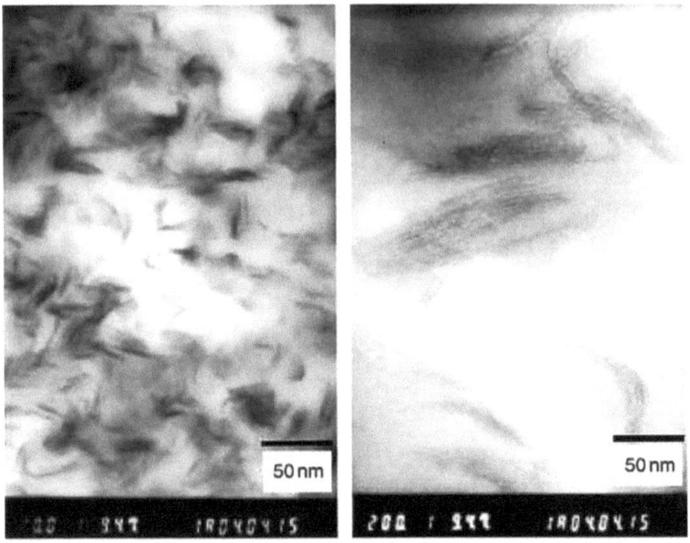

(a) (b)

Figure 8 TEM photographs of BR/organoclay composite [71].

particles was studied. Figure 8 shows the sandwich microstructures consisting of parallel alternating clay layers and BR matrix observed by TEM.

4.2 3-D Printing Materials

3-D printing is the process of joining materials to make 3-D objects, and it is also known as "additive manufacturing." Current manufacturing processes can be better described as "subtractive." Subtractive manufacturing, or machining, removes material to create objects: raw material is cut into shape by abrading, cutting, grinding, and so forth. 3-D printing creates objects on demand with less waste since in most cases additive manufacturing uses only as much material as is necessary to create an object. In fact, 3-D printing is the process of being able to print any object layer by layer [72].

3-D printing can perform by some methods, and photopolymerization has an important role in some methods of this technology. In PolyJet photopolymer, curing by light is the process by which the solid forms from a liquid. A photopolymerization method is also used in stereolithography, and the 3-D model can be created by radiation of a UV light [73,74].

4.3 Photopolymerization Kinetic of Scattering Materials

In the case of thick materials, the decay of UV light intensity through the material thickness can induce a heterogeneous reaction in the medium. This

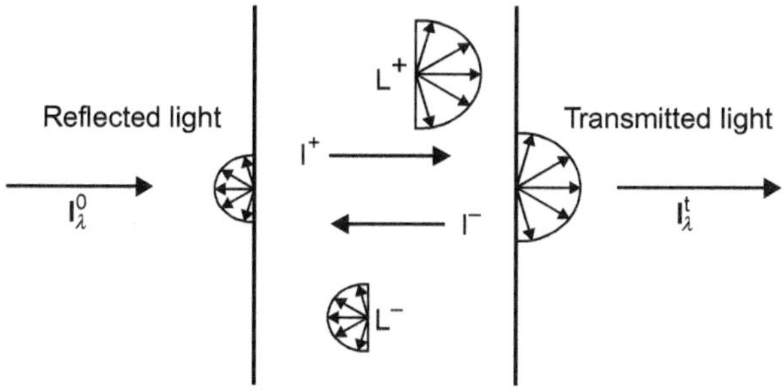

Figure 9 Schematic representation of the four contributions of the radiation field inside an inhomogeneous sample [75,76].

heterogeneous reaction is a drawback when the final product must present uniform properties, but in fact, this can also be used to create deliberately a polymerization gradient and consequently to synthesize a property gradient material. Whatever the objective, the knowledge of UV light intensity variation through the material is important to perfectly control the reaction evolution.

There are some approaches to describe the photopolymerization of scattering media. Modification of the Beer-Lambert law as well as the four flux theories can be used for the kinetic study of these systems. Figure 9 shows the schematic four contributions of the radiation field inside an inhomogeneous sample.

5. SUMMARY AND FUTURE OUTLOOK

UV-curable nanocomposites and materials are a rapidly expanding technology because of their advantages. UV-curable nanocomposites can be prepared by the incorporation of nanoparticles into a UV-curable matrix or *in situ* preparation of inorganic phase in a UV-curable phase via a sol-gel method.

The structure of these systems can be characterized by NMR spectroscopy, FTIR spectroscopy, and electron microscopy as well as by X-ray techniques. UV cross-linking process and UV-curing process can also be characterized by real-time FTIR spectroscopy or photo-DSC techniques. Partly integrated model, fully integrated model, and autocatalytic model can be used to describe UV-curing reaction.

The future outlook leans mostly towards studying the various aspects of material selection, processing, and application for UV-curable nanocomposites. Material selection will be carried out according to a few deciding factors:

price, availability, durability, and safety in terms of more efficient photoinitiators, dual curing systems, 3-D curing, curing by LEDs, 3-D printing, protective coatings, biological materials such as biodegradable systems, functional and smart materials, and finally sandwich structures of great interest.

TERMINOLOGY

k_t^b	bimolecular termination rate coefficient
k_t^m	monomolecular termination rate coefficient
t	time
P	conversion
ΔH_T	total enthalpy
H	enthalpy
ΔH_T^{theory}	theoretical total enthalpy
[M]	concentration of double bonds
$[P^{\cdot}]$	radical concentration
R_p	rate of propagation reactions
R_t^b	rate of bimolecular termination reactions
k_t^m	rate coefficient of monomolecular termination reactions
k_t^b	rate coefficient of bimolecular termination reactions
p_d	degree of conversion of double bonds during the dark polymerization
k_1, k_2	Arrhenius rate constants
n	reaction degree
m	autoacceleration component

REFERENCES

[1] Z.W. Wicks, Organic Coatings: Science and Technology, third ed., Wiley-Interscience, Hoboken, NJ, 2007, pp. 574–584.

[2] E. Zarshenas, S. Bastani, M. Pishvaei, Curing behavior study of UV-curable coatings containing nanosilica and different multifunctional monomers via depth profiling assessment, Ind. Eng. Chem. Res. 52 (2013) 16110–16117.

[3] S. Bastani, S. Moradian, UV-curable powder coatings containing interpenetrating polymer networks (IPNs), Prog. Org. Coat. 56 (2006) 248–251.

[4] P. Kardar, M. Ebrahimi, S. Bastani, M. Jalili, Using mixture experimental design to study the effect of multifunctional acrylate monomers on UV cured epoxy acrylate resins, Prog. Org. Coat. 64 (2009) 74–80.

[5] G. Mashouf, M. Ebrahimi, S. Bastani, UV curable urethane acrylate coatings formulation: experimental design approach, Pigm. Resin Technol. 43 (2014) 61–68.

[6] J.V. Koleske, first ed., in: Radiation Curing of Coatings, vol. 1, ASTM; Radtech International, West Conshohocken, PA; Bethesda, MD; North America, 2001, pp. 1–6.

[7] E. Andrzejewska, Photopolymerization kinetics of multifunctional monomers, Prog. Polym. Sci. 26 (2001) 605–665.

[8] K. Dušek, M. Dušková-Smrčková, Network structure formation during crosslinking of organic coating systems, Prog. Polym. Sci. 25 (2000) 1215–1260.

[9] Y. Yagci, S. Jockusch, N.J. Turro, Photoinitiated polymerization: advances, challenges, and opportunities, Macromolecules 43 (2010) 6245–6260.

[10] P. Kardar, M. Ebrahimi, S. Bastani, Influence of temperature and light intensity on the photocuring process and kinetics parameters of a pigmented UV curable system, J. Therm. Anal. Calorim. 118 (2014) 541–549.

[11] A. Labbani Motlagh, S. Bastani, M.M. Hashemi, Investigation of synergistic effect of nano sized Ag/TiO_2 particles on antibacterial, physical and mechanical properties of UV-curable clear coatings by experimental design, Prog. Org. Coat. 77 (2014) 502–511.

[12] M. Mohseni, S. Bastani, A. Jannesari, Influence of silane structure on curing behavior and surface properties of sol-gel based UV-curable organic-inorganic hybrid coatings, Prog. Org. Coat. 77 (2014) 1191–1199.

[13] W. Tao, F. Fei, W. Yue-Chuan, Structure and thermal properties of titanium dioxide-polyacrylate nanocomposites, Polym. Bull. 56 (2006) 413–426.

[14] F. Li, S. Zhou, L. Wu, Preparation and characterization of UV-curable MPS-modified silica nanocomposite coats, J. Appl. Polym. Sci. 98 (2005) 2274–2281.

[15] Y. Pan, Y. Xu, L. An, H. Lu, Y. Yang, W. Chen, S. Nutt, Hybrid network structure and mechanical properties of rodlike silicate/cyanate ester nanocomposites, Macromolecules 41 (2008) 9245–9258.

[16] M. Martin-Gallego, R. Verdejo, M.A. Lopez-Manchado, M. Sangermano, Epoxy-graphene UV-cured nanocomposites, Polymer 52 (2011) 4664–4669.

[17] F.M. Uhl, S.P. Davuluri, S.-C. Wong, D.C. Webster, Organically modified montmorillonites in UV curable urethane acrylate films, Polymer 45 (2004) 6175–6187.

[18] M. Mohseni, S. Bastani, A. Jannesari, Effects of silane precursors on curing behavior of UV-curable hybrid coatings, J. Therm. Anal. Calorim. 199 (2015) 515–526.

[19] H. Yahyaei, M. Mohseni, S. Bastani, Using Taguchi experimental design to reveal the impact of parameters affecting the abrasion resistance of sol-gel based UV curable nanocomposite films on polycarbonate, J. Sol-Gel Sci. Technol. 59 (2011) 95–105.

[20] F. Li, S. Zhou, L. Wu, Effects of preparation method on microstructure and properties of UV-curable nanocomposite coatings containing silica, J. Appl. Polym. Sci. 98 (2005) 1119–1124.

[21] Y. De Hazan, J. Heinecke, A. Weber, T. Graule, High solids loading ceramic colloidal dispersions in UV curable media via comb-polyelectrolyte surfactants, J. Colloid Interface Sci. 337 (2009) 66–74.

[22] L. Fogelström, P. Antoni, E. Malmström, A. Hult, UV-curable hyperbranched nanocomposite coatings, Prog. Org. Coat. 55 (2006) 284–290.

[23] E.J.A. Pope, S. Sakka, L.C. Klein, Sol-gel science and technology, in: Ceramic Transactions, American Ceramic Society, Westerville, OH, 1995, pp. ix, 402 p.

[24] Y.Z. Lahijani, M. Mohseni, S. Bastani, Utilizing Taguchi design of experiment to study the surface treatment of a nanosilica with an acrylic silane coupling agent and revealing the dispersibility of particles in a urethane acrylate resin, J. Coat. Technol. Res. 10 (2013) 537–547.

[25] H. Yahyaei, M. Mohseni, Mechanically controlled, morphologically determined sol-gel derived UV curable hybrid nanocomposites: SAXS and DMTA studies, J. Sol-Gel Sci. Technol. 66 (2013) 187–192.

[26] C.J. Brinker, G.W. Scherer, Sol-Gel Science : The Physics and Chemistry of Sol-Gel Processing, Academic Press, Boston, 1990.

[27] S. Sakka, first ed., in: Handbook of Sol–Gel Science and Technology: Processing, Characterization, and Applications, vol. 1, Kluwer Academic Publishers, Boston, 2005, pp. 529–538.

[28] T.C. Chang, Y.T. Wang, Y.S. Hong, Y.S. Chiu, Organic–inorganic hybrid materials. V. Dynamics and degradation of poly(methyl methacrylate) silica hybrids, J. Polym. Sci. A Polym. Chem. 38 (2000) 1972–1980.

[29] Z. Zong, M.D. Soucek, C. Xue, Unusual inorganic phase formation in ultraviolet-curable organic-inorganic hybrid films, J. Polym. Sci. A Polym. Chem. 43 (2005) 1607–1623.
[30] F.M. Uhl, D.C. Webster, S.P. Davuluri, S.-C. Wong, UV curable epoxy acrylate-clay nanocomposites, Eur. Polym. J. 42 (2006) 2596–2605.
[31] C. Decker, K. Moussa, A new method for monitoring ultra-fast photopolymerizations by real-time infra-red (RTIR) spectroscopy, Die Makromol. Chem. 189 (1988) 2381–2394.
[32] B.-S. Chiou, S.A. Khan, Real-time FTIR and in situ rheological studies on the UV curing kinetics of thiol-ene polymers, Macromolecules 30 (1997) 7322–7328.
[33] T. Scherzer, U. Decker, Real-time FTIR–ATR spectroscopy to study the kinetics of ultrafast photopolymerization reactions induced by monochromatic UV light, Vib. Spectrosc. 19 (1999) 385–398.
[34] C. Esposito Corcione, M. Frigione, A. Maffezzoli, G. Malucelli, Photo – DSC and real time – FT-IR kinetic study of a UV curable epoxy resin containing *o*-Boehmites, Eur. Polym. J. 44 (2008) 2010–2023.
[35] J.D. Cho, J.W. Hong, Photo-curing kinetics for the UV-initiated cationic polymerization of a cycloaliphatic diepoxide system photosensitized by thioxanthone, Eur. Polym. J. 41 (2005) 367–374.
[36] J. He, A. Nebioglu, Z. Zong, M.D. Soucek, K.M. Wollyung, C. Wesdemiotis, Preparation of a siloxane acrylic functional siloxane colloid for UV-curable organic-inorganic hybrid films, Macromol. Chem. Phys. 206 (2005) 732–743.
[37] H. Tan, D. Yang, M. Xiao, J. Han, J. Nie, Preparation of silica/polyurethane nanocomposites by UV-induced polymerization from surfaces of silica, J. Appl. Polym. Sci. 111 (2009) 1936–1941.
[38] E. Amerio, M. Sangermano, G. Colucci, G. Malucelli, M. Messori, R. Taurino, P. Fabbri, UV curing of organic-inorganic hybrid coatings containing polyhedral oligomeric silsesquioxane blocks, Macromol. Mater. Eng. 293 (2008) 700–707.
[39] H. Lu, L.G. Lovell, C.N. Bowman, Exploiting the heterogeneity of cross-linked photopolymers to create high-Tg polymers from polymerizations performed at ambient conditions, Macromolecules 34 (2001) 8021–8025.
[40] J.E. Elliott, C.N. Bowman, Monomer functionality and polymer network formation, Macromolecules 34 (2001) 4642–4649.
[41] K.A. Berchtold, B. Hacioğlu, L. Lovell, J. Nie, C.N. Bowman, Using changes in initiation and chain transfer rates to probe the kinetics of cross-linking photopolymerizations: effects of chain length dependent termination, Macromolecules 34 (2001) 5103–5111.
[42] K.A. Berchtold, T.M. Lovestead, C.N. Bowman, Coupling chain length dependent and reaction diffusion controlled termination in the free radical polymerization of multivinyl (meth)acrylates, Macromolecules 35 (2002) 7968–7975.
[43] J.E. Elliott, C.N. Bowman, Effect of primary cyclization on free radical polymerization kinetics: modeling approach, Macromolecules 35 (2002) 7125–7131.
[44] A.K. O'Brien, C.N. Bowman, Modeling thermal and optical effects on photopolymerization systems, Macromolecules 36 (2003) 7777–7782.
[45] T.M. Lovestead, C.N. Bowman, A modeling investigation of chain length dependent termination during multivinyl free radical chain photopolymerizations: accounting for the gel, Macromolecules 38 (2005) 4913–4918.
[46] T.M. Lovestead, K.A. Berchtold, C.N. Bowman, An investigation of chain length dependent termination and reaction diffusion controlled termination during the free radical photopolymerization of multivinyl monomers, Macromolecules 38 (2005) 6374–6381.
[47] A.K. O'Brien, C.N. Bowman, Impact of oxygen on photopolymerization kinetics and polymer structure, Macromolecules 39 (2006) 2501–2506.

[48] P.M. Johnson, J.W. Stansbury, C.N. Bowman, Alkyl chain length effects on copolymerization kinetics of a monoacrylate with hexanediol diacrylate, J. Comb. Chem. 9 (2007) 1149–1156.
[49] H. Kilambi, J.W. Stansbury, C.N. Bowman, Deconvoluting the impact of intermolecular and intramolecular interactions on the polymerization kinetics of ultrarapid mono (meth)acrylates, Macromolecules 40 (2007) 47–54.
[50] H. Kilambi, S.K. Reddy, C.N. Bowman, Kinetic and mechanistic studies of photopolymerizations of acrylates in the presence of iniferters, Macromolecules 40 (2007) 6131–6135.
[51] P.M. Johnson, J.W. Stansbury, C.N. Bowman, Kinetic modeling of a comonomer photopolymerization system using high-throughput conversion data, Macromolecules 41 (2007) 230–237.
[52] K.A. Berchtold, J. Nie, J.W. Stansbury, C.N. Bowman, Reactivity of monovinyl (meth)acrylates containing cyclic carbonates, Macromolecules 41 (2008) 9035–9043.
[53] K.A. Berchtold, B. Hacioglu, J. Nie, N.B. Cramer, J.W. Stansbury, C.N. Bowman, Rapid solid-state photopolymerization of cyclic acetal-containing acrylates, Macromolecules 42 (2009) 2433–2437.
[54] B.D. Fairbanks, T.F. Scott, C.J. Kloxin, K.S. Anseth, C.N. Bowman, Thiol-yne photopolymerizations: novel mechanism, kinetics, and step-growth formation of highly cross-linked networks, Macromolecules 42 (2008) 211–217.
[55] K.M. Schreck, D. Leung, C.N. Bowman, Hybrid organic/inorganic thiol-ene-based photopolymerized networks, Macromolecules 44 (2011) 7520–7529.
[56] S. Ye, N.B. Cramer, C.N. Bowman, Relationship between glass transition temperature and polymerization temperature for cross-linked photopolymers, Macromolecules 44 (2011) 490–494.
[57] N.M. Mohsen, R.G. Craig, F.E. Filisko, Effects of curing time and filler concentration on curing and postcuring of urethane dimethacrylate composites: a microcalorimetric study, J. Biomed. Mater. Res. 40 (1998) 224–232.
[58] P. Kardar, M. Ebrahimi, S. Bastani, Curing behaviour and mechanical properties of pigmented UV-curable epoxy acrylate coatings, Pigm. Resin Technol. 43 (2014) 177–184.
[59] D. Pananakis, E.W. Abel, A comparison of methods for the deconvolution of isothermal DSC data, Thermochim. Acta 315 (1998) 107–119.
[60] A. Palanisamy, B.S. Rao, Photo-DSC and dynamic mechanical studies on UV curable compositions containing diacrylate of ricinoleic acid amide derived from castor oil, Prog. Org. Coat. 60 (2007) 161–169.
[61] G.R. Tryson, A.R. Shultz, A calorimetric study of acrylate photopolymerization, J. Polym. Sci. Polym. Phys. Ed. 17 (1979) 2059–2075.
[62] H.-D. Hwang, C.-H. Park, J.-I. Moon, H.-J. Kim, T. Masubuchi, UV-curing behavior and physical properties of waterborne UV-curable polycarbonate-based polyurethane dispersion, Prog. Org. Coat. 72 (2011) 663–675.
[63] Z. Dogruyol, N. Arsu, S.K. Dogruyol, O. Pekcan, Critical phenomenon during photoinitiated gelation at different temperatures: a photo-DSC study, Prog. Org. Coat. 72 (2011) 763–768.
[64] D.S. Esen, F. Karasu, N. Arsu, The investigation of photoinitiated polymerization of multifunctional acrylates with TX-BT by photo-DSC and RT-FTIR, Prog. Org. Coat. 70 (2011) 102–107.
[65] E. Andrzejewska, M.B. Bogacki, Monomolecular and bimolecular termination in the polymerization of di(meth)acrylates, Macromol. Chem. Phys. 198 (1997) 1649–1664.
[66] D.L. Kurdikar, N.A. Peppas, A kinetic model for diffusion-controlled bulk crosslinking photopolymerizations, Macromolecules 27 (1994) 4084–4092.
[67] A. Nebioglu, M.D. Soucek, Reaction kinetics and network characterization of UV-curing polyester acrylate inorganic/organic hybrids, Eur. Polym. J. 43 (2007) 3325–3336.

[68] Y. Zamani Ketek Lahijania, M. Mohseni, S. Bastani, Characterization of mechanical behavior of UV cured urethane acrylate nanocomposite films loaded with silane treated nanosilica by the aid of nanoindentation and nanoscratch experiments, Tribol. Int. 69 (2014) 10–18.
[69] P. Kardar, M. Ebrahimi, S. Bastani, Study the effect of nano-alumina particles on physical-mechanical properties of UV cured epoxy acrylate via nano-indentation, Prog. Org. Coat. 62 (2008) 321–325.
[70] O.T. Thomsen, E. Bozhevolnaya, A. Lyckegaard, Sandwich structures 7 advancing with sandwich structures and materials, in: Proceedings of the 7th International Conference on Sandwich Structures, Aalborg University, Aalborg, Denmark, 29-31 August 2005, 2005.
[71] S. Wang, Y. Zhang, W. Ren, Y. Zhang, H. Lin, Morphology, mechanical and optical properties of transparent BR/clay nanocomposites, Polym. Test. 24 (2005) 766–774.
[72] J.D. Prince, 3D printing: an industrial revolution, J. Electron. Resour. Med. Libr. 11 (2014) 39–45.
[73] R. Bogue, 3D printing: the dawn of a new era in manufacturing? Assem. Autom. 33 (2013) 307–311.
[74] M.D. Symes, P.J. Kitson, J. Yan, C.J. Richmond, G.J.T. Cooper, R.W. Bowman, T. Vilbrandt, L. Cronin, Integrated 3D-printed reactionware for chemical synthesis and analysis, Nat. Chem. 4 (2012) 349–354.
[75] N. Désilles, C. Gautrelet, L. Lecamp, P. Lebaudy, C. Bunel, Effect of UV light scattering during photopolymerization on UV spectroscopy measurements, Eur. Polym. J. 41 (2005) 1296–1303.
[76] V. Azan, L. Lecamp, P. Lebaudy, C. Bunel, Simulation of the photopolymerization gradient inside a pigmented coating: influence of TiO_2 concentration on the gradient, Prog. Org. Coat. 58 (2007) 70–75.

CHAPTER 8

Advancements in Diamond-Like Carbon Coatings

Robert Bogdanowicz
Department of Metrology and Optoelectronics, Faculty of Electronics, Telecommunications and Informatics, Gdansk University of Technology, Gdansk, Poland

Contents

1. INTRODUCTION

Diamonds are valuable not only as the most expensive jewelry material but also as the material whose optic, dielectric, mechanical, and thermal properties cannot be substituted by other substances [1]. The physical and chemical properties of diamond, resulting in its potential uses, are presented in Table 1.

Natural, monocrystalline diamond is an allotropic carbon C form with the highest density. In nature, diamonds form under a pressure above 60 GPa and at a temperature of around 1000 K. These conditions allow for creating strong covalent carbon bonds with sp^3 hybridization. Natural conditions for diamond synthesis occur ca. 160 km beneath the Earth's surface. Nowadays, diamond and diamond-like carbon (DLC) materials can be synthesized thanks to the technological processes of

- high-pressure high temperature (HPHT),
- low-pressure chemical vapor deposition (CVD),
- magnetron sputtering (MS)—physical vapor deposition (PVD),
- detonation nanodiamond (DND).

Handbook of Nanoceramic and Nanocomposite Coatings and Materials
http://dx.doi.org/10.1016/B978-0-12-799947-0.00008-0

Table 1 Physical and chemical properties of diamond

High hardness (ca. 90 GPa)
High bulk modulus (1.2×10^{12} N m^{-2})
Low compressibility (8.3×10^{-13} m^2 N^{-1})
High heat conductivity (2×10^3 W m^{-1} K^{-1})
Low thermal expansion (1×10^{-6} K)
Wide range of high optical transmittance (0.2-6 μm)
The highest speed of sound propagation (17.5 km s^{-1})
Good insulator (surface resistivity of ~10^{16} Ω cm at 20°)
Semiconductor with a wide energy gap of ~5.4 eV
Resistance to chemicals and corrosion
Biocompatibility
Low-friction coefficient

In the HPHT process, which imitates the conditions of natural diamond synthesis, small defected crystals are obtained. These crystals are mostly used as emery/sand material. Moreover, the HPHT process is conducted under diamond-stable conditions.

In the low-pressure CVD process enhanced by plasma or hot filament, it is possible to create polycrystalline thin diamond or DLC films, which are contaminated with graphite and/or amorphous phase due to the pressure and temperature inside the CVD chamber. In recent years, the DND particles have started to play a very important role. The unique features of nanodiamonds and their low-cost and high-scale production resulted in unprecedented performance in various fields. Furthermore, recent studies indicate that the pulse plasma sintering (PPS) has a growing potential as an alternative method for diamond fabrication.

This chapter will focus on plasma-based processes, including various sputtering techniques (PVD) and CVD. Figure 1 illustrates the deposition pressure and temperature for specific PVD and CVD techniques. In both aforementioned techniques, the species (atoms and molecules) exist in the vapor state, while condensation on the substrate requires a chemical reaction (CVD) or a conversion to form a solid material (PVD) [2]. The main principles of PVD and CVD deposition process are shown in Figure 2. Plasma-enhanced chemical vapor deposition (PECVD) is a process by which solid films are synthesized from a chemically reactive gas mixture. The energy needed to support the chemical reactions is supplied by high electric fields known as cold plasma or electric discharge. An electric discharge is the most flexible means of creating and sustaining low-temperature plasma.

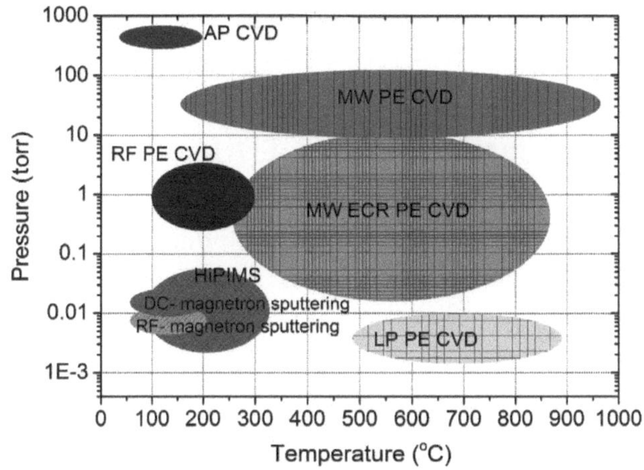

Figure 1 Diagram of deposition pressure and temperature used in the synthesis of specific diamond and diamond-like carbon thin films.

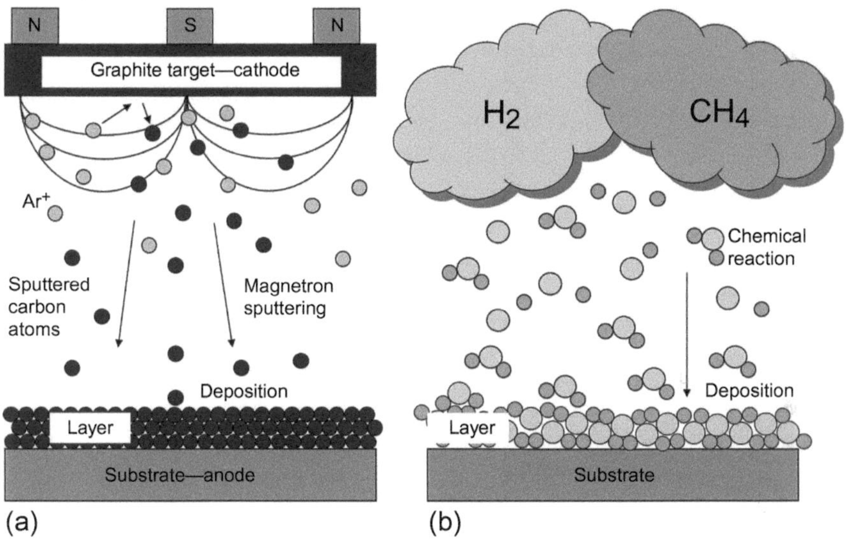

Figure 2 Principles of physical (a) and chemical (b) vapor deposition of diamond.

Both direct current (DC) and AC power sources have been developed to generate plasma. Among many, the radio frequency (RF)-PECVD at 13.56 MHz) and microwaves (MW-PECVD at 2.45 GHz) [3] are the most efficient and widespread configurations used to ionize plasma.

The other widely used methods are DC arcjet, combustion flame, MS, and laser-assisted CVD [4]. These methods have greatly reduced the unit production cost of diamond and increased the prospects for large-scale industrial applications in tribological and other fields. In fact, several commercial companies presently offer diamond-coated products at a reasonable price.

2. RADIO-FREQUENCY PLASMA-ENHANCED CVD PLASMA FOR DLC COMPOSITE SYNTHESIS

Diamond, DLC, and diamond-like nanocomposite thin films are based on amorphous carbon films. Due to their fundamental and technological importance, all these thin films have generated a great interest among academia. At present, researchers make efforts to fabricate various metal [5] or silica nanoparticle [6] composites in DLC matrix.

Amorphous DLC coatings consist of a mixture of sp^3 and sp^2 carbon structures, where sp^2-bonded graphite-like clusters are embedded in an amorphous sp^3-bonded carbon matrix [7]. They can be roughly divided into hydrogenated and hydrogen-free DLC coatings [8,9]. Both types of DLC can be metal-doped (Me-DLC), possibly forming carbide particles (MeC-DLC). It was observed that the addition of metallic elements such as Ti and W or nonmetallic elements such as N and Si decreases the compressive stress of the film and thus improves its adhesion to the substrate [10].

Dwivedi et al. [11] reported that the addition of Ag dots and Ag interlayers to DLC and nitrogen-doped DLC (DLC:N) films changes their bulk amorphous carbon material to micro- and nanostructured morphologies and enhances the sp^2 bonding. This suggests that low-threshold field emission can be achieved from DLC films by the inclusion of Ag, while N doping is sufficient to change the sp^2 phase in the films. The sp^3 content and the value of elastic modulus were reduced with increasing Ag content, possibly due to the compensation of compressive stress, which is required for sp^3 formation [12]. Another investigation [13] showed that silver-doped tetrahedral amorphous DLC (ta-C:DLC) on titanium substrate is more effective as an antibacterial agent than undoped ta-C:DLC coating, or the bare substrate.

Composite titanium or copper/DLC multilayer films could be deposited using a hybrid system combining RF sputtering and RF-PECVD techniques. Ti/hydrogenated amorphous carbon (a-C:H) films were found to have excellent nanomechanical properties with low internal stress and maximum hardness of 41.2 GPa. The recorded spectrum revealed the high optical transparency in the visible and NIR regions, which was close to over 90% despite the presence of a very thin metallic Ti layer in the multilayer structure [14].

The amorphous nature of DLC opens the possibility of incorporating elements such as Si, F, P, Ag, and N, which imparts improved properties to DLC [15]. Kim et al. [16] described the effect of Si incorporation on the wear-corrosion properties. These authors demonstrated that the aforementioned properties improve when the bonding ratio of sp^3/sp^2 is increased by raising the content of Si from 0 to 2 at.%. Maguire et al. [17] reported that Si incorporation enhances the barrier properties of DLC.

Tang and Wang [18] investigated the effects of silicon incorporation on the microstructure and nanomechanical behavior of RF-PECVD-deposited DLC films. Silicon doping suppresses the formation of sp^2 graphitic clusters and increases the sp^3-bonding content in DLC films, which is associated with a reduction in film hardness. Because C—C bond is more stable than C—Si bond as the bonding energy of C—C (610 kJ/mol) is higher than that of C—Si (451.5 kJ/mol), C—C bonds form preferentially. As a result, the film hardness decreases from 23.1 to 13.7 GPa and Si ratio increases from 0% to 19.73%.

Furthermore, SiO_x-containing diamond-like carbon (DLC:SiO_x) composites were developed in the early 1990s [19]. DLC:SiO_x films received considerable attention due to their substantially reduced stress level, increased thermal stability, high hardness, and low friction. Dielectric permittivity and refractive index of SiO_x-containing DLC are lower than the respective parameters of undoped DLC, and they are comparable with those of SiO_2, while optical transparency is higher [6]. Capacitively coupled RF-PECVD is the most frequently used technique for DLC:SiO_x deposition [20].

DLC:SiO_x films can be synthesized by using different silicon- and oxygen-containing precursors. These precursors might be gaseous (silane (SiH_4) and oxygen (O_2) mix and tetramethylsilane and O_2 mix) or liquid (hexamethyldisiloxane, tetraethyl orthosilicate, and hexamethyldisilazane precursors) [6]. The liquid precursors are more commonly used because they are easier to handle and offer a variety of choices. Raman spectroscopy was successfully used to analyze DLC:SiO_x films. Raman spectra of DLC:SiO_x films are similar to those of DLC films, that is, the main feature is a broad peak in 1100-1700 cm^{-1} range [21]. In comparison with DLC films, a characteristic shift of the G-peak to the lower wave numbers was observed in DLC:SiO_x films [22]. Moreover, the toughness of DLC:SiO_x films is higher than that of hydrogenated DLC films. However, in most cases, it is achieved at the cost of decreased hardness and lower Young's modulus. The hardness and Young's modulus of DLC:SiO_x can be increased by setting the appropriate deposition conditions (ion energy and plasma density) and by choosing the correct structure and chemical composition. Present industrial applications of DLC:SiO_x films include low-friction wear-resistant coatings for increased lifetime of car engines, plastic mold components, optical disk mold parts, and textile machinery parts. The use of these films for different lithography techniques, for MEMS fabrication, and in biomedicine is considered.

Kaczorowski et al. [23] reported the deposition of DLC coatings by dual MW-PECVD/RF-PECVD on a poly(ether ether ketone) (PEEK) substrates. Such a composite enabled a decrease in the coefficient of friction from the level of 0.4 (typical for PEEK with no modification) up to 0.26, and a ca. 35-fold decrease in the specific wear rate. DLC coatings deposited in MW/RF plasma exhibited hardness of about 2 GPa with a relatively low content of sp^3 phase and surface roughness (RMS ~4-10 nm). These coatings were specifically developed to improve the surface of the artificial cardiac valve disks.

3. GROWTH OF DIAMOND NANOSTRUCTURES VIA MW PLASMA-ASSISTED CVD

MW plasma-assisted (2.45 GHz) CVD is a powerful technique used for a wide range of diamond materials (e.g., DLC, polycrystalline diamond, and single-crystal diamond) for growth. Kamo et al. [24] pioneered the use of MW plasma to synthesize diamond films. In this work, the authors used a fused silica tube with a flow of hydrogen and 1-3% methane admixture under 60 torr pressure. The substrate heated to 1000 °C was placed at the cross point with MW waveguide.

For the activation of particles, a low-temperature plasma excited at the MW frequency of 2.45 GHz was used. Despite many years of research, the mechanism of diamond synthesis and the process control are not fully controlled. Nucleation takes place under very specific conditions as it depends on the partial pressure of gas precursors (hydrogen and methane) and the appropriate excitation of these particles in plasma [25–27].

Boron-doping technology was developed by Derjaguin et al. and confirmed by Fujimori et al. [28,29]. Boron is an excellent acceptor and it can be relatively easily incorporated into the structure of DLC and diamond. Diborane (B_2H_6) was successfully used for doping, which resulted in the process of diamond synthesis in the standard CH_4/H_2 mixture via activation process with both hot filament and MW irradiation [30]. Hartman et al.

studied in detail the influence of boron impurities on the properties of diamond layers [31]. A small addition of boron in the gas phase (<500 ppm B/C) improves the quality of the layers and increases the growth rate, while the share of boron greater than 4000 ppm causes defects in the grid and the graphitization of diamond layers [32].

Boron-doped diamond electrodes exhibit interesting electrochemical properties, which include a wide electrochemical potential window (see Figure 3), stability in aqueous and nonaqueous media, low background current density, and high resistance against chemical and electrochemical corrosion [33,34]. BDD thin films are an interesting candidate for the substrate of third-generation biosensors [35,36]. Figure 3 presents linear voltammetry curves registered in 0.5 M Na_2SO_4. It is apparent that the BDD layer, used as a working electrode, exhibits the wider voltage potential range compared to GC and Pt electrodes. In addition to electroanalytical applications, diamond electrodes have been used for treating polluted water [37] and the precise detection of contaminants in the natural environment [38], food [39], and pharmaceutical industry [40]. Most electrochemical studies of diamond have been performed on highly boron-doped p-type material in aqueous solvents.

Liao et al. [41] investigated the influence of boron concentration on the structure of diamond thin films. The carrier density ranged from 3.4×10^{17} to 1.8×10^{21} holes/cm^{-3}. The average grain size of the film decreased from 5 to 1 μm with increasing boron concentration. Boron introduces

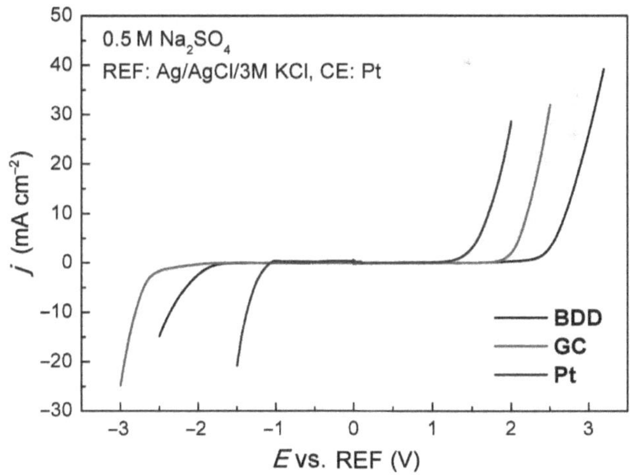

Figure 3 Linear voltammetry curves recorded for boron-doped diamond, glassy carbon (GC), and platinum working electrodes immersed in 0.5 M Na_2SO_4.

Figure 4 SEM microimages of BDD films deposited with different [B]/[C] ratios in plasma, namely, 1000, 2000, and 10,000 ppm. Magnification of 10,000 ×. SEM image of undoped CVD diamond is presented for comparative purposes.

renucleation, which results in the creation of smaller crystallites on the higher primary diamond crystals.

Si/BDD electrodes were synthesized in an MW-plasma-assisted chemical vapor deposition (PACVD) system on p-type Si wafers with (111) orientation (see Figure 4). The substrates were seeded by sonication in nanodiamond suspension (crystallite size of 5-10 nm) for 2 h [42–44]. The substrate temperature was kept at 1000 °C during the deposition process. Excited plasma was ignited by MW radiation (2.45 GHz) [45,46]. The plasma MW power, optimized for diamond synthesis, was kept at 1300 W. The gas mixture ratio was 1% of the molar ratio of CH_4/H_2 at gas volume of 300 sccm of total flow rate. The base pressure was about 10^{-6} torr and the process pressure was kept at 50 torr. All samples were doped by using diborane (B_2H_6) dopant precursor. [B]/[C] ratios in plasma were 1000, 2000, and 10,000 ppm. The time of polycrystalline layer growth was 6 h, which provided the thickness of deposited films of approx. 2 μm [46,47].

Wang et al. [48] examined the internal stress and strain in heavily doped BDDs. The source of boron was trimethylboron ($B(CH_3)_3$) and the volume ratio in the reaction gas mixture varied between 0% and 4.5%. The total internal stress increased with increasing concentration of dopant. Feng et al. [49] investigated, *inter alia*, electrocatalytic performance in phenol solution. With increasing level of boron in a BDD electrode, the electrolysis time decreased and the anodic peak shifted to a lower potential range. However, the properties of the electrode kept improving up to a certain level of dopant; for heavily doped electrodes, these properties deteriorated.

Moreover, information about electrochemical response of BDD electrodes with different boron-doping levels can be found in literature. Salazar-Banda et al. [50] examined the changes in electrochemical behavior of boron-doped diamond surfaces in relation to time after cathodic pretreatment in the electrodes containing different boron levels. However, the study focused on low-doped electrodes.

Currently, there is a tremendous interest in fabricating nanostructured diamond composites due to their extraordinary mechanical, electrical, and optical properties as predicted by theory for quasi one-dimensional sp^3 nanostructures [51,52]. The investigated three-dimensional diamond shapes include nanowire, nanorod [53–55], teepee-shaped [56], various carbon nanowall [57–59] (see structure of samples in Figure 5), porous [60], and nanostructured honeycomb [61] structures. Unfortunately, synthesizing or fabricating these nanostructures experimentally proves to be very challenging. To date, only several attempts have been reported, using a top-down approach, either by etching diamond using e-beam lithography or by reactive ion etching (RIE) [54,62–64]. Krauss et al. [65] deposited a diamond film on silicon tips. Krauss et al. [66] exploited RIE to etch diamond films and form diamond tips. Ternyak et al. [67] chose a photolithography process to produce a silicon mold of concave pyramidal shape and then deposited diamond onto the concave shape to form diamond tips after removing the silicon. Additionally, nanostructured diamond honeycomb films have been prepared by etching through a porous anodic alumina mask [61]. Zou et al. [68] reported the fabrication of nanopillar arrays by using self-aligned Au nanodots as etching mask in a bias-assisted RIE and by applying a hydrogen-argon plasma.

However, it is difficult to produce horizontally aligned diamond structures with relatively small width and large length without any residual stress. These parameters are essential for better performance in nanoelectronic devices due to high surface-to-volume ratio. In addition, the RIE process often raises concerns in relation to diamond fabrication since possible diamond degradation and the etching mask residues can occur afterward.

The bottom-up approach using a CVD method was reported by Zou et al. [56]. The authors described a method for fabricating diamond-coated

Figure 5 Microimages of carbon nanowall composite structures deposited in microwave plasma on p-Si substrates. Magnification of 5000× and 25,000×.

carbon nanotubes (CNTs), which results in tepee-shaped structures that are ideal for FE applications. The dense arrays of vertically aligned multiwalled carbon nanotubes (MWCNTs) were seeded with a nanodiamond suspension in methanol by using electrospray deposition. This treatment caused the tips consisting of groups of 20-40 MWCNTs to stick together forming structures resembling "tepees." The before-growth formation of diamond seed patterns is one of the solutions that does not require any aftergrowth structuring [69,70]. Adopting the strategy of before-growth structuring followed by CVD is often called selected-area deposition (SAD). Direct distribution of diamond seeds by ink-jet printing has also been applied as an alternative technique [71]. The most popular SAD technique is based on the removal of diamond nuclei (seeds) from the predefined areas [72,73]. Furthermore, with a selective seeding technique [62], it is now possible to selectively grow ultrananocrystalline diamond nanowires patterned without the use of RIE on the deposited ultrananocrystalline diamond.

Diamond films made by CVD are probably the most promising, owing to their negative electron affinity surfaces with low effective work functions [57,66,74]. The field emission properties of diamond can be enhanced by reducing the grain size [66], doping it with boron to increase its conductivity [75], and/or fabricating the diamond into sharp tips to increase the field enhancement factor [56].

4. APPLICATION OF HIPIMS FOR DEPOSITING DIAMOND AND METAL-DIAMOND COMPOSITES

The MS is a popular and well-known technique for the deposition of metal and metal oxide films. In the MS technique, atoms or molecules are sputtered from the surface of a target made from appropriate material by bombarding it with high-energy ions in a vacuum. The magnetic field is used to confine electrons near the target to sustain plasma. Mainly argon is used as an inert gas. The sputtering process can be generated in either the DC or RF (13.56 MHz) modes [76]. The films are deposited at low substrate temperatures up to 300 °C [77].

MS is probably the most commonly used industrial method for depositing DLC films. In the unbalanced configuration, the substrate is bombarded with Ar ions, thus increasing the sp^3 content of the films. Ar-H mixtures allow the production of hydrogenated DLC films. With the mass-selected ion-beam technique, it is possible to obtain ta-C films that contain up to 80% sp^3-hybridized C by varying the acceleration, mass, and energy selection of a C ion beam that is condensing onto the substrate. Recently, the novel pulsed techniques, such as high-power impulse magnetron sputtering (HiPIMS), have been developed [78].

By loading the voltage in pulses, researchers generated higher voltage on the electrodes and achieved higher current per square centimeter than with standard MS [79,80]. Such a design results in a meaningful film densification and the avoidance of metal target poisoning by oxygen [81]. These advantageous features have been applied to increase the density and hardness of DLC coatings as well. The sputtering sources are mounted into the vacuum chamber (see Figure 6a). The vacuum chamber can be equipped with several sputtering sources, for example, for the deposition of multistructural film, composites, or alloys. Stranak et al. [82–84] proposed to charge the capacitors from the DC source during the idle part of the pulses. The stored energy is released in pulses activated by a handmade power switch based on insulated-gate bipolar transistors. The high-power switch transistors are controlled by an external multichannel pulse-delay generator. Most of the other scientific groups use a combination of DC power supply loading a large capacitor bank of a pulsing unit directly connected to the magnetron cathode [78,85,86]. The typical time evolution of peak discharge currents and voltage waveforms in HiPIMS mode is presented in Figure 6b.

Such an approach produces higher ionization of the sputtering plasma, together with higher amounts of activated radicals [87]. This generally produces denser amorphous coating materials compared to the conventional nonpulsed, less dense, less ionized, and dissociated plasma coating processes

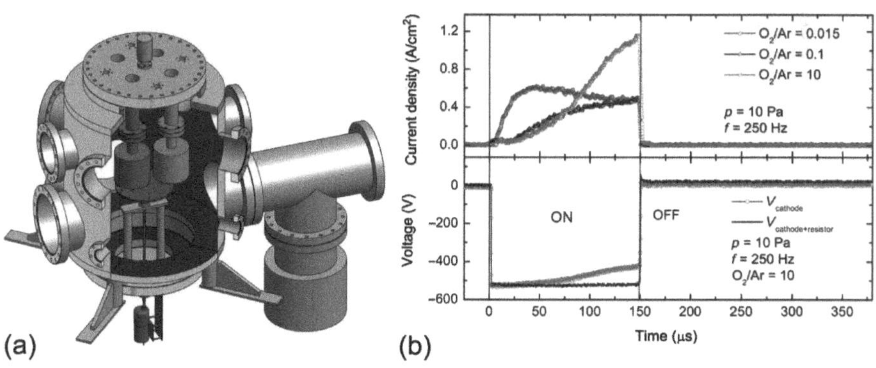

Figure 6 Schematic of sputtering sources (red- and green-colored parts) mounted into the vacuum chamber (a). Time evolution of peak discharge currents (upper panel) and voltages V_{C+R} and V_c (b).

(PVD and PACVD). HiPIMS produces coatings with lower stress as a consequence of the period where the plasma power is off and the condensing activated radicals and neutrals can relax; the stress produced during the period the power is on in the preceding working phase produces an intense bombardment of ions, activated particles, and neutrals [82]. No matter whether harder and better quality amorphous DLC material with good adherence (because of the lower stress) is produced, the hardness remains limited (30 GPa) compared to previous filtered arc devices (when the process is adjusted correctly). This is caused by the lower ion flux to neutral flux ratio.

Aijaz et al. [88] implemented a sputtering-based synthesis route for the synthesis of hydrogenated amorphous carbon (a-C:H) thin films that entails coupling a hydrocarbon gas (acetylene) with high-density discharges generated by the superposition of HiPIMS and direct current magnetron sputtering (DCMS). The authors achieved conditions capable of providing a 10-fold increase in the deposition rate of a-C:H film growth compared to HiPIMS Ar discharge [89]. The deposition of DLC films by using Ar in HiPIMS results in the growth of amorphous carbon films with sp^3 bond fraction up to 45% and density up to 2.2 g cm^{-3}. Nevertheless, the use of Ar–C_2H_2 atmosphere gives mass densities of a-C:H films in the order of 2.32 g cm^{-3}, with the hardness exceeding 25 GPa and H content lower than 10 at.%. Aijaz et al. [90] have also presented an approach of increasing carbon ionization in a single-pulse HiPIMS discharge by using Ne as the sputtering gas. An increase in the degree of carbon ionization can be achieved in a HiPIMS discharge (via electron temperature enhancement) by using Ne, a gas with higher ionization energy than Ar. The direct consequence of the ionization enhancement on carbon growth was demonstrated by the resulting film density, which was found to reach 2.8 g cm^{-3} [91].

Recent work presented by Lin et al. [92] showed an alternative HiPIMS technique that uses large voltage oscillation packets to achieve high-power pulses for sputtering, which results in the smoother surface of DLC films. In comparison with the outcome of the study by Aijaz et al. [90], the films exhibited the maximum hardness value of 35 GPa and the lowest dry sliding coefficient of friction of 0.09.

The metal carbide TiC/a-C:H nanocomposite could be reactively deposited by employing HiPIMS and DCMS [93] in an Ar-C_2H_2 atmosphere of various compositions. The growth of TiC and hydrogenated amorphous carbon (a-C:H) nanocomposites, nanocrystalline TiC, or Ti/TiC depends strongly on the C/Ti ratio in the plasma. Samuelsson et al. [93]

found that TiC films grown by HiPIMS show the C/Ti ratio close to 1 for a wide C_2H_2 flow range (4-15 sccm), with free C ranging from 0% to 20%. Thus, the deposited films exhibit near stoichiometric single-phase TiC to TiC/a-C:H structure.

5. PULSED PLASMA SINTERING AND DETONATION DIAMOND NANOCOMPOSITES

PPS, a method developed in the 1960s and refined by Japanese scientists at the end of the twentieth century, made it possible to fabricate composite materials with novel characteristics in an approach based on passing high-power currents through the sample. This approach is employed primarily in the fabrication of ceramic materials and composites of refractory alloys for cutting tools and corrosion- and wear-resistant coatings. The PPS technique was found to be instrumental in sintering composites based on ceramic/metal and polymer/metal combinations and in the development of tools fabricated by sintering cobalt and bronze compounds, crushed and bulk rocks, cutting blades, etc. A number of research groups try to achieve mechanically stable and thermally efficient diamond composites via sintering DND with various metals. DND, often also called ultradispersed diamond, is a diamond that originates from a detonation. When an oxygen-deficient explosive mixture of TNT/RDX is detonated in a closed chamber, diamond particles with a diameter of ca. 5 nm are formed at the front of detonation wave in the span of several microseconds [94].

Ritasalo et al. [95] reported the dispersion strengthening of copper by nano- and submicron-sized diamonds (DND and SMD). Mechanical alloying in an argon atmosphere by high-energy vibratory ball mill resulted in composite powders with diamond particles strongly adhering onto the surfaces of Cu particles. The pulsed electric current-sintered Cu composite had a high density (>97% of T.D.), with quite evenly distributed diamonds, and a higher hardness than that of pure submicron-sized copper.

Senyut and Mosunov [96] studied the dependence of the microhardness of composites made from DND and DND modified by cobalt and titanium on sintering temperature at a pressure of 7 GPa. The authors observed that at 1300 °C, the density and microhardness of DND samples were 2.73 g cm^{-3} and 20 GPa, respectively, while at 1600-1800 °C, the values of these parameters reached 2.85-2.90 g cm^{-3} and 30-35 GPa, respectively. Shulzhenko et al. [97] investigated the dependence of the strength and hardness of polycrystalline DND on sintering conditions (i.e., 1150-1600 °C at 8 GPa) and the prior surface modification of DND particles. The authors found that heat pretreatment in a vacuum prior to sintering at $T = 500$ °C efficiently removes the gases adsorbed on DND surface. Due to this treatment, the density of composites increased up to 2.86 g cm^{-3}, while their strength reached as high as 28 GPa. Kidalov et al. [98] measured the thermal conductivity of DND sintered at a high pressure (6.5 GPa) and found it to range from 5 W/(m K) to 50 W/(m K), depending on the actual sintering temperature.

In order to manage the high-power density dissipation in microelectronic devices and thermal management problems, new materials of high-thermal conductivity are needed. They should also be characterized by the thermal expansion coefficient fitting microelectronic devices. These requirements can be satisfied by the diamond/metal composites with a metal matrix of high-thermal conductivity, such as Cu. Cu/diamond composites have been considered as a next generation of thermal management material for electronic packages and heat sink applications. The major challenge in the development of these composites is to obtain a well-bonded interface between copper and diamond.

Recently, Rosinski et al. [99] studied copper/diamond composites with 50 vol.% of diamond particles under conditions of thermodynamic instability of diamond by consolidating the composite using the PPS method. The process temperature was kept at 900 °C, the pressure was 80 MPa, and the sintering time was 10 min. Cu/diamond PPS consolidated composites had a theoretical density of 96% and the diamond particles were distributed uniformly within the copper matrix.

Schubert et al. [100] produced Cu-diamond composites (copper matrix with the carbide-forming element chromium) with the density close to theoretical value and a thermal conductivity of 600 W/mK. Without alloying, a low thermal conductivity of the composite (200 W/mK) was measured, indicating a high-thermal boundary resistance. The improvements in thermophysical properties of the composites were achieved by using copper alloy with chromium to increase the interfacial bonding in Cu/diamond composites by a thin nanosized Cr_3C_2 layer. The microscopic observations (see Figure 7a) revealed the nanocrystalline structure of sintered composite [101].

Also, experiments performed by Michalski and Rosinski [102] have demonstrated that PPS method might be a prospective technology for manufacturing WCCo/diamond sinters on cemented carbide substrates without the graphitization of diamond. Figure 7b presents the SEM image

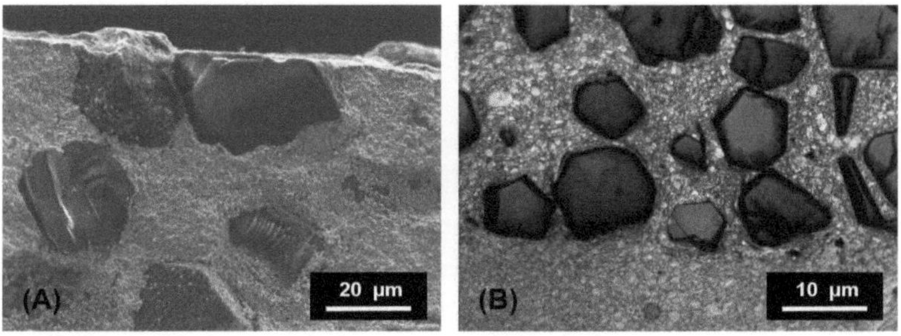

Figure 7 SEM microimages of the fracture surface in WCCo/diamond composite (a); Cu/diamond composite (b) [101].

of the fracture surface in WCCo/diamond composite [101]. No graphite precipitates can be seen around the diamond particle, while the particle itself remains in a very good contact with WCCo matrix.

Moreover, Shen et al. [103] and Zhang et al. [104] demonstrated that MWCNTs transform into micrometer-sized diamonds at 1500 °C and only 80 MPa during SPS in the absence of any additional catalyst. The resulting diamonds are single or agglomerated crystalline particles, with diameters up to 100 μm, and are sheathed with an amorphous carbon layer and a residue of a few CNT layers. Spark plasmas are believed to play a key role in the transition of MWCNTs to diamond; the mechanism involves the breakage of some C—C bonds, the formation of carbon nano-onions, and the nucleation and growth of the diamond phase within the onion cores [105].

Detonation-synthesis nanodiamonds were first synthesized by researchers from the former USSR in 1963 by explosive decomposition of high-explosive mixtures with negative oxygen balance in a nonoxidizing medium [106]. Recently, the DND particles started to play a very important role in biomedicine, thermal management in electronics, energy storage, field emission displays, and other advanced applications [107]. Superior mechanical properties, rich surface chemistry, and good biocompatibility of DND nanoparticles make them attractive in biomaterial applications.

Zhang et al. [108] investigated a composite bone scaffold material produced by utilizing a biodegradable polymer, poly(L-lactic acid) (PLLA), and octadecylamine-functionalized nanodiamond (ND-ODA). The mechanical properties of ND-ODA/PLLA composites were improved dramatically with the addition of ND-ODA, that is, 1 wt% of ND-ODA increased the hardness of the composites by the factor of 4 and 10 wt% of ND-ODA resulted in almost an order of magnitude higher hardness, three times higher

Young's modulus, and reduced creep. Cytotoxicity and biocompatibility experiments demonstrated that both ND and ND-ODA are nontoxic to osteoblasts and support cell proliferation and differentiation *in vitro*. Increased mechanical properties and enhanced biomineralization of ND-ODA/PLLA composites make them a promising material for bone tissue engineering, bone surgical fixation devices, and regenerative medicine [109].

Hydroxyapatite/DND composites were studied by Pramatarova et al. [110] on silica glass. The composite growth was carried out through the addition of DND particles to supersaturated simulated body fluid (SBF) solutions. The applied methods provide evidence that the functional groups on nanodiamond surface interact strongly with the biological solutions used. The DND surface is chemically multifunctional (surface OH, CO—H, C=C, C—O—C, and C=O groups exist), so that hydroxyapatite is grown by both physical adsorption and chemical interaction. The OH^- groups are regarded to play an important role in the growth of hydroxyapatite on the diamond surface in SBF medium, as their negative charge attracts Ca^{2+} ions, which in turn attract PO_4^{3-} ions, and thus, apatite nuclei are formed.

Neitzel et al. [111] studied hot-pressed epoxy-ND composites with ND content, which displayed an up to 470% increase in Young's modulus and up to 300% increase in hardness compared to neat epoxy. These high values have never been achieved before for any other epoxy with a nanoparticle filler and could be potentially further improved by increasing the packing density of ND. Scratch and sliding tests showed increased scratch resistance, with 35% less removed material, a 48% reduction of piled up material, and a 40% decrease in the coefficient of friction for 25 vol.% ND composites in comparison with the neat epoxy samples. The authors also demonstrated that the thermal conductivity of high concentration epoxy-ND composites improved, thus adding more value to this material. Due to the small size of only 5 nm (see Figure 8), the distance between ND particles is significantly smaller (particle density is higher) at the same concentration when compared to other nanofillers, such as nanosilica with a particle size of 10 nm.

The nanodiamond is an attractive material for the investigation of relations in the world of carbon nanomaterials and a candidate for valuable technological applications. It can be produced in bulk quantities, can be functionalized noncovalently and covalently, and, so far, did not show any biohazardous effects. This makes it a prospective material for a variety of biomedical applications, such as labeling or targeted drug delivery. On the

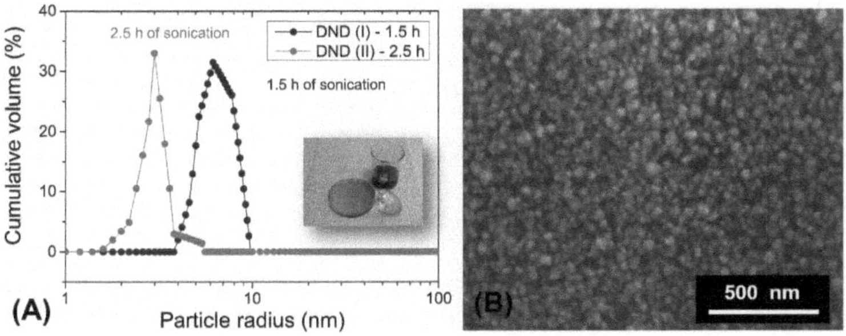

Figure 8 Grain size distribution of deagglomerated DND suspension (a); SEM microimage of DND-polymer (PVA) composite (b). Inset shows the photograph of nanodiamond suspension.

other hand, its mechanical and electronic properties are attractive for the production of various composite materials, with either polymer or metallic matrices [112].

REFERENCES

[1] J.E. Field, Properties of Natural and Synthetic Diamond, first ed., Academic Press, London, San Diego, 1992, 710 p.
[2] K. Seshan, Handbook of Thin Film Deposition Techniques, Processes, and Technologies, Elsevier, Amsterdam, 2012.
[3] S.J. Harris, A.M. Weiner, T.A. Perry, Measurement of stable species present during filament-assisted diamond growth, Appl. Phys. Lett. 53 (17) (1988) 1605–1607.
[4] J. Asmussen, D. Reinhard, Diamond Films Handbook, first ed., CRC Press, New York, 2002, 656 p.
[5] J.C. Sánchez-López, A. Fernández, Doping and alloying effects on DLC coatings, in: C. Donnet, A. Erdemir (Eds.), Tribology of Diamond-Like Carbon Films [Internet], Springer, USA, 2008, pp. 311–338, Available from: http://link.springer.com/chapter/10.1007/978-0-387-49891-1_12 [cited 2014 Aug 25].
[6] Š. MešKinis, A. Tamulevičlenė, Structure, properties and applications of diamond like nanocomposite (SiOx containing DLC) films: a review, Mater. Sci. 17 (4) (2011), Available from: http://matsc.ktu.lt/index.php/MatSc/article/view/770 [cited 2014 Aug 23].
[7] E.G. Spencer, P.H. Schmidt, D.C. Joy, F.J. Sansalone, Ion-beam-deposited polycrystalline diamondlike films, Appl. Phys. Lett. 29 (2) (1976) 118–120.
[8] Y. Lifshitz, Diamond-like carbon—present status, Diam. Relat. Mater. 8 (8–9) (1999) 1659–1676.
[9] J. Robertson, Diamond-like amorphous carbon, Mater. Sci. Eng. R. Rep. 37 (4-6) (2002) 129–281.
[10] D. Galvan, Y.T. Pei, J.T.M. De Hosson, Influence of deposition parameters on the structure and mechanical properties of nanocomposite coatings, Surf. Coat. Technol. 201 (3-4) (2006) 590–598.
[11] N. Dwivedi, S. Kumar, J.D. Carey, R.K. Tripathi, H.K. Malik, M.K. Dalai, Influence of silver incorporation on the structural and electrical properties of diamond-like carbon thin films, ACS Appl. Mater. Interfaces 5 (7) (2013) 2725–2732.

[12] D. Batory, M.C. Reczulska, L. Kolodziejczyk, W. Szymanski, Gradient titanium and silver based carbon coatings deposited on AISI316L, Appl. Surf. Sci. 275 (2013) 303–310.
[13] N.M. Chekan, N.M. Beliauski, V.V. Akulich, L.V. Pozdniak, E.K. Sergeeva, A.N. Chernov, et al., Biological activity of silver-doped DLC films, Diam. Relat. Mater. 18 (5-8) (2009) 1006–1009.
[14] N. Dwivedi, S. Kumar, H.K. Malik, Nanostructured titanium/diamond-like carbon multilayer films: deposition, characterization, and applications, ACS Appl. Mater. Interfaces 3 (11) (2011) 4268–4278.
[15] R. Asakawa, S. Nagashima, Y. Nakamura, T. Hasebe, T. Suzuki, A. Hotta, Combining polymers with diamond-like carbon (DLC) for highly functionalized materials, Surf. Coat. Technol. 206 (4) (2011) 676–685.
[16] H.-G. Kim, S.-H. Ahn, J.-G. Kim, S.J. Park, K.-R. Lee, Effect of Si-incorporation on wear-corrosion properties of diamond-like carbon films, Thin Solid Films 482 (1-2) (2005) 299–304.
[17] P.D. Maguire, J.A. McLaughlin, T.I.T. Okpalugo, P. Lemoine, P. Papakonstantinou, E.T. McAdams, et al., Mechanical stability, corrosion performance and bioresponse of amorphous diamond-like carbon for medical stents and guidewires, Diam. Relat. Mater. 14 (8) (2005) 1277–1288.
[18] J. Tang, Y. Wang, Characterization of DLC-Si films prepared by RF-PECVD, in: 2012 International Conference on Optoelectronics and Microelectronics (ICOM), 2012, , pp. 431–433.
[19] V.F. Dorfman, Diamond-like nanocomposites (DLN), Thin Solid Films 212 (1-2) (1992) 267–273.
[20] V. Buršíková, V. Navrátil, L. Zajíčková, J. Janča, Temperature dependence of mechanical properties of DLC/Si protective coatings prepared by PECVD, Mater. Sci. Eng. A 324 (1-2) (2002) 251–254.
[21] W.J. Yang, Y.-H. Choa, T. Sekino, K.B. Shim, K. Niihara, K.H. Auh, Structural characteristics of diamond-like nanocomposite films grown by PECVD, Mater. Lett. 57 (21) (2003) 3305–3310.
[22] M. Veres, M. Koós, S. Tóth, M. Füle, I. Pócsik, A. Tóth, et al., Characterisation of a-C:H and oxygen-containing Si:C:H films by Raman spectroscopy and XPS, Diam. Relat. Mater. 14 (3-7) (2005) 1051–1056.
[23] W. Kaczorowski, W. Szymanski, D. Batory, P. Niedzielski, Tribological properties and characterization of diamond like carbon coatings deposited by MW/RF and RF plasma-enhanced CVD method on poly(ether-ether-ketone), Plasma Process. Polym. 11 (9) (2014) 878–887.
[24] M. Kamo, Y. Sato, S. Matsumoto, N. Setaka, Diamond synthesis from gas phase in microwave plasma, J. Cryst. Growth 62 (3) (1983) 642–644.
[25] P.W. May, Diamond thin films: a 21st-century material, Philos. Trans. R. Soc. Lond. Ser. Math. Phys. Eng. Sci. 358 (1766) (2000) 473–495.
[26] J. Robertson, Comparison of diamond-like carbon to diamond for applications, Phys. Status Solidi A 205 (9) (2008) 2233–2244.
[27] R.S. Sussmann, CVD Diamond for Electronic Devices and Sensors, John Wiley and Sons, London, UK, 2009, 603 p.
[28] H. Shiomi, Y. Nishibayashi, N. Fujimori, Characterization of boron-doped diamond epitaxial films, Jpn. J. Appl. Phys. 30 (Part 1, No. 7) (1991) 1363–1366.
[29] B.V. Spitsyn, L.L. Bouilov, B.V. Derjaguin, Vapor growth of diamond on diamond and other surfaces, J. Cryst. Growth 52 (Part 1(0)) (1981) 219–226.
[30] W. Gajewski, P. Achatz, O.A. Williams, K. Haenen, E. Bustarret, M. Stutzmann, et al., Electronic and optical properties of boron-doped nanocrystalline diamond films, Phys. Rev. B 79 (4) (2009) 045206.

[31] P. Wurzinger, P. Pongratz, P. Hartmann, R. Haubner, B. Lux, Investigation of the boron incorporation in polycrystalline CVD diamond films by TEM, EELS and Raman spectroscopy, Diam. Relat. Mater. 6 (5-7) (1997) 763–768.
[32] J. Zhang, J.W. Zimmer, R.T. Howe, R. Maboudian, Characterization of boron-doped micro- and nanocrystalline diamond films deposited by wafer-scale hot filament chemical vapor deposition for MEMS applications, Diam. Relat. Mater. 17 (1) (2008) 23–28.
[33] E. Brillas, C.A.M. Huitle, Synthetic Diamond Films: Preparation, Electrochemistry, Characterization and Applications, John Wiley & Sons, Hoboken, NJ, 2011, 679 p.
[34] R.G. Compton, Electroanalysis at diamond-like and doped-diamond electrodes, Electroanalysis 15 (17) (2003) 1349–1363.
[35] C.E. Nebel, N. Yang, H. Uetsuka, E. Osawa, N. Tokuda, O. Williams, Diamond nano-wires, a new approach towards next generation electrochemical gene sensor platforms, Diam. Relat. Mater. 18 (5-8) (2009) 910–917.
[36] A. Qureshi, W.P. Kang, J.L. Davidson, Y. Gurbuz, Review on carbon-derived, solid-state, micro and nano sensors for electrochemical sensing applications, Diam. Relat. Mater. 18 (12) (2009) 1401–1420.
[37] N. Spataru, T.N. Rao, D.A. Tryk, A. Fujishima, Determination of nitrite and nitrogen oxides by anodic voltammetry at conductive diamond electrodes, J. Electrochem. Soc. 148 (3) (2001) E112–E117.
[38] J.K. Zak, J.E. Butler, G.M. Swain, Diamond optically transparent electrodes: demonstration of concept with ferri/ferrocyanide and methyl viologen, Anal. Chem. 73 (5) (2001) 908–914.
[39] A. Manivannan, M.S. Seehra, D.A. Tryk, A. Fujishima, Electrochemical detection of ionic mercury at boron-doped diamond electrodes, Anal. Lett. 35 (2) (2002) 355–368.
[40] E. Popa, H. Notsu, T. Miwa, D.A. Tryk, A. Fujishima, Selective electrochemical detection of dopamine in the presence of ascorbic acid at anodized diamond thin film electrodes, Electrochem. Solid-State Lett. 2 (1) (1999) 49–51.
[41] X.Z. Liao, R.J. Zhang, C.S. Lee, S.T. Lee, Y.W. Lam, The influence of boron doping on the structure and characteristics of diamond thin films, Diam. Relat. Mater. 6 (2-4) (1997) 521–525.
[42] R. Bogdanowicz, M. Śmietana, M. Gnyba, M. Ficek, V. Straňák, Ł. Goluński, et al., Nucleation and growth of CVD diamond on fused silica optical fibres with titanium dioxide interlayer, Phys. Status Solidi A 210 (10) (2013) 1991–1997.
[43] R. Bogdanowicz, M. Śmietana, M. Gnyba, Ł. Gołunski, J. Ryl, M. Gardas, Optical and structural properties of polycrystalline CVD diamond films grown on fused silica optical fibres pre-treated by high-power sonication seeding, Appl. Phys. A 116 (4) (2014) 1927–1937.
[44] A. Fabiańska, T. Ossowski, R. Bogdanowicz, J. Czupryniak, M. Gnyba, T. Odzga, et al., Electrochemical oxidation of ionic liquids at highly boron doped diamond electrodes, Phys. Status Solidi A 209 (9) (2012) 1797–1803.
[45] A. Fabiańska, R. Bogdanowicz, P. Zięba, T. Ossowski, M. Gnyba, J. Ryl, et al., Electrochemical oxidation of sulphamerazine at boron-doped diamond electrodes: influence of boron concentration, Phys. Status Solidi A 210 (10) (2013) 2040–2047.
[46] J. Czupryniak, A. Fabiańska, P. Stepnowski, T. Ossowski, R. Bogdanowicz, M. Gnyba, et al., Application of BDD thin film electrode for electrochemical decomposition of heterogeneous aromatic compounds, Cent. Eur. J. Phys. 10 (5) (2012) 1183–1189.
[47] R. Bogdanowicz, A. Fabiańska, L. Golunski, M. Sobaszek, M. Gnyba, J. Ryl, et al., Influence of the boron doping level on the electrochemical oxidation of the azo dyes at Si/BDD thin film electrodes, Diam. Relat. Mater. 39 (2013) 82–88.
[48] W.L. Wang, M.C. Polo, G. Sánchez, J. Cifre, J. Esteve, Internal stress and strain in heavily boron-doped diamond films grown by microwave plasma and hot filament chemical vapor deposition, J. Appl. Phys. 80 (3) (1996) 1846–1850.

[49] Y. Feng, J. Lv, J. Liu, N. Gao, H. Peng, Y. Chen, Influence of boron concentration on growth characteristic and electro-catalytic performance of boron-doped diamond electrodes prepared by direct current plasma chemical vapor deposition, Appl. Surf. Sci. 257 (8) (2011) 3433–3439.
[50] G.R. Salazar-Banda, L.S. Andrade, P.A.P. Nascente, P.S. Pizani, R.C. Rocha-Filho, L.A. Avaca, On the changing electrochemical behaviour of boron-doped diamond surfaces with time after cathodic pre-treatments, Electrochim. Acta 51 (22) (2006) 4612–4619.
[51] A.S. Barnard, S.P. Russo, I.K. Snook, Ab initio modeling of diamond nanowire structures, Nano Lett. 3 (10) (2003) 1323–1328.
[52] O. Shenderova, D. Brenner, R.S. Ruoff, Would diamond nanorods be stronger than fullerene nanotubes? Nano Lett. 3 (6) (2003) 805–809.
[53] N. Yang, H. Uetsuka, E. Osawa, C.E. Nebel, Vertically aligned diamond nanowires for DNA sensing, Angew. Chem. Int. Ed. 47 (28) (2008) 5183–5185.
[54] B.J.M. Hausmann, M. Khan, Y. Zhang, T.M. Babinec, K. Martinick, M. McCutcheon, et al., Fabrication of diamond nanowires for quantum information processing applications, Diam. Relat. Mater. 19 (5-6) (2010) 621–629.
[55] M. Karlsson, P. Forsberg, F. Nikolajeff, From hydrophilic to superhydrophobic: fabrication of micrometer-sized nail-head-shaped pillars in diamond, Langmuir 26 (2) (2010) 889–893.
[56] Y. Zou, P.W. May, S.M.C. Vieira, N.A. Fox, Field emission from diamond-coated multiwalled carbon nanotube "teepee" structures, J. Appl. Phys. 112 (4) (2012) 044903.
[57] C.Y. Cheng, M. Nakashima, K. Teii, Low threshold field emission from nanocrystalline diamond/carbon nanowall composite films, Diam. Relat. Mater. 27-28 (2012) 40–44.
[58] C. Terashima, K. Arihara, S. Okazaki, T. Shichi, D.A. Tryk, T. Shirafuji, et al., Fabrication of vertically aligned diamond whiskers from highly boron-doped diamond by oxygen plasma etching, ACS Appl. Mater. Interfaces 3 (2) (2011) 177–182.
[59] A.T.H. Chuang, B.O. Boskovic, J. Robertson, Freestanding carbon nanowalls by microwave plasma-enhanced chemical vapour deposition, Diam. Relat. Mater. 15 (4-8) (2006) 1103–1106.
[60] N.A. Braga, C.A.A. Cairo, M.R. Baldan, V.J. Trava-Airoldi, N.G. Ferreira, Optimal parameters to produce high quality diamond films on 3D porous titanium substrates, Diam. Relat. Mater. 20 (1) (2011) 31–35.
[61] H. Masuda, M. Watanabe, K. Yasui, D. Tryk, T. Rao, A. Fujishima, Fabrication of a nanostructured diamond honeycomb film, Adv. Mater. 12 (6) (2000) 444–447.
[62] X. Wang, L.E. Ocola, R.S. Divan, A.V. Sumant, Nanopatterning of ultrananocrystalline diamond nanowires, Nanotechnology 23 (7) (2012) 075301.
[63] M. Larsson, K.N. Dinyari, H. Wang, Composite optical microcavity of diamond nanopillar and silica microsphere, Nano Lett. 9 (4) (2009) 1447–1450.
[64] K.A. Lister, B.G. Casey, P.S. Dobson, S. Thoms, D.S. Macintyre, C.D. W. Wilkinson, et al., Pattern transfer of a 23 nm-period grating and sub-15 nm dots into CVD diamond, Microelectron. Eng. 73-74 (2004) 319–322.
[65] A.R. Krauss, O. Auciello, D.M. Gruen, A. Jayatissa, A. Sumant, J. Tucek, et al., Ultrananocrystalline diamond thin films for MEMS and moving mechanical assembly devices, Diam. Relat. Mater. 10 (11) (2001) 1952–1961.
[66] A.R. Krauss, O. Auciello, M.Q. Ding, D.M. Gruen, Y. Huang, V.V. Zhirnov, et al., Electron field emission for ultrananocrystalline diamond films, J. Appl. Phys. 89 (5) (2001) 2958–2967.
[67] O. Ternyak, R. Akhvlediani, A. Hoffman, Evolution and properties of adherent diamond films with ultra high nucleation density deposited onto alumina, Diam. Relat. Mater. 14 (2) (2005) 144–154.

[68] Y.S. Zou, Y. Yang, W.J. Zhang, Y.M. Chong, B. He, I. Bello, et al., Fabrication of diamond nanopillars and their arrays, Appl. Phys. Lett. 92 (5) (2008) 053105.
[69] O. Babchenko, E. Verveniotis, K. Hruska, M. Ledinsky, A. Kromka, B. Rezek, Direct growth of sub-micron diamond structures, Vacuum 86 (6) (2012) 693–695.
[70] Y. Fu, H. Du, J. Miao, Patterning of diamond microstructures on Si substrate by bulk and surface micromachining, J. Mater. Process. Technol. 132 (1-3) (2003) 73–81.
[71] N.A. Fox, M.J. Youh, J.W. Steeds, W.N. Wang, Patterned diamond particle films, J. Appl. Phys. 87 (11) (2000) 8187–8191.
[72] K. Hirabayashi, Y. Taniguchi, O. Takamatsu, T. Ikeda, K. Ikoma, N. Iwasaki-Kurihara, Selective deposition of diamond crystals by chemical vapor deposition using a tungsten-filament method, Appl. Phys. Lett. 53 (19) (1988) 1815–1817.
[73] E. Scorsone, S. Saada, J.C. Arnault, P. Bergonzo, Enhanced control of diamond nanoparticle seeding using a polymer matrix, J. Appl. Phys. 106 (1) (2009) 014908.
[74] H. Zanin, P.W. May, M.H.M.O. Hamanaka, E.J. Corat, Field emission from hybrid diamond-like carbon and carbon nanotube composite structures, ACS Appl. Mater. Interfaces 5 (23) (2013) 12238–12243.
[75] P.M. Koinkar, S.S. Patil, T.-G. Kim, D. Yonekura, M.A. More, D.S. Joag, et al., Enhanced field emission characteristics of boron doped diamond films grown by microwave plasma assisted chemical vapor deposition, Appl. Surf. Sci. 257 (6) (2011) 1854–1858.
[76] K. Wasa, S. Hayakawa, Handbook of Sputter Deposition Technology: Principles, Technology, and Applications, Noyes Publications, Park Ridge, NJ, USA, 1992.
[77] M.-J. Lee, S.I. Kim, C.B. Lee, H. Yin, S.-E. Ahn, B.S. Kang, et al., Low-temperature-grown transition metal oxide based storage materials and oxide transistors for high-density non-volatile memory, Adv. Funct. Mater. 19 (10) (2009) 1587–1593.
[78] N. Brenning, I. Axnäs, M.A. Raadu, D. Lundin, U. Helmerson, A bulk plasma model for dc and HiPIMS magnetrons, Plasma Sources Sci. Technol. 17 (4) (2008) 045009.
[79] K. Macák, V. Kouznetsov, J. Schneider, U. Helmersson, I. Petrov, Ionized sputter deposition using an extremely high plasma density pulsed magnetron discharge, J. Vac. Sci. Technol. A 18 (4) (2000) 1533–1537.
[80] J.T. Gudmundsson, J. Alami, U. Helmersson, Evolution of the electron energy distribution and plasma parameters in a pulsed magnetron discharge, Appl. Phys. Lett. 78 (22) (2001) 3427–3429.
[81] V. Stranak, M. Quaas, R. Bogdanowicz, H. Steffen, H. Wulff, Z. Hubicka, et al., Effect of nitrogen doping on TiOxNy thin film formation at reactive high-power pulsed magnetron sputtering, J. Phys. Appl. Phys. 43 (28) (2010) 285203.
[82] V. Stranak, A.-P. Herrendorf, S. Drache, M. Cada, Z. Hubicka, M. Tichy, et al., Highly ionized physical vapor deposition plasma source working at very low pressure, Appl. Phys. Lett. 100 (14) (2012) 141604.
[83] S. Drache, V. Stranak, A.-P. Herrendorf, M. Cada, Z. Hubicka, M. Tichy, et al., Time-resolved Langmuir probe investigation of hybrid high power impulse magnetron sputtering discharges, Vacuum 90 (2013) 176–181.
[84] V. Straňák, M. Quaas, H. Wulff, Z. Hubička, S. Wrehde, M. Tichý, et al., Formation of TiOx films produced by high-power pulsed magnetron sputtering, J. Phys. Appl. Phys. 41 (5) (2008) 055202.
[85] A.P. Ehiasarian, R. New, W.-D. Münz, L. Hultman, U. Helmersson, V. Kouznetsov, Influence of high power densities on the composition of pulsed magnetron plasmas, Vacuum 65 (2) (2002) 147–154.
[86] V. Kouznetsov, K. Macák, J.M. Schneider, U. Helmersson, I. Petrov, A novel pulsed magnetron sputter technique utilizing very high target power densities, Surf. Coat. Technol. 122 (2-3) (1999) 290–293.

[87] V. Straňák, M. Čada, M. Quaas, S. Block, R. Bogdanowicz, Š. Kment, et al., Physical properties of homogeneous TiO_2 films prepared by high power impulse magnetron sputtering as a function of crystallographic phase and nanostructure, J. Phys. Appl. Phys. 42 (10) (2009) 105204.

[88] A. Aijaz, K. Sarakinos, M. Raza, J. Jensen, U. Helmersson, Principles for designing sputtering-based strategies for high-rate synthesis of dense and hard hydrogenated amorphous carbon thin films, Diam. Relat. Mater. 44 (2014) 117–122.

[89] K. Sarakinos, A. Braun, C. Zilkens, S. Mráz, J.M. Schneider, H. Zoubos, et al., Exploring the potential of high power impulse magnetron sputtering for growth of diamond-like carbon films, Surf. Coat. Technol. 206 (10) (2012) 2706–2710.

[90] A. Aijaz, K. Sarakinos, D. Lundin, N. Brenning, U. Helmersson, A strategy for increased carbon ionization in magnetron sputtering discharges, Diam. Relat. Mater. 23 (2012 Mar) 1–4.

[91] U. Helmersson, Ionautics | The Excellence in Thin Film Coatings [Internet], Available from: http://www.ionautics.com/hipims-coatings/dlc-coatings/ [cited 2014 Aug 27].

[92] J. Lin, W.D. Sproul, R. Wei, R. Chistyakov, Diamond like carbon films deposited by HiPIMS using oscillatory voltage pulses, Surf. Coat. Technol. [Internet], Available from: http://www.sciencedirect.com/science/article/pii/S0257897214005738 [cited 2014 Aug 25].

[93] M. Samuelsson, K. Sarakinos, H. Högberg, E. Lewin, U. Jansson, B. Wälivaara, et al., Growth of Ti-C nanocomposite films by reactive high power impulse magnetron sputtering under industrial conditions, Surf. Coat. Technol. 206 (8-9) (2012) 2396–2402.

[94] Detonation Nanodiamond [Internet], Wikipedia, the free encyclopedia, 2014, Available from: http://en.wikipedia.org/w/index.php?title=Detonation_nanodiamond&oldid=621906155 [cited 2014 Aug 26].

[95] R. Ritasalo, U. Kanerva, Y. Ge, S.-P. Hannula, Mechanical and thermal properties of pulsed electric current sintered (PECS) Cu-diamond compacts, Metall. Mater. Trans. B 45 (2) (2014) 489–496.

[96] V.T. Senyut, E.I. Mosunov, Physical-mechanical properties of nanocrystalline materials based on ultrafine-dispersed diamonds, Phys. Solid State 46 (4) (2004) 767–769.

[97] A.A. Shulzhenko, I.Y. Ignatyeva, A.S. Osipov, T.I. Smirnova, Peculiarities of interaction in the Zn-C system under high pressures and temperatures, Diam. Relat. Mater. 9 (2) (2000) 129–133.

[98] S.V. Kidalov, F.M. Shakhov, A.Y. Vul, Thermal conductivity of sintered nanodiamonds and microdiamonds, Diam. Relat. Mater. 17 (4-5) (2008) 844–847.

[99] M. Rosinski, L. Ciupinski, J. Grzonka, A. Michalski, K.J. Kurzydlowski, Synthesis and characterization of the diamond/copper composites produced by the pulse plasma sintering (PPS) method, Diam. Relat. Mater. 27-28 (2012) 29–35.

[100] T. Schubert, Ł. Ciupiński, W. Zieliński, A. Michalski, T. Weißgärber, B. Kieback, Interfacial characterization of Cu/diamond composites prepared by powder metallurgy for heat sink applications, Scr. Mater. 58 (4) (2008) 263–266.

[101] A. Michalski, WC/Diamond Composites [Internet], 2014, Available from: http://www.sintering.pl/.

[102] M. Rosinski, A. Michalski, WCCo/cBN composites produced by pulse plasma sintering method, J. Mater. Sci. 47 (20) (2012) 7064–7071.

[103] J. Shen, F.M. Zhang, J.F. Sun, Y.Q. Zhu, D.G. McCartney, Spark plasma sintering assisted diamond formation from carbon nanotubes at very low pressure, Nanotechnology 17 (9) (2006) 2187.

[104] F. Zhang, C. Mihoc, F. Ahmed, C. Lathe, E. Burkel, Thermal stability of carbon nanotubes, fullerene and graphite under spark plasma sintering, Chem. Phys. Lett. 510 (1-3) (2011) 109–114.

[105] F. Zhang, F. Ahmed, J. Bednarcik, E. Burkel, Diamond synthesis through the generation of plasma during spark plasma sintering, Phys. Status Solidi A 209 (11) (2012) 2241–2246.
[106] V.V. Danilenko, On the history of the discovery of nanodiamond synthesis, Phys. Solid State 46 (4) (2004) 595–599.
[107] L. Pramatarova, E. Radeva, E. Pecheva, T. Hikov, N. Krasteva, R. Dimitrova, et al., The advantages of polymer composites with detonation nanodiamond particles for medical applications, in: L. Pramatarova (Ed.), On Biomimetics, InTech, 2011, Available from: http://www.intechopen.com/books/howtoreference/on-biomimetics/the-advantages-of-polymer-composites-with-detonation-nanodiamond-particles-for-medical-applications [cited 2014 Aug 26].
[108] Q. Zhang, V.N. Mochalin, I. Neitzel, I.Y. Knoke, J. Han, C.A. Klug, et al., Fluorescent PLLA-nanodiamond composites for bone tissue engineering, Biomaterials 32 (1) (2011) 87–94.
[109] Q. Zhang, V.N. Mochalin, I. Neitzel, K. Hazeli, J. Niu, A. Kontsos, et al., Mechanical properties and biomineralization of multifunctional nanodiamond-PLLA composites for bone tissue engineering, Biomaterials 33 (20) (2012) 5067–5075.
[110] L. Pramatarova, R. Dimitrova, E. Pecheva, T. Spassov, M. Dimitrova, Peculiarities of hydroxyapatite/nanodiamond composites as novel implants, J. Phys. Conf. Ser. 93 (1) (2007) 012049.
[111] I. Neitzel, V. Mochalin, I. Knoke, G.R. Palmese, Y. Gogotsi, Mechanical properties of epoxy composites with high contents of nanodiamond, Compos. Sci. Technol. 71 (5) (2011) 710–716.
[112] A. Krueger, The structure and reactivity of nanoscale diamond, J. Mater. Chem. 18 (13) (2008) 1485–1492.

CHAPTER 9

Vapor-Phase Synthesis of Free-Standing and Stable Colloidal Magnetoplasmonic Nanoparticles Encapsulated by an Amorphous Silicon Shell

Maria Benelmekki, Mukhles Sowwan
Okinawa Institute of Science and Technology Graduate University, Onna-son, Japan

Chapter Contents

Handbook of Nanoceramic and Nanocomposite Coatings and Materials
http://dx.doi.org/10.1016/B978-0-12-799947-0.00009-2

1. INTRODUCTION

1.1 Magnetoplasmonic Nanoparticles

Nanoparticles (NPs) are considered the starting point for many "bottom-up" approaches to prepare nanostructured materials and devices. Hybrid NPs that contain both magnetic and optically active components have emerged as attractive candidates for advanced biomedical applications, such as multimodal bioimaging, targeted drug delivery, and magnetic hyperthermia. In particular, dumbbell-like NPs composed of magnetite (Fe_3O_4) and Au or Ag are attractive because of their biocompatibility, plasmonic activity, and magnetic properties [1–7]. A broad range of techniques in chemistry and physics have been explored for the preparation of such multicomponent NPs. Further details of synthesis, properties, and applications of dumbbell-like NPs can be found in several review articles [1,8,9].

In the last decade, physical vapor deposition techniques have been developed for the production of hybrid inorganic NPs. However, only a few attempts have been made to produce heterogeneous, dumbbell-shaped multicomponents [10,11], other than the widely reported core-shell nanostructures [12,13]. In light of the present work, it is noteworthy to mention reports on the preparation of Fe-rich-core/Si-rich-shell NPs using double-glow discharge sources [14,15]. A recent study by Elsukova et al. [16] also reported the behavior of Ag-Fe magnetoplasmonic nanoparticles (MPNPs) prepared by magnetron sputtering, with subsequent in-flight annealing, and demonstrated the instability of these NPs under atmospheric conditions. On the other hand, the use of high-moment magnetic cores such as Fe and Co still faces several limitations, such as the oxidation and degradation of these metals when exposed to the environment.

Processes for harvesting and transferring NPs to stable colloidal suspensions need to be addressed. Attempts were made to deposit simultaneously generated NPs, with a beam of water vapor, methanol, or isopropanol, onto nitrogen-cooled substrates [17,18]. However, the stability of the resulting suspensions was not reported.

1.2 Magnetron Sputtering-Based Inert Gas Condensation Methods

Physical methods offer good alternatives to chemical methods for the generation of tailored NPs. Of the physical techniques, magnetron sputtering-based inert gas condensation has emerged as one of the most flexible and successful techniques, as it allows control of the growth process by varying parameters such as the magnetron power, the length of the so-called *aggregation zone*, and the inert gas pressure. Also, the size of NPs can be monitored, and the composition can be altered *in situ* by the introduction of gases such as oxygen and nitrogen, in order to form oxidized and nitrided NPs with tunable stoichiometries.

Since their initial development, inert gas aggregation sources have been based on the formation of a vapor of the desired elements that aggregate to form NPs [19,20]. These sources differ in the methods used to generate elemental vapors: arc discharge, pulsed microplasma, laser vaporization, Joule heating, and sputtering [21–24]. The aim of this chapter is not to review the state-of-the-art inert gas aggregation sources for NP generation. For more details on this subject, the reader can refer to various reviews [25–29]. This chapter focuses on the gas aggregation source, based on magnetron sputtering, as the most popular technique for generating NPs in both academia and industry. The widespread acceptance of magnetron sputtering may be a consequence of its ability to generate a large proportion of ionized NPs that can be mass-/charge-selected by attaching a quadrupole to the source (quadrupole aggregation sources became commercially available in 2001) [30].

2. EXPERIMENTAL

2.1 Deposition of FeAg@Si MPNPs

NPs were synthesized by combining inert gas condensation with cosputtering of three independent (1 in. in diameter) targets (Figure 1). During deposition, targets were DC-sputtered in an Ar atmosphere ($P1 \sim 10^{-1}$ mbar working pressure), and a saturated vapor of metal atoms was generated on the target surfaces. The generated vapor condenses into nanoclusters along its propagation path through the aggregation zone (Figure 1a). Nanoclusters are extracted and accelerated by the differential pressure between the aggregation zone and the deposition chamber ($P2 \sim 10^{-5}$ mbar), where they are deposited on the desired substrate (Figure 1c).

One inch in diameter pure Si (99.999%), Ag (99.99%), and Fe (99.9%) targets were used as ion sources for NP formation. Power was applied to the Si, Fe, and Ag targets at 64, 10, and 4.8 W, respectively. The Ar flux in the aggregation zone was set at 80 sccm. NPs were deposited on flat silicon substrates (5×5 mm^2), glass slide substrates (76 mm × 26 mm), and carbon-coated copper grids. Working distances from target surfaces to the exit slit (Figure 1b) and from the exit slit to the substrate were ~90 and ~20 cm, respectively.

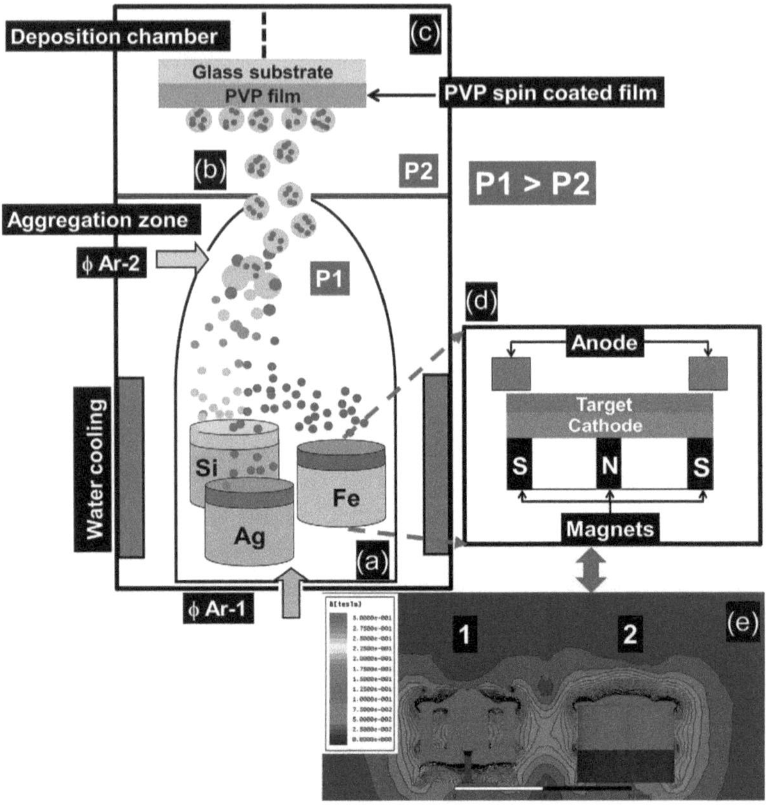

Figure 1 Sketch of the nanoparticle (NP) source used in this work. (a) The aggregation zone. (b) The differential pumping aperture. (c) The main chamber where particles are deposited on substrates. (d) The cathode magnetron geometry. (e) The magnetic field configuration in the vertical cross section of the cathodes (1) with and (2) without an Fe target (the magnetic field was calculated using the finite element method: Maxwell v14).

2.2 Characterization

TEM analysis was performed using both a Cs-corrected FEI Titan 80-300 kV operating at 300 kV and a JEOL JEM-1230 operating at 100 kV. SEM images were collected using an FEI Quanta FEG 250 system. For XPS analysis, a Kratos Axis Ultra^{DLD} 39-306 electron spectrometer equipped with a mono AlKα source operated at 300 W was used. The field dependence of magnetization (M-H loops) was measured with in-plane sample configuration using a quantum design cryogen-free Physical Property Measurement System. UV-visible absorption spectra of each sample suspension were recorded using a Thermo Scientific Multiskan GO UV/Vis microplate spectrophotometer.

Hydrodynamic radii of nanoclusters were determined with dynamic light scattering (DLS) (Zetasizer Nano ZSP, Malvern Instruments).

2.3 Harvesting and Transferring MPNPs to Aqueous Suspensions

To transfer NPs to a stable aqueous suspension, a glass slide substrate (76 mm × 26 mm, polished and precleaned) was thoroughly cleaned in absolute methanol for 10 min under ultrasonication and dried with N_2 gas. Then, 10 mg of polyvinylpyrrolidone (PVP) (Sigma-Aldrich, St. Louis, USA) was dissolved in 250 μL of methanol solution and gently dispensed onto the cleaned glass substrate. A thin polymer film was formed with a spin coater (MS-A150, MIKASA, Japan) at 3000 rpm for 30 s, and NPs were deposited as described above. For harvesting, NP/PVP/glass samples were immersed in methanol and sonicated for 15 min, followed by a separation step to remove excess PVP polymers using a centrifuge at 100,000 rpm for 60 min. After washing precipitated NPs with methanol, the NPs were redispersed in ultrapure water from a Milli-Q system (Nihon Millipore K.K., Tokyo, Japan) using 0.1 μm filters.

3. RESULTS AND DISCUSSION

3.1 Morphology, Structure, and Composition of FeAg@Si NPs

About 65% of the resulting MPNPs (hereafter MPNPs-1) were 10–20 nm in size, and 20% were 20–30 nm in size (Figure 2a and b). This size distribution can be correlated with the number of cores per NP, as reported in our previous work [31]. An HRTEM image of a representative four-core NP (Figure 2e) shows crystalline structure differences within the dumbbell-like shaped cores embedded in an amorphous shell. An STEM image of the same NP (Figure 2c) shows the difference in contrast (due to the atomic number of constituents) within the cores, indicating the position of Ag NPs (brighter contrast). For further information on the composition of the cores, EELS analysis was performed along the NP in the direction of the arrow as indicated in Figure 2c. Figure 2f shows the most representative spectra selected from the line scan (more than 80 points) collected along the NP (Figure 2d). At the beginning of the line scan (Figure 2e, point 1 spectrum), only the O-K edge (532 eV) was observed (coming from the oxidation of the Si shell when exposed to air). Progressing across the dumbbell-like NP, the Fe-L edge (710 eV) and the O-K edge were observed (point 20). Continuing with the line scan, both Fe and Ag edges are observed (point 40 spectrum).

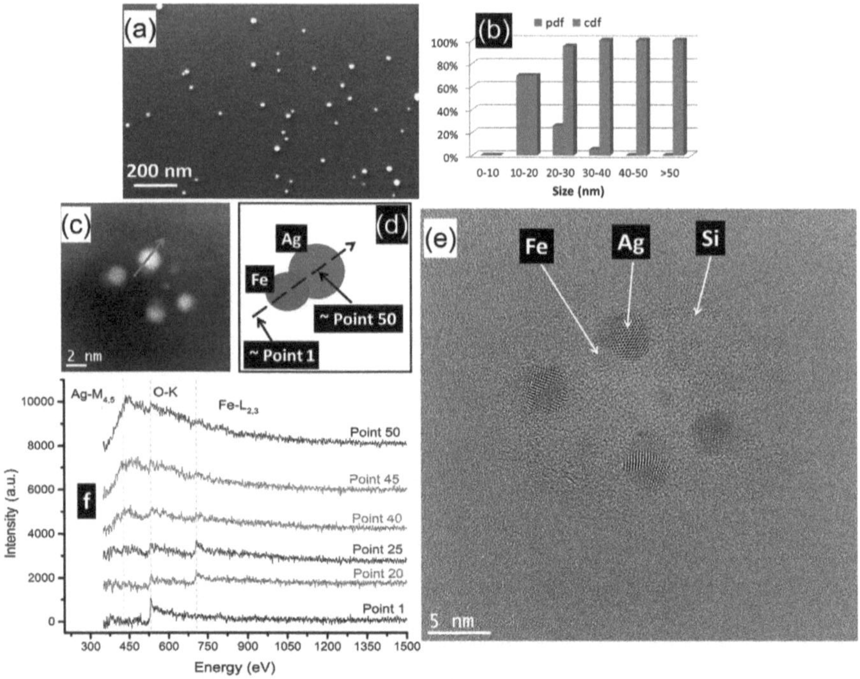

Figure 2 (a) SEM image and (b) size probability distribution function (pdf) and cumulative distribution function (cdf) of MPNPs-1, calculated from SEM images. (c) STEM image and (d) sketch of the scanned NPs. (e) HRTEM images of the same NP. (f) Selected EEL spectra acquired along the scan line across the dumbbell-like NP, as indicated by the arrows in (c) and (d).

This position corresponds to the Fe/Ag interface. Continuing with the line scan, Fe-L edge intensity is considerably reduced and finally, only Ag-$M_{4,5}$ is detected (point 50 spectrum). In all cases, EEL spectra show a Si edge at 1860 eV (not shown) corresponding to a complete Si shell [31,32].

From HRTEM images (Figure 2d) and associated FFTs (not shown), the Ag-rich region reveals the {111} Ag planes with an interplanar distance of 0.238 nm, while Fe-rich and Si regions showed an amorphous structure. Thus, HRTEM and STEM imaging and EELS analysis confirm the core-shell structure of the nanoclusters, with cores made of FeAg dumbbell-like NPs covered with Si shells.

3.2 Morphological Tuning of MPNPs

3.2.1 Effects of MPNP Residence Time in the Aggregation Zone

By introducing Ar (flux of 50 sccm) at the differential pumping aperture level (Figure 1b, ϕAr-2), the pressure in the deposition chamber increased from 7.5×10^{-4} to 3.9×10^{-3} mbar, resulting in the formation of MPNPs termed MPNPs-2 (Figure 3a). In comparison with MPNPs-1, only 50% of MPNPs-2 were below 30 nm in size, 20% were 30-50 nm, and 30% exceeded 50 nm (Figure 3b). These results demonstrate that by reducing the differential pressure between the aggregation zone and the deposition chamber, the residence time of the NPs in the aggregation zone increased, and the coalescence process was enhanced, leading to the formation of larger NPs with a higher number of cores and thicker Si shells (Figure 3c and d) [32]. EELS profiles were intended to determine the position of each component within the MPNPs-2; however, due to the Si shell thickness, resulting profiles were noisy and, therefore, inconclusive (see Figure 5c and d).

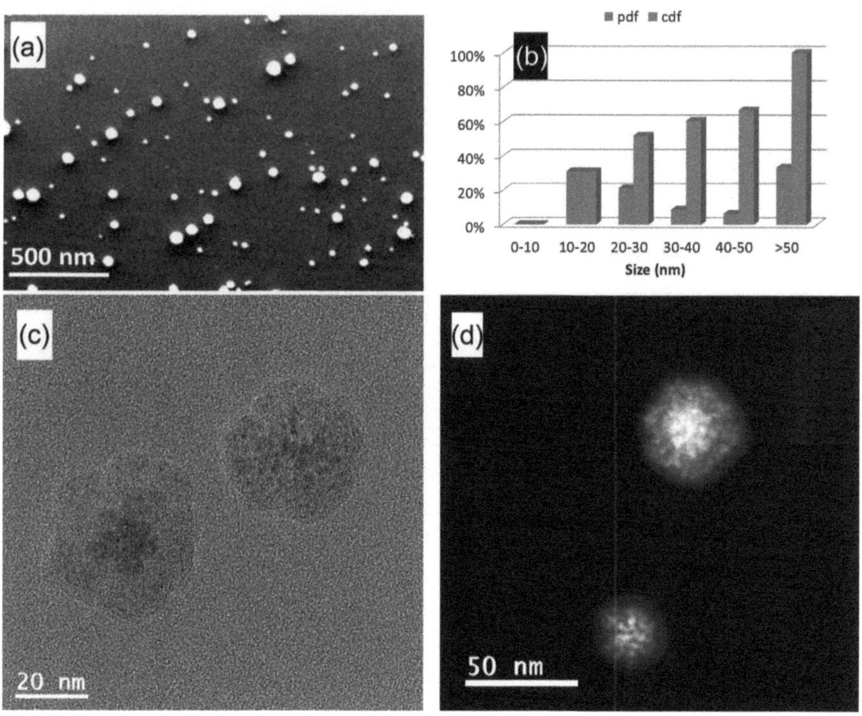

Figure 3 (a) SEM image and (b) size probability distribution function (pdf) and cumulative distribution function (cdf) of MPNPs-2, calculated from the SEM image. (c) TEM image and (d) STEM image showing the multicore-shell morphology of resulting NPs.

3.2.2 Effects of Discharge Powers on Ag Target

In this experiment, the power on Ag target was increased from 4.8 to 6.7 W, while the other parameters were kept constant as described in Section 2.1.

The resulting MPNPs (hereafter MPNPs-3) clearly show a mixture of two types of structures, namely, core-shell and core-satellite structures (Figure 4a and b). Core-shell NPs have irregular shapes with well-defined cores of one material encapsulated with another material. Core-satellite NPs have a roughly spherical core, with a number of smaller clusters of other material deposited on their surfaces. The observed satellite morphology suggests that, because of the higher content of Ag in the aggregation zone, Ag clusters deposit on the surface of the formed NPs while in-flight in

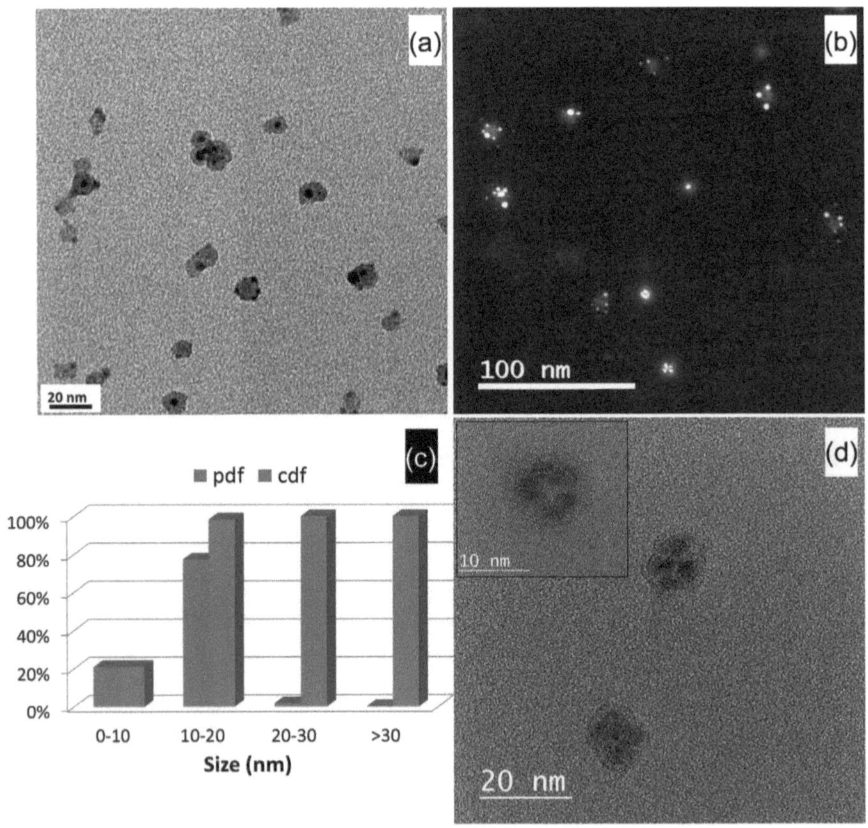

Figure 4 Low-magnification TEM (a) and STEM (b) images of MPNPs-3 showing both core-shell and satellite morphologies. (c) The size probability distribution (pdf) and cumulative distribution functions (cdf) of MPNPs-3 calculated from TEM images. (d) High-magnification TEM image showing the proximity of cores inside a Si shell.

the aggregation zone. Size distributions were obtained from TEM images. 80% of the NPs were 10-20 nm in size, clearly smaller than MPNPs-1 (Figure 4c). From high-magnification TEM images (Figure 4d), cores appear to be closer to each other compared with MPNPs-1.

3.3 Correlation of Morphology with Oxidation State of MPNPs

3.3.1 EELS Intensity Profile Study

Scanning across one NP of each sample demonstrated the dumbbell-like and core-shell morphologies for MPNPs-1 and MPNPs-3, respectively (Figure 5). The line scan through the particle corresponding to the original MPNPs-1 sample, in the direction of the arrow as indicated in Figure 5a, shows an increment of the Ag signal that reaches a maximum peak around 4 nm and then decreases to a minimum around 7.5 nm. In the case of the Fe signal, a defined minimum at ~4 nm was observed, followed by a distinct maximum at ~7.5 nm. This behavior confirms the dumbbell-like structure of the core, in agreement with the study presented in Section 3.1. Regarding MPNPs-3, EELS line scans were performed across the particle, in the direction of the arrow (Figure 5e and f). Both the Ag and Fe intensity profiles show maxima at ~4.5 and ~9 nm positions, respectively, indicating a core-shell structure in both core NPs. The narrow shoulders of the Fe signal, appearing at ~7.5 and ~10.5 nm, indicate the depletion of Fe during the growth process. These results suggest that by increasing the power on the Ag target, the formed Ag NPs have enough energy to mix with iron and segregate to form Fe-Ag core-shell NPs [11]. In the case of MPNPs-2, due to the thickness of the Si shell and the larger size of the NPs, it is difficult to determine whether the cores have a core-shell structure or a dumbbell-like shape (Figure 5c and d).

3.3.2 Oxidation States of MPNPs

XPS measurements were performed after the exposure of the samples to the atmosphere. Fe $2p_{3/2}$ fitting peaks suggest that the iron cores in MPNPs-1 are dominated by the Fe^{3+} state (711.12 eV), while in the cases of MPNPs-2 and MPNPs-3, the iron is present in both metallic (707.12 eV) and oxidized (711.33 eV) states [33,34] (Figure 6). For Si 2p fitting peaks (Figure 6a, c, and e), the atomic ratios between the Si2p@Si (99.11 eV) and Si2p@SiOx (102.10 eV) are ~20% for MPNPs-1 and ~43% for MPNPs-2, respectively. In the case of the MPNPs-3, the Si shell is nearly fully oxidized to SiOx. These results suggest that in the case of MPNPs-2, since the Si shell is thicker than that of MPNPs-1 and MPNPs-3, a gradual oxidation occurred from the

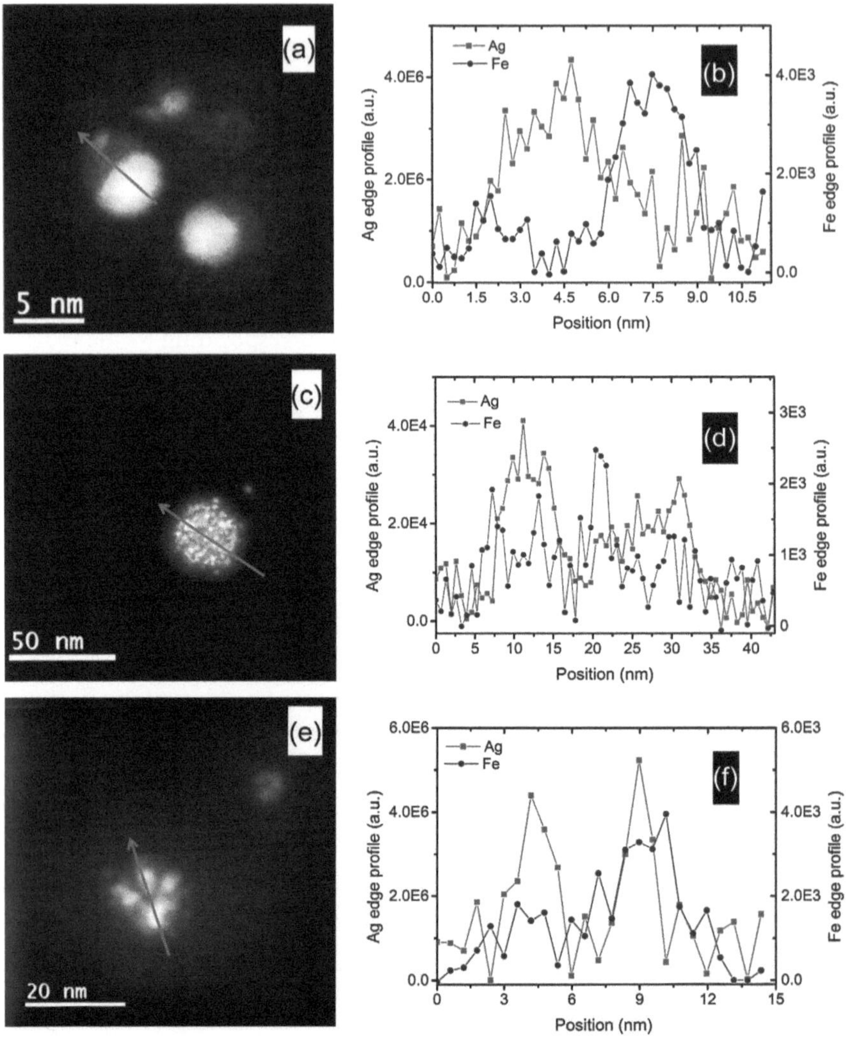

Figure 5 STEM images (a, c, e) and their corresponding EELS line scans (b, d, f) of MPNPs-1, MPNPs-2, and MPNPs-3, respectively. Green arrows in STEM images represent the path where the beam was scanned.

surface of the Si shell to the interface between the Si and the FeAg cores, providing better protection against oxidation of the Fe cores. In the case of MPNPs-3, despite the full oxidation of the Si shell, Fe is present in both its metallic and oxidized states. This behavior is in agreement with EELS profiles, confirming that the iron is protected by the Ag shell.

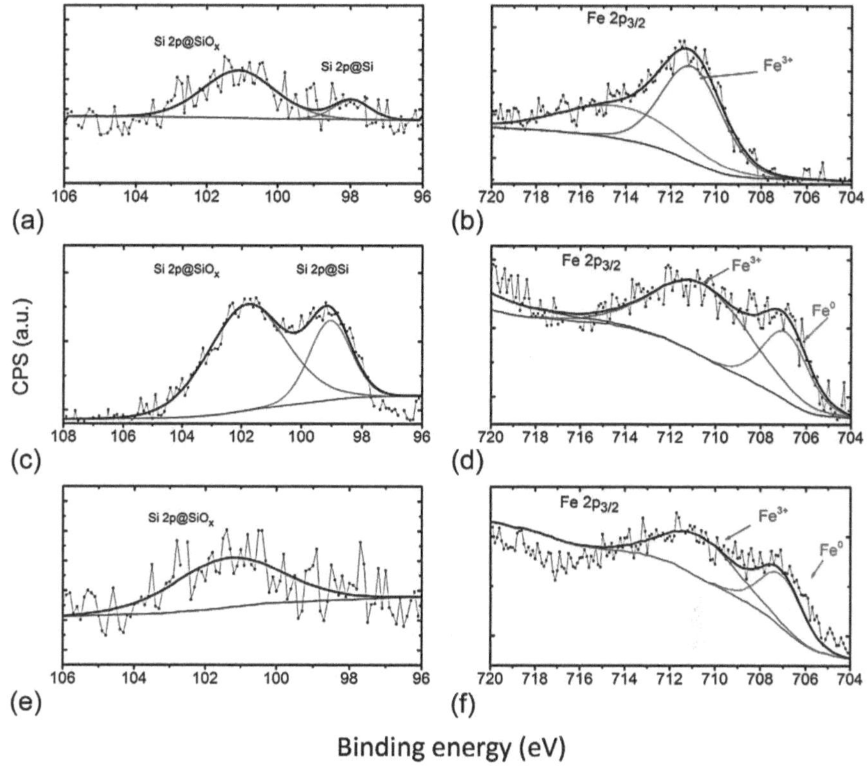

Figure 6 XPS spectra corresponding to Si 2p and Fe 2p core levels from MPNPs-1 (a, b), MPNPs-2 (c, d), and MPNPs-3 (e, f).

3.4 Mechanism of MPNPs Formation

When the Ag target is being sputtered, due to its high sputtering yield (~1.20 under our experimental conditions), electrons and ions are highly concentrated in the plasma region, inducing high-density plasma near the surface of the Ag target. Under these conditions, nucleation and growth of NPs are completed before leaving the plasma zone. In the aggregation zone, the density of inert gas is sufficient to quench their high energy. As a result, particles retain their high-temperature phase, and larger nanoclusters with a monocrystalline structure are formed (Figure 7, step 1).

For Fe NPs, the estimated sputtering yield is ~0.45. Moreover, the Fe target provides a path for the magnetic flux generated by the permanent magnet of the cathode (Figure 1e). Flux passes through the Fe target (Figure 1e) and only a small portion escapes and reaches the surface, contributing to the generation of

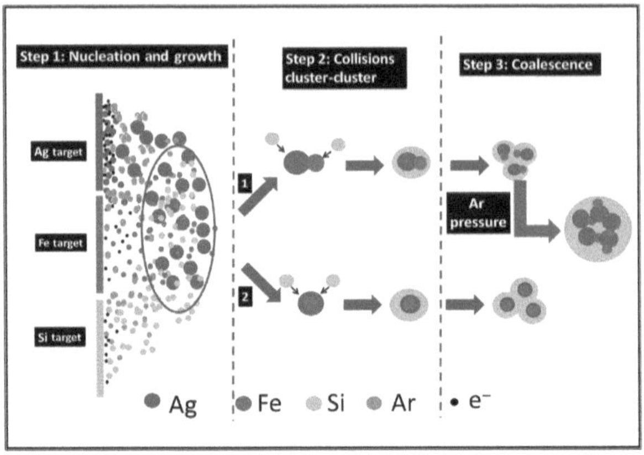

Figure 7 Suggested mechanism of MPNP formation. Step 1 shows the nucleation and growth phase in the plasma zone near the surface of the sputtered targets. Step 2-1 depicts cluster-cluster collisions. Fe and Ag nanoclusters (NCs) juxtapose, forming dumbbell-like shaped NPs. Step 2-2 Ag and Fe nanoclusters mix and then segregate in core-shell structures. Then, Si NCs cover the surface of FeAg NCs resulting in a core-shell structure. The resulting core-shell NPs coalesce in-flight forming multicore-shell NPs (step 3).

the magnetron plasma. Since the density and the distribution of the plasma are strongly correlated with the magnetic field strength and magnetic flux distribution [35,36], during Fe sputtering, only a low concentration of electrons near the surface is achieved. As a result, the ionization rate is lower than that of the Ag target, leading to a lower density of plasma. In this case, nucleation and growth of NPs occur during the flight of atoms into the aggregation zone. There, Fe atoms lose their energy rapidly and form small, amorphous Fe nanoclusters. Regarding Si nucleation and growth, due to the low sputtering yield (~0.29), plasma density on the surface of the Si target is also low, similarly leading to the formation of small amorphous Si nanoclusters. In all cases, sputtering yields were estimated using the simple sputtering yield calculator [37].

In addition, since the surfaces of the nanoclusters are not oxidized, and the fraction of surface atoms per unit volume is very large, in the aggregation zone, nanoclusters collide and coalesce by diffusion of the constituent atoms at their contact interfaces, creating larger NPs. Due to the large, positive free energy of Fe and Ag mixing, segregation of the elements forming core-shell or dumbbell-like structures would be expected [11,16]. However, from our TEM study, an almost *exclusive* formation of dumbbell-like structures was observed in MPNPs-1. A plausible explanation for this behavior is that when Fe and Ag NPs collide in the aggregation zone, their energy is not sufficient to induce complete coalescence so that only dumbbell-like structures can be formed. In the case of MPNPs-3, due to the higher current discharge applied to the Ag target, the preformed Ag NPs seem to retain enough energy to collide and mix with the Fe NPs and then segregate to form an Fe/Ag core-shell structure (Figure 7, step 2). Further, due to the low surface energy of amorphous silicon ($1.05\ J\ m^{-2}$) [38], Si clusters eventually cover the surface of FeAg NPs. Assisted by the lower melting temperature associated with their amorphous structure, these clusters are able to premelt and wet the surface of FeAg NPs, resulting in core-shell structures with shells of rather uniform thicknesses. Finally, during their flight through the aggregation zone, FeAg/Si core-shell NPs collide and partially coalesce with each other, forming multicore MPNPs.

By reducing the differential pressure between the aggregation zone and the main chamber, the residence time of MPNPs in the aggregation zone is increased, increasing the probability of coalescence, resulting in larger NPs with a higher number of cores and thicker Si shells.

3.5 Colloidal Stability and Magneto-Optic Behavior of MPNPs

Following the process described in Section 3.3, MPNPs were harvested and transferred to an aqueous suspension. From DLS measurements (Table 1), PVP-stabilized MPNPs-1 and MPNPs-3 show a narrow size distribution, while MPNPs-2 shows a broad distribution, in agreement with TEM and SEM results. The zeta potential of the MPNPs was found to be around -21 ± 3.24 mV at pH ~6. Both DLS and zeta potential data were acquired from 12 runs per measurement and showed uniform distributions between runs, corresponding to a high-quality criterion. The sample also retained outstanding colloidal stability in water (>6 months), indicating that the particles are thoroughly wrapped and stabilized with PVP after harvesting from the substrate.

Table 1 Dynamic light scattering (DLS) measurements showing the size distribution of MPNPs encapsulated with PVP in aqueous suspension

	Peak 1			Peak 2		
	Size (nm)	Intensity (%)	Width (nm)	Size (nm)	Intensity (%)	Width (nm)
MPNPs-1	27.70	100	9.84	-	-	-
MPNPs-2	30.39	80.55	12.22	184	19.15	61.28
MPNPs-3	23.80	100	7.33	-	-	-

Compared with nonencapsulated metallic Ag NPs prepared under the same experimental conditions, which present a typical surface plasmon resonance (SPR) peak at 410 nm (not shown), MPNPs show redshifted absorption bands at ~600 nm for MPNPs-3, ~650 nm for MPNPs-1, and ~700 nm for MPNPs-2 (Figure 8a). This significant redshift can be attributed to the strong near-field coupling between the Ag cores and the dielectric shell. This redshift from ~600 to ~700 nm can be correlated with the oxidation state of the shell of each sample. In fact, as demonstrated from the XPS study, Si shell presents different degrees of oxidation, depending on the sample. The MPNPs-3 shell is fully oxidized while those of MPNPs-1 and MPNPs-2 show partial oxidation. This variation in oxidation induces sizable increments in the local refractive index of the dielectric shell [39,40].

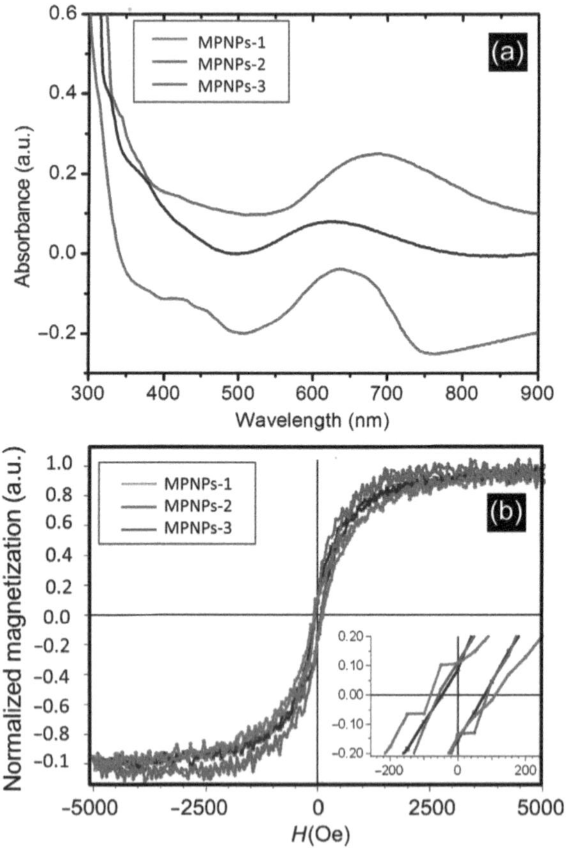

Figure 8 (a) UV-visible absorbance spectra showing the redshifted SPR peak and (b) normalized hysteresis loops of MPNPs at 300 K.

MPNPs exhibit typical ferromagnetic behavior with a coercivity of ~102 Oe at 300 K (Figure 8b). MPNPs-2 and MPNPs-3 show slightly lower coercivities and reach their saturation magnetization at relatively lower magnetic fields. The observed coercivities in all samples can be attributed to the anisotropy at the FeAg interface. In fact, when Ag participates in a metallic alloy, hybridization can induce holes in the 4d states of Ag; therefore, a magnetic moment can be induced in the Ag sites. As reported recently by Alonso et al. [41], the coupling between the Fe magnetic moment and the magnetic moment induced in the Ag atoms is ferromagnetic, and as a consequence, the amorphous FeAg interfacial region affects the magnetic response of the whole sample, giving rise to ferromagnetic behavior.

4. SUMMARY

Multicomponent hybrid nanostructures that contain two or more nanometer-scale components continue to attract much attention, owing to the synergistic properties induced by interactions between these different nanometer-scale objects. Here, we developed a one-step method to produce hybrid NPs composed of magnetoplasmonic FeAg cores encapsulated with Si shell. The size of the NPs and the shape of the cores in these NPs can be modulated by controlling the deposition parameters as follows: (1) By adjusting the discharge current on each target, we control the plasma density on the surface of each target and thus the size and crystallinity of the formed clusters. (2) By reducing the differential pressure between the aggregation zone and the deposition chamber, we increase the residence time of the NPs inside the aggregation zone, enhancing the coalescence process, and therefore, larger NPs with a higher number of cores are formed. (3) The immiscibility of Fe and Ag and the low surface energy of the amorphous Si play important roles in the position of each material within the hybrid NPs.

The Si shell provides an excellent way to bind the Fe to the Ag and to inhibit particle-particle magnetic interactions. When these NPs are exposed to the atmosphere, the Si surface is oxidized to SiOx, providing an easy path to functionalize their surface for a wide range of bioapplications. Moreover, the thickness and the oxidation state of the Si shell seem to play an important role in the optical properties of the NPs.

5. FUTURE OUTLOOK

Magnetron sputtering-based inert gas condensation method is a powerful tool to synthesize multifunctional NPs. This method offers accurate control on the morphology, chemical composition, and size distribution of the NPs, opening new horizons for theranostic nanocomposites. However, for a better understanding of NP formation, further improvements in characterization methods of aggregation zone conditions are needed. In addition, there is a growing focus on yield improvement and harvesting processes of the HNPs in order to make profitable the scale-up of this method.

REFERENCES

[1] C. Wang, C. Xu, H. Zeng, S. Sun, Recent progress in syntheses and applications of dumbbell-like nanoparticles, Adv. Mater. 21 (2009) 3045–3052.

[2] I. Parsina, C. DiPaola, F. Baletto, A novel structural motif for free CoPt nanoalloys, Nanoscale 4 (2012) 1160–1166.

[3] H. Yu, M. Chen, P.M. Rice, S.X. Wang, R.L. White, S. Sun, Dumbbell-like bifunctional Au-Fe3O4 nanoparticles, Nano Lett. 5 (2005) 379–382.

[4] H. Gu, Z. Yang, J. Gao, C.K. Chang, B. Xu, Heterodimers of nanoparticles: formation at a liquid-liquid interface and particle-specific surface modification by functional molecules, J. Am. Chem. Soc. 127 (2005) 34–35.

[5] L. Zhang, Y.H. Dou, H.C. Gu, Synthesis of Ag-Fe_3O_4 heterodimeric nanoparticles, J. Colloid Interface Sci. 297 (2006) 660–664.

[6] Y. Lu, H. Xiong, X. Jiang, Y. Xia, Asymmetric dimers can be formed by dewetting half-shells of gold deposited on the surfaces of spherical oxide colloids, J. Am. Chem. Soc. 125 (2003) 12724–12725.

[7] X. Meng, H.C. Seton, L.T. Lu, I.A. Prior, N.T.K. Thanh, B. Song, Magnetic CoPt nanoparticles as MRI contrast agent for transplanted neural stem cells detection, Nanoscale 3 (2011) 977–984.

[8] H. Zeng, S. Sun, Syntheses, properties, and potential applications of multicomponent magnetic nanoparticles, Adv. Funct. Mater. 18 (2008) 391–400.

[9] A. Perro, S. Reculusa, S. Ravaine, E. Bourgeat-Lami, E. Duguet, Design and synthesis of Janus nanoparticles, J. Mater. Chem. 15 (2005) 3745–3760.

[10] J. Bai, Y.H. Xu, J. Thomas, J.P. Wang, (FeCo)3Si–SiOx core–shell nanoparticles fabricated in the gas phase, Nanotechnology 18 (2007) 065701.

[11] Y.H. Xu, J.P. Wang, Direct Gas-Phase Synthesis of heterostructured nanoparticles through phase separation and surface segregation, Adv. Mater. 20 (2008) 994–999.

[12] V. Skumryev, S. Stoyanov, Y. Zhang, G. Hadjipanayis, D. Givord, J. Nogués, Beating the superparamagnetic limit with exchange bias, Nature 423 (2003) 850–853.

[13] M. Kaur, J.S. McCloy, W. Jiang, Q. Yao, Y. Qiang, Size dependence of inter-and intracluster interactions in core–shell iron–iron oxide nanoclusters, J. Phys. Chem. C 116 (2012) 12875–12885.

[14] K. Sumiyama, R. Katoh, S. Kadowaki, T. Hihara, Fe–Si core/Si-shell clusters prepared by double glow discharge sources, J. Nanopart. Res. 12 (2012) 2589–2596.

[15] N. Tanaka, K. Sumiyama, R. Katoh, T. Hihara, K. Sato, T.J. Konno, K. Mibu, Core-shell formation and juxtaposition in fe and si hybrid clusters prepared by controlling the collision stages, Materials Transactions 51 (2010) 1990–1996.

[16] A. Elsukova, Z.A. Li, C. Möller, M. Spasova, M. Acet, M. Farle, M. Kawasaki, P. Ercius, T. Duden, Structure, morphology, and aging of Ag–Fe dumbbell nanoparticles, Phys. Status Solidi A 208 (2011) 2437–2442.

[17] K.V. Haeften, C. Binns, A. Brewer, O. Crisan, P.B. Howes, M.P. Lowe, C. Sibbley-Allen, S.C. Thornton, A novel approach towards the production of luminescent silicon nanoparticles: sputtering, gas aggregation and co-deposition with H_2O, Eur. Phys. J. D 52 (2009) 11–14.

[18] G. Galinis, H. Yazdanfar, M. Bayliss, M. Watkins, K.V. Haeften, Towards biosensing via fluorescent surface sites of nanoparticles, J. Nanopart. Res. 14 (2012) 1019–1024.
[19] C.G. Granqvist, R.A. Buhrman, Ultrafine metal particles, J. Appl. Phys. 47 (1976) 2200–2219.
[20] K. Sattler, J. Mühlbach, E. Recknagel, Generation of metal clusters containing from 2 to 500 atoms, Phys. Rev. Lett. 45 (1980) 821–824.
[21] P. Milani, W.A. deHeer, Improved pulsed laser vaporization for production of intense beams of neutral and ionized clusters, Rev. Sci. Instrum. 61 (1990) 1835–1838.
[22] H.R. Siekmann, Ch. Lüder, J. Faehrmann, H.O. Lutz, K.H. Meiwes-Broer, The pulsed arc cluster ion source, Z. Phys. D: At. Mol. Clusters 20 (1991) 417–420.
[23] H. Haberland, M. Karrais, M. Mall, A new type of cluster and cluster ion source, Z. Phys. D: At. Mol. Clusters 20 (1991) 413–415.
[24] H. Haberland, M. Karrais, M. Mall, Y. Thurner, Thin films from energetic cluster impact: a feasibility study, J. Vac. Sci. Technol. A 10 (1992) 3266–3271.
[25] C. Xirouchaki, R.E. Palmer, Deposition of size-selected metal clusters generated by magnetron sputtering and gas condensation: a progress review, Philos. Trans. R. Soc. A 362 (2004) 117–124.
[26] J. Bansmann, S.H. Baker, C. Binns, J.A. Blackman, et al., Magnetic and structural properties of isolated and assembled clusters, Surf. Sci. Rep. 56 (2005) 189–275.
[27] K. Wegner, P. Piseri, H. Vahedi-Tafreshi, P. Milani, Cluster beam deposition: a tool for nanoscale science and technology, J. Phys. D: Appl. Phys. 39 (2006) R439–R459.
[28] C. Binns, K.N. Trohidou, J.A. Bansmann, et al., The behaviour of nanostructured magnetic materials produced by depositing gas-phase nanoparticles, J. Phys. D: Appl. Phys. 38 (2005) R357–R379.
[29] M.T. Swihart, Vapor-phase synthesis of nanoparticles, current opinion in colloid and interface science 8 (2003) 127–133.
[30] Oxford Applied Research, http://www.oaresearch.co.uk.
[31] M. Benelmekki, M. Bohra, J.H. Kim, R.E. Diaz, J. Vernieres, P. Grammatikopoulos, M. Sowwan, Facile single-step synthesis of ternary multicore magneto-plasmonic nanoparticles, Nanoscale 6 (2014) 3532–3535.
[32] M. Benelmekki, J. Vernieres, J.H. Kim, R.E. Diaz, P. Grammatikopoulos, M. Sowwan, On the formation of ternary metallic-dielectric multicore-shell nanoparticles by inert-gas condensation method, Mater. Chem. Phys. 151 (2015) 275–281.
[33] T. Yamashita, P. Hayes, Analysis of XPS spectra of Fe^{2+} and Fe^{3+} ions in oxide materials, Applied surface science 254 (2008) 2441–2449.
[34] Moulder J F, Stickle W F, Sobol P E, Bomben K D (1992) "Handbook of X-ray photoelectron spectroscopy" ISBN 0-9627026-2-5 ED Jill Chastain. Pub. Perkin Elmer Corporation.
[35] K. Nanbu, S. Kondo, Analysis of three-dimensional DC Magnetron Discharge by the Particle-in-Cell/Monte Carlo Method, Jpn. J. Appl. Phys. 36 (1997) 4808–4814.
[36] S. Ido, M. Kashiwawagi, M. Takahashi, Computational studies of plasma generation and control in a magnetron sputtering system, Jpn. J. Appl. Phys. 38 (1999) 4450–4454.
[37] www.iap.tuwien.ac.at/www/surface/sputteryield, " A simple sputter yield calculator", accessed June, 06, 2014.
[38] S. Hara, S. Izumi, T. kumagai, S. Sakai, Surface energy, stress and structure of well-relaxed amorphous silicon: a combination approach of ab initio and classical molecular dynamics, Surf. Sci. 585 (2005) 17–24.
[39] Y. Chen, N. Gao, J. Jiang, Surface Matters: Enhanced bactericidal property of core–shell Ag–Fe_2O_3 nanostructures to their heteromer counterparts from one-pot synthesis, Small 9 (2013) 3242–3246.
[40] Refractive index database, http://refractiveindex.info. Accessed June 06, 2014.
[41] J. Alonso, M.L. Fdez-Gubieda, et al., Interfacial magnetic coupling between Fe nanoparticles in Fe–Ag granular alloys, Nanotechnology 23 (2012) 025705.

CHAPTER 10

Thermally Sprayed Nanoceramic and Nanocomposite Coatings

Pierre Fauchais, Armelle Vardelle, Michel Vardelle
SPCTS UMR 7315, CEC, University of Limoges, Limoges, France

Chapter Contents

Handbook of Nanoceramic and Nanocomposite Coatings and Materials
http://dx.doi.org/10.1016/B978-0-12-799947-0.00010-9

1. INTRODUCTION: HOW TO PRODUCE NANOSTRUCTURED THERMALLY SPRAYED COATINGS?

Nanomaterials (i.e., materials with structure at least one dimension smaller than 100 nm, following the definition of the US National Nanotechnology Initiative, USNNI) are at the frontier of materials science due to their remarkable and, in some cases, novel properties [1]. Nanostructured coatings should exhibit better properties than conventional microstructured coatings because of a high volume fraction of internal interfaces. Thus, ceramic coatings should exhibit lower thermal diffusivity, higher toughness due to small grain sizes, and also better wear resistance. The latter are explained by a higher toughness, grain sliding plasticity, and change in the fracture and material-removal mechanism due to the ultrafine particle sizes.

Since the mid-1990s, a large body of works has been devoted to the use of thermal spray technologies for the deposition of finely structured or nanometer-structured coatings [2–4]. The two main routes to achieve such coatings are the following:

- Spraying agglomerated nanometer-sized particles, melting them only partially, and embedding the unmolten nanostructures within fully melted ones, forming the "cement" of the coating, which, therefore, exhibits a bimodal structure [5–7]. Only ceramic particles have been sprayed that way and a few industrial applications have been developed.
- Spraying a liquid feedstock consisting either of submicrometer-sized or nanometer-sized particles dispersed in a solvent (suspension) or of chemical precursors (solution) [3,4]. For suspensions, once the liquid carrier has been fragmented and vaporized by the plasma flow or the high-velocity oxygen fuel (HVOF) flame, the solid particles contained in solvent droplets are, in principle, accelerated, heated, and melted before impacting onto the substrate. For solutions, after fragmentation and partial evaporation of the liquid, precipitation or gelation and pyrolysis occur, followed by sintering and impacting the melted particles onto the substrate.

2. NANOCERAMIC COATINGS BUILT UP FROM NANOSIZED AGGLOMERATED PARTICLES

This thermal spray method uses ceramic powders (in the few tens of micrometers) and powder feeders used in conventional thermal spraying. However, it requires a close control of the in-flight processing of the sprayed particles [3–5] as they must be melted only partially: The melted part, with a micrometer structure after spraying, is the "cement" of the coating, while the unmelted nanometer particles keep their nanostructure. The result is a coating with a bimodal (nano- and micro-)structure.

2.1 Spray Conditions

The particles are made of nanosized ceramic particles, which are agglomerated via the spray drying process into micrometer-sized particles and then calcinated or sintered [5]. The densification process must be performed very carefully to avoid partial or total loss of the nanostructural character of the powder by particle coarsening and sintering effects. These agglomerated particles can be either dense or rather porous.

A typical coating of yttria (8 wt%) partially stabilized zirconia (YPSZ) with a bimodal structure is shown in Figure 1a where lighter-colored and darker-colored zones can be distinguished [5]. The darker zones, as shown in Figure 1b, correspond to the semimolten nanostructured agglomerated YPSZ particles embedded in the coating microstructure (about 35 vol.% according to image analysis). The key parameter is the density of the nanosized zones [5], which depends on thermal processing, spraying conditions, and feedstock characteristics (e.g., particle diameter and powder density). Compared to conventional coatings, these coatings exhibit significantly higher crack propagation resistance or relative toughness.

The nanometer zones that form the coating can be either porous (see Figure 1b), as the original feedstock, or much denser. The porous nanometer-sized zones are obtained when the molten part of the semimolten particles does not fully infiltrate their unmolten core. Thus, these coatings can be engineered to exhibit different properties and microstructures

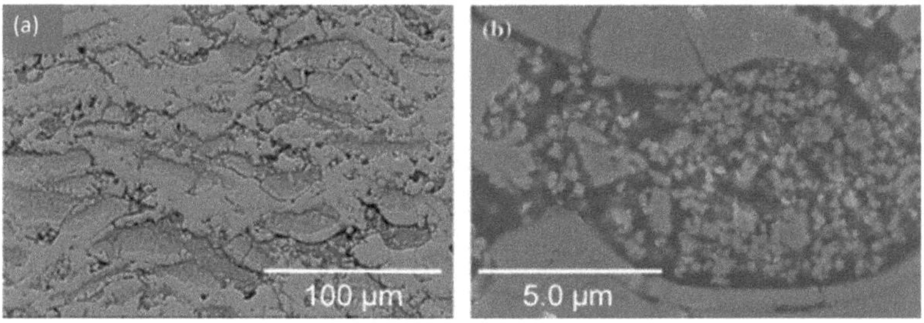

Figure 1 (a) Microstructure of the bimodal-structured zirconia-yttria coating made from agglomerated feedstock. (b) Darker-colored region containing the semimolten feedstock particles [5].

[3–5], such as antiwear and abradable coatings. Very porous nanostructured agglomerated particles help obtaining porous nanosized zones. The dense nanosized zones probably occur when the molten part of the agglomerated semimolten particle infiltrates into the small capillaries of its unmolten core. Two spray routes are used: atmospheric plasma spraying APS and HVOF flame. With APS, large particles with a wide particle size distribution (e.g., 40–160 μm) have to be used, while with HVOF, getting dense nanosized zones embedded in the coating microstructure requires to employ small particles with a narrow size distribution. Especially with APS, the in-flight temperatures of particles are one of the key parameters; they must be kept around the melting point, with approximately 50% of them above it and 50% below it. Lima and Marple [5] had delineated some general and practical engineering guidelines for the production of bimodal-structured coatings based on previous experimental results; they have summarized the spray conditions in [3–5].

2.2 Applications

- Bimodal-structured wear-resistant coatings, mainly of titania or alumina-titania, have better abrasion or sliding wear performance than coatings sprayed with conventional spray techniques. If the hardness of both types of coatings is about the same, the bimodal-structured coatings exhibit significantly higher crack propagation resistance or relative toughness when compared to conventional coatings [3–5]; for example, the nanostructured titania coatings provide superior resistance against abrasive and erosive wear for ball valves destined for high-pressure acid leach service; the thermal shock behavior of nanostructured Al_2O_3-13 wt%TiO_2 plasma-sprayed coatings is much higher than that of conventional coatings [6].
- Abradable ceramic coatings are obtained by embedding a significant amount of semimolten porous particles in the microstructure, the coating becoming thus friable [3–5].
- Much work has been performed on plasma-sprayed nanostructured thermal barrier coatings (TBC) [3–5]. Basically, the porous nanometer-sized zones reduce significantly the increase of the elastic modulus and thermal conductivity values of these coatings when exposed at high temperatures by counteracting densification effects via differential sintering, which are highly desired characteristics for TBC applications. After heating at 1400 °C during 20 h, the coating porosity increases [7], because

the nanometer-sized zones sinter at much faster rate than the coating matrix, thereby reducing the growth rate of thermal conductivity governed by sintering.

- Biomedical applications use mainly hydroxyapatite (HA) and titania. HVOF-sprayed nanosized TiO_2-HA composite coatings exhibit bond strength levels of at least 2.5 times that of conventional thermally sprayed HA coatings [3–5].

3. SUSPENSION OR SOLUTION SPRAYING

In conventional thermal spraying, the particles of the coating material are injected into the hot gas jet (flame or plasma) thanks to a gas carrier. However, when shifting from particles 40 μm in diameter to particles 0.4 μm in diameter, the particle mass is 10^6 times smaller and the carrier gas flow rate to achieve the same acceleration at the injector exit would result in a cold flow cutting through the hot gas flow and preventing a contact between the particles and the hot gases. Therefore, the only way to introduce submicrometer-sized particles in the high-energy gas flow is to use a liquid carrier, which has a mass of about 1000 times higher than that of gas. Two routes have been suggested [3,4] for liquid feedstock spraying:

- Spraying submicrometer-sized or nanometer-sized ceramic or cermet particles via a suspension that is a mixture of fine particles and solvent. Once the suspension drops have been fragmented and vaporized by the plasma flow or the HVOF flame, particles contained in the droplets either agglomerate or disperse in the gas flow where they are heated and usually melted.
- Spraying chemical solutions of coating material precursors. As with suspensions, the liquid injected in the gas flow undergoes rapid fragmentation and evaporation followed by precipitation or gelation, pyrolysis, and, for particles traveling in the hot jet core, sintering and melting, resulting in molten particles between a few tenths and few micrometers.

Compared to conventional thermal spraying, these two processes are by far more complex because the coating properties and process efficiency depend strongly on the suspensions and solution preparation, liquid injection into the gas jet, interaction of the liquid with the hot gases, processing of the small solid particles in hot gases, and elimination or not of the small particles poorly treated (too low temperatures and velocities). Moreover, the in-flight measurement of the size of single droplets or particles is not feasible yet for dimensions below 4-5 μm [8]. Also, the measurement of coating porosity for pore size below hundreds of nanometers requires instruments such as ultrasmall-angle X-ray scattering (USAXS) [9].

3.1 Suspensions and Solutions Preparation

Suspensions are obtained from slurries consisting of particles and a solvent, with particle sizes varying from a few tens of nanometers to micrometers. The most frequently used solvents are alcohols (ethanol or isopropanol), water, or a mixture of both [10]. The load of particles can vary from 5 to 30 wt%. However, in most cases, it is below 20 wt% as the viscosity of the suspension increases and the injection of the liquid in the gas jet is more difficult as well as an efficient treatment of the suspension. After stirring, the suspension stability can be tested by sedimentation. Generally, a dispersant, which adsorbs on the particle surfaces, separating them by electrostatic, steric, or electrosteric repulsions is used. The percentage of dispersant must be adjusted in such a way the viscosity of the suspension is close to that of solvent. The suspension has a shear-thinning behavior, and the main property for its injection in the hot gas jet is its surface tension that partly controls its fragmentation [4]. Different chemicals can be added to the liquid phase to modify its surface tension and/or its viscosity [4].

However, the question is what happens to particles contained in a droplet when the liquid has been evaporated? When observing the resulting splats, it seems that very often, they result from the impact and spreading of melted agglomerates [4]. The possible agglomeration of particles depends to a great extent, on their manufacturing process.

Most suspensions are made with ceramic particles. To our best knowledge, cermets have not been studied yet except WC-Co, probably because of the different acid/base properties of the corresponding oxides present at particle surfaces. Solutions with a mixing at the molecular level of the constituent chemicals allow an excellent chemical homogeneity [11]. Several liquid precursors, such as solution/sol/polymeric complexes, have been evaluated for different oxide systems [12]. The success in forming the phase required for a given system depends on the decomposition characteristics of the different precursors. They include (a) the mixture of nitrates in water/ethanol solution, (b) mixtures of nitrates and metal-organics in isopropanol (hybrid sol), (c) mixed citrate/nitrate solution (polymeric complex), and (d) coprecipitation followed by peptization (gel dispersion in water/ethanol). The aqueous solutions allow higher precursor concentrations than organic solutions; they are less expensive to produce and safer to store and handle. Volumetric precipitation, when the droplets are heated within the plasma jet, depends on the solute initial concentration and its value relative to the equilibrium saturation [13,14]. With solute value close to equilibrium saturation, liquid pyrolysis occurs within the whole volume, while with lower concentration, it occurs in the periphery of the droplet. The concentration of precursors results in almost no variation in the solution specific mass and surface tension, but large variations in the solution viscosity while mixing the solution with the solvent modify its viscosity, surface tension, and specific mass. At last, solutions behave as the solvent, that is, as a Newtonian fluid. As for suspensions, the liquid vaporization consumes energy, and moreover, often, precursors undergo significant endothermic processes in the early stages of heating [15]. It is then possible to add a component presenting an exothermic decomposition under heating. For example, Muoto et al. [15] had added ammonium acetate to their solution of nitrates to increase the exothermicity of the decomposition process. Acetates can oxidize in the air that is present during spraying, and nitrate can act as an oxidizer for the acetate in the acetate-nitrate combinations. Coatings obtained with the ammonium acetate were denser. Generally, the pyrolysis process occurring in solution plasma spraying results in particles that impact bigger than those contained in suspensions.

3.2 Liquid-Hot Gas Interactions

3.2.1 Liquid Injection

The liquid injection depends on the injection method. For suspension or solution thermal spraying, spray atomization and mechanical injection are both used [3,4,10,12].

The different techniques of atomization used in liquid plasma spraying are described in the papers of Jordan et al. [16] and Marchand et al. [17]. In the atomization process, the expanding atomizing gas fragments the low-velocity liquid, and thus, the liquid jet or drops break up into droplets [18]. For liquid viscosity, μ, such as $0.1 < \mu < 50\text{-}60$ mPa s, the breakup mechanism depends on the Weber number (see Eq. 1) that is a measure of the relative importance of the gas cross flow inertia compared to the injected liquid surface tension. For the Newtonian liquids in subsonic gas flows, these mechanisms are generally classified as bag breakup ($12 < We < 100$). deformation of the drop as a bag-like structure that is stretched and swept off in the flow direction; stripping breakup ($100 < We < 350$), drawing of thin sheets from the periphery of the deforming droplets; and catastrophic breakup ($We > 350$), a multistage breaking process [19]. The Weber number is defined as

$$We = \frac{\rho_g \times u_r^2 \times d_l}{\sigma_l} \quad (1)$$

where ρ_g is the gas mass density (kg/m^3), u_r the relative velocity between gas and liquid (m/s), d_l the drop or liquid jet diameter (m), and σ_l (N/m) the surface tension of the liquid.

For a fluid with higher viscosity, the Ohnesorge number (Oh) (see Eq. 2) that relates the viscous force to inertial and surface tension forces must also be considered [10,17]:

$$Oh = \frac{\mu_l}{\sqrt{\rho_l \sigma_l d_l}} \quad (2)$$

where ρ_l is the mass density of the liquid and μ_l its molecular viscosity (Pa s).

The typical droplet sizes, achieved by atomization, range between 2 and 100 μm and the corresponding velocities vary between 60 and 5 m/s [17]. However, the velocities and diameters of droplets cannot be controlled independently [17], while a distribution of droplet trajectories as narrow as possible results in the best thermal treatment of the liquid in the hot core of the plasma jet.

Mechanical injection can be achieved with pressurized reservoir feeding the liquid to a nozzle at the backside of which a magnetostrictive rod superimposes pressure pulses at frequencies up to a few tens of kilohertz [20], breaking up the liquid stream into monosized and equidistant droplets. However, in this system, the size and velocity of droplets cannot be controlled separately. Another system consists in the liquid pressurized in a reservoir and forced through a nozzle of a given diameter d_i [21,22]. A liquid jet is produced with a velocity v_l and a diameter of about 1.5-1.8 × d_i. The square of the liquid velocity varies linearly with the pressure imparted to the liquid in the reservoir and with a 50 μm injector diameter to get the same liquid flow rate, as that obtained at 0.5 MPa with the 150 μm nozzle requires multiplying the pressure by 81.

3.2.2 Hot Gas-Liquid Interactions

In conventional spraying, two injection modes are used: For HVOF spraying (generally called high-velocity suspension flame spraying, HVSFS), the feedstock is injected axially in the combustion chamber, and for most plasma spray torches, it is injected radially at the anode-nozzle exit, except for the axial plasma torch presented in Section 3.4.2.3.

When the liquid is injected in the combustion chamber of an HVOF gun, according to the low gas velocity in this chamber, droplets are essentially vaporized. The injector must provide a liquid pressure high enough to overcome that of the combustion chamber. Dongmo et al. [23] were the first to model the phenomena occurring in a Top Gun G torch: ethanol evaporation followed by nonpremixed combustion of gaseous ethanol with excess oxygen from the fuel gas (propane/oxygen) and interactions between gas and liquid as well as between gas and particles. The liquid vaporization cools the flame with a rapid decrease of its temperature, even with a combustible solvent, and increases the pressure of the chamber [23], which can modify the combustion stability and make the use of acetylene impossible. A longer combustion chamber [23] permitted the heating of particles contained in droplets after their release in the high-temperature combustion chamber and not in the nozzle expansion where gas temperature decreases. Other authors have proposed to limit the liquid feed rate to avoid this phenomenon [4].

In plasma torches, when the liquid is injected radially, it will be fragmented according to the Weber number (see Eq. 1) before being vaporized. The vaporization time is at least two orders of magnitude longer than that of fragmentation [21]. Compared to conventional plasma spraying, where particles are heated and melted without noticeable change in their diameter, the initial drop diameter rapidly decreases.

An efficient liquid penetration requires [24] that the dynamic pressure of the liquid flow is higher than that of the gas flow:

$$\rho_l v_l^2 \gg \rho_g v_g^2 \tag{3}$$

where the subscript l corresponds to the suspension at the injector exit and g to the plasma flow, ρ and v varying with the radial and axial positions in the hot jet. As the liquid jet fragmentation depends strongly on the Weber number (*We*), the solvent surface tension, σ_l, plays a key role. For example, shifting from ethanol to water, all the other parameters being the same, reduces *We* by a factor of 3.2, ratio of the corresponding surface tensions. In Figure 2, a water jet injected within two different plasma flows is shown [25]: pure argon ($We=41$) and a mixture of Ar-He-H_2 ($We=495$), the primary fragmentation of the liquid jet being by far more important with the latter gas mixture. The droplets formed are then fragmented farther downstream. More tiny droplets are formed 15 mm downstream of the nozzle exit with the Ar-He-H_2 plasma gas mixture compared to the argon plasma.

With the axial plasma torch of Mettech (see Section 5.4), the shear produced by the three converging plasma jets also induces liquid fragmentation followed by vaporization.

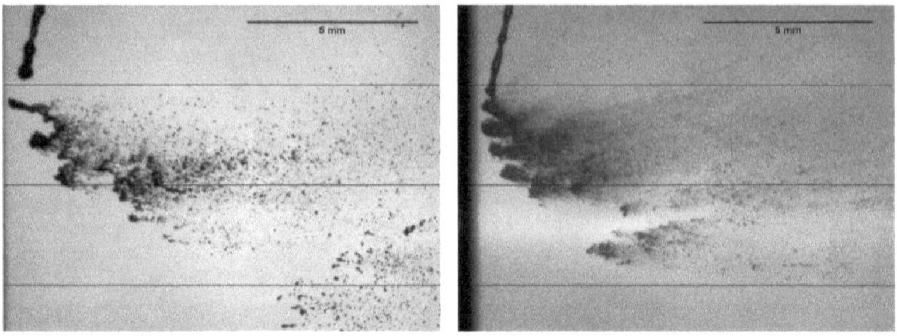

Figure 2 Water jet breakup by the cross plasma flow: (a) Ar plasma jet, $We=41$; (b) Ar-He-H_2 plasma jet, $We=495$ [25].

At last, it must be pointed out that with plasma jets, the combustion of the solvent, when combustible, can only occur when the hot gas temperature is below the combustion temperature, which occurs in the plasma plume at distances generally farther than the substrate location for suspension or solution spraying.

3.3 Behavior of Small (Less Than Few Micrometer) Droplets or Particles in Hot Gases

3.3.1 Heat and Momentum Transfers

Particles or droplets with diameters below a few micrometers have a very low inertia and thus reach rapidly a velocity close to that of the flow, which, for plasma jets, decreases very fast with the surrounding air entrainment and mixing [4,5]. When the particle size is below 5 μm [26], the Knudsen effect limits more or less their acceleration and heating and results in lower velocity and temperature. It means that, when assuming the hot gas flow velocity constant, the particle will reach the velocity it would have had with no Knudsen effect farther downstream [25]. For example, the predicted velocity of a 0.1 μm yttria-stabilized zirconia (YSZ) particle, 40 mm downstream of its injection point, reaches, at most, about 200 m/s in an Ar-H_2 (20 vol.%) plasma jet of 35 kW produced with a 6 mm internal diameter anode nozzle, while the gas velocity reaches 2500 m/s at nozzle exit [27]. In most cases, the final velocities of particles contained in suspension droplets, or formed from the solution, are mostly those gained by their origin droplets before vaporization or pyrolysis occurs. For example, Joulia et al. [28], in an Ar-He plasma jet (anode-nozzle i.d. = 6 mm, $h=14$ MJ/kg, and mean $We=400$) where an ethanol suspension of YSZ particles (<1 μm) was injected, had measured, by using a particle imaging velocity technique, velocity distribution between a few and 300 m/s at 10 mm downstream of the nozzle exit against 100-400 m/s at 30 mm. Beyond 30 mm, the particle or droplet velocities decrease.

Also small particles, when traveling in areas where temperature gradients are high, which is the case of the plasma jet core with temperatures higher than 6000 K, experience thermophoretic force [26] that repels them in the direction of decreasing temperature.

With the Axial III torch (see Section 3.4.2.3), the problem is quite different. After the interchangeable water-cooled nozzle, where the three separate plasma jets converge and liquid is injected, an extension nozzle delays the mixing of the plasma with the surrounding air and results in a much lower velocity decrease. Betoule [29], for conventional d.c. plasma torches, had experimentally observed the lower decrease of the jet velocity, at the extension exit even for long extensions (50 mm): 1700 m/s instead of 700 m/s without nozzle extension. With the Axial III torch working at power levels between 57 and 118 kW and suspensions of alumina/YSZ particles with a mean diameter either of 1.4 μm or between 27 and 43 nm, the particle velocities were measured between 590 and 840 m/s with temperatures between 2600 and 3160 °C [30] thanks to the high electric power dissipated and nozzle extension.

In the high-velocity flame spray process, the Knudsen effect is generally negligible. According to the long acceleration in the divergent part of the nozzle and the barrel, gas velocities between 1200 and 1900 m/s were observed at the barrel exit. For example, Oberste-Berghaus et al. [31], using a DJ-2700 HVOF gun working with 85 slpm propylene and 279 slpm O_2, had sprayed an ethanol suspension of alumina (60 wt%) and YSZ (40 wt%) particles (Al_2O_3 particle diameters ranging between 27 and 43 nm and YSZ particle diameters between 29 and 68 nm). Under these conditions, the particle impact velocities were about 740 m/s and particle temperatures about 2814 °C.

3.3.2 Stokes' Effect

The small particles can follow the parietal gas streamlines developing parallel to the substrate surface and so do not impact on it, if their velocities are below those corresponding to Stokes' numbers, *St*, below 1 [32]:

$$St = \frac{\rho_p d_p^2 v_p}{\mu_g l_{BL}} \tag{4}$$

where the indexes p and g are related to particle and gas, respectively, ρ is the specific mass (kg/m^3), d is the particle diameter (m), v_p is its velocity (m/s),

μ is the molecular viscosity (Pa s), and l_{BL} is the thickness of the flow boundary layer, BL, in front of the substrate (m). This thickness varies as the reverse of the square root of the gas velocity close to the substrate and can be smaller than 0.1 mm [21]. Stokes' effect helps to eliminate the poorly accelerated particles as they follow the parietal gas flow. However, these particles can also create defects initiated by the substrate asperities:

- The first type of defect is that observed by VanEvery et al. [33] in YSZ coatings formed by deposition of 0.1 μm particles on a substrate that exhibited surface peaks about 8 μm high corresponding to an average roughness (R_a) of about 1 μm. Coarse columnar microstructures [33] are formed, with a cauliflower aspect, due to particle trajectories close to the substrate bringing about a shadow effect for particles with $St < 1$.
- The second type of defect is the formation of stacking defects, called speckles, appearing as soon as the ratio $R_t/d_{50} > 40$–50 (R_t is about $8 \times R_a$) [34].

3.4 Spray Torches Used

3.4.1 High-Velocity Suspension Flame Spraying Guns

To overcome the solvent evaporation problems (see Section 3.2.2), Dongmo et al. [23] and Gadow [35] modified the design of the combustion chamber. Gozali et al. [36] modeled the suspension injection in the Diamond Jet (DJ) 2700 from Sulzer Metco, working with propane, and showed that the initial mass flow rate of liquid feedstock mainly controls the HVSFS process. An optimum amount of liquid mass must be injected in the gun to achieve the maximum evaporation of the solvent inside the combustion chamber, while the cooling of the gas flow and reduction of its velocity are minimized.

In most cases, ethanol is used as solvent. Sometimes, a low water percentage is added to ethanol to reduce the temperature in the combustion chamber. Isopropanol is also used for the high enthalpy released during combustion. According to combustion temperatures in combustion chamber, up to 3300 K with propane, and low liquid mass flow rate, the HVSFS gun is well adapted to ceramic materials. For YSZ suspensions, ethene is used as fuel to achieve higher flame temperature.

The substrate standoff distances with suspensions vary generally between 100 and 150 mm from the nozzle exit.

3.4.2 Plasma Guns

Most plasma torches operate in a current control mode, that is, the current is kept constant. The current set point and the plasma gas flow rate and composition are the independent parameters and are chosen by the operator. In suspension or solution spraying, three types of torches are used: conventional plasma torches with a stick-type cathode and radial injection of the liquid feedstock at the anode-nozzle exit; Triplex-type torch (from Sulzer Metco) with three stick-type cathodes and one anode separated by neutrodes, as in cascaded arc torches, with radial injection of the liquid feedstock; and the Axial III plasma torch (from Mettech) consisting of three torches arranged to have the three plasma jets converging and surrounding the central feedstock injector.

3.4.2.1 Conventional Torches

They have a single cathode and a concentric anode nozzle with internal diameters between 5 and 10 mm. Roughly, the plasma flow velocity at the nozzle exit is inversely proportional to the square of the nozzle internal diameter. Ar-H_2 or N_2-H_2 plasma gas mixtures enable not only the highest power levels (up to 60 kW about) but also the highest voltage fluctuations due to arc root fluctuations on the anode-nozzle wall. These fluctuations can be characterized by the ratio $\Delta V/V_m$, where ΔV stands for the dispersion of the instant arc voltage distribution and V_m for the mean arc voltage. This ratio can be close to 1. The voltage fluctuations bring about variation in the enthalpy input to the plasma jet and thus variation in its velocity and temperature. When $\Delta V/V_m$ is close to 1, the gas velocity v varies by a factor up to 3 and the dynamic pressure of the plasma jet (ρv^2) varies even more [24], modifying drastically the instant liquid fragmentation (see Section 3.5.4). When using Ar-He mixtures, fluctuations are reduced below 0.3 and liquid fragmentation is more controllable. It is the same if the H_2 percentage in Ar is below 5 vol.% [28]. The spray distances for solutions and suspensions to achieve particle velocities of about 300 m/s are between 30 and 60 mm, resulting in very high heat fluxes to the substrate (up to 40 MJ/kg).

3.4.2.2 Triplex Torch

The Triplex torches from Sulzer Metco are based on the use of three cathodes insulated between them and supplied by independent sources and one anode separated from the cathodes by neutrodes (cascaded arc). The electrical energy is distributed through three parallel arcs striking at the single anode. This torch works with Ar-He gas mixtures with relatively low arc current (<500 A) and high voltage (>100 V). With this plasma-forming gas mixture, $\Delta V/V_m < 0.15$. The spray distances range generally between 50 and 70 mm.

3.4.2.3 Axial III Torch

The Axial III torch, from Northwest Mettech Corp, is made of three conventional plasma torches operated by three independent power supplies. $Ar-N_2-H_2$ or $Ar-N_2-He$ gas mixtures are used with electric power levels up to 120 kW. The three separate plasma jets converge in an interchangeable water-cooled nozzle where the liquid is injected axially between the three plasma jets. This axial injection improves the momentum and heat transfers between the injected suspension and the converging plasma jets and also makes its penetration in the plasma gas medium easier. As the three plasma torches fluctuate independently, voltage fluctuations have lower effects on the axially injected particles or liquid. At last, a nozzle extension (typically 9 mm in i.d.), following the interchangeable water-cooled nozzle, allows high acceleration of particles contained in suspension [4], and particle velocities can vary between 500 and 800 m/s at the substrate location, values for which $St > 1$. As for other plasma torches used for suspensions, the spray distances are shorter (50-80 mm) than distances used when spraying particles with sizes in the few tens micrometers (about 120 mm).

3.5 Coating Formation

3.5.1 General Remarks

Whatever may be the thermal spray process, it is impossible to avoid the impact of poorly treated particle onto the substrate. When spraying oxide particles with rather short spray distances and thus high heat fluxes to substrates, they have the tendency to stick on the hot coating under formation and create defects. Figure 3 illustrates the trajectories of fully treated and partially treated particles (a) and the resulting impacts on the substrate (b) with a conventional plasma spray torch.

The incorporation of poorly treated particles in the coating results in a rather porous microstructure. The use of wind jets as in conventional powder plasma spraying [4] can help to reduce this incorporation but with a noticeable decrease (up to 50%) in deposition efficiency.

3.5.2 Suspensions

As pointed out in Sections 3.2.2 and 3.3.1, especially for suspensions that are not axially injected (see Section 3.4), the velocity of the solid particles depends strongly on that of their mother droplets. Thus, the liquid injection and fragmentation steps play a key role in the coating buildup. Other key parameters are the particle characteristics (morphology and size distributions) and their mass load (5-20 wt%) in the solvent [32]. The particle

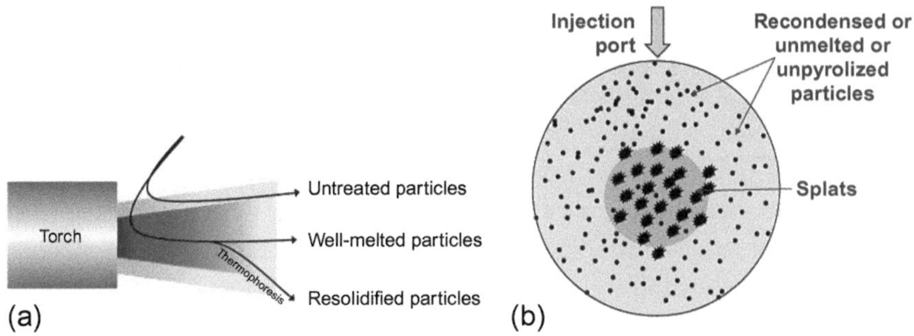

Figure 3 (a) Trajectories of well-molten, untreated, and resolidified particles within a conventional plasma jet. (b) Schematic distribution of splats and particles collected on a disk-shaped substrate.

agglomeration or aggregation in the suspension is impeded or seriously limited thanks to dispersants. However, the behavior of particles in suspensions when the liquid is evaporating is not yet clear.

According to the Young-Laplace equation, and the very important internal pressure of the flattening fully melted particles (below 1 μm), the particle flattening degree at impact on the substrate should be lower than 2. For example, Delbos [27] had collected splats of YSZ ethanol-based suspensions where the solid particles on the one hand were prepared by a chemical route (Tosoh powder) with diameters between 40 nm and 3 μm, called "wide powder" (see the distribution in Figure 4a) and on the other hand resulted

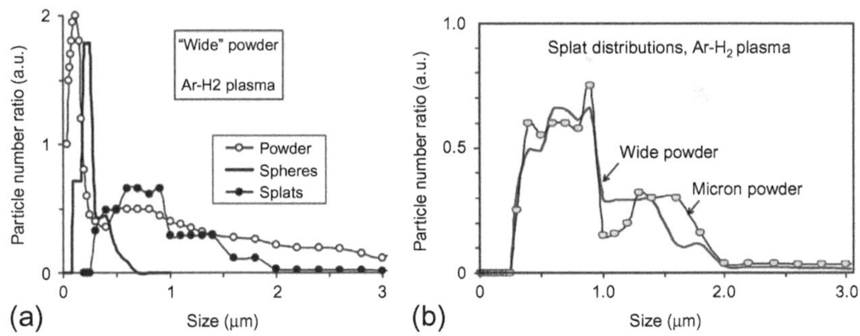

Figure 4 (a) For sprayed suspensions of Tosoh powder (10-3000 nm) called "wide powder" the following distributions are obtained: that of initial powder, that of sherical powders, which correspond to molten particles cooled slightly below melting temperature before impact because they travelled in the jet fringes and that of the resulting splats, (b) Splat distributions of the "wide powder" and "micron powder" resulting from agglomerates and aggregates of the initial Tosoh powder. Spray conditions Ar-H2 plasma: 45-15 slpm, hplasma = 17.9 MJ/kg [27].

from agglomerates and aggregates of the same powder, roughly between 0.2 and 3 μm, called "micron powder." Figure 4 shows that both types of particles result in almost the same splat size distribution. A similar splat distribution is also obtained when attrition milled particles with a diameter distribution of 0.2-3 μm are sprayed under the same spray conditions.

During the vaporization of the liquid containing the agglomerates, the liquid between agglomerated particles can evaporate either slowly, keeping the agglomerate, which then melts, or fast enough to separate the agglomerate in solid particles or small agglomerates of them. The large initial agglomerates keep their trajectory, melt rather rapidly, and form large splats (few tenths to a few micrometers in diameter) on the substrate. On the contrary, "exploded" agglomerates result in smaller agglomerates or single particles with highly dispersed trajectories, which can be melted or not. When melted, the corresponding small, below 1 μm, spherical particles (see Figure 4a) cool down very fast and keep their spherical shape.

Tarasi et al. [37] had used the Axial III torch to deposit suspensions of Al_2O_3-ZrO_2 powders. Two suspensions were sprayed with the same injection conditions: one with solid particles having average sizes of about 1.4 micrometer, the resulting coating being presented in Fig. 5a and the other with average sizes of 50 nm, the resulting coating being presented in Fig. 5b.

The picture of coating shown in Figure 5a reveals that the particles impacting on the substrate were melted and their flattening was improved, while the picture of coating presented in Figure 5b reveals lack of well-bonded splats. Amorphous and crystalline phases can be observed within the as-sprayed coatings. It seems that particle in-flight melting followed

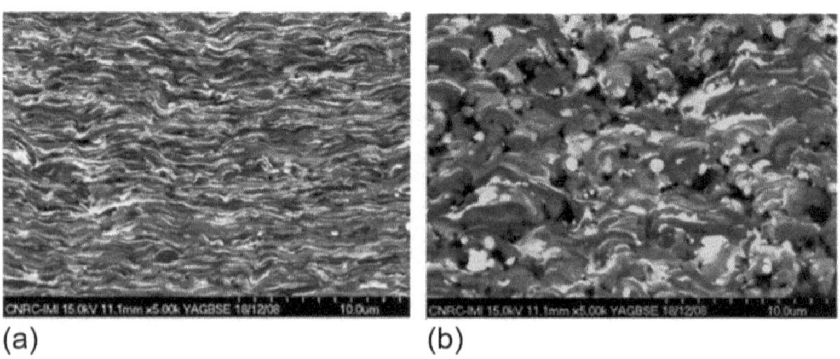

Figure 5 SPS coatings from (a) loose micron powder mixture 8 mol% YSZ (average size 1.5 μm) + 3 mol% YSZ (average size 2 μm) + alumina 95% pure (average size 1.4 μm) and (b) loose nanopowder mixture (8 mol% YSZ+3 mol% YSZ) (30-60 nm) + alumina (23-47 nm) [37].

by the mixing of alumina and zirconia is a crucial process for amorphous formation, which is increased by lower impact velocity. The amorphous phase in the coating seems to be composed of two compositional ranges, one with high alumina and the other one with high zirconia.

3.5.3 Solutions

As in the numerical models of Ozturk and Cetegen [38] and Basu et al. [39], the conservation equations of solute mass and energy have to be considered besides the momentum and heat transfer equations describing the interactions between solution drops or droplets and plasma. Such models show that the solution drops are first fragmented by the hot gas flow. Then, the heating of the formed droplets results in the precipitation of the solute as a shell and the droplet size is thus fixed by the outer diameter of the precipitate shell. The interior of the droplet is so divided into three zones: solid shell, liquid core, and vapor between them. Depending on the solute and solvent characteristics and thermophysical conditions controlling the solid precipitation, the shell will be more or less porous. If the pressure of the vapor inside it is too high, the shell will fracture. The precipitate formation depends on droplet sizes and solute initial mass fraction. In small droplets (5 μm in diameter), precipitation, whatever may be the initial solute level, encompasses the whole droplet, thus creating solid particles [39]. For larger particles (only particles up to 40 μm in diameter were considered in the model), a shell precipitate is formed and subsequently fragments, depending on the vapor pressure within the droplet. To summarize, Saha et al. [40] showed that the particle morphology, resulting from droplet processing, is sensitive to the solute chemistry, mass diffusivity, solute solubility, droplet size, thermal history, liquid injection mode, and velocity [40].

As for suspensions, the liquid droplet trajectories must be optimized to achieve a distribution as narrow as possible in the hot core of the plasma jet and obtain homogeneous droplet processing.

The surface of the spray beads deposited onto the substrate can be divided into adherent deposits (bead central part) and powdery deposits (bead edges) that correspond to particle trajectories in the hot and cold regions of the plasma jet, respectively. Four deposition mechanisms, corresponding to splat formation or particle deposition, have been identified [12,41]:

i. Deposition of fine droplets that undergo further heating on the substrate until fully melting: they recrystallize upon impact and form ultrafine (0.5-2 μm average diameter) splats, as illustrated in Figure 6a.

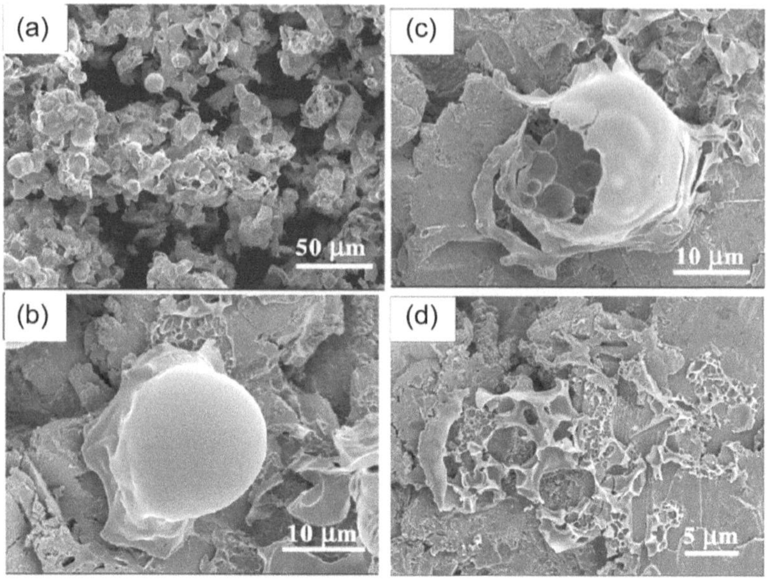

Figure 6 Deposition modes of SPPS droplets: (a) fine splats, (b) crystallized spheres, (c) ruptured shell, and (d) vapor-deposited film [12].

ii. At certain spray distances, droplets undergo resolidification and crystallization before impacting on the substrate. They form fine crystallized spheres, as illustrated in Figure 6b.

iii. Deposition of droplets entrained in the cold regions of the thermal jet where they experience sufficient heating to cause solute evaporation leading to the formation of a gel phase, deposited on the substrate. Some droplets also form a pyrolyzed shell containing nonpyrolyzed solution that fractures during deposition, as illustrated in Figure 6c.

iv. Some precursor solution droplets can reach the substrate in liquid form, having undergone none of the aforementioned processes, as illustrated in Figure 6d.

Finally, compared to suspensions, bigger solid particles are obtained with size higher than a few tenths of micrometers especially when pyrolysis occurs in the whole droplet volume.

3.5.4 Coating Generation

More or less nonpyrolyzed particles (solution spraying) or poorly treated particles (suspension spraying) can be embedded in the coating during deposition. The key parameters that control coating properties are the following [3,4]:

(i) The close control of liquid penetration into the plasma jet hot core.
(ii) For solutions, a precursor concentration close to equilibrium saturation, promoting droplet volume precipitation [13].
(iii) For suspensions, the size distribution and morphology of particles. Both are linked to their manufacturing process [3,4].
(iv) The solvent type. Solvents with high surface tension and high boiling point experience incomplete evaporation if the enthalpy of the gas jet in which the liquid feedstock is injected is not high enough. For suspensions, the solid particles contained in liquid droplets that are incompletely evaporated are poorly treated. For solutions, mud-like cracked film on the substrate is obtained, resulting in a porous coating.
(v) The use of plasma jets that have a low level of fluctuations ($\Delta V/V_m < 0.3$), especially with conventional plasma torches. To illustrate the effect of plasma jet fluctuations on coating microstructure, coatings formed from an ethanol suspension of YSZ (4.5 mol% of Y_2O_3) with particles of diameter less than 1 μm sprayed with Ar-He plasma ($h = 14$ MJ/kg and $\Delta V/V_m = 0.15$) are presented in Figure 7a, while coatings sprayed with Ar-He-H_2 plasma ($h = 20$ MJ/kg and $\Delta V/V_m = 1$) are shown in Figure 7b [28].
(vi) The substrate and coating temperature during spraying. First, substrates must be preheated over the transition temperature to get rid of adsorbates and condensates in the 250-300 °C temperature range. For solutions, nonpyrolyzed materials are pyrolyzed at the substrate surface and form aggregates when the substrate temperature is above the precursor pyrolysis temperature. This pyrolysis also occurs during the formation of the coating [12,14] and results in porous coatings.

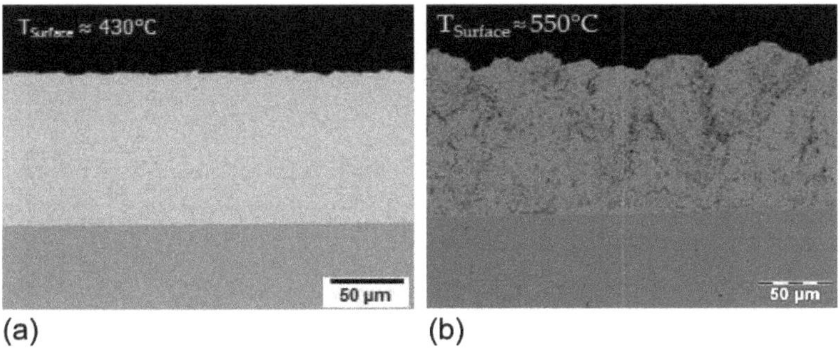

Figure 7 Cross sections of plasma-sprayed coatings deposited with Ar-He (a) and Ar-He-H_2 (b) plasmas using YSZ suspensions with particles below 1 μm in diameter [28].

For suspensions, the high temperature of substrate and coating, as well as the heat flux brought by the hot gas, can delay the splat cooling, resulting in the recoil of the flattening droplet before the onset of solidification. This brings about coatings with granular structures [27]. With short spray distances, especially with plasma spraying (see Section 3.4.2) that yields high heat fluxes (tens of MW/m^2) to substrates and with ceramic particles that have low thermal conductivities, the surface temperature of the coating under construction can reach very high values (>800 °C). The latter can be limited by using air jets and/or an adapted spray pattern to increase the time between two successive impacts of hot gas jet and particles at the same location. High temperatures favor the sticking of poorly heat-treated particles contained in the suspension that brings about defects between the successive sprayed beads [3].

Also, high heat flux input to coatings with low thermal conductivity results in high-temperature gradients within coatings and induces tensile stresses that are released by vertical cracking [42]. For example, in YSZ TBCs made by suspension plasma spraying, the segmentation cracks survive during thermal cycling test, but sintering occurs quickly during the first hours of exposure at 1400 °C, limiting the coating strain tolerance. On the contrary, TBCs deposited from solutions presented excellent durability in furnace cycling tests, that is, 2.5 times the spallation life of air plasma coatings APS and 1.5 times the life of electron beam physical vapor deposition coatings [43]. These vertical cracks are due to the pyrolysis of unmolten material, contained in bead wings, when heated by either subsequent plasma torch passes or postdeposition treatment. Tensile stresses are then generated, driving the formation of vertical cracks [44].

(vii) With the Axial III plasma torch, resulting in particle velocities over 500 m/s, coatings can be deposited on substrates with rough surfaces. Figure 8 shows the image of a solid oxide fuel cell (SOFC), presented by Waldbillig and Kesler [45]. Three layers of the cell were deposited with the Axial III plasma torch. The electrolyte was made from a water-based suspension of YSZ with a particle mean diameter of 1.6 μm. Starting from a porous ferritic stainless steel, the cathode surface was rather rough and the sprayed suspension followed well the asperities. The electrolyte-specific resistance was below 0.1 Ω cm^2 at 750 °C for a thickness <20 μm.

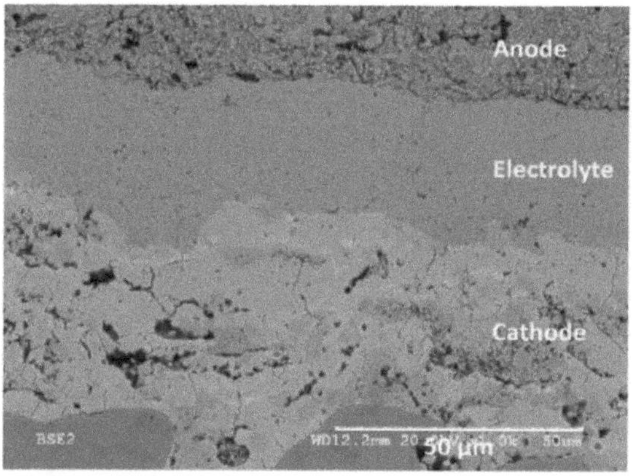

Figure 8 BSE SEM images of typical SOFC microstructure at 1000 × magnification [45].

3.6 Examples of Coatings

3.6.1 Suspensions

3.6.1.1 HVSFS-Sprayed Coatings

(i) *Alumina*: Bolelli et al. [46] deposited coatings using two nanosized Al_2O_3 powders and a micron-sized one, the latter resulting in denser and more wear-resistant coatings.

(ii) *Alumina-zirconia:* Oberste-Berghaus et al. [31] sprayed with HVSFS and plasma (Mettech III torch) alumina-zirconia coatings, which exhibited higher hardness and fracture toughness and lower thermal conductivity than pure alumina or zirconia coatings.

(iii) *Bioglasses:* Bioactive glasses were frequently reported to exhibit higher bioactivity than HA. Bolelli et al. [47] processed by HVSFS two different oxide-based bioactive materials. Bellucci et al. [48] sprayed with HVSFS processed a suspension of water and isopropanol containing a novel bioactive glass composition. With this process, the glass transition temperature of the system was never attained, and the coating consisted of layers with uniformly distributed micrometric pores. A TiO_2 bond coat on the Ti substrate improved their adhesion. Altomare et al. [49] showed that HVSFS-deposited 45S5 bioglass coatings were highly bioactive and had potentials as replacement of conventional HA for implants.

(iv) *Hydroxyapatite:* Bolelli et al. [50] had HVSFS-sprayed a suspension of nanometer-sized powders. Coatings were dense, 27-37 μm thick, and with some transverse microcracks.

(v) *Mullite:* Oberste-Berghaus and Marple [51] used HVSFS to spray suspension of micron-sized ($d_{50}=1.8$ µm) mullite ($3Al_2O_3$-$2SiO_2$) particles. They showed that the crystalline phases were retained in the coating only when the latter was gradually heated during deposition up to temperatures above 400 °C.

(vi) *Oxides:* Gadow et al. [35] sprayed ceramic suspensions (Al_2O_3, YSZ, Cr_2O_3, and TiO_2) by using HVSFS. They used isopropanol as a solvent and powders whose particle size was either micronic, or submicronic, or nanometric. Rauch et al. [52] performed a similar work except for TiO_2 particles. The optical micrographs of coatings showed that they were rather dense, with a porosity in the range of 5% for Al_2O_3, 5-10% for Cr_2O_3, and <1% for the YSZ. The wear rate of HVSF-sprayed Al_2O_3 coating was nearly one order in magnitude lower than that of standard APS and HVOF coatings. HVSFS-sprayed Cr_2O_3 coatings exhibited the highest hardness values (1446 Berkovich) and elastic modulus (206 GPa).

(vii) *Samaria-doped ceria (SDC):* Oberste-Berghaus et al. [53] sprayed SDC electrolyte for SOFC and showed that for the optimized spray conditions, the defect density in the resulting coating was reduced.

(viii) *Titania:* The HVSFS process allows tailoring the coating properties (thickness, porosity, anatase content, hardness, etc.) according to the desired objective: for example, for photocatalytic activity, the coating anatase content (up to 30 wt%) and porosity can be varied [54]. The work of Yuan et al. showed that hybrid micro-/nanostructured TiO_2 coatings exhibited superhydrophilic performances (contact angle close to 0°) [55].

3.6.1.2 Plasma-Sprayed Coatings

(i) *Alumina:* Most studies, with conventional plasma torches, were devoted to achieve coatings with a lower friction coefficient than that of APS coatings [56–58]. The addition of SiC in the alumina matrix permitted to decrease the coating friction coefficient [59]. Suspension coatings deposited by conventional plasma and HVSFS spray processes were also compared [60]. Brousse et al. [61] used aluminum and alumina suspensions and checked how a metallic matrix and very few alumina inclusions modified the reflectance and resistivity of the coatings.

(ii) *Alumina-zirconia and composites:* The suspension plasma spraying was used with the Axial III plasma torch for the deposition of pseudoeutectic composition of alumina-YSZ as potential TBC [29,60,62]. The particle velocity was found as a crucial parameter dictating the crystalline structure of the resulting coatings. With conventional plasma torch, finely structured layers of Al_2O_3-ZrO_2 with a large variety of architectures were produced [58].

(iii) *Alumina-titania:* Darut et al. [63] showed that coatings from suspensions with submicrometer-sized Al_2O_3-TiO_2 particles exhibited higher performances than that of coatings from suspensions with micrometer-sized particles. When the TiO_2 content was increased, the content of Al_xTiO_y or Al_2TiO_5 compounds was also increased.

(iv) *Bioglass:* Bioactive glass coatings deposited via suspension plasma spraying were studied to improve the adhesion between orthopedic implants and bones [64,65].

(v) *Hydroxyapatite:* $Ca_{10}(PO_4)_6(OH)_2$ powder was formulated with water and alcohol to obtain suspension, which was plasma-sprayed onto titanium substrate [66]. Huang et al. [67] used a novel liquid precursor plasma spraying process to deposit Si, Mg, and carbonate HA coatings (alone and cosubstituted) onto Ti6Al4V substrates.

(vi) *Hot extrusion tool:* Different oxide ceramics were mixed and prepared in one suspension and plasma-sprayed to produce coatings with low friction coefficient under dry sliding conditions [68]. Titania and chromia showed a different behavior during tribological testing on aluminum and copper-based extrudable alloys.

(vii) *SOFC:* Many works have been performed to deposit by using plasma suspension spraying mainly the electrolyte [45,69,70] and also the anode [70,71] and cathode [72] of SOFCs. For the electrolyte when the Axial III plasma torch was used [45,69], dense coatings were obtained (see Section 3.5.4.vii). However, the control of the coating during spraying was important to avoid vertical cracks due to the high heat flux brought by the plasma jet [69]. Conventional torches have also been used [72], but the electrolyte coatings were less dense. The anode (Ni-YSZ) has been sprayed with the Axial III plasma torch [70,71]. Carbon black addition to the suspension controlled the coating total porosity as well as the suspension feed rate. The size of the NiO particles in the suspension affected both the composition and the total porosity of coatings [71]. Cuglietta et al. [70] injected axially a multicomponent aqueous suspension of CuO, Co_3O_4, and NiO and radially a dry powder of SDC. Depending on the processing conditions, the SDC and the metal phases could be either very well mixed or layered. Wang et al. [72] sprayed with conventional plasma torch a suspension of $La_{0.8}Sr_{0.2}MnO_3$ (LSM) and obtained a specific surface resistance of 0.45 Ω cm^2 at 950 °C.

(viii) *TBC:* The results of Guignard et al. [42] for YSZ thermal barriers sprayed with Triplex torch have been presented in Section 3.5.4.vi. Gong et al. [73] sprayed with conventional plasma torch suspensions of YSZ powder in ethanol where rare-earth nitrates were dissolved into prior to plasma spraying.

(ix) *Titania:* Titania is used for its wear resistance (rutile phase) or photocatalytic properties (anatase phase). Dense coatings were obtained starting from aqueous suspensions, and porous coatings were elaborated with alcoholic suspension, but with the latter, the transformation from anatase to rutile was important [74]. Mauer et al. [75] added TiN to the anatase powder to enhance the photocatalytic activity. Unfortunately, to dissolute the nitrogen in the TiO_2 lattice, hot spray parameters are necessary and initial anatase is mainly transformed into rutile.

(x) *YSZ:* Kozerski et al. [76] with a conventional plasma torch deposited coatings with thermal conductivities below 1 W/m K. Darut et al. [56], also with a conventional plasma torch, showed that the tribological performance was improved in comparison with conventional coatings. Addition of SiC particles in the suspension resulted in a higher wear resistance. Waldbillig and Kesler [77] with the Axial III plasma torch optimized the spray conditions to achieve dense coatings with good deposition efficiency. Vert et al. [78] showed that a thin (<100 μm) nanostructured layer of YSZ sprayed with a conventional torch on smooth (R_a ~0.5 μm) and thin sheets of Haynes 230 (1 mm thick) allowed depositing conventional YSZ coating with an acceptable tensile bond strength, between 12 and 24 MPa, depending on the spraying conditions and SPS layer thickness.

(xi) *YAG:* Zhao et al. [79] sprayed with the Axial III plasma torch a suspension of Y_2O_3 and Al_2O_3 with a mole ratio of Y-Al=5:3 in distilled water. They showed that high torch power and short standoff distance resulted in high density and high hardness coating.

(xii) *Yttria:* Yttrium oxide (Y_2O_3) coatings, prepared by axial suspension plasma spraying, exhibited high density, uniform structure, high hardness, high plasma erosion resistance, and retention of smoother surface after plasma erosion [80].

3.6.2 Solutions

3.6.2.1 HVSFS-Sprayed

Cetegen and Basu [81] had modeled the thermophysical processes in liquid ceramic precursor droplets injected into an HVOF flame jet. Chen et al. [82] had used the HVSFS process to deposit Al_2O_3-ZrO_2 ceramic coatings. The as-sprayed coating was composed of mixed nanocrystalline ZrO_2 and γ-Al_2O_3 as well as amorphous phases. Polished cross section showed that the coating was quite dense with a thickness of 40 μm.

3.6.2.2 Plasma-Sprayed

Ozturk and Cetegen [83] modeled droplets containing a ceramic precursor injected axially or transversely into a plasma jet. Axial injection was found to result in more rapid heat-up and precipitation than transverse injection.

(i) *Al_2O_3 α-phase:* Sivakumar et al. [84] sprayed organic precursors with a conventional plasma torch. Fully molten Al_2O_3 splats, as well as extremely fine particles due to secondary pyrolytic action, were observed.

(ii) *SOFC:* Works were devoted on the one hand to cathodes sprayed with conventional plasma torches with either LSM [85] or $Sm_{0.7}Sr_{0.3}Co_3Ld$ [86] and on the other hand to anodes [71]. In the latter case, it was found that the addition of citric acid to the aqueous solution delayed resolidification of NiO particles, improving the deposition efficiency and coating adhesion.

(iii) *TBC:* The solution precursor plasma spray (SPPS) process offered the prospect of depositing highly durable TBCs of low thermal conductivity. All studies were performed with conventional plasma torches [12,44].

(iv) *TiO_2:* Porous titania coatings [87] were deposited from aqueous solution containing titanium isopropoxide. A dense titania (TiO_2) coating was deposited from an ethanol-based solution containing titanium isopropoxide [88].

(v) *Zn ferrite:* Deposition of pure spinel phase, photocatalytic zinc ferrite films on SS-304 substrates by SPPS has been demonstrated [89].

(vi) *Metastable ceramics* (ZrO_2-Al_2O_3): Vasiliev et al. [90] used this process to deposit 75% dense coatings of metastable ceramics in the ZrO_2-10 mol% Al_2O_3 binary system. Vasiliev and Padture [91] deposited 75% dense, metastable coatings in the following ternary systems (mol%): (i) $10Al_2O_3$-$86.4ZrO_2$-$3.6Y_2O_3$ (10AlZrY) and (ii) $20Al_2O_3$-$76.4ZrO_2$-$3.6Y_2O_3$ (20AlZrY).

4. CONCLUSIONS

The interest for nanostructured coatings has been leading to the development of many works over the last decade and the emerging of specific thermal spray techniques. These techniques are based on either conventional spray processes using powders with particles in the tens of micrometers size range (called "conventional coatings" in the following) or new processes using liquid feedstocks, either suspensions of submicrometer solid particles or solutions of chemical precursors (called "coatings from liquid" in the following).

Conventional coatings

- The spraying of agglomerated nanometer-sized particles with a close control of their heating in the hot gas jet in order to melt them only partially and embed the unmolten nanostructures within fully melted ones, forming the "cement" of the coating, results in the formation of a coating with a bimodal structure. Only ceramic particles have been sprayed that way and a few industrial applications are developed.
- To conclude such, coatings are relatively easy to implement in industry because they require the same equipments as those used for conventional coatings.

Coatings from liquids

- Suspension thermal spraying and solution precursor thermal spraying allow achieving finely structured layers with a thickness varying from a few micrometers to a few hundreds of micrometers. However, they are more complex than conventional thermal spray processes:
- They can be achieved only with plasma or HVOF (then called HVSFS torches).
- The liquid injection (radially or axially) in the gas flow and their interactions result in the liquid fragmentation by the plasma jet, followed by droplet vaporization with either the acceleration and the melting of solid particles contained in suspension droplet or the pyrolyzation, sintering, and melting of solution droplets.
- Droplets or particles with sizes below a few micrometers are submitted to the Knudsen effect, which delays their acceleration and heating. Moreover, they must have a sufficient velocity to impact the substrate (Stokes' effect) and also avoid creating defects due to substrate roughness relatively to impacting particles' diameters. If this condition can be

achieved with HVSFS or the axial plasma torch of Mettech, it is not necessarily the case with conventional plasma spray torches.

- All the preceding remarks mean that the spray conditions of suspensions or solutions are more difficult to optimize than those of conventional thermal spray techniques.
- Moreover, the diagnostic techniques and sensors required to monitor the process have to be adapted or developed in order to observe the in-flight behavior of droplets and particles with sizes below 5 μm. The measuring techniques for coating properties have also to be adapted, for example, measuring the coating porosity with pores below a few hundreds of micrometers implies using USAXS.
- To better control the plasma process, the devices used to inject the liquid have to be optimized. In particular, the injectors should produce liquid jets or drops with controlled narrow distributions of sizes, if possible below 50 μm, and velocities, both varying independently. Commercial liquid feedstock feeders begin to appear on the market. The spraying of liquid feedstock also requires higher power levels than those used to spray conventional powders and plasma torches with low arc voltage fluctuations. Thus, plasma torches must be adapted or specific torch developed. The deposition rate, which is about 10–30% that of conventional spraying, should also be increased.
- At last, no standards have been defined yet for both the spray process and the solution or suspension preparation.
- However, many potential applications have been described in the literature [12,37,47–91] and could emerge in the short or medium term at an industrial level.

REFERENCES

[1] M.F. Ashby, P.J. Ferreira, D.L. Schodek, Nanomaterials properties, Nanomaterials, Nanotechnologies and Design, Elsevier Ltd, Kidlington, UK, 2009, pp. 199–255.
[2] V. Viswanathan, T. Laha, K. Balani, A. Agarwal, S. Seal, Challenges and advances in nanocomposite processing techniques, Mater. Sci. Eng. R54 (2006) 121–285.
[3] P. Fauchais, G. Montavon, R.S. Lima, B.R. Marple, Engineering a new class of thermal spray nano-based microstructures from agglomerated nanostructured particles, suspensions and solutions: an invited review, J. Phys. D 44 (2011) 093001.
[4] P. Fauchais, J. Heberlein, M. Boulos, Thermal Spray Fundamentals, Springer, New York, 2014, p. 1560.
[5] R.S. Lima, B.R. Marple, Thermal spray coatings engineered from nanostructured ceramic agglomerated powders for structural, thermal barrier and biomedical applications: a review, J. Therm. Spray Technol. 16 (1) (2007) 40–63.

[6] L.L. Shaw, D. Goberman, R. Ren, M. Gell, S. Jiang, Y. Wang, T.D. Xiao, P.R. Strutt, The dependency of microstructure and properties of nanostructured coatings on plasma spray conditions, Surf. Coat. Technol. 130 (2000) 1–8.
[7] R.S. Lima, B.R. Marple, Toward highly sintering-resistant nanostructured ZrO_2-7wt% Y_2O_3 coatings for TBC applications by employing differential sintering, J. Therm. Spray Technol. 17 (5-6) (2008) 846–852.
[8] P. Fauchais, A. Vardelle, Innovative and emerging processes in plasma spraying: from micro- to nanostructured coatings, J. Phys. D Appl. Phys. 44 (2011) 194011 (14 pp).
[9] A. Bacciochini, J. Ilavsky, G. Montavon, A. Denoirjean, F. Ben-ettouil, S. Valette, P. Fauchais, K. Wittmann-Teneze, Quantification of void network architectures of suspension plasma-sprayed (SPS) yttria-stabilized zirconia (YSZ) coatings using ultra-small-angle-X-ray-scattering (USAXS), Mater. Sci. Eng. A528 (2010) 91–102.
[10] R. Rampon, C. Filiatre, G. Bertrand, Suspension plasma spraying of YPSZ coatings: suspension atomization and injection, J. Therm. Spray Technol. 17 (1) (2008) 105–114.
[11] B.G. Ravi, S. Sampath, R. Gambino, P.S. Devi, J.B. Parise, Plasma spray synthesis from precursors: progress, issues and considerations, J. Therm. Spray Technol. 15 (4) (2006) 701–707.
[12] M. Gell, E.H. Jordan, M. Teicholz, B.M. Cetegen, N. Padture, L. Xie, D. Chen, X. Ma, J. Roth, Thermal barrier coatings made by the solution precursor plasma spray process, J. Therm. Spray Technol. 17 (1) (2008) 124–135.
[13] D. Chen, E.H. Jordan, M. Gell, Effect of solution concentration on splat formation and coating microstructure using the solution precursor plasma spray process, Surf. Coat. Technol. 202 (2008) 2132–2138.
[14] D. Chen, E.H. Jordan, M. Gell, The solution precursor plasma spray coatings: influence of solvent type, Plasma Chem. Plasma Process. 30 (2010) 111–119.
[15] C.K. Muoto, E.H. Jordan, M. Gell, M. Aindow, Identification of desirable precursor properties for solution precursor plasma spray, J. Therm. Spray Technol. 20 (4) (2011) 802–816.
[16] E.H. Jordan, M. Gell, P. Bonzani, D. Chen, S. Basu, B. Cetegen, F. Wu, X. Ma, Making dense coatings with the solution precursor plasma spray process, in: B.R. Marple, et al. (Eds.), Thermal Spray 2007: Global Coating Solution, ASM International, Materials Park, OH, 2007, pp. 463–470.
[17] O. Marchand, L. Girardot, M.P. Planche, P. Bertrand, Y. Bailly, G. Bertrand, An insight into suspension plasma spray: injection of the suspension and its interaction with the plasma flow, J. Therm. Spray Technol. 20 (6) (2011) 1310–1320.
[18] P. Fauchais, A. Joulia, S. Goutier, C. Chazelas, M. Vardelle, A. Vardelle, S. Rossignol, Suspension and solution plasma spraying, J. Phys. D Appl. Phys. 46 (2013) 224015 (14 pp).
[19] C. Marchand, A. Vardelle, G. Mariaux, P. Lefort, Modelling of the plasma spray process with liquid feedstock injection, Surf. Coat. Technol. 202 (2008) 4458–4464.
[20] P. Blazdell, S. Kuroda, Plasma spraying of submicron ceramic suspensions using a continuous ink jet printer, Surf. Coat. Technol. 123 (2000) 239–246.
[21] J. Fazilleau, C. Delbos, V. Rat, J.-F. Coudert, P. Fauchais, B. Pateyron, Phenomena involved in suspension plasma spraying part 1: suspension injection and behaviour, Plasma Chem. Plasma Process. 26 (4) (2006) 371–391.
[22] P. Fauchais, R. Etchart-Salas, C. Delbos, M. Tognovi, V. Rat, J.F. Coudert, T. Chartier, Suspension and solution plasma spraying of finely structured coatings, J. Phys. D Appl. Phys. 40 (2007) 2394–2406.
[23] E. Dongmo, R. Gadow, A. Killinger, M. Wenzelburger, Modelling of combustion as well as heat, mass, and momentum transfer during thermal spraying by HVOF and HVSFS, J. Therm. Spray Technol. 18 (5-6) (2009) 896–908.

[24] R. Etchart-Salas, V. Rat, J.F. Coudert, P. Fauchais, N. Caron, K. Wittman, S. Alexandre, Influence of plasma instabilities in ceramic suspension plasma spraying, J. Therm. Spray Technol. 16 (5-6) (2007) 857–865.
[25] D. Damiani, Elaboration, Fitting in Shape and Characterization of Materials for Energy by the Interaction Radiation-Matter, HdR University of Limoges, France, 2012.
[26] E. Pfender, Particle behavior in thermal plasmas: I, Plasma Chem. Plasma Process. 9 (1989) 167S–194S.
[27] C. Delbos, J. Fazilleau, V. Rat, J.F. Coudert, P. Fauchais, B. Pateyron, Phenomena involved in suspension plasma spraying part 2: zirconia particle treatment and coating formation, Plasma Chem. Plasma Process. 26 (2006) 393–414.
[28] A. Joulia, Contribution to the development of thermal barriers for space propulsion by plasma spraying of liquid feedstock, PhD Thesis, Univ. of Limoges France Dec. 19 th (2013).
[29] O. Betoule, Influence of velocity and temperature distributions of d.c. plasma jets and alumina particles in flight onto coating properties, PhD Thesis, University of Limoges, Limoges, France, 1994.
[30] F. Tarasi, M. Medraj, A. Dolatabadi, J. Oberste-Berghaus, C. Moreau, Phase formation and transformation in alumina/YSZ nanocomposite coating deposited by suspension plasma spray process, J. Therm. Spray Technol. 19 (4) (2010) 787–795.
[31] J. Oberste Berghaus, J.-G. Legoux, C. Moreau, F. Tarasi, T. Chraska, Mechanical and thermal transport properties of suspension thermal-sprayed alumina-zirconia composite coatings, J. Therm. Spray Technol. 17 (1) (2008) 91–104.
[32] P. Fauchais, R. Etchart-Salas, V. Rat, J.F. Coudert, N. Caron, K. Wittmann-Téenèze, Parameters controlling liquid plasma spraying: solutions, sols, or suspensions, J. Therm. Spray Technol. 17 (2008) 31–59.
[33] K. VanEvery, M.J.M. Krane, R.W. Trice, Parametric study of suspension plasma spray processing parameters on coating microstructures manufactured from nanoscale yttria-stabilized zirconia, Surf. Coat. Technol. 206 (2012) 2464–2473.
[34] E. Brousse-Pereira, Elaboration by thermal spray of finely structured elements of a high temperature electrolyser for hydrogen production: processes, structures and characteristics, PhD Thesis, University of Limoges, France, September 21 2010.
[35] R. Gadow, A. Killinger, J. Rauch, Introduction to high-velocity suspension flame spraying (HVSFS), J. Therm. Spray Technol. 17 (5-6) (2008) 655–661.
[36] E. Gozali, S. Kamnis, S. Gu, Numerical investigation of combustion and liquid feedstock in high velocity suspension flame spraying process, Surf. Coat. Technol. 228 (2013) 176–186.
[37] F. Tarasi, M. Medraj, A. Dolatabadi, J. Oberste-Berghaus, C. Moreau, Amorphous and crystalline phase formation during suspension plasma spraying of the alumina-zirconia composite, J. Eur. Ceram. Soc. 31 (2011) 2903–2913.
[38] A. Ozturk, B. Cetegen, Modeling of axially and transversely injected precursor droplets into a plasma environment, J. Heat Mass Transfer 48 (21-22) (2005) 4367–4383.
[39] S. Basu, E.H. Jordan, B.M. Cetegen, Fluid mechanics and heat transfer of liquid precursor droplets injected into high temperature plasmas, J. Therm. Spray Technol. 15 (4) (2006) 576–581.
[40] A. Saha, S. Seal, B. Cetegen, E. Jordan, A. Ozturk, S. Basu, Thermo-physical processes in cerium nitrate precursor droplets injected into high temperature plasma, Surf. Coat. Technol. 203 (2009) 2081–2091.
[41] L. Xie, X.C. Ma, E.H. Jordan, N.P. Padture, D.T. Xiao, M. Gell, Deposition mechanisms of thermal barrier coatings in the solution precursor plasma spray process, Surf. Coat. Technol. 177-178 (2004) 103–107.
[42] A. Guignard, G. Mauer, R. Vaßen, D. Stöver, Deposition and characteristics of submicrometer-structured thermal barrier coatings by suspension plasma spraying, J. Therm. Spray Technol. 21 (3-4) (2012) 416–424.

[43] L. Xie, D. Chen, E.H. Jordan, N.P. Padture, A. Ozturk, F. Wu, X.C. Ma, B. M. Cetegem, M. Gell, Formation of vertical cracks in solution-precursor plasma-sprayed thermal barrier coatings, Surf. Coat. Technol. 201 (2006) 1058–1064.
[44] M. Gell, L. Xie, X. Ma, E.H. Jordan, N.P. Padture, Highly durable thermal barrier coatings made by the solution precursor plasma spray process, Surf. Coat. Technol. 177-178 (2004) 97–102.
[45] D. Waldbillig, O. Kesler, Electrochemical testing of suspension plasma sprayed solid oxide fuel cell electrolytes, J. Power Sources 196 (2011) 5423–5431.
[46] G. Bolelli, B. Bonferroni, V. Cannillo, R. Gadow, A. Killinger, L. Lusvarghi, J. Rauch, N. Stiegler, Wear behaviour of high velocity suspension flame sprayed (HVSFS) Al_2O_3 coatings produced using micron- and nano-sized powder suspensions, Surf. Coat. Technol. 204 (2010) 2657–2668.
[47] G. Bolelli, N. Stiegler, D. Bellucci, V. Cannillo, R. Gadow, A. Killinger, L. Lusvarghi, A. Sola, Deposition mechanisms in high velocity suspension spraying: case study for two bioactive materials, Surf. Coat. Technol. 210 (2012) 28–45.
[48] D. Bellucci, G. Bolelli, V. Cannillo, R. Gadow, A. Killinger, L. Lusvarghi, A. Sola, N. Stiegler, High velocity suspension flame sprayed (HVSFS) potassium-based bioactive glass coatings with and without TiO_2 bond coat, Surf. Coat. Technol. 206 (2012) 3857–3868.
[49] L. Altomare, D. Bellucci, G. Bolelli, B. Bonferroni, V. Cannillo, L. De Nardo, R. Gadow, A. Killinger, L. Lusvarghi, A. Sola, N. Stiegler, Microstructure and in vitro behaviour of 45S5 bioglass coatings deposited by high velocity suspension flame spraying (HVSFS), J. Mater. Sci. Mater. Med. 22 (2011) 1303–1319.
[50] G. Bolelli, D. Bellucci, V. Cannillo, L. Lusvarghi, A. Sola, N. Stiegler, P. Müller, A. Killinger, R. Gadow, L. Altomare, L. De Nardo, Suspension thermal spraying of hydroxyapatite: microstructure and in vitro behavior, Mater. Sci. Eng. C 34 (2014) 287–303.
[51] J. Oberste Berghaus, B.R. Marple, High-velocity oxy-fuel (HVOF) suspension spraying of mullite coatings, J. Therm. Spray Technol. 17 (5-6) (2008) 671–678.
[52] J. Rauch, G. Bolelli, A. Killinger, R. Gadow, V. Cannillo, L. Lusvarghi, Advances in high velocity suspension flame spraying (HVSFS), Surf. Coat. Technol. 203 (2009) 2131–2138.
[53] J. Oberste Berghaus, J.-G. Legoux, C. Moreau, R. Hui, C. Decès-Petit, W. Qu, S. Yick, Z. Wang, R. Maric, D. Ghosh, Suspension HVOF spraying of reduced temperature solid oxide fuel cell electrolytes, J. Therm. Spray Technol. 17 (5-6) (2008) 700–707.
[54] E. Bemporad, G. Bolelli, V. Cannillo, D. De Felicis, R. Gadow, A. Killinger, L. Lusvarghi, J. Rauch, M. Sebastiani, Structural characterization of high velocity suspension flame sprayed (HVSFS) TiO_2 coatings, Surf. Coat. Technol. 204 (2010) 3902–3910.
[55] J. Yuan, Q. Zhan, Q. Lei, Si. Ding, H. Li, Fabrication and characterization of hybrid micro/nano-structured hydrophilic titania coatings deposited by suspension flame spraying, Appl. Surf. Sci. 258 (2012) 6672–6678.
[56] G. Darut, H. Ageorges, A. Denoirjean, G. Montavon, P. Fauchais, Effect of the structural scale of plasma-sprayed alumina coatings on their friction coefficients, J. Therm. Spray Technol. 17 (5-6) (2008) 788–795.
[57] F.-L. Toma, L.-M. Berger, C.C. Stahr, T. Naumann, S. Langner, Microstructures and functional properties of suspension-sprayed Al_2O_3 and TiO_2 coatings: an overview, J. Therm. Spray Technol. 19 (1-2) (2010) 262–274.
[58] O. Tingaud, P. Bertrand, G. Bertrand, Microstructure and tribological behavior of suspension plasma sprayed Al_2O_3 and Al_2O_3-YSZ composite coatings, Surf. Coat. Technol. 205 (2010) 1004–1008.

[59] G. Darut, F. Ben-Ettouil, A. Denoirjean, G. Montavon, H. Ageorges, P. Fauchais, Dry sliding behavior of sub-micrometer-sized suspension plasma sprayed ceramic oxide coatings, J. Therm. Spray Technol. 19 (1-2) (2010) 275–285.
[60] P. Müller, A. Killinger, R. Gadow, Comparison between high-velocity suspension flame spraying and suspension plasma spraying of alumina, J. Therm. Spray Technol. 21 (6) (2012) 1120–1129.
[61] E. Brousse-Pereira, K. Wittmann-Teneze, V. Bianchi, J.L. Longuet, L. Del Campo, Optical and electrical properties of heterogeneous coatings produced by aluminum powder and boehmite suspension plasma spraying, J. Therm. Spray Technol. 21 (10) (2012) 1110–1121.
[62] F. Tarasi, M. Medraj, A. Dolatabadi, J. Oberste-Berghaus, C. Moreau, Effective parameters in axial injection suspension plasma spray process of alumina-zirconia ceramics, J. Therm. Spray Technol. 17 (5-6) (2008) 685–691.
[63] G. Darut, El. Klyatskina, S. Valette, P. Carles, A. Denoirjean, G. Montavon, H. Ageorges, F. Segovia, M. Salvador, Architecture and phases composition of suspension plasma sprayed alumina-titania sub-micrometer-sized coatings, Mater. Lett. 67 (2012) 241–244.
[64] A. Cattini, L. Łatka, D. Bellucci, G. Bolelli, A. Sola, L. Lusvarghi, L. Pawłowski, V. Cannillo, Suspension plasma sprayed bioactive glass coatings: effects of processing on microstructure, mechanical properties and in-vitro behavior, Surf. Coat. Technol. 220 (2013) 52–59.
[65] Y. Xiao, L. Song, X. Liu, Y. Huang, T. Huang, Y. Wu, J. Chen, F. Wu, Nanostructured bioactive glass-ceramic coatings deposited by the liquid precursor plasma spraying process, Appl. Surf. Sci. 257 (2011) 1898–1905.
[66] H. Podlesak, L. Pawlowski, R. Haese, J. Laureyns, T. Lampke, S. Bellayer, Advanced microstructural study of suspension plasma sprayed hydroxyapatite coatings, J. Therm. Spray Technol. 19 (3) (2010) 657–664.
[67] T. Huang, Y. Xiao, S. Wang, Y. Huang, X. Liu, F. Wu, Z. Gu, Nanostructured Si, Mg, CO_3^{2-} substituted hydroxyapatite coatings deposited by liquid precursor plasma spraying: synthesis and characterization, J. Therm. Spray Technol. 20 (4) (2011) 829–836.
[68] M. Erne, D. Kolar, C. Hübsch, M. Möhwald, Fr.-W. Bach, Synthesis of tribologically favorable coatings for hot extrusion tools suspension plasma spraying, J. Therm. Spray Technol. 21 (3-4) (2012) 668–675.
[69] A. Macwan, D.L. Chen, M. Marr, O. Kesler, Residual stresses in suspension plasma sprayed electrolytes in metal-supported solid oxide fuel cell half cells, J. Power Sources 221 (2013) 397–405.
[70] M. Cuglietta, J. Kuhn, O. Kesler, A novel hybrid axial-radial atmospheric plasma spraying technique for the fabrication of solid oxide fuel cell anodes containing Cu, Co, Ni, and samaria-doped ceria, J. Therm. Spray Technol. 22 (5) (2013) 609–621.
[71] C. Metcalfe, J. Kuhn, O. Kesler, Characterization of Ni-YSZ anodes for solid oxide fuel cells fabricated by suspension plasma spraying with axial feedstock injection, J. Power Sources 243 (2013) 172–180.
[72] X.-M. Wang, C.-X. Li, C.-J. Li, G.-J. Yang, Effect of microstructures on electrochemical behavior of $La_{0.8}Sr_{0.2}MnO_3$ deposited by suspension plasma spraying, Int. J. Hydrogen Energy 35 (2010) 3152–3158.
[73] S. Gong, K. VanEvery, H. Wang, R.W. Trice, Microstructure and thermal properties of inflight rare-earth doped thermal barriers prepared by suspension plasma spray, J. Eur. Ceram. Soc. 34 (5) (2014) 1243–1253.
[74] F.-L. Toma, G. Bertrand, S. Begin, C. Meunier, O. Barres, D. Klein, C. Coddet, Microstructure and environmental functionalities of TiO_2-supported photocatalysts obtained by suspension plasma spraying, Appl. Catal. B Environ. 68 (2006) 74–84.

[75] G. Mauer, A. Guignard, R. Vaßen, Plasma spraying of efficient photoactive TiO_2 coatings, Surf. Coat. Technol. 220 (2013) 40–43.
[76] S. Kozerski, L. Łatka, L. Pawlowski, F. Cernuschi, F. Petit, C. Pierlot, H. Podlesak, J. P. Laval, Preliminary study on suspension plasma sprayed ZrO_2 + 8 wt.% Y_2O_3 coatings, J. Eur. Ceram. Soc. 31 (2011) 2089–2098.
[77] D. Waldbillig, O. Kesler, Effect of suspension plasma spraying process parameters on YSZ coating microstructure and permeability, Surf. Coat. Technol. 205 (2011) 5483–5492.
[78] R. Vert, P. Carles, E. Laborde, G. Mariaux, E. Meillot, A. Vardelle, Adhesion of ceramic coating on thin and smooth metal substrate: a novel approach with a nanostructured ceramic interlayer, J. Therm. Spray Technol. 21 (6) (2012) 1128–1134.
[79] D. Zhao, T.W. Coyle, K. Chien, Phase composition and microstructure of yttrium aluminum garnet (YAG) coatings prepared by suspension plasma spraying of Y_2O_3-Al_2O_3 powders, Surf. Coat. Technol. 235 (2013) 303–309.
[80] J. Kitamura, Z. Tang, H. Mizuno, K. Sato, A. Burgess, Structural, mechanical and erosion properties of yttrium oxide coatings by axial suspension plasma spraying for electronics applications, J. Therm. Spray Technol. 20 (1-2) (2011) 170–185.
[81] B.M. Cetegen, S. Basu, Review of modeling of liquid precursor droplets and particles injected into plasmas and high-velocity oxy-fuel (HVOF) flame jets for thermal spray deposition applications, J. Therm. Spray Technol. 18 (5-6) (2009) 769–793.
[82] D. Chen, E.H. Jordan, M. Gell, Solution precursor high-velocity oxy-fuel spray ceramic coatings, J. Eur. Ceram. Soc. 29 (2009) 3349–3353.
[83] A. Ozturk, B.M. Cetegen, Modeling of axially and transversely injected precursor droplets into a plasma environment, Int. J. Heat Mass Transfer 48 (2005) 4367–4383.
[84] G. Sivakumar, R.O. Dusane, S.V. Joshi, A novel approach to process phase pure α-Al_2O_3 coatings by solution precursor plasma spraying, J. Eur. Ceram. Soc. 33 (2013) 2823–2829.
[85] X.-M. Wang, C.-X. Li, C.-J. Li, G.-J. Yang, Microstructure and polarization of $La_{0.8}Sr_{0.2}MnO_3$ cathode deposited by alcohol solution precursor plasma spraying, Int. J. Hydrogen Energy 37 (2012) 12879–12885.
[86] Li Cheng-Xin, Shuai Liu, Yang Zhang, Chang-Jiu Li, Characterization of the microstructure and electrochemical behavior of $Sm_{0.7}Sr_{0.3}Co_3Ld$ cathode deposited by solution precursor plasma spraying, Int. J. Hydrogen Energy 37 (2012) 13097–13102.
[87] D. Chen, E.H. Jordan, M. Gell, Porous TiO_2 coating using the solution precursor plasma spray process, Surf. Coat. Technol. 202 (2008) 6113–6119.
[88] D. Chen, E.H. Jordan, M. Gell, M. Wei, Apatite formation on alkaline-treated dense TiO_2 coatings deposited using the solution precursor plasma spray process, Acta Biomater. 4 (2008) 553–559.
[89] R. Dom, G. Sivakumar, N.Y. Hebalkar, S.V. Joshi, P.H. Borse, Deposition of nanostructured photocatalytic zinc ferrite films using solution precursor plasma spraying, Mater. Res. Bull. 47 (2012) 562–570.
[90] A.L. Vasiliev, N.P. Padture, X. Ma, Coatings of metastable ceramics deposited by solution-precursor plasma spray: I. Binary ZrO_2-Al_2O_3 system, Acta Mater. 54 (2006) 4913–4920.
[91] A.L. Vasiliev, N.P. Padture, Coatings of metastable ceramics deposited by solution-precursor plasma spray: II. Ternary ZrO_2-Y_2O_3-Al_2O_3 system, Acta Mater. 54 (2006) 4921–4928.

CHAPTER 11

Microarc Oxidation: A Novel Process for Preparing Nanocrystallized Ceramic Metal Oxide Coating

Xudong Jiang[1,2], Chunxu Pan[1]
[1]School of Physics and Technology, Wuhan University, Wuhan, PR China
[2]Hubei Provincial Museum, Wuhan, PR China

Chapter Contents

1. INTRODUCTION

Short for microarc oxidation, MAO is also called microplasma oxidation, anode spark electrolysis, plasma electrolytic anode treatment, and plasma

Handbook of Nanoceramic and Nanocomposite Coatings and Materials
http://dx.doi.org/10.1016/B978-0-12-799947-0.00011-0

electrolytic oxidation, which is a kind of surface treatment through an electrochemical and plasma-chemical process. The process occurs at potentials above the breakdown voltage of the anode oxidation on the surface of valve metals including Al, Mg, Ti, Zr, Ta, and Nb and their alloys, which are immersed in the given electrolyte. By the coaffection of thermochemistry, plasma chemistry, and electrochemistry, the micro spark discharge occurs, which provides a possibility for "*in situ*" formation of a ceramic oxide coating upon their surface [1].

The discharge phenomena associated with electrolysis had been discovered more than a century ago. Günterschultze and Betz [2] also studied this phenomenon in detail in the 1930s. The discharge was thought to cause destruction on the ceramic oxide coating and should be avoided by their researches. However, in the 1960s, McNeill and Gruss [3] used a spark discharge to deposit cadmium niobate onto a cadmium anode in an Nb-containing electrolyte. Then, the practical benefits were first exploited. In the 1970s, Markov [4] studied the oxide deposition on an aluminum anode under an arc discharge condition. This technique was further improved and named "microarc oxidation."

The MAO technique has many advantages, such as being a simple process and toxic-free, which can be potentially in accordance with the requirements of the clean technology. This technique can be applied to any sample of irregular shape, especially for those with holes and welding seams. The ceramic coating prepared by MAO exhibits many excellent properties [5].

Our group has been forcing on the study of MAO for several years. In this chapter, combined with our recent research results, we explained the process of the spark discharge of MAO and its influencing factors and analyzed the structure of ceramic coating synthesized by MAO in micro- and nanoscale. Then, the mechanical properties, corrosion resistance, and photocatalytic properties were reviewed in detail.

2. FUNDAMENTALS AND PROCESSES

Al, Mg, Ti, Zr, Ta, and Nb and their alloys can shut down the oxidized electric current in electrolytes, like electrochemical valves. So they are all named valve metals [6]. When they are acted as anode and immersed in given electrolytes, compact oxide coatings with insulation will grow on their surface immediately at first. The current will be significantly reduced as the oxide coating grows. As the voltage grows up and exceeds the critical value, the spark arc microdischarges happen at the weak points. The spark arc microdischarges rapidly develop and extinguish within 10^{-4}-10^{-5} s [7]. At the same time, the local temperature and pressure inside the discharge channel can reach 10^3-10^4 K and 10^2-10^3 MPa, respectively, which are high enough to give rise to plasma thermochemical interactions between the substrate and the electrolyte. These interactions result in the formation of melt-quenched high-temperature oxides and complex compounds on the surface, composed of oxides of both the substrate material and electrolyte-borne modifying elements [1]. The melted discharge channels are preserved by a surrounding electrolyte, which results in the microporous structural ceramic coating. The typical scanning electron microscope (SEM) morphologies of MAO ceramic coatings are shown in Figure 1. This unique "*in situ*" growing process of the MAO coating has great influence on its microstructure, compositions and properties. The following section will introduce them in detail.

Figure 2 illustrates the schematic diagram of the MAO setup. A valve metal is used as an anode and immersed in stirred electrolyte. A stainless steel container is selected as an electrolyte cell and cathode. The electric current is applied by an MAO power supply. The temperature of the electrolyte is controlled below 40 °C by condensed water.

Generally, according to the voltage characteristics, the process of MAO is composed of four stages in sequence, as shown in Figure 3 [1].

(1) Anodic oxidation: The course in Figure 3 represents a system where oxide coating formation occurs. At relatively low voltages, the kinetics of the electrode processes for the system conforms to Faraday's laws, and the current-voltage characteristics of the cell vary according to Ohm's law. Thus, an increase in voltage leads to a proportional rise

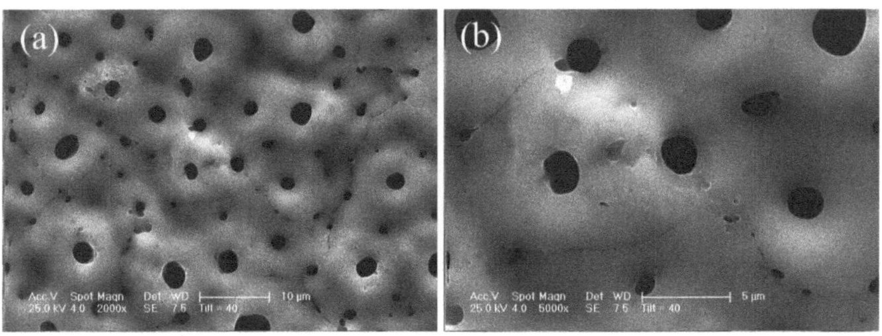

Figure 1 SEM morphologies of MAO coatings. (a) Low magnification; (b) high magnification.

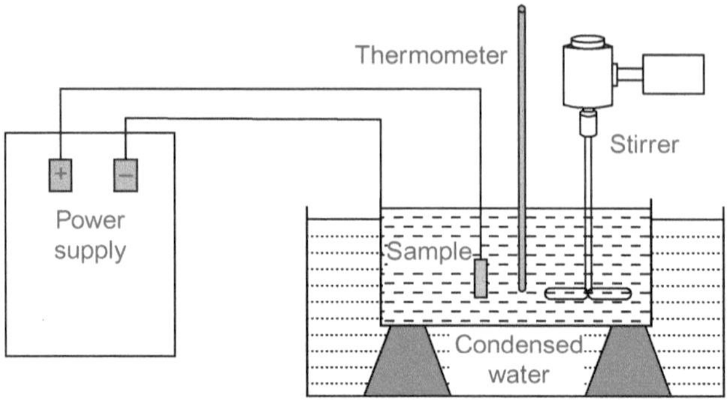

Figure 2 Schematic diagram of microarc oxidation experimental installation.

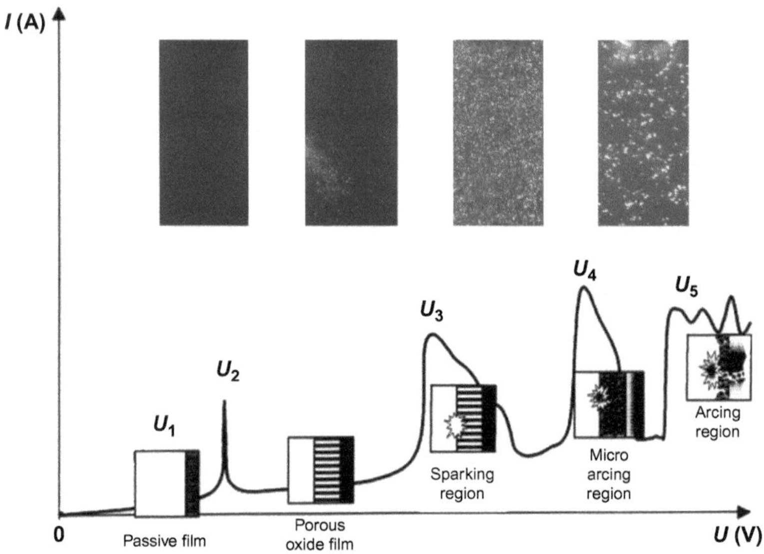

Figure 3 Current-voltage diagram for the process of MAO [1].

in the current (region "0-U_1"). However, beyond a certain critical voltage, the behavior of a particular system may change significantly. The passive coating previously formed begins to dissolve at point U_2, which, in practice, corresponds to the corrosion potential of the material. Then, in the region of repassivation, U_2-U_3, a porous oxide coating, grows, across which most of the voltage drops now occur. The reaction from U_1 to U_3 is shown in inset (anodic oxidation).

(2) Sparking region: At point U_3, the electric field strength in the oxide coating reaches a critical value beyond which the coating is broken through due to impact or tunneling ionization. So, small luminescent sparks are observed to move rapidly across the surface of the oxide coating, facilitating its continued growth.
(3) Microarcing region: At point U_4, the mechanism of impact ionization is supported by the onset of thermal ionization processes, and slower, larger arc discharges arise. In the region U_6–U_7, thermal ionization is partially blocked by negative charge buildup in the bulk of the thickening oxide coating, resulting in discharge decay shorting of the substrate. This effect determines the relatively low power and duration of the resultant arc discharges. Owing to microarcing, the coating is gradually fused and alloyed with elements contained in the electrolyte.
(4) Arcing region: Above the point U_7, the arc microdischarges occurring throughout the coating penetrate through to the substrate and (since negative charge blocking effects can no longer occur) transform into powerful arcs that may cause destructive effects, such as thermal cracking of the coating. So, this process should be avoided in coating growth.

3. INFLUENCE FACTORS

3.1 Electric Parameters

The electric parameters, especially the voltage (U) [8] and current density (J) [9], have great importance on the morphology and properties of MAO coatings. Generally, the properties of coatings prepared by pulsed power supplies are better than common power sources, because the frequency (f) and duty cycle (δ) [10] provided by them can influence the spark characteristics. The thickness of coatings is usually determined by the oxidation time (t) [11]. Therefore, the properties of MAO coatings can be optimized by electric parameters, which are concluded in Table 1.

3.1.1 Voltage [8]

The spark discharges only happen on the surface of metals, when the voltage exceeds their valve values. So, the voltage of MAO is closely related to the thickness of coatings. The value of spark discharge voltage is mainly determined by different metals, component of the electrolyte, and concentration of the electrolyte, while the current density, shape of the cathode, and mode of increasing voltage have little influence on it. The thickness

Table 1 The influence of electric parameters on ceramic coatings of MAO

Influence factors	Growth rate	Pore sizes	Porosity	Roughness
U	P	P	N	P
J	P	P	N	P
δ	P	P	N	P
t	N	P	N	P
f	N	N	P	N

P, positive effect; N, negative effect.

of the MAO coating grows as the oxidation voltage increases. The voltage also influences the morphology of the coating, because the amount of micropores increases and the micropore sizes grow larger as the voltage arises. The coating also turns rough due to the increase of remains on the edge of micropores.

3.1.2 Current Density [9]

The current density also can influence the thickness, growth rate, compositions, and morphologies of MAO coatings. It shows that the current density decrease leads to its growth rate to drop under the voltage constant mode. While on the current constant mode, the thickness of the coating and oxidation time has a linear relationship. As the current density increases, in the preparation of the MAO coating on the surface of Ti alloy, the growth rate arises.

3.1.3 Frequency and Duty Ratios [10]

The frequency and duty ratios are both influence factors of a single pulse discharge, which determine the energy of a single discharge. In order to reveal the effect of the alternate pulse duty cycle and frequency on MAO coatings, Chen and his coauthors prepared ceramic coatings using an alternate pulse MAO method on Ti6A14V alloy by a constant current regime of MAO. The results showed that with the increase of the duty cycle, the growth rates of MAO coatings increased. With the increase of frequency, the growth rates of MAO coatings decreased, and the roughness of the coating surface decreased gradually.

3.1.4 Oxidation Time [11]

The oxidation time affects the morphologies, growth rate, surface roughness, and surface hardness of MAO coatings. It shows that, with the increasing oxidation time, the number of micropores and the growth rate of MAO coatings decrease, while both the size of the micropores and the thickness of MAO coatings increase. The average surface hardness increases at the initial stage of the oxidation process and then decreases.

3.2 Electrolyte [12]

MAO coatings of various compositions, microstructures, and properties can be prepared by choosing the proper electrolyte. According to the pH value, electrolyte systems are generally divided into neutral electrolyte (which contains silicate, phosphate, carbonate, and aluminosilicate), alkaline electrolyte (including sodium hydroxide and potassium hydroxide), and acidic electrolyte (sulfuric acid). The affection of concentration, compositions, pH value, and additives of electrolytes on MAO coatings will be explained in the following sections.

3.2.1 Concentration [13]

The concentration of the MAO electrolyte has a significant effect on the arc starting voltage, morphology, compositions, and properties of coatings. Zhang and his coauthors studied the effect of different phosphate electrolyte concentrations on the microstructure. The results showed that as the solution conductivity increased, the arc starting voltage fell down, the thickness of the ceramic coating represented linear growth, and the number of micropores on the coating surface decreased obviously, while the micropore size increased.

3.2.2 Compositions [12]

The compositions of electrolytes influence the properties of MAO coatings significantly. Bulyshev et al. [14] studied on the adsorption of ions by MAO ceramic coatings and thought that the adsorption activity of different ions followed the descending order $SiO_3{}^{2-}$, $PO_4{}^{3-}$, VO_4^{3-}, $MoO_4{}^{2-}$, $B_4O_7{}^{2-}$, and $CrO_4{}^{2-}$. Lv and coworkers prepared MAO coatings on the titanium alloy in Na_3PO_4, Na_2SiO_3, and NaOH electrolyte solutions and studied the effects of electrolytes on the phase components and surface morphology of coatings. The research showed that the critical arc starting voltage was highest in phosphate electrolyte solution and lowest in sodium hydroxide electrolyte solution. The coating prepared in Na_3PO_4 electrolyte solution consisted of dominant anatase phases, the size of the discharge pore was smaller, and the pore number was infrequent, whereas the coating prepared in NaOH electrolyte solution consisted of dominant rutile phases, the coating surface was porous, and the pore size was larger.

3.2.3 pH Value [15]

The effect of pH value of the MAO electrolyte on the component of MAO ceramic coatings had been also studied by Gao and his coworkers [15]. Three kinds of TiO_2 coatings, pure anatase-type TiO_2, mixed phase of rutile-type TiO_2 and anatase-type TiO_2, and pure rutile-type TiO_2, could be obtained in acidic, neutral, and alkaline electrolyte solutions, respectively.

3.2.4 Additives

In order to obtain MAO coatings with superior properties, some additives [16], for example, fluoride, glycerin, molybdate, and chromate, are usually added into electrolyte solutions, which can promote their growth rate, wear resistance, and corrosion resistance.

The traditional additives are soluble. However, our research team tried to add some insoluble particles, like YAG:Ce^{3+} and Eu_2O_3, into the MAO electrolyte solution. The coating compounded with additives can be synthesized because of the spark arcing and molten and rapid solidification processes. So this technique has expanded the additives from soluble to insoluble, and the significant advantage of this method is suitable for any insoluble particles, which are expected to be compounded on the surface of MAO coatings.

When the YAG:Ce^{3+} [17] powders were added into the electrolyte, it was homogeneously distributed under vigorous stirring. During MAO, when the molten TiO_2 erupted from the discharge channels, YAG:Ce^{3+} particles would be mixed within the TiO_2. Under high temperature from the molten TiO_2, the YAG:Ce^{3+} particles were melted partly and had a short-range diffusion around the interface. Therefore, the original YAG:Ce^{3+} polyhedron particles transformed into roundish particles in the TiO_2 coating without obvious boundary between the YAG:Ce^{3+} particle and the TiO_2 substrate. Figure 4 shows the SEM morphologies of pure TiO_2 coating and TiO_2 coating compounded with the YAG:Ce^{3+} semiconductor, respectively. In the YAG:Ce^{3+} particles, the rare earth metals with incompletely occupied 4f and empty 5d orbits could restrain the growth of crystalline TiO_2, which makes the grain size much smaller. The heterojunction between the YAG:Ce^{3+} particle and the TiO_2 phase provided a superior photocatalytic effect.

We also tried to add the Eu_2O_3 into the electrolyte solution and obtained TiO_2/Eu_2O_3 composite coating by MAO technique, as shown in Figure 5 [18]. The ion Eu^{3+} in Eu_2O_3 had an incomplete 4f orbital track and an empty 5d orbital track, which tended to produce a multielectron

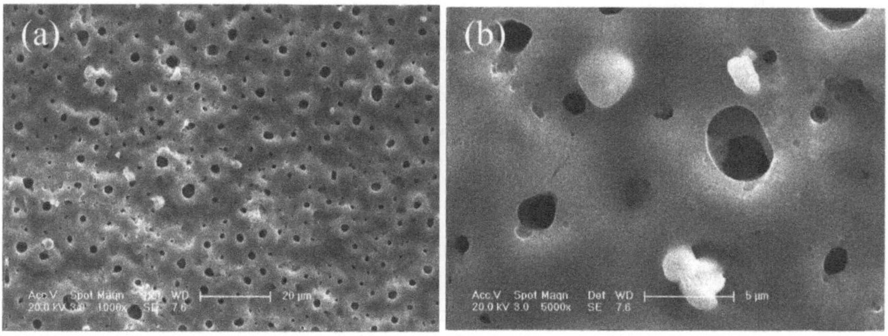

Figure 4 SEM morphologies of $TiO_2/YAG:Ce^{3+}$ compounded coating. (a) Low magnification; (b) high magnification.

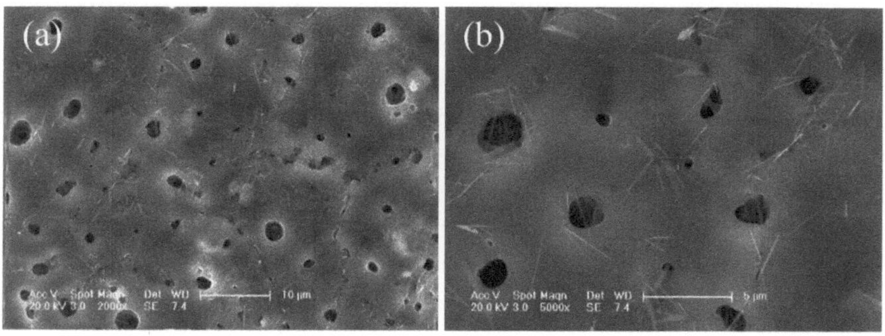

Figure 5 SEM morphologies of TiO_2/Eu_2O_3 compounded coating. (a) Low magnification; (b) high magnification.

configuration, and therefore could effectively inhibit the recombination of photogenerated electrons and holes. In addition, the ground-state energy was close to the excited-state energy in the ion Eu^{3+}, which enhanced the transition of f-electrons from the ground-state to the excited-state under visible light irradiation and then exhibited a higher absorption in the visible light range. Therefore, the photocatalytic performance of the TiO_2/Eu_2O_3 composite coating exhibited twice higher than that of the coating without the Eu_2O_3 additive.

4. MICROSTRUCTURE OF MAO COATING

Generally, the thickness of MAO coatings ranges from several micrometers on the titanium substrate to tens of micrometers on the aluminum substrate. From previous reports, the MAO coating was usually divided into three layers from substrate to surface: bonding layer, dense layer, and loose layer.

The bonding layer connects the substrate and coating, which are combined with each other like a sawtooth state in the micro area. Therefore, the ceramic coating is firmly bonded to the substrate. The dense layer, between the bonding layer and loose layer, contains fewer micropores, which is composed of density materials to be mainly part of MAO coatings. The loose layer, on the coating exterior, consists of crystalline with large grain sizes and many micropores, which lead to the MAO coating that presents porous morphology. There is no obvious boundary between every layer. Therefore, the MAO coating exhibits good adhesion to the substrate, excellent chemical durability, high abradability, and high-temperature resistance.

From the cross section, the MAO coating generally consists of three typical portions with variant microstructures, that is, (1) a loose layer beneath the surface containing many micropores with thickness of several micrometers, (2) a dense layer in the middle of the coating consisting of only few pores and defects with thickness of dozens of micrometers, and (3) a transitional zone adjacent to the Ti substrate consisting of only polycrystalline particles with a few micrometers.

The chemical compositions of MAO coatings are generally composed of elements from the substrate and electrolyte. The distribution of elements diffuses from the substrate to the surface of coating: The substrate elements decrease, while the electrolyte elements increase. Figure 6 illustrates Ti, P, and O distribution curves in the MAO coating obtained by titanium substrate in the electrolyte of Na_3PO_4. From the substrate to the surface of coating (right side to left side in Figure 6), the contents of Ti decreases, while the P and O increase gradually. This is in agreement with previous reports.

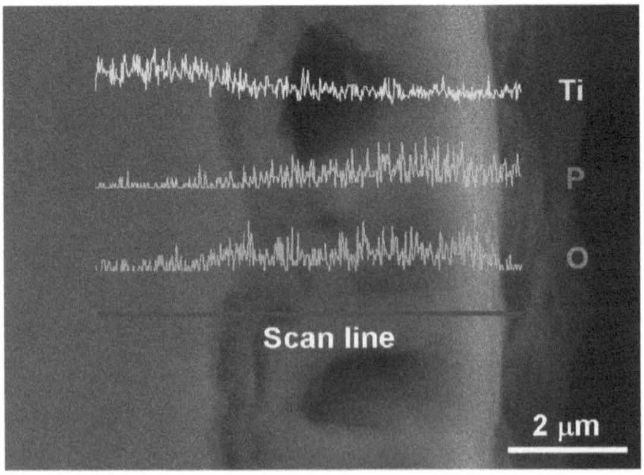

Figure 6 Cross-sectional chemical composition distributions of MAO coatings.

Referring to the former researches, the microstructure of MAO coatings was generally observed by using a SEM in microscale. Recently, we studied the microstructures of MAO coatings on titanium, magnesium, and aluminum substrates in nanoscale by using high-resolution transmission electron microscopy (HRTEM) and firstly proposed that the ceramic coatings synthesized by MAO technique were a new kind of nanocrystalline coating [19].

Figure 7 shows the HRTEM micrographs of the TiO_2 coating from MAO on titanium. Obviously, there is a layer of full TiO_2 amorphous phase with thickness around 10-20 nm beneath the surface. After the amorphous layer, an intermediate zone consisting of amorphous phase and crystallized anatase and few rutile TiO_2 grains is observed. The most grains in the intermediate zone are in the average size of 12 nm. By Scherrer's formula,

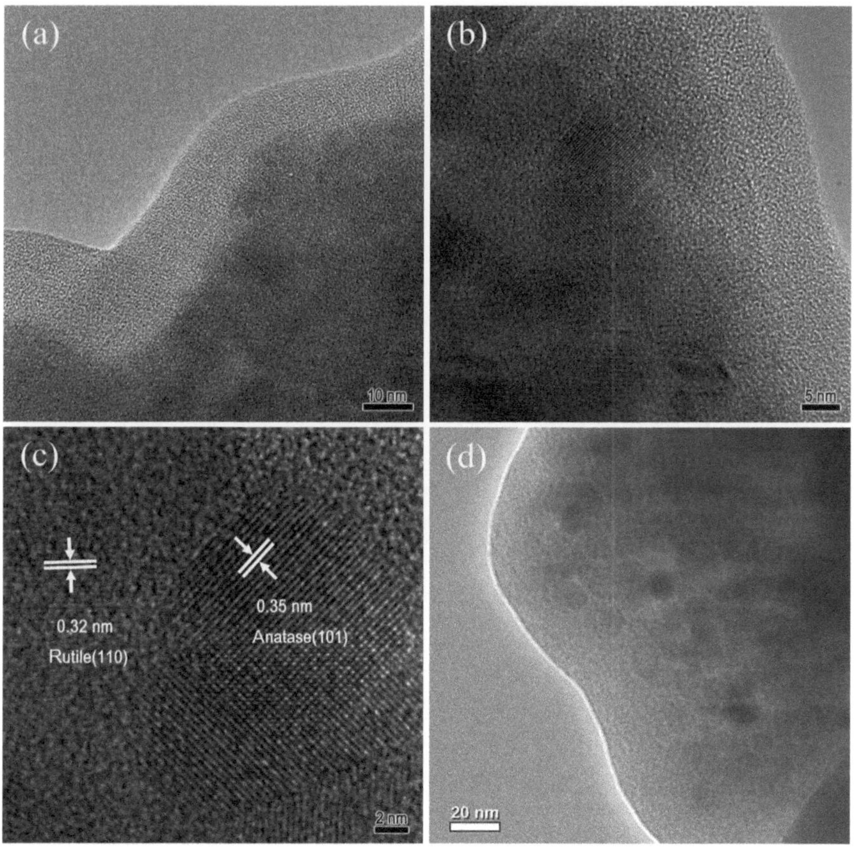

Figure 7 HRTEM micrographs of the TiO_2 coating by MAO. (a) Low magnification; (b) high magnification; (c) TiO_2 lattice image; (d) nanocrystalline TiO_2.

the grain size of the as-prepared TiO_2 coating was calculated to be about 20 nm, which proves that the average grain size of the TiO_2 coating is significantly larger than that in the intermediate zone as observed from HRTEM.

A layer of MgO amorphous phase with thickness around 10 nm beneath the surface is also observed in the HRTEM morphologies of the MgO coating by MAO on the substrate of magnesium, as shown in Figure 8. Amorphous phase and crystallized magnesium oxide with the size of 5 nm composes the intermediate zone. The average grain size of the MgO coating is calculated to be 26 nm by Scherrer's formula, which is also larger than that in the intermediate zone.

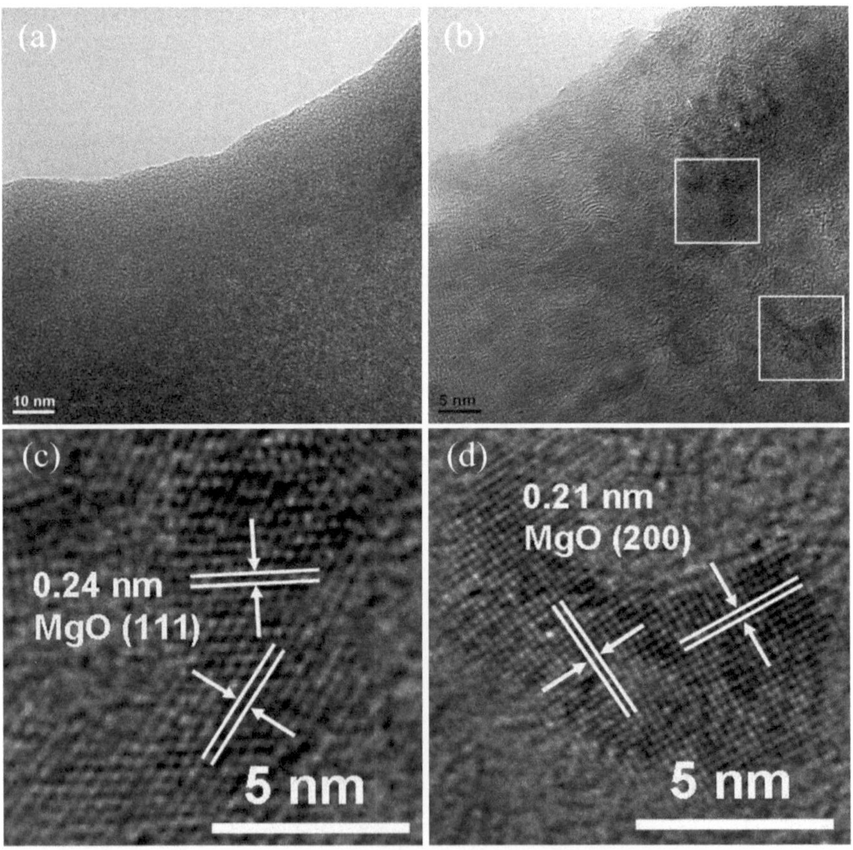

Figure 8 HRTEM morphologies of the MgO coating by MAO. (a) Low magnification; (b) high magnification; (c) lattice image of (111) plane; (d) lattice image of (200) plane.

Similar results are also obtained in Al_2O_3 coating via the MAO technique, as shown in Figure 9. A layer of amorphous phases with thickness around 10 nm, and crystallized α-Al_2O_3 beneath the amorphous phase are observed. The grain size of α-Al_2O_3 is about 20 nm. The average grain size of the Al_2O_3 coating is calculated to be about 43 nm by Scherrer's formula.

The grain sizes of ceramic coatings obtained by MAO are listed in Table 2, which indicates that the ceramic coating is of specific nonhomogeneous microstructures, especially the microstructural variation from the surface into the interior regions. In fact, this microstructural variation is associated with the MAO process. It has been reported that the temperature and pressure in the discharge channels can reach to 2×10^4 K and 10^2 MPa

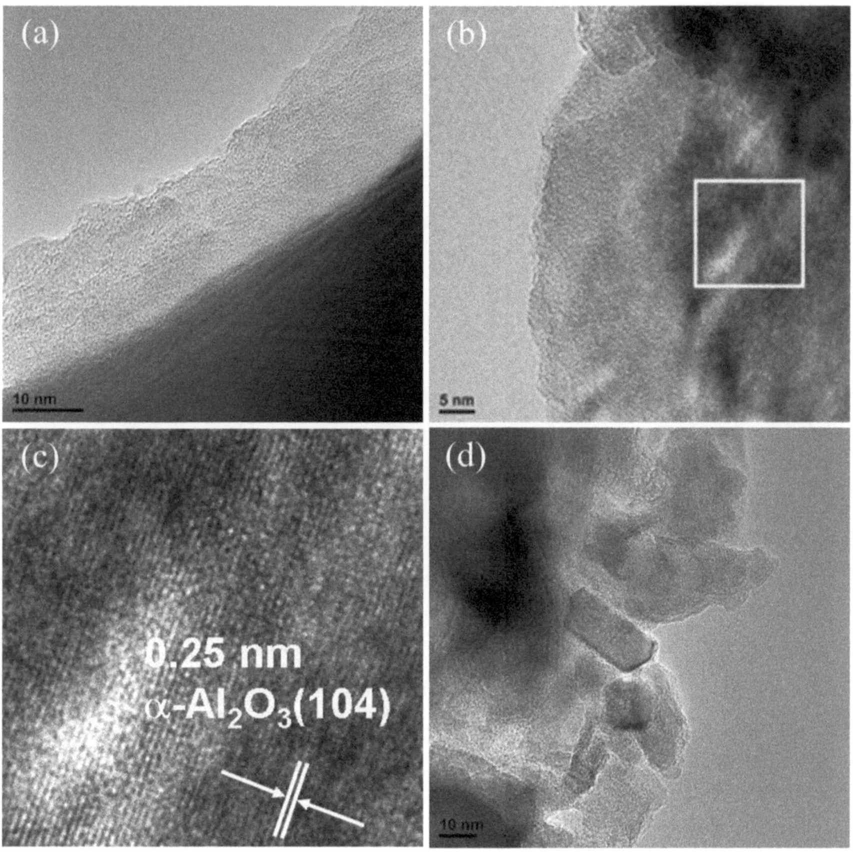

Figure 9 HRTEM morphologies of the Al_2O_3 coating by MAO. (a) Low magnification; (b) high magnification; (c) lattice image of α-Al_2O_3 (104) lattice plane; (d) Al_2O_3 grain.

Table 2 The grain size in different layers of MAO nanocrystalline coating

MAO coating	TiO_2	MgO	Al_2O_3
Adjacent to surface	Amorphous	Amorphous	Amorphous
Intermediate zone (nm)	12	5	20
Interior portion (nm)	20	26	43

during MAO [2], which promotes the conversion of the substrate into molten oxide temporarily. Then, the molten oxide erupts from the discharge channels and is cooled rapidly by the surrounding electrolyte, which results in the formation of crater and porous morphologies. Anyway, due to the extremely rapid cooling rate and thermal isolation, the ceramic MAO coating is composed of nanocrystalline layer. Therefore, the ceramic coating obtained via the MAO technique is a kind of new nanocrystalline coating.

5. PROPERTIES

5.1 Mechanical Behavior

The hardness and elastic modulus are primary parameters to evaluate the mechanical properties of films and coatings. Zhang and his coworkers [20] studied the mechanical properties of the ceramic coating on LY12 aluminum alloy fabricated by the MAO technique. The maximum hardness and elastic modulus values of the ceramic coating were 25 GPa and 300 GPa, respectively, compared with 2 GPa and 80 GPa of the substrate of LY12 aluminum alloy. After undergoing 40 times heat shock circulation from 300 to 15 °C, neither cracks nor breaking-off was found. The friction coefficient of the ceramic coating against bearing steel in air was in the range of 0.25-0.37, showing a good wear resistance.

Wu et al. [21] synthesized the ceramic coatings via MAO on the substrates of pure Al, LC9, and LY12. They found that the coatings were composed of α-Al_2O_3, γ-Al_2O_3, a large amount of γ-Al_2O_3, and a little α-Al_2O_3 on the surfaces of pure Al, LC9, and LY12, respectively. Along the section of the coating, hardness of the coating at both sides was lower than that in the middle, and the maximum hardness of the coating obtained on pure Al, LC9, and LY12 was 33.4 GPa, 22.15 GPa, and 16.8 GPa, respectively. The anticorrosion property of the coating on LY12 was a little better than that on LC9, while the coating on pure Al had the worst anticorrosion property.

Figure 10 The cover of the "One S" obtained by MAO.

Relying on its superior mechanical properties, the MAO coating has been applied to instead the traditional cover material of mobile phones and appeared on the "One S" by HTC Corporation, as shown in Figure 10.

5.2 Corrosion Resistance

Mg, Al, and Ti have been widely used in aerospace, automobiles, chemical industry, etc. However, their applications were largely restrained for the high chemical activity and poor corrosion resistance. Fortunately, the MAO coating on their surface can be used to improve their corrosion resistance.

Our group evaluated the corrosion resistance of MAO coatings on Mg by total immersion test and salt spray test and found that both phosphate and silicate coatings could enhance the corrosion resistance of AZ91D alloy significantly. However, the corrosion resistance of silicate coating was superior to that of phosphate coating [22].

Wang [23] obtained the ceramic coating on the surface of pure aluminum in a silicate electrolyte by MAO technique and compared its corrosion resistance with aluminum substrate by their Tafel curves. The corrosion potential of the MAO coating was enhanced to 0.4 V, and the corrosion current density decreased by three orders of magnitude, which indicated that the MAO coating could protect the aluminum substrate well.

5.3 Photocatalytic Activity

TiO_2 has been chosen as an excellent photocatalyst and extensively studied due to its chemical stability, biological compatibility, and antiphotocorrosion ability. However, TiO_2 powder is hard to be recycled in polluted water and the depth of photoradiation is limited. TiO_2 powder fossilization and the preparation of TiO_2 coating have become hot research directions recently. When Ti or Ti alloys are used as a substrate, an excellent adhesive TiO_2 coating will be made on its surface via MAO technique, which exhibits super photocatalytic activity.

Our research group has been studying on the photocatalyst of TiO_2 coating via the MAO technique for several years. Firstly, TiO_2/YAG:Ce^{3+} [17] and TiO_2/Eu_2O_3 [18] compound coatings were obtained by adding YAG:Ce^{3+} and Eu_2O_3 powders into the electrolyte, as shown in Figures 4 and 5. Compared with the pure TiO_2 coating, the compounded coating exhibited enhanced absorption in the visible light and higher photogenerated current density, which improved the photocatalytic property markedly. Therefore, it is expected that MAO will be a simple, economic, and promising approach to preparing various compound coatings of superior photocatalytic property.

Referring to previous reports, the doping TiO_2 coating is usually synthesized by adding the soluble compound containing doping elements into the MAO electrolyte. Our recent research indicated that the elements in the electrolyte were not doped into the TiO_2 lattice while they existed in the amorphous states. Figure 11 shows the HRTEM morphology of TiO_2 obtained on the surface of the Ti substrate by MAO in a sodium phosphate solution. The TiO_2 crystalline grains are surrounded by amorphous states. "A" has been added in Figure 11a, it contains Ti, P, and O elements. By annealing to 700-800 °C, some new XRD patterns ($Ti(PO_4)$ and $Ti(PO_3)_3$) appeared, as shown in Figure 12. $Ti(PO_4)$ and $Ti(PO_3)_3$ are similar

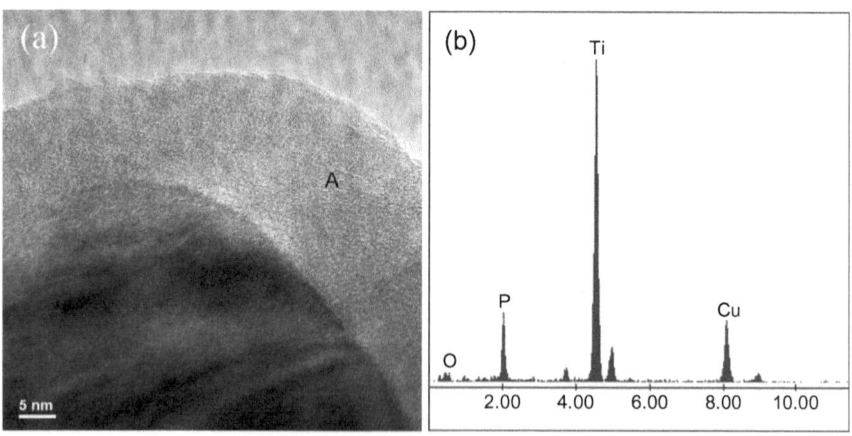

Figure 11 HRTEM morphologies of the MAO coating (a) and ESD analysis on point A (b).

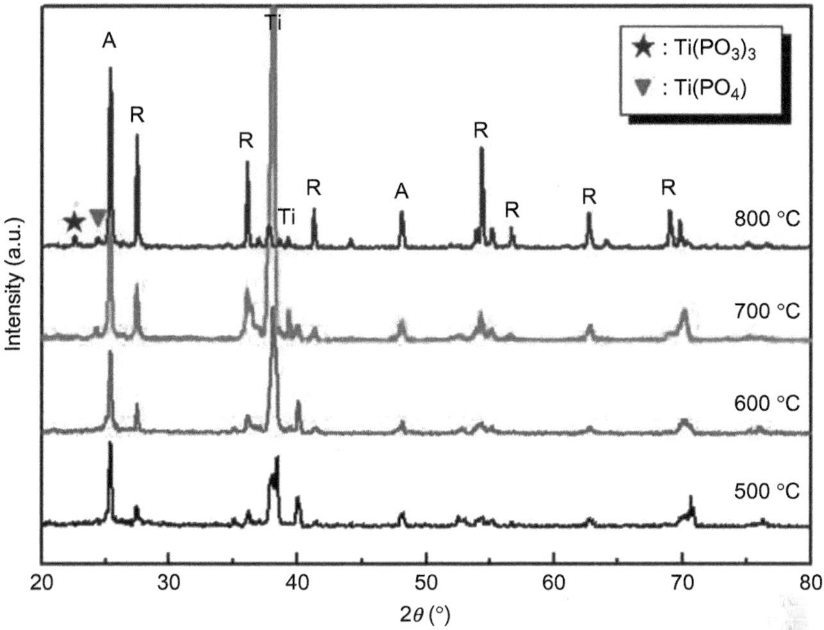

Figure 12 XRD patterns of MAO coatings annealed in different temperatures.

to the compounded structures of TiO_2/PO_2 and $TiO_2/P_2O_5/PO_2$, which is in agreement with previous reports.

Therefore, it is different to synthesize the doping TiO_2 coating by conventional methods, such as adding doping elements in the MAO electrolyte. We propose a novel and valid process for preparing a high-concentration nonmetal element-doped TiO_2 photocatalytic coating, that is, firstly, the titanium substrate is thermochemically treated in the atmosphere containing doping elements and secondly, doping TiO_2 coating is synthesized on the surface of the treated substrate by the MAO technique [24]. For example, the titanium substrate is firstly plasma nitride to make a layer of TiN_x on its surface. And then, the nitride substrate is oxidized by the MAO technique. The rapid process of oxidation can restrict the N atoms escaping from the substrate and obtain a substitutional N-doped TiO_2 coating with higher nitrogen doping concentration up to 3.21 at.% and x to 0.11 in $TiO_{2-x}N_x$, compared with $x < 0.04$ in the regular powders and coatings. The N-doped TiO_2 coating exhibits advantages including a redshift in the band gap transition, narrow band gap of 2.6 eV, high photogenerated charge carrier density, and superior photocatalytic property. Furthermore, it is expected

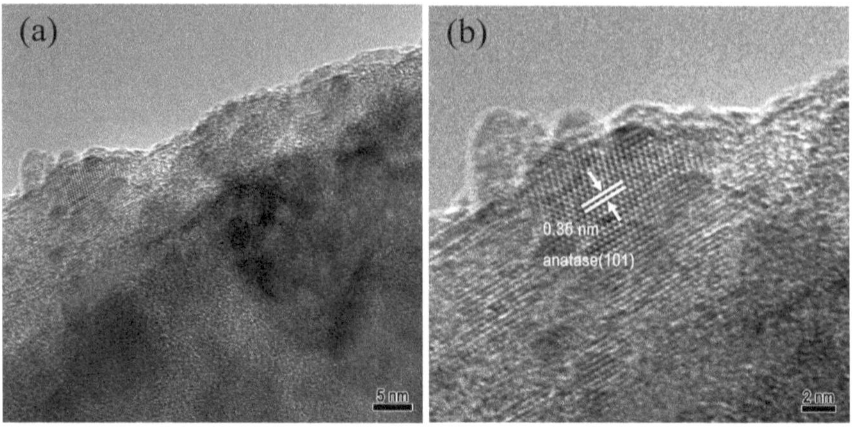

Figure 13 HRTEM micrographs of the annealed TiO_2 coating. (a) High magnification; (b) TiO_2 lattice image.

that MAO combines with nitriding or other chemical heat treatments (carburizing, boronizing, etc.) will be a simple, economical, and a promising approach for preparing multiple doped TiO_2 coatings even in a larger area.

Furthermore, the MAO coating is covered by an amorphous layer with thickness around 10-20 nm from the previous introduction. It is well known that the amorphous TiO_2 has detrimental effects on the photocatalytic activity, because it contains more defects, which act as recombination centers of photogenerated electrons and holes. By annealing, the surface amorphous phase transforms into a nanosized anatase TiO_2, as shown in Figure 13. The photocatalytic activity of the postannealed TiO_2 coating is enhanced to more than twice that of the as-prepared TiO_2 coating, as shown in Figure 14. Therefore, we propose that the crystallization of amorphous phase beneath the surface plays a key role for the improvement of its photocatalytic property [19].

6. CONCLUSION

As a novel process for preparing nanocrystallized ceramic metal oxide coatings, MAO technique is simple, convenient, and controllable, which has been widely used in the areas of aerospace, machinery, environmental protection, electronic products, and decoration. It's expected that more products of MAO will appear in more industry and daily life areas with the rapid development of researches in the future.

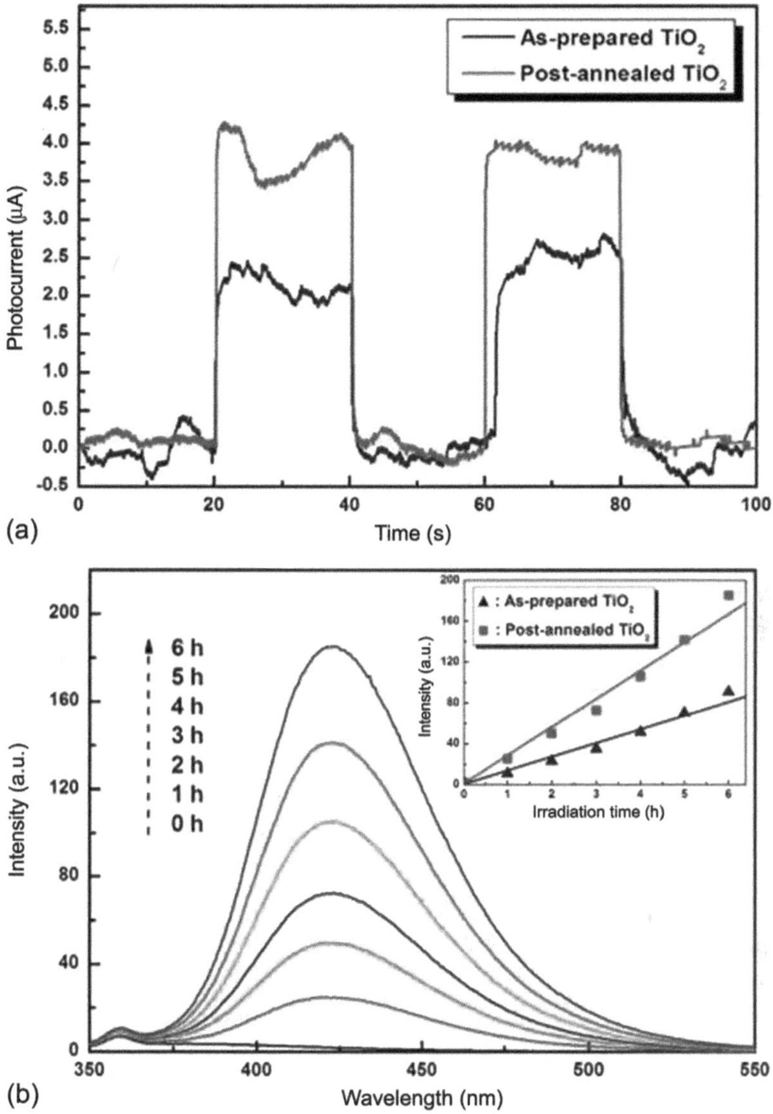

Figure 14 Photogenerated current of as-prepared and postannealed TiO_2 coatings (a) and fluorescence spectra of the postannealed TiO_2 at different irradiation times (b). Inset is the time dependences of the fluorescence intensity of TiO_2 coatings.

REFERENCES

[1] A. Yerokhin, X. Nie, A. Leyland, A. Matthews, et al., Plasma electrolysis for surface engineering, Surf. Coat. Technol. 122 (2) (1999) 73–93.
[2] A. Güntershulze, H. Betz, Elektroliticheskie kondensatory (electrolytic capacitors), Obomizidat, Moscow, 1938.
[3] W. McNeill, L.L. Gruss, US Patent 3 293 158 (1966).
[4] G.A. Markov, G.V. Markova, USSR Patent 526 961, Bul. Inv. 32 (1976).
[5] A. Yerokhin, A. Leyland, A. Matthews, Kinetic aspects of aluminium titanate layer formation on titanium alloys by plasma electrolytic oxidation, Appl. Surf. Sci. 200 (1) (2002) 172–184.
[6] X. Xi, Review on micro-arc oxidation, Surf. Technol. 36 (4) (2007) 66–68.
[7] J. Li, L. Wan, J. Feng, Micro arc oxidation of S-containing TiO_2 films by sulfur bearing electrolytes, J. Mater. Process. Technol. 209 (2) (2009) 762–766.
[8] B. Guo, J. Liang, J. Tian, et al., The effect of anodic voltage on the performance of micro-arc oxidation film on titanium alloy, Plat. Finish. 27 (3) (2005) 1–4.
[9] Z. Liu, Z. Xiang, Z. Zhang, et al., Effect of current density on titanium alloy micro-arc oxidation coating, Light Met. (1) (2008) 48–51.
[10] N. Chen, Q. Zhao, Z. Zhang, Effect of the alternate pulse parameter on the micro-arc oxidation coatings on TC4 titanium alloy, Surf. Technol. 36 (3) (2007) 43–45.
[11] B. Guo, J. Liang, J. Chen, et al., Effects of oxidation time on microstructure and performance of micro-arc oxidation coatings formed on T-i 6A-l 4 V alloy, Chin. J. Nonferrous Met. 15 (6) (2005) 981–986.
[12] X. Lv, H. Wu, J. Wang, et al., The effects of electrolytical parameters on phase component and microstructure of titania films by MAO, J. Yunnan Univ. 27 (5A) (2005) 583–586.
[13] Y. Zhang, Micro-arc oxidation on surface of marine titanium alloy, Dev. Appl. Mater. 31 (3) (2006) 26–29.
[14] S.I. Bulyshev, V.A. Fedorov, The kinetic of coating formation in microarc oxidation process, J. Fiz. Kim. obrob. Mater. 17 (1993) 93–95.
[15] Y. Gao, L. Yan, S. Zhang, et al., Structural characteristics of TiO_2 ceramic coating by micro-plasma oxidation, Chin. Surf. Eng. (6) (2003) 35–37.
[16] L. Wan, J. Li, J. Feng, et al., Anatase TiO_2 films with 2.2 eV band gap prepared by micro-arc oxidation, Mater. Sci. Eng. B 139 (2) (2007) 216–220.
[17] X. Jiang, Y. Wang, C. Pan, Micro-arc oxidation of TC4 substrates to fabricate TiO_2/YAG:Ce^{3+} compound films with enhanced photocatalytic activity, J. Alloys Compd. 509 (2011) L137–L141.
[18] Y. Wang, X. Jiang, C. Pan, 'In-situ' preparation of a TiO_2/Eu_2O_3 composite film upon Ti alloy substrate by micro-arc oxidation and its photo-catalytic property, J. Alloys Compd. 538 (2012) 16–20.
[19] X. Jiang, A. Shi, Y. Wang, et al., Effect of surface microstructure of TiO_2 film from micro-arc oxidation on its photocatalytic activity: a HRTEM study, Nanoscale 3 (9) (2011) 3573–3577.
[20] T. Zhang, H. Yuan, Y. Yan, et al., Microstructure and properties of ceramic coating by microarc oxidation on aluminum alloy, Sci. Technol. Eng. 11 (001) (2011) 154–158.
[21] Z. Wu, Z. Jiang, Z. Yao, et al., Analysis of properties of micro arc oxidation ceramic coating for aluminum and its alloys, Rare Metal Mater. Eng. 35 (A02) (2006) 148–151.
[22] L. Wang, C. Pan, Q. Cai, et al., Corrosion failure mechanism of micro-arc oxidation coatings formed on AZ91D magnesium alloy, J. Chin. Soc. Corros. Protect. 28 (4) (2008) 219–224.
[23] Z. Wang, Y. Sun, K. Ding, et al., Structure and electrochemistry corrosion behaviors of micro-arc oxidation on aluminum, Trans. China Weld. Inst. 29 (12) (2008) 74–76.
[24] X. Jiang, Y. Wang, C. Pan, High concentration substitutional N-doped TiO_2 film: preparation, characterization and photocatalytic property, J. Am. Ceram. Soc. 94 (2011) 4078–4083.

CHAPTER 12

Development of Nanostructured Composite Coatings on Metallic Surfaces

Tahir I. Khan*
Department of Mechanical & Industrial Engineering, Qatar University, Doha, Qatar

Chapter Contents

1. INTRODUCTION

Engineering components are designed to fulfill certain functional requirements, which are normally based on their bulk mechanical properties such as yield strength, tensile strength, and fracture/impact toughness. However, it is often the surface properties of the component that determine the life and failure of the part when exposed to service environments and external loads. As a result, the engineering of surfaces has been practiced by industries to protect and extend the wear life of engineering parts.

The engineering of surfaces is normally associated with metal processing or the automotive industry. However, the petrochemical, oil, and gas industries also suffer from costs associated with failures linked to degradation

*On leave from Department of Mechanical & Manufacturing Engineering, University of Calgary, Calgary, Alberta, Canada.

Handbook of Nanoceramic and Nanocomposite Coatings and Materials
http://dx.doi.org/10.1016/B978-0-12-799947-0.00012-2

through corrosion, erosion, and abrasive wear of surfaces. The material failures that result from this lead to costs accrued in revenue loss due to downtime and due to repair/replacement of parts.

Furthermore, in the pursuit of hydrocarbon resources, energy companies have to explore reservoirs that lie in deeper waters and farther offshore. The harsher environment means that traditional metal surfacing and coatings cannot be used for these applications. In parts of North America, such as Alberta in Canada, there are large reserves of oil sands that contain "heavy oil" in the form of bitumen. This bitumen must be extracted from the ground and separated from the sand. Therefore, thick surface coatings need to be developed to protect the outer surfaces of pumps, gate valves, and pipelines. The coatings that are available commercially tend to be microstructured and may contain a ceramic dispersion (e.g., Al_2O_3 and TiO_2) as a way of increasing the hardness and surface wear resistance of the coating [1]. However, it is known that as the grain structure of a material is reduced from micron- to nanoscale, the size of the inherent defects is also reduced [2]. Furthermore, the micromechanisms of plastic deformation change and the presence of grain boundaries and triple junctions become very large compared to conventional materials [2,3]. This has a significant effect on the mechanical properties resulting in an increase in the fracture toughness and strength of the material.

The use of thermal spraying technology to produce nanostructured coatings for use in other industries such as the oil and gas sectors or the oil sands industry in North America is a recent development and has grown to considerable importance. It is evident that coatings developed for severe erosive and abrasive wear environments will need to be thick and the only techniques currently available for deposition of such coatings are based on thermal processes, for example, flame spraying, electric arc spraying, detonation gun (D-Gun), and plasma spraying process. The feedstock powder can be exposed to very high temperatures (5000–10,000 °C). At these high temperatures, the feedstock powders melt during the spraying process, and this results in the original nanostructure being lost during the spraying step so that the solidified coat suffers from not only oxidation but also excessive grain growth. A comparison of some spraying techniques and parameters is shown in Table 1. A relatively low flame temperature and high gaseous velocity can be conditions that are more favorable for producing and retaining the nanostructure of cermet-based coatings.

Therefore, based on this comparison, the best option is to use the high velocity oxy-fuel (HVOF) spraying process as a technique for producing

Table 1 A comparison of thermal spraying techniques and their parameters [4]

Spraying technique	Temperature (°C)	Impact velocity (m/s)	Oxide content (%)	Adhesive strength
Flame wire	3000	180	4	D
Electric arc	5500	240	0.5-3	C
Detonation gun	4000	900	0.2	B
Plasma arc	5500	240	0.5-1	C
HVOF	2500	700-1200	0.1	A

Adhesive strength: A, best; D, worst.

these specialized coatings. This chapter reviews the use of the HVOF technique as a method of producing nanostructured coatings and highlights the important parameters that affect the spraying process. Furthermore, a comparison is also made with a low-temperature spraying process known as the cold gas dynamic spraying (CGDS) technique.

2. PRODUCTION OF POWDER FEEDSTOCK

In order to produce a nanostructured coating, the powder feedstock must possess a component of it that is nanostructured. Several methods have been used to produce a nanostructured feedstock, including gas condensation/atomization, mechanical alloying (MA), sol-gel processing, and vapor deposition methods [5] to name a few. However, the actual thermal spraying of nanosized powders is difficult due to agglomeration of these fine particles. The subsequent "clogging" of powder feed flow lines into the spray gun, and a decrease in the size of particles to within the nanometer range means that a much higher particle injection force would be necessary. This could only be achieved by increasing the cold carrier gas flow rate to such an extent that the high-enthalpy jet flow within this flame would be affected [6,7]. Therefore, producing a powder feedstock that contains the nanosized particles in such a way that the sprayed powder is within the micron range will help to solve this problem. MA process is one such technique that has the capability to produce sufficient quantities of powder for commercial spraying. Furthermore, this process has the potential to produce composite powder mixtures within the micron range and suitable for thermal spraying in a carrier gas.

As a technique, the MA process was developed in 1966 by the International Nickel Company to produce oxide dispersion-strengthened superalloys [8]. The MA process uses a variety of milling equipment, for example, vibratory mill, planetary mill, uniball mill, and attritor mill, but the latter has the highest capacity of powder charge [9]. Therefore, the attritor mill is employed to synthesize feedstock powders because a minimum quantity of 500 g is necessary for thermal spraying.

The attritor mill consists of a cylindrical stainless steel tank, which contains ceramic balls with a ball-to-powder ratio of 20:1. Several horizontal impellers are used to generate motion of the balls (Figure 1a). As the balls collide against each other, they trap and deform the powder by impact, and this in turn results in significant grain refinements. The atmosphere of the attritor mill is controlled with the use of argon/nitrogen in order to decrease oxidation effects and prevent an increase in temperature. To control powder size and morphology, a liquid environment of liquid nitrogen, methanol, acetone, or hexane may be used [8]. The effect of atmosphere and milling time can be seen in Figure 2b. The graph shows that as milling time increases, the average particle size of Cr_3C_2-25 (Ni,Cr) powder decreases to a value of 5 μm. In contrast, the Inconel 625 increases in particle size to 84 μm. In both cases, the most significant change is seen within the first 4 h of milling.

3. THERMAL SPRAYING TECHNIQUES FOR NANOSTRUCTURED COATINGS PRODUCTION

There are a number of thermal spraying processes that have the potential to be used to form nanostructured coatings and they include plasma spraying, D-Gun spraying, HVOF spraying, and cold spraying techniques.

It is known that the highest particle velocity can be obtained by particles sprayed using a gaseous flame with the highest velocity [9]. Furthermore, the temperature of the gaseous flame is determined by the type of spraying process and this directly affects the temperature to which the particles are exposed. A comparison of the flame temperature for different thermal spraying methods and particle velocity range achieved can be seen in Figure 2.

It was reported that the size of the feedstock powder influences particle heating and acceleration. Work by Li and Christofides showed that the heating rate and acceleration of powdered particles increase as particle size decreases [9]. This observation can detrimentally affect the decarburization/oxidation of carbide particles (used in the form of WC dispersions in coatings) during the thermal spraying process.

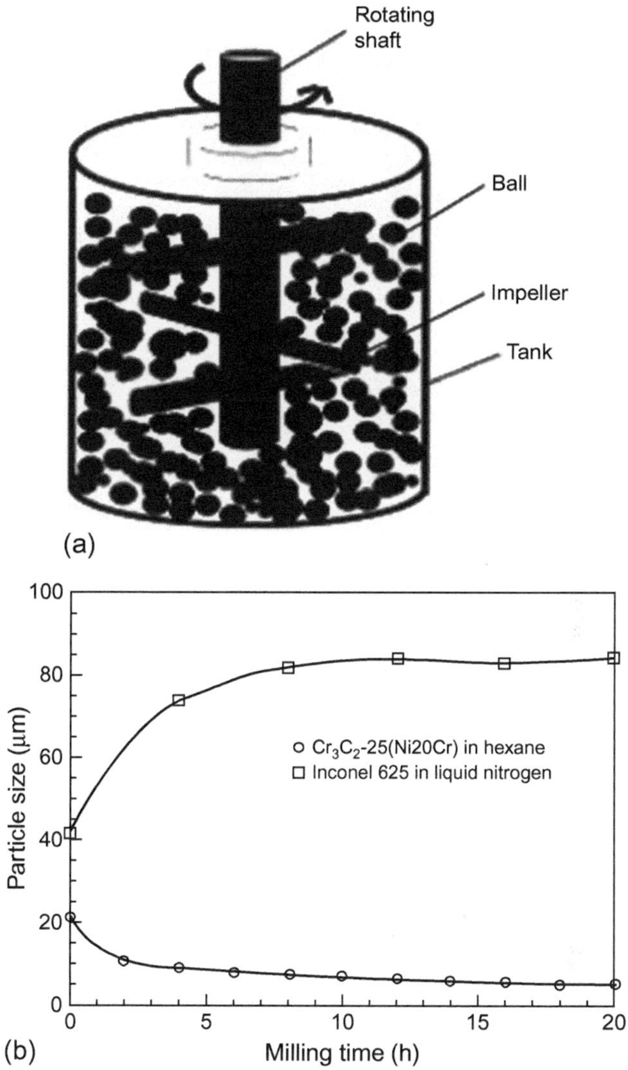

Figure 1 Schematic showing (a) attritor mill and (b) graph showing dependence of the average particle size of attritor-milled powders on milling time [9].

Although a number of thermal spraying techniques have been developed to spray conventional coatings, the HVOF spraying process has received the most attention. The HVOF technique is particularly attractive for developing nanostructured coatings because a high particle velocity is expected resulting in better coating adhesive strengths, lower % porosity (i.e., denser coatings), and lower oxide content (Table 1).

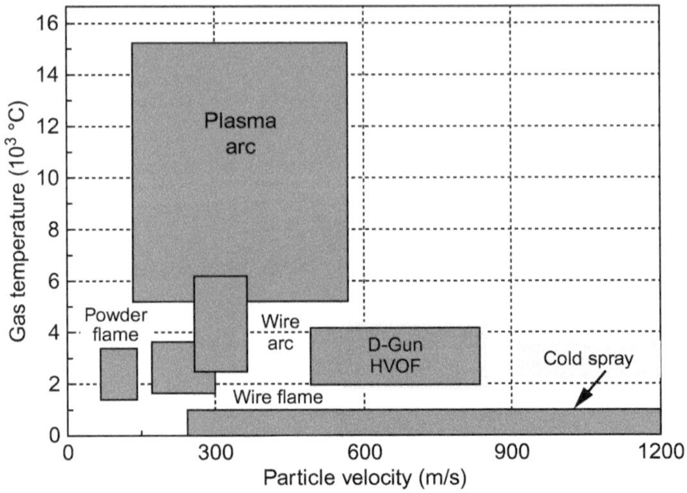

Figure 2 Flame temperature of different thermal spray methods and obtainable particle velocity range [9].

The principles of the HVOF process were developed by Schoop in 1910 [10]. His experimentation involved pouring liquid metal into a gas stream so that molten drops could be carried and projected onto a substrate surface. The basic principles of Schoop's methods have remained unaltered in almost a century of work on thermal spray processes. The HVOF spraying technique was further developed by Browning in the 1980s, and his JetKote process was the first HVOF gun to be used to deposit metal carbides for wear-resistant coatings [11].

As a spraying method, the "Diamond Jet" spray gun developed by Sulzer Metco consists of a water cooled gun that uses two fuel gases such as oxygen and propylene or methane (Figure 3). The gun brings together the fuel and oxygen in the correct stoichiometric ratio, and this mixture is ignited by an

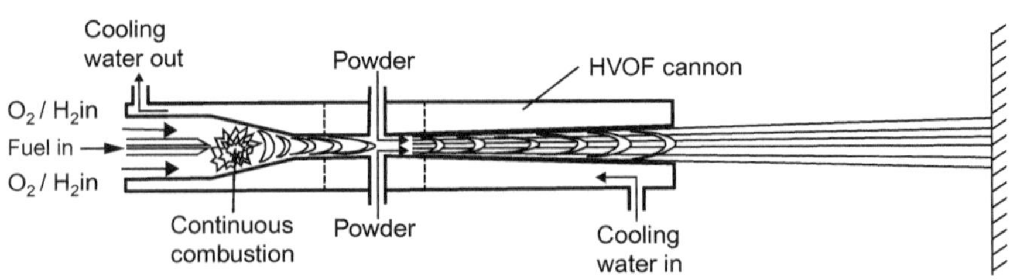

Figure 3 Schematic showing the HVOF spray gun.

arc current that creates a supersonic flame at a temperature of 2500–3000 °C. A hopper powder feed unit provides the powder feedstock, which is carried into the jet flame via a carrier gas such as nitrogen. The powder particles are injected axially into the jet gas, heated, and propelled towards the substrate.

3.1 Influence of HVOF Spraying Parameters

3.1.1 Feedstock Powders

The nanostructured WC-Co powders have a spherical morphology that is superficially indifferent to microcrystalline powders (Figures 4 and 5). The particle size distribution is near-Gaussian with a range from 0.5 to 50 μm (Figure 6) for the micro- and nanostructured powders. The WC dispersion in a Co matrix was prepared by Powdermet Inc. (Ohio, USA) to a size between 100 and 300 nm. Micrograph in Figure 4a shows the duplex

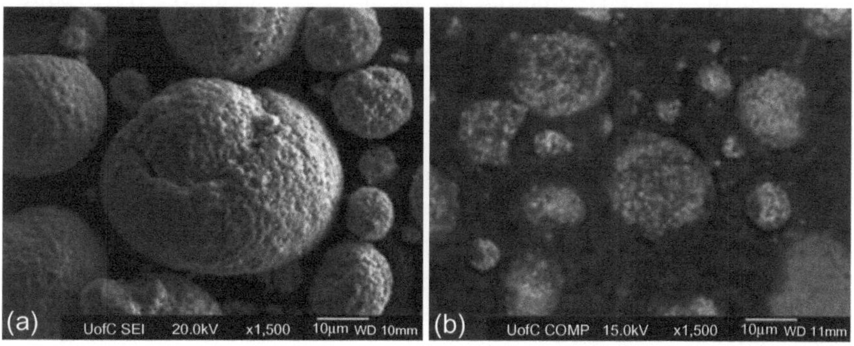

Figure 4 SEM micrograph showing (a) nanostructured duplex WC-17%Co powder for spraying and (b) cross sections of the nanostructured powder (backscattered image) [12].

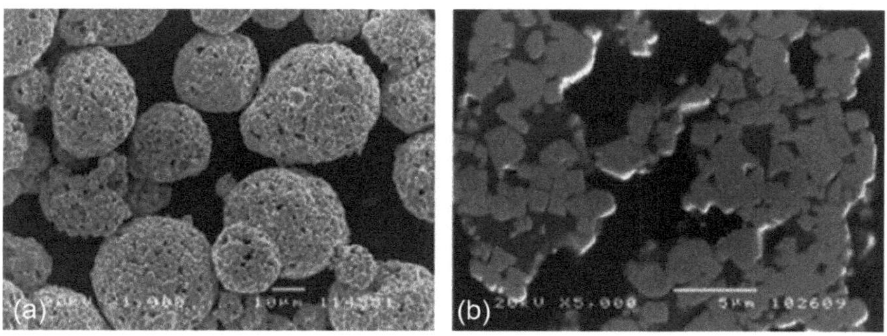

Figure 5 SEM micrograph showing (a) microstructured WC-17%Co powder for spraying and (b) cross sections of microstructured powder [13].

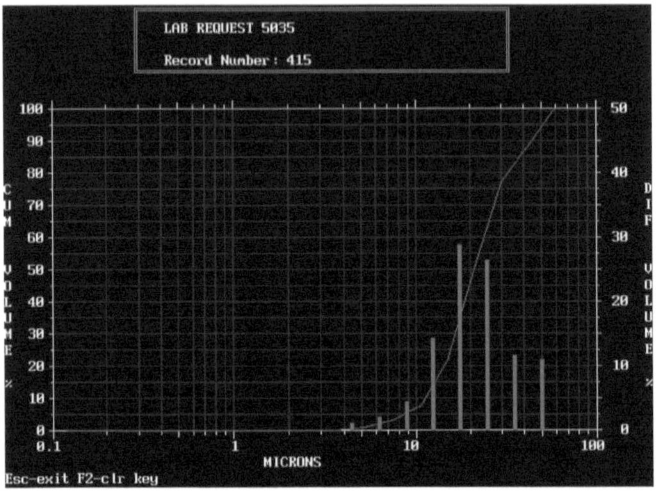

Figure 6 SEM micrograph showing particle size distribution for the WC-17%Co powder.

Co-coated structure of the powder. The duplex coating technique is a relatively new method introduced to prevent decarburization observed during the HVOF spraying process. Furthermore, the Co coating prevents agglomeration of the powder during the HVOF spraying process.

It was reported that the nanostructured sprayed powder tends to be more homogeneous when compared with microstructured sprayed powders [12,13]. Both coatings have low porosity and the individual WC particles can be clearly identified and are held by the Co matrix as shown in Figure 7.

Work by Skandan et al. on the spraying of nanostructured WC-Co system showed that the dissolution of the WC particles occurs in the Co matrix [14]. This leads to a hardened matrix that also holds the WC particles in position. Guilemany et al. compared the dissolution behavior of nanostructured

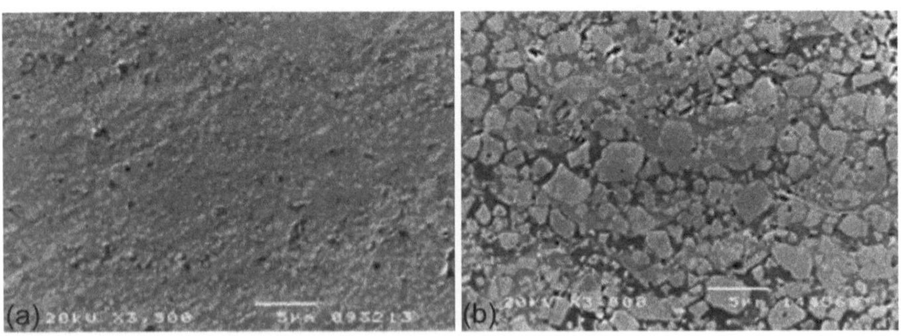

Figure 7 SEM cross-sectional images of sprayed powders.

powders [13]. In their observations, it was found that decarburization was higher for nanostructured than conventional microstructured powders [13]. Earlier He et al. [15] reviewed the mechanisms of WC dissolution and considered that the formation of non-WC phases depends on the composition and structure of the matrix phase [10]. The decarburization transformation from WC to W_2C and W occurred as a result of the direct oxidation of the surface of solid WC due to the interaction of the fuel gases. However, according to transmission electron microscopic studies by Stewart et al., it was suggested that the Co matrix melted during the spraying process and the WC suffered dissolution [16]. A schematic showing the two processes leading to dissolution of WC particles is given in Figure 7.

3.1.2 Fuel Chemistry

The type of fuel used in HVOF spraying and the fuel ratio can also affect the behavior of nanostructured coatings. Research work by He et al. showed that the flame temperature was influenced by the fuel/O_2 ratio that affected the extent of WC decarburization [15]. A summary of their results is shown in Table 2. The flame temperature increased from 1246 to 1567 °C with an increase in the H_2/O_2 ratio from 1.98 to 3.46. This increase resulted in a greater W_2C vol% forming within the coating showing a greater extent of decarburization from WC to W_2C. Similar observations were recorded for propylene/O_2 ratio. XRD spectra taken from nanostructured WC-18%Co coatings sprayed using different fuel ratios show a distinct peak for the WC phase. The intensity of this peak increases with an increase in the H_2/O_2 ratio (Figure 8).

Furthermore, the type of fuel used during the spraying process can influence the extent of decarburization. For instance, when the propylene/O_2 ratio is increased from 0.242 to 0.367, the flame temperature also increases

Table 2 Influence of fuel chemistry and fuel ratios on nanostructured W-18%Co coatings [15]

Coating code	Fuel	Fuel/ O_2	Temperature (°C)	W_2C (vol%)	Porosity (vol%)
MH45	H_2	1.98	1246	1.57	25.3
MH62	H_2	2.8	1534	3.8	11.98
MH75	H_2	3.46	1567	4.68	9.5
MP32	C_3H_6	0.242	1857	3.79	2.08
MP40	C_3H_6	0.304	2018	6.71	1.99
MP48	C_3H_6	0.367	1950	5.69	1.98

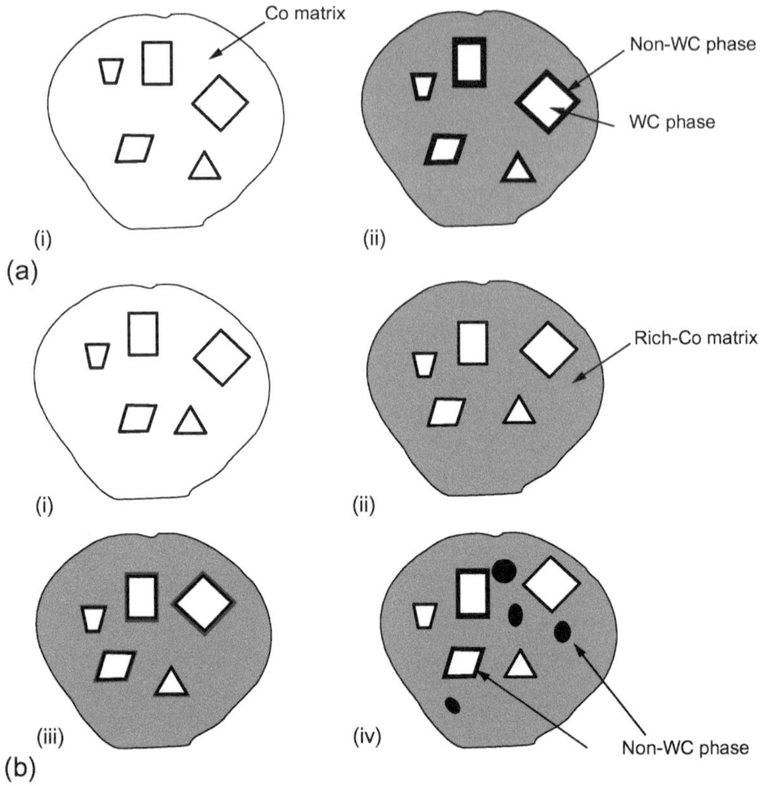

Figure 8 Schematic indicating the two mechanisms responsible for the formation of non-WC phases. Light areas denote WC particles or Co matrix, gray areas denote matrix containing W and C, and dark areas denote non-WC phases [17]. (a) Non-WC phases form by direct oxidation on the surface of WC phase: (i) initial stage and (ii) non-WC phases formed on the surface of the WC phase. (b) Melting of the binder phase leads to the formation of non-WC phases: (i) initial stage, (ii) Co matrix melted and WC particles dissolved, (iii) further dissolution of WC on periphery, and (iv), after impingement with substrate, precipitation of non-WC phases in binder.

from 1857 to 1950 °C; however, the decomposition of WC was on the same level as that for coatings sprayed using H_2 fuel, as shown in Table 2. These results correspond to the XRD peaks that show lower intensity peaks for the formation of W_2C (at $2\Theta = 40°$) with increasing propylene/O_2 ratio as shown in Figure 8b. Characterization of the sprayed coatings produced with different fuel mixtures shows that the propylene/O_2 ratio results in coatings with the lowest % porosity. This suggests that the propylene/O_2 mixture is more suitable for spraying WC-Co nanostructured powders than H_2/O_2 mixture.

3.2 Mechanical Properties of Coatings

The steels used for components in the oil sands are heat-treated so that they have a tempered martensitic structure as shown in Figure 9a. The typical surface hardness of this type of steel before heat treatment is 230 VHN, and after heat treatment, it increases to 348 VHN [18]. In comparison, the HVOF-sprayed WC-Co coating gives a value close to 1084 VHN, which is a threefold increase in hardness over the hardened steel plate currently employed in the industry.

The hardness of a coating is an important performance characteristic and gives information on the wear resistance of a coating. Research has shown that the average microhardness of nanostructured Cr_3C_2-25(Ni,Cr) coatings increases from a value of 846 VHN for microstructured coatings to 1020 VHN [19]. A similar study done by Kear and McCandlish [20] has also indicated WC-23%Co with a higher hardness compared to conventional coatings of the same composition. The high hardness of Cr_3C_2-25(Ni,Cr) coatings has been attributed to the uniformity of microstructures due to the synthesis process of the feedstock powder and the intrinsic high hardness of the nanostructured phases.

The increase in surface hardness and corresponding increase in wear resistance are highlighted in Figure 10. In contrast, the HVOF-sprayed micro- and nanostructured WC-Co-based coating provides excellent abrasive wear resistance. In this particular study, a duplex Co-coated powder was used in order to prevent decarburization and dissolution of the nanosized WC phase in the Co matrix.

Similar results showing the strong relationship between HVOF-sprayed coat hardness and abrasive wear resistance were observed by Li et al. [17] and are shown in Figure 11 [17].

3.3 CGDS of WC-Co Coatings

Low-temperature spraying methods such as the "cold" gas dynamic spraying (CGDS) of WC-Co coatings have also been investigated in order to reduce the decarburization of WC particles. The CGDS technique is relatively new compared with the HVOF spraying process and was developed in the 1980s at the Institute of Theoretical and Applied Mechanics in Russia and then introduced to North America in the 1990s [21,22]. Research using the CGDS process has shown that micro- and nanostructured WC-Co coatings were successfully used to deposit thick coatings with low internal porosity [22]. However, metallographic analysis using scanning electron microscopy

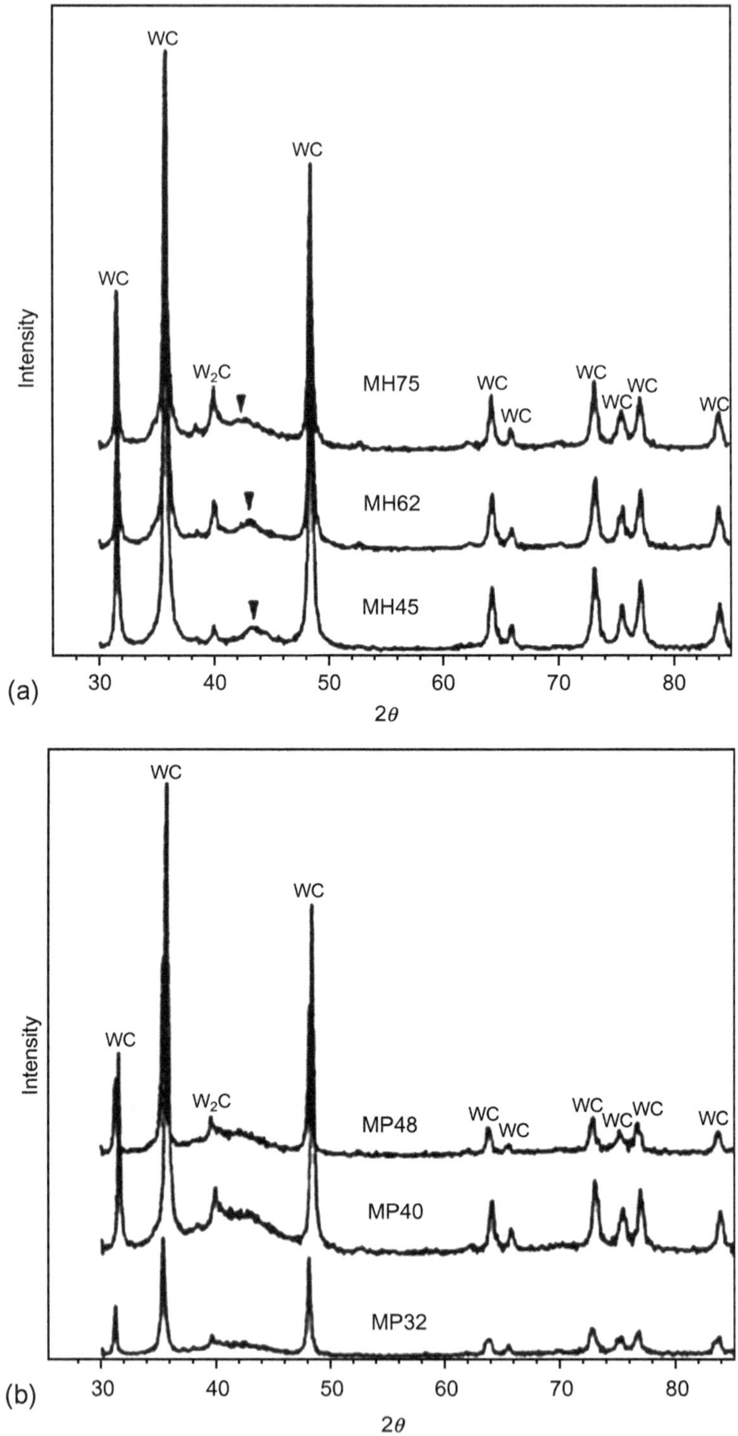

Figure 9 Influence of fuel ratio on XRD of near nanostructured WC-18%Co coatings: (a) hydrogen/oxygen ratios, MH45—$H_2/O_2=1.98$; MH62—$H_2/O_2=2.80$; and MH75—$H_2/O_2=3.46$; (b) propylene/oxygen ratios, MP32—$C_3H_6/O_2=0.242$; MP40—$C_3H_6/O_2=0.304$; and MP48—$C_3H_6/O_2=0.367$ [21].

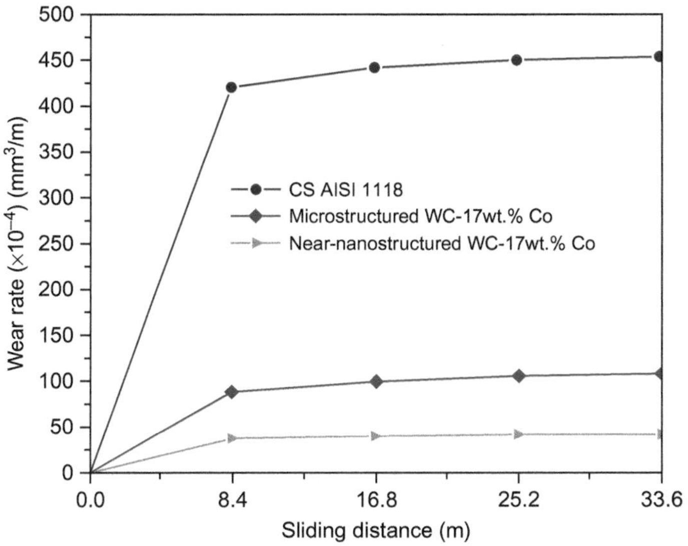

Figure 10 Volume loss as a function of sliding distance showing a comparison between ceramic-metal composite coating, a hardened C-Mn steel (AISI 1118), and micro- and nanostructured coatings [18].

revealed microcracks close to the coating/substrate interface [23,24]. Research work by Ang et al. used cold gas spaying with a converging-diverging Laval-type nozzle of diameter 2 mm; helium was used as a carrier gas for feedstock powder at a pressure of 2-2.5 MPa [25]. The gas temperature in the prechamber was set to 650 °C. A spraying standoff distance of 20 mm and a spraying speed of 0-400 mm/s were used. A comparison of WC-Co coatings containing different WC dispersion sizes ranging from WC of 2 to 3 μm to nanostructured WC (40-800 nm) showed that WC greater than a critical size prevented a buildup of the coating on the substrate surface. In comparison, the nanostructured coating builds up to a thickness greater than 500 μm. This difference in behavior was attributed to a lack of WC particles for the cold spraying process. As a result, the WC particles act as erosive medium on the first layer of WC-Co coating deposited on the steel substrates. Therefore, it was suggested that it becomes more difficult for WC-Co powders to deform on the deposited WC-Co coatings. Interestingly, XRD analysis of the feedstock powder before and after spraying showed some refinement of WC particles. In all the cold spraying processes, the % porosity was recorded between 11% and 38% that is significantly higher compared with porosity observed in HVOF coatings (less than 1%).

Unlike the thermal spraying processes, which use elevated temperatures for deposition, in cold spraying, the hardness of the substrate surface can also

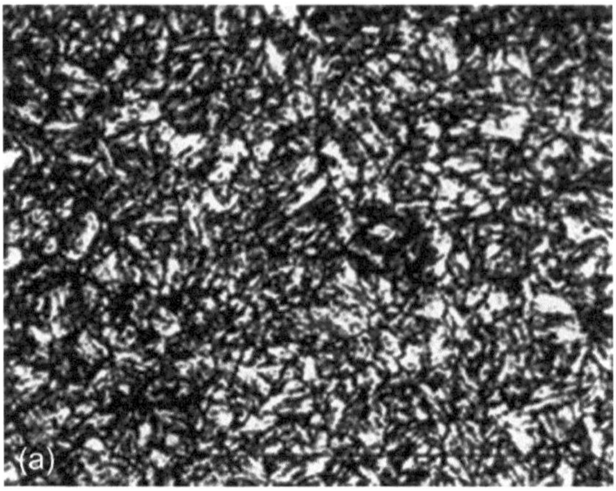

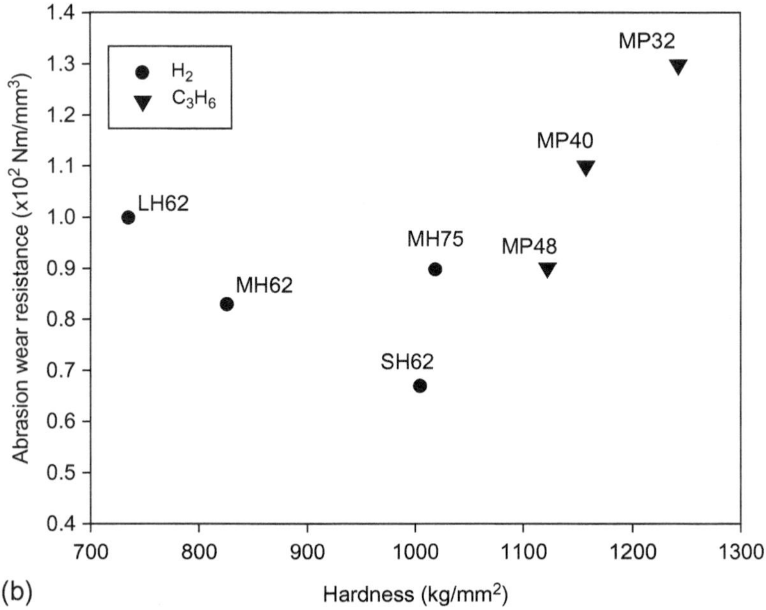

Figure 11 (a) Microstructure of hardened C-Mn steel plate (AISI 1118); (b) abrasive wear resistance of nanostructured WC-18%Co coatings as a function of their hardness [17].

influence the extent of coat deposition and buildup on a surface. The work by Gao et al. showed that a WC-12Co coating could form on a stainless steel of hardness 201 VHN, but the coating failed to deposit to a reasonable thickness on harder Ni alloy surfaces [26]. However, because elevated temperatures are not used in the cold spraying process, the decarburization of WC into the brittle W_2C formed and limited the oxidation effects within the coating because no phase changes are expected below a temperature of 1250 °C [27,28]. Research shows cold spraying offers the potential to deposit nanostructured powder feedstock as an alternative to HVOF techniques. However, work is still necessary in order to optimize process conditions and powder compositions to reduce internal porosity and increase the quality of coatings.

4. SUMMARY AND FUTURE TRENDS

Thick nanostructured coatings can be produced onto steel substrates using the HVOF spraying technique and CGDS process. In both cases, the starting powder feedstock is nanostructured containing nanosized WC dispersion in a Co powder matrix. However, the careful control of spraying parameters such as the type of fuel employed in HVOF spraying or the fuel to oxygen ratio can reduce the extent of decarburization of the WC dispersions to form brittle W_2C. The nanosized WC has a tendency to decarburize faster than the micron-sized WC particles. The only way to truly eliminate decarburization of these fine dispersions is to use "engineered powders" that have an outer Co layer to protect the WC particles during the spraying process. Research has shown that this method offers considerable potential in maintaining the nanostructure of the final WC-Co coatings.

The use of CGDS also offers potential in retaining the original composition and microstructure of the feedstock powders. The low temperatures used in the process can eliminate decarburization, but there is an increase in defects such as microcracking and greater porosity, which compromise the final mechanical properties of the coatings.

Research is also necessary in order to fully exploit the use of duplex Co-coated powders to other more inert systems compared to the WC-Co composition. The development of coatings more suited for applications in the oil sands industry is necessary, and the use of Ni/Cr coatings containing a nanosized dispersion of Al_2O_3 or Y_2O_3 can offer greater wear and corrosion resistance in erosive-corrosive environments.

REFERENCES

[1] T.I. Khan, S. Miller, Surface modification of an aluminium 2124 composite by eutectic alloying, J. Mater. Sci. 36 (5) (2001) 1307–1311.
[2] R.W. Siegel, Nanostructured materials, Adv. Top. Mater. Sci. Eng. 3 (1993) 273–288.
[3] H.G. Jiang, M. Ruhle, et al., On the applicability of the x-ray diffraction line profile analysis in extracting grain size and microstrain in nanocrystalline materials, J. Mater. Res 14 (1999) 549–559.

[4] M.L. Thorpe, Thermal spraying, Industry in Transition, Adv. Mater. Proc. 5 (1993) 50–61.
[5] S. Suryanarayana, Nanocrystalline materials, Int. Mater. Rev. 40 (2) (1995) 41–64.
[6] P. Fauchais, R. Etchart-Salas, et al., Parameters controlling liquid plasma spraying: solutions, sols, or suspensions, J. Therm. Spray Technol. 17 (2008) 31–59.
[7] P. Fauchais, G. Montavon, et al., Developments in direct current plasma spraying, Surf. Coat. Technol. 5 (2006) 1908–1921.
[8] J.S. Benjamin, Mechanical alloying—a perspective, Met. Powder Rep. 45 (2) (1990) 122–127.
[9] M. Li, P.D. Christofides, Computational study of particle in-flight behavior in HVOF thermal spraying process, Chem. Eng. Sci. 61 (2006) 6540–6552.
[10] M.U. Schoop, Apparatus for spraying molten metal and other fusible substances, 30th March 1915, Patent No. 819722.
[11] J.A. Browning, Method and apparatus for ultra high velocity dual flame spraying, 25th January 1983, Patent No. 4370 538.
[12] T.I. Khan, G.C. Saha, et al., Nanostructured composite coatings for oil sands applications, Surf. Eng. 26 (2010) 540–545.
[13] J.M. Guilemany, S. Dosta, et al., The enhancement of the properties of WC-Co HVOF coatings through the use of nanostructured and microstructured feedstock powders, Surf. Coat. Technol. 201 (2006) 1180–1190.
[14] G. Skandan, R. Yao, et al., Multimodal powders: a new class of feedstock material for thermal spraying of hard coatings, Scr. Mater. 44 (8-9) (2001) 1699–1702.
[15] J. He, Y. Liu, et al., Near nanostructured WC-18pctCo coatings with low amounts of non-WC carbide phase: part 1. Synthesis and characterization, Metall. Mater. Trans. A 33A (2002) 145–157.
[16] D.A. Stewart, P.H. Shipway, et al., Microstructural evolution in thermally sprayed WC-Co coatings: comparison between nanocomposite and conventional starting powders, Acta Mater. 48 (2000) 1593–1604.
[17] Y. Liu, Y. Qiao, Near-nanostructured WC-18 pct Co coatings with low amounts of non-WC carbide phase: part II. Hardness and resistance to sliding and abrasive wear, Metall. Mater. Trans. A 33 (2002) 159–164.
[18] G.C. Saha, T.I. Khan, Development of wear resistant nanostructured duplex coatings by high velocity oxy-fuel process for use in oil sands industry, J. Nanosci. Nanotechnol. 9 (7) (2009) 4316–4323.
[19] J. He, M. Ice, et al., Synthesis of nanostructured Cr_3C_2-25(Ni_20Cr) coatings, Metall Mater. Trans. A 31A (2000) 555–564.
[20] B.H. Kear, L.E. McCandlish, Chemical processing and properties of nanostructured WC-Co materials, Nanostruct. Mater. 3 (1993) 19–30.
[21] A.P. Alkhimov, S.V. Kosarev, et al., Gas dynamic method for applying coating, 1994, US Patent No. 5302414.
[22] C.J. Li, W.Y. Li, Recent advances in coating development by cold spraying, in: X.Y. Liu, P.K. Chu, C.X. Ding (Eds.), Surface Modification of Materials by Coatings and Films, Research Signpost, Kerala, India, 2009, pp. 1–107.
[23] B. Jodoin, P. Richer, et al., Pulsed gas dynamic spraying: process analysis, development and selected coated examples, Surf. Coat. Technol. 16 (2007) 7544–7551.
[24] M. Yandouzi, E. Sansoucy, et al., WC-based cermet coatings produced by cold gas dynamic and pulsed gas dynamic spraying processes, Surf. Coat. Technol. 202 (2) (2007) 382–390.
[25] A.S.M. Ang, C.C. Berndt, et al., Deposition effects of WC particle size on cold sprayed WC-Co coatings, Surf. Coat. Technol. 205 (2011) 3260–3267.
[26] P.-H. Gao, C.-J. Li, et al., Influence of substrate hardness transition on built-up of nanostructured WC-12Co by cold spraying, Appl. Surf. Sci. 256 (2010) 2263–2268.
[27] N.M. Melendez, Development of WC-based metal matrix composite coatings using low-pressure cold gas dynamic spraying, Surf. Coat. Technol. 214 (2013) 101–109.
[28] C.J. Li, W.Y. Li, Deposition characteristics of titanium coatings in cold spraying, Surf. Coat. Technol. 167 (2003) 278–283.

CHAPTER 13

Plasma-Induction Spraying of Nanocrystalline Hydroxyapatite Coatings Obtained on Titanium Intraosseous Implants

Aleksandr Aleksandrovich Fomin
Saratov State Technical University, Saratov, Russia

Chapter Contents

Handbook of Nanoceramic and Nanocomposite Coatings and Materials
http://dx.doi.org/10.1016/B978-0-12-799947-0.00013-4

1. INTRODUCTION

Modern metal constructions of transosteal implants, intraosseous implants, and endoprostheses of large joints are a set of materials varying in their properties [1,2]. These devices are characterized by a biocompatible layer (near-surface layer, film, or coating) having the following distinctive features:

- Necessary structural parameters, in particular roughness [3,4], porosity [5], and the presence of macro-, micro-, and nano-sized grains and their agglomerates [6–10]
- Biocompatible chemical and phase composition determining the presence of bioinert [9,11] or bioactive [12–17] interaction with bone tissue, sufficient corrosion resistance [18], and bactericidal [19,20] and other properties of the coating [8,10,21,22]
- High physicomechanical properties, particularly adhesive strength [18,23,24], hardness *H* combined with the required value of modulus of elasticity *E* [14,25,26], fracture toughness, resistance to elastic deformation of fracture (index of plasticity) and material resistance to plastic deformation [27], wear resistance [28] and biomechanical characteristics of the structure in general

1.1 Surface Treatment and Deposition of Biocompatible Coatings on Medical Metal Items

This set of properties required for the implant surface and construction in general is practically impossible to provide by a single method of obtaining a biocompatible interface layer. Conventionally, to obtain the required surface characteristics, various chemical [29], electrochemical [30–32], mechanical [3], physical [33–35], and vacuum-condensation methods [36] are applied. For the surface treatment of intraosseous implants, complex methods are widely used, including preparatory mechanical microtexturing and/or etching [3,4], functional layer deposition [14,17,21,32,35,36], and in some cases subsequent thermal [3,37] or chemical treatment, particularly biomineralization with the deposition of bioactive ceramics [9,38]. The latter treatment can also be applied as a test method for assessing the potential for osteoinduction [16]. On the other hand, a number of methods for the production of coatings can meet the criteria of bioactivity in terms of biochemistry and morphological heterogeneity of indicators of micro- and nanostructures [7,39,40], but their mechanical properties are not sufficient for the implants exposed to extremely high loads of hard bone pieces when installed into the bone bed. Under these conditions, numerous "openwork" elements are subjected to friction, which leads to the inevitable wear, and in some cases, they are completely separated from the metal substrate of the implant [1]. Several approaches can prevent this process.

(1) Lowering of the parameters of surface morphology and porosity, which is especially well manifested when PVD and CVD methods are used [36]. These methods are widely used in microelectronics; however, their application is appropriate in obtaining nanostructured surfaces combined with other methods of macro- and microtexturing.
(2) Strengthening of porous materials (blocks and coatings) with special fillers of submicron- and nano-sized bioceramic powders of various oxides (Al_2O_3, ZrO_2, and TiO_2) [6,41,42], carbides [43], nitrides (TiN) [23], multiwalled carbon nanotubes [44], and whiskers [11].
(3) Sublayer deposition enhancing the adhesion of the main functional layer [23,24,36].
(4) Production of high-strength coatings of matrix type for the subsequent introduction of the bioactive filler [32,45].
(5) Application of subsequent heat treatment, which also enhances the crystallinity of hydroxyapatite bioceramic coatings [35,46].
(6) Thermal activation during obtaining of the coating. The use of which is limited due to a number of technological features [47].

In most cases, additional treatment complicates the technology of obtaining coatings and increases labor costs and the cost of biomedical products. Highly efficient methods of coatings production are gas-thermal methods, such as plasma spraying [48,49] and High Velocity Oxygen Fuel (HVOF) [50]. The main disadvantage of gas-thermal methods is significant phase-structural change in the initial composition of the deposited powder particles. It affects significantly the biomechanical compatibility of the coating and bone.

During the interaction between the molten particles of bioceramics and the metal substrate, the cooling rate reaches 10^5–10^7 K/s [13]. Under these conditions, splats solidify almost immediately before they can crystallize. Numerous crystal nuclei are formed but their subsequent growth is limited. During plasma heating of hydroxyapatite (HAp) particles, their thermal destruction occurs with the formation of phases: oxyhydroxyapatite (OHAp), oxyapatite (OAp), tricalcium phosphate (TCP), tetracalcium phosphate (TTCP), and amorphous calcium phosphate (ACP) [1,51]. Resorption rate of the constituent phases of such bioceramic coating in biological fluids is represented by the following increasing sequence: $\mathrm{HAp} \ll \beta\text{-TCP} < \mathrm{OHAp/OAp} < \alpha\text{-TCP} < \mathrm{CaO} < \mathrm{TTCP} \ll \mathrm{ACP}$. The proportion of these phases does not exceed a certain limit, since their presence accelerates coating resorption and might cause the loss of the implant [1].

Amorphization of HAp coating caused by ultrahigh cooling rate is one of the reasons for the accelerated resorption of bioceramic coatings, reduction of mechanical properties, and subsequent rejection of the whole implant [2,3]. Therefore, this research describes the results of the study of phase-structural state, chemical composition, and physicomechanical properties of HAp coatings obtained by plasma-induction spraying. Activation of titanium implant comprises the application of high-speed and noncontact inductive heating by generating high-frequency electromagnetic waves.

1.2 Temperature Influence on the Structure and Mechanical Properties of Plasma-Sprayed Biocompatible Coatings

1.2.1 Coating Formation

Structure formation is accompanied by the following phenomena: the formation of interphase boundaries of new composition (the appearance of oxides and other compounds), porous microstructure production, changes in the initial phase composition of powder particles (amorphization, start of phase decomposition), and the occurrence of stresses leading to fracture formation [51–53].

During plasma spraying, the HAp particle has an amorphous phase on its surface. At the moment of impact with the cold substrate in the contact area, its proportion is still increasing; that is why the adhesive strength decreases. Several studies show that amorphous phase in the coating is virtually absent during spraying onto the substrate heated to 800 °C; however, additional phases of TCP and TTCP are formed [13,51]. Thus, the crystallinity can be enhanced by special heat treatment.

The adhesive strength between the splat and the substrate depends on physicochemical parameters that are well known in the theory of gas-thermal spraying:

$$\sigma_{\mathrm{AD}} = \left[1 - \exp\left(-\frac{\nu\,\tau}{\exp\left(\frac{E_{\mathrm{a}}}{kT_{\mathrm{SPS}}}\right)}\right)\right] \times \sigma_{\mathrm{U}} \tag{1}$$

where σ_{AD} is the adhesion strength, MPa; σ_{U} is the tensile strength of the material of powder particle, MPa; $\nu \approx 10^{13}$ Hz is the frequency of natural oscillations of substrate atoms; $\tau = 10^{-7} - 10^{-4}$ s is the characteristic values of the interaction time of molten particle (splat) and the substrate material; k is the Boltzmann constant; E_{a} is the activation energy of interatomic interaction; and T_{SPS} is the contact temperature in the zone of interaction between the splat and the substrate, K.

From expression (1), it follows that to improve the adhesive strength of the coating, it is necessary to raise the contact temperature T_{SPS}, which, in its turn, depends on the temperature of the particles, the substrate, and properties of the materials:

$$T_{SPS} = \frac{b_S T_S + b_P T_P}{b_S + b_P} \tag{2}$$

where b_S and b_P are the coefficients of heat accumulation of materials of the substrate and the particles, respectively. In general, coefficient of heat accumulation b depends on thermal conductivity coefficient λ, specific heat capacity c, and density ρ:

$$b = \sqrt{\lambda c \rho} \tag{3}$$

1.2.2 Crystallization Modeling of Splats

Structure formation of plasma coatings can be considered in terms of the theory of crystallization of microsized ingots. It includes the stages of nucleation and nuclei growth [54,55]. Thus, depending on the cooling conditions/solidification of sprayed particles (splats), amorphous or crystalline structure is formed of nanograins and larger entities, such as submicron or microsized agglomerates [56].

Mathematical model of the initial stage of solidification involves probabilistic Kolmogorov equation:

$$X(t) = 1 - \exp\left\{ -\frac{4\pi}{3} \int_0^t J(T(\theta)) \cdot \left[R_C + \int_\tau^t U(T(\theta)) \mathrm{d}\theta \right]^3 \mathrm{d}\theta \right\} \tag{4}$$

where t is the crystallization duration, s; θ is the current time coordinate, s; T is the temperature, K; X is the volume fraction of crystallized substance; J is the frequency of the critical nuclei formation per volume unit, $\mathrm{m}^{-3}\,\mathrm{s}^{-1}$; U is the speed of grain growth, m/s; and R_C is the critical radius of the nucleus in the crystalline phase, m.

Supercritical growth stage in differential form is described by the expression [57]

$$\frac{\mathrm{d}R}{\mathrm{d}t} = \frac{m\gamma}{R} \tag{5}$$

Thus, in Equation (5), crystal growth rate U is described by the product of empirical values of grain boundary mobility m and surface grain energy γ. Integration of this equation leads to the following expression:

$$R^n - R_C^n = k_n t \tag{6}$$

where R is the grain radius, m; $k_n = nm\gamma R^{n-2}$ is the structural characteristics of the material; n is the empirical exponent for polycrystalline materials in the range from 2 to 4; m is the empirical parameter characterizing the grain boundary mobility, $m^4/(J\ s)$; and γ is the grain boundary energy, J/m^2.

Let us rewrite expression (6):

$$R = R_C + \sqrt[n]{nm\gamma R^{n-2} t} \tag{7}$$

In the simplest case, when $n=2$, the expression for determining the average grain radius takes the form

$$R = R_C + \sqrt{2m\gamma t} \tag{8}$$

In the case of a spherical critical nucleus the radius R_C containing N_C of atoms, we get the following expression:

$$N_C = \frac{4\pi R_C^3}{3V_C} \tag{9}$$

Let us assume that the lattice volume V_C is proportional to the cube of the average lattice constant a^3. For HAp, the average value of $a = 9.42$ Å is determined by the reference data of "JCPDS" base (card 09-0432, 24-0033, and 34-0010). Thus, when calculating the value of HAp critical nucleus from Equation (9), we get $R_C = 2.9$–4.0 nm (the number of atoms in the nucleus of about 30–80 [54,55]).

Let us assume that grain growth is isotropic; then, we will get the following estimate for the growth rate of grains:

$$m\gamma = \frac{a^2}{t_{LS}} \tag{10}$$

where t_{LS} is the duration of the transition of atoms from the melt to the crystalline phase.

Considering expressions (8) and (10) and the duration of cooling $t = t_{ref}$, the expression for determining of the grain size R takes the following form:

$$R = R_C + a\sqrt{\frac{2t_{ref}}{t_{LS}}} \tag{11}$$

The faster atoms move from the melt to the volume of crystalline phase, the more intense is the linear growth of grain size R. The duration of cooling limits the grain growth; in this case, two stages are observed:

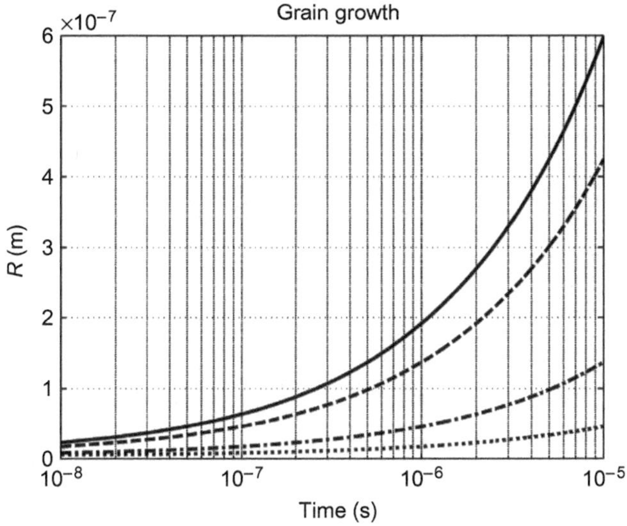

Figure 1 Growth kinetics of nano- and submicron grains at different values of t_{LS}: —, 5×10^{-11} s [55]; - - -, 10^{-10} s; –·–·–, 10^{-9} s; ······, 10^{-8} s.

1. Intensive grain growth at $t_{ref} \ll t_{LS}$
2. Elimination of the influence of grain growth factor at $t_{ref} \to t_{LS}$

Dependency of grain size R on cooling duration t_{ref} at different values of t_{LS} is shown in Figure 1.

1.3 Key Processing Parameters of Plasma-Induction Spraying

The experimental work involved the use of plasma spraying equipment "VRES 744.3227.001." The scheme of plasma spraying with preliminary induction heat treatment (IHT) [58] of metal substrates, for example, of titanium implants, is shown in Figure 2. The spraying parameters are presented in Table 1.

The laboratory apparatus for IHT includes power supplies with rectified high (up to +300 V) and stable low (+15 V) voltages, a generator unit (GU), and an induction heating unit. When selecting the induction heating system configuration, the typical power conversion flow is used. It includes an electromagnetic compatibility (EMC) filter, an AC-DC converter, a capacitor filter, a DC-AC converter, and load. A block diagram of the IHT device also shows an auxiliary power unit with low (+12 V) voltage for air-cooling devices (fans) and a closed liquid-cooling system (Figure 3a).

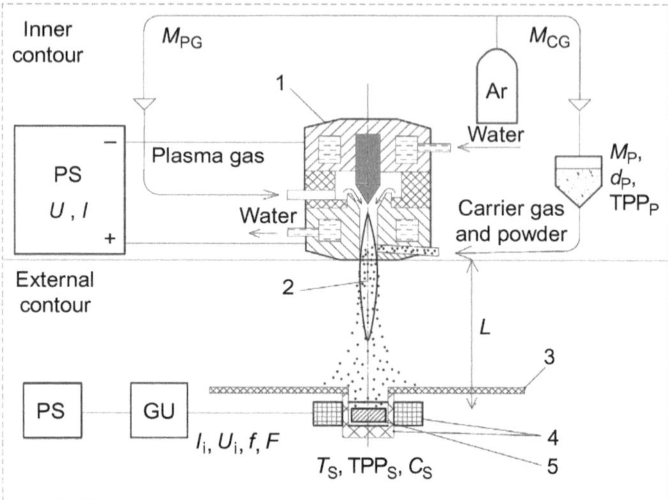

Figure 2 Plasma-induction spraying scheme. Internal circuit: U, I, voltage and current of plasma torch 1 from the power supply (PS); M_{PG} and M_{CG}, flow rate of plasma and carrier gas (argon); M_P, d_P, TPP_P, flow rate, particle size, and thermophysical properties of the powder. Outer circuit: L, spraying distance; T_S, TPP_S, C_S, temperature, thermophysical properties, and structural features of substrate 5; 3, protective screen against the plasma jet 2 of the inductor 4 supplied by voltage U_i and current I_i with frequency f of the induction heating device comprising the main PS and a GU; F, function depending on the treated item geometry.

Table 1 Regimes of plasma spraying with IHT

Parameter	**Value**
Electric power of IHT device (kW)	0.5
Current frequency of inductor (kHz)	100 ± 20
Preheating temperature (°C)	20–1000
Spraying distance (mm)	90–130
Arc current of plasma torch (A)	540
Voltage (V)	30–35
Size of powder (μm)	<90
Plasma gas flow rate (Ar 100%) (L min^{-1})	55–60
Carrier gas flow rate (Ar 100%) (L min^{-1})	5–7

The generating unit includes an oscillator of RC type with the adjustable frequency in the range of 100 ± 20 kHz and a half-bridge output of MOSFETs (metal-oxide-semiconductor field-effect transistors). Frequency control in the selected range ensures the desirable parameters for the heating of small-size items, such as titanium implants. In a low-power version (no more

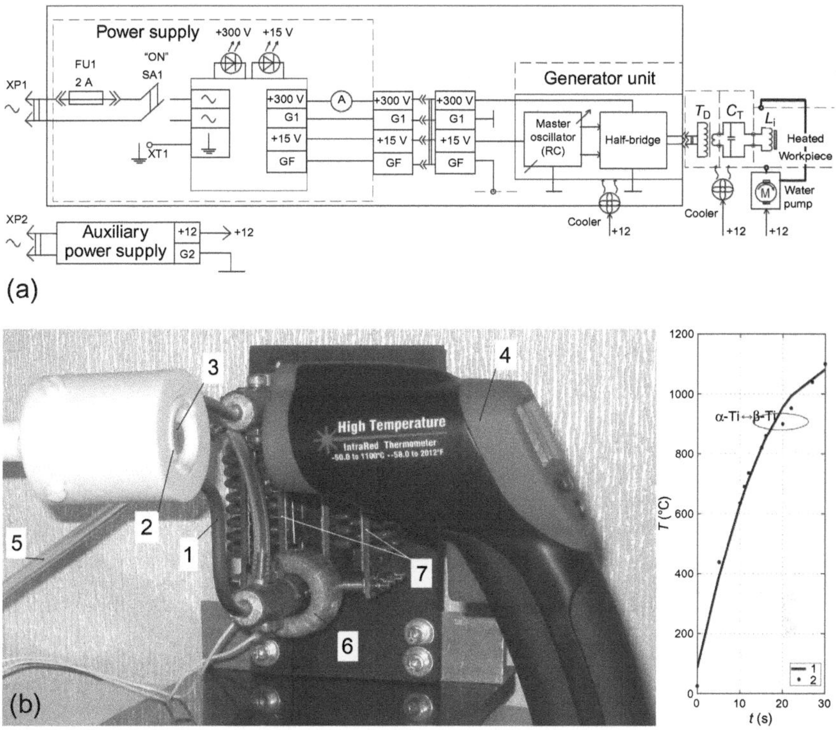

Figure 3 Laboratory apparatus for induction heat treatment: (a) flowchart; (b) temperature measuring of the samples (1, inductor; 2, ceramic muffle; 3, metal sample; 4, pyrometer; 5, water cooling pipe; 6, primary winding of a down transformer; 7, capacitor tank).

than 0.5 kW) of the IHT device, the power is controlled by varying the magnitude of the supply AC voltage using a laboratory autotransformer. A down transformer T_D is included into the output diagonal of half-bridge scheme of the IHT device generator. Further, the IHT device scheme has an oscillating circuit of LC type, which includes a capacitor tank with a constant capacitance C_T and an induction-type heater (inductor) L_i, the geometry of which depends on the heated item geometry.

The temperature and heating rate values were obtained by noncontact temperature measuring with an infrared pyrometer DT-8828 (Figure 3b).

In Figure 3b, at the maximum current value and heating duration of 18-20 s, the deviation increase of the calculated curve from the experimental mean is shown, the physical meaning of which is associated with a polymorphous transformation α-Ti↔β-Ti.

2. HYDROXYAPATITE (HAp) COATINGS ON TITANIUM IMPLANTS OBTAINED BY PLASMA-INDUCTION SPRAYING

2.1 Materials and Methods

2.1.1 Samples

The samples are 2 mm thick plates, which are made of commercially pure (CP) titanium VT1-00 (analog of titanium CP-3 grade 2). The main preparatory operations are corundum sandblasting abrasive with an average particle size of 200-400 μm, which provides the necessary parameters of the surface microrelief and further removal of various impurities in detergent solutions and in ethanol when titanium samples are put into an ultrasonic bath. Plasma spraying of HAp coatings was carried out at the spraying distance of 90 and 130 mm, arc current of 540 A, and voltage of 30 V. HAp was synthesized using calcium nitrate and diammonium phosphate in ammonia solution according to the well-known reactions of liquid-phase synthesis. Further, HAp unit was crushed and sieved into powder with an average size of 50 ± 20 μm.

IHT was carried out immediately before plasma spraying, according to the scheme shown in Figure 2. The heating rate of the samples at the maximum power of IHT device ranged from 30 to 70 °C/s. The obtained samples of coatings were given two-digit numbering: the first number indicates the spraying distance (1-90 and 2-130 mm) and the second one indicates the preheating temperature of substrate (0 to no heating, 1-100, 2-200, ... 10-1000 °C). The deposition duration was 3-5 s per sample, which corresponds to the obtained coating thickness of 25-35 μm.

2.1.2 Characterization

Phase-structural state of CaP bioceramic coatings was determined by X-ray diffraction (XRD) on "Gemini/Xcalibur" diffractometer using X-ray tube with a copper anode Cu-Kα radiation. To determine the crystal structure of coatings, "Crystallography Open Database" was used. Test sample preparation for XRD comprised the obtaining of a conglomerate of sprayed coating fragments. Delamination from the titanium substrate was performed by scratching; then, the microparticles were bound into the globular conglomerate using epoxy resin with an amorphous structure. Attachment of the obtained sample was performed on glass fiber.

Microstructure and nanostructure of HAp coatings were studied using scanning electron microscopy (SEM). Processing of SEM images was performed using hardware and software for the analysis of geometric parameters of micro-objects "AGPM-6M" and program "Metallograph." As a result of the morphological analysis of the coating surface images, the following parameters were determined: average linear dimension of microrelief elements (grains/pores and protrusions/cavities), their dispersion in size, and the number of microrelief elements in the view field.

Mechanical properties (modulus of elasticity and hardness) of HAp coatings were evaluated by nanoindentation using mechanical properties tester "NANOVEA Ergonomic Workstation." The selected load applied to Berkovich indenter varied from 50 to 100 mN and provided its penetration into the coating within the thickness of 1-3 splats (about 0.5-2.5 μm).

Experimental data were processed statistically with the calculated mean values and confidence interval; approximation of least squares was performed. As a result, summarized dependencies of the structural parameters, chemical composition, and mechanical properties were obtained. Mathematical processing was performed in "MATLAB 6.0."

Preliminary biocompatibility checking of the studied coating samples was carried out *in vitro*. Human dermal fibroblasts (HDFs culture) were seeded in tissue culture flasks (Falcon BD) and cultivated in Dulbecco Modified Eagle Medium (DMEM) (Biolot, Russia) containing 10% of fetal bovine serum (FBS, HyClone, the United Kingdom), 0.1% gentamicin (Sigma-Aldrich), and 4.5 g/L D-glucose (Sigma-Aldrich). Fibroblasts were used as a test model for toxicity of the studied materials. Fibroblasts were seeded on the surfaces with an average density of 5×10^4 in 1 mL of cell culture medium per well and incubated for 7 days. Media was changed every 3 days, and cells were maintained in a humidified incubator at 37 °C with 5% of CO_2. The samples were deposited in 24-well cell culture plates (Falcon BD) after sterilization procedure (70% ethanol for 30 min). Then, samples were air-dried (under sterile conditions) and washed three times with sterile PBS. After cultivation, the samples were washed with PBS twice and fixed in 4% paraformaldehyde for 15 min at 4 °C. SEM was used to observe and analyze the cell adhesion and attachment on the sample surfaces.

2.2 Results and Discussion

2.2.1 XRD Analysis

Phase composition of coatings significantly depends on the temperature of preliminary IHT (Figure 4). The initial composition of the sprayed bioceramic powder corresponds to HAp (all peaks correspond to the data card "9001233") [59].

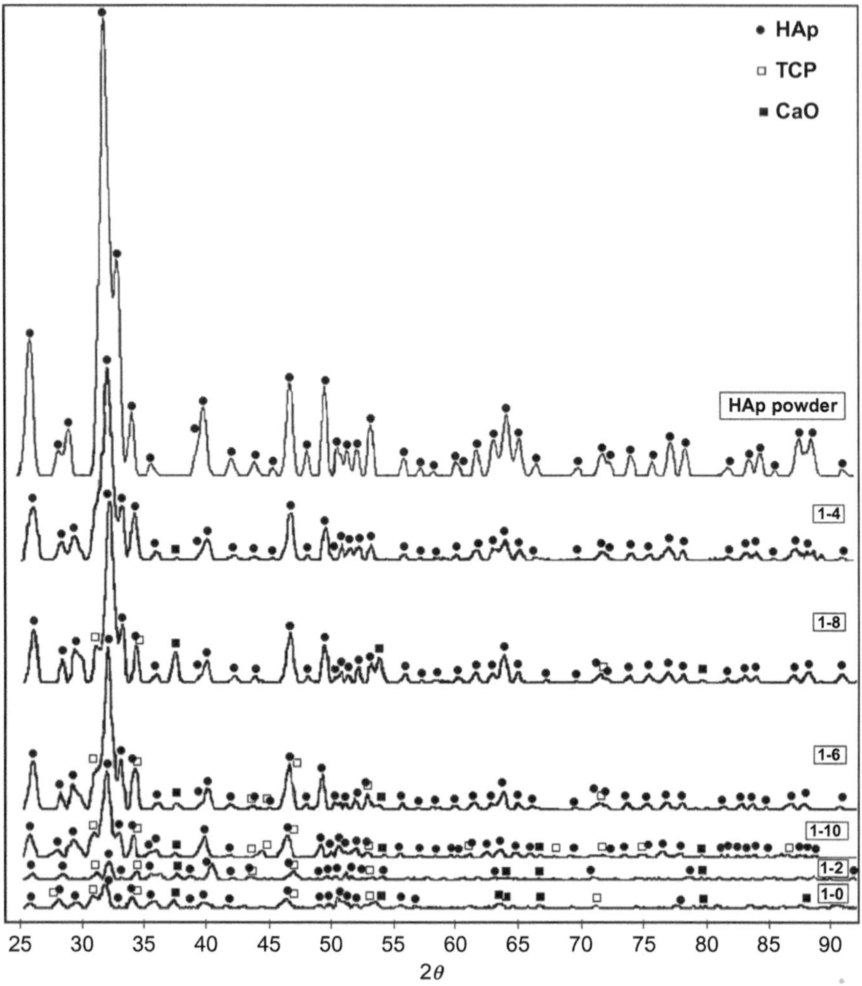

Figure 4 XRD of HAp powder and coating samples obtained at the spraying distance of 90 mm: •—HAp; □—TCP; ▪—CaO.

Analysis of initial diffractograms (no correction of amorphous phase) of the obtained coatings shows that the initial and the main component HAp remained in sufficient quantity on regimes 1-4, 1-6, and 1-8. Its proportion ranges from 40 to 60%, and the maximum of the crystalline phase equaling 70% is at the preheating temperature of about 600 °C. A slightly lower amount of the crystalline phase of calcium phosphate compounds (HAp with 10% addition of TCP and 3-7% of CaO impurity) is observed in the range of IHT over 600 °C and up to 1000 °C. The samples of coatings obtained on regimes 1-0 and 1-2 are characterized by lower crystallinity (about 10-20%) and HAp amount of 10-12% with a slight addition of TCP and CaO, not more than 5% each. In coatings obtained on these regimes, the increased amount of ACP (up to 80-90%) is observed, which can be a significant drawback of the coatings. TCP, CaO, and ACP are the phases of HAp decomposition; they also have increased resorption rate and lower mechanical strength compared to the initial HAp.

2.2.2 Surface Morphology

2.2.2.1 Microstructure

Morphological heterogeneity MH_S of HAp coatings is a quantitative characterization of the size distribution of protrusions and/or cavities on the surface of splats and their structural components. The microstructure of HAp coatings has significant differences depending on plasma spraying regimes, particularly on IHT temperature (Figure 5). Average diameter dependency for microprotrusions of the coating D_S on the temperature of preliminary IHT of titanium substrate is parabolic (Table 2).

Without preheating (regime 1-0), parameter D_S takes a minimum value. Maximum is observed at the preheating temperature of 400-600 °C (regimes 1-4 and 1-6). With further increase of temperature T_S to 800-1000 °C, parameter D_S decreases (regimes 1-8 and 1-10).

Analysis of morphological heterogeneity MH_S was based on the density of splat protrusions in the studied area, whereas MH_S varies according to the periodic law. MH_S minimum corresponds to the temperature of 400 and 800 °C. The same values are observed on regime 1-0, which corresponds to conventional plasma spraying (Figure 5a). Small spherical particles sprayed at this spraying distance interact weakly with the surface of titanium substrate and large splats forming the coating. MH_S increases from 2 to 4 times on regimes 1-2, 1-6, and 1-10. On regime 1-2, the cohesion of fine spherical particles and dust HAp powder particles improves (Figure 5b). Increased fluidity caused by viscosity decrease smoothens the microrelief (regime 1-4); however, at the impact of liquid droplets, the emerging surges influence the microgeometry of HAp coatings on regime 1-6 (Figure 5c and d). With further temperature increase of preliminary IHT (regime 1-8), morphological heterogeneity is reduced due to the lengthy stay of sprayed particles of all size modes in viscoplastic or liquid state (Figure 5e). Splat thickness decreases from 5-15 μm to the minimum value of about 1-2 μm, which leads to a sharp increase in the cooling rate up to the supercritical value of 10^5-10^7 °C/s and, as a result, to a noticeable cracking (Figure 5f).

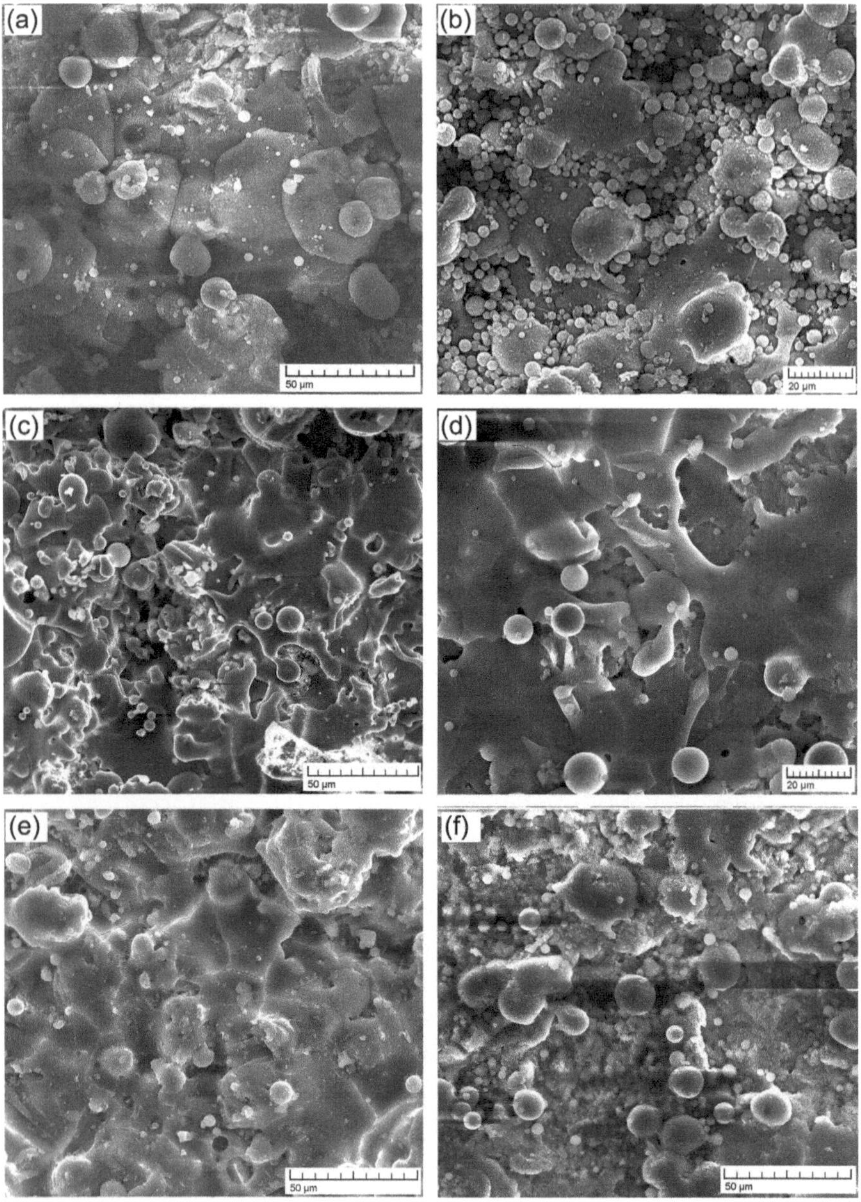

Figure 5 Microstructure of HAp coatings obtained on regimes: (a) 1-0; (b) 1-2; (c) 1-4; (d) 1-6; (e) 1-8; (f) 1-10.

At the upper limit of the studied range of IHT, that is, above the phase transition α-Ti$\leftrightarrow$$\beta$-Ti temperature (regime 1-10), the formation of a titania TiO_2 film is observed, which is characterized by increased morphological heterogeneity (it is discussed in detail in Chapter 19). The presence of

Table 2 Morphology indicators of HAp coatings

Regime	Morphological heterogeneity MH_S (mm^{-2})	Average splat size D_S and confidence interval ΔD_S (μm)
1-0	5821	5.67 ± 3.32
1-2	21,826	8.79 ± 4.53
1-4	6447	12.43 ± 5.30
1-6	11,245	12.88 ± 5.13
1-8	4210	10.22 ± 6.49
1-10	18,918	6.37 ± 4.07
Ti substrate after sandblasting	99,251	4.92 ± 2.84

intermediate metal oxide sublayer reduces the cracking of bioceramic HAp layer, enhances the attachment of small spherical splats, and limits the spreading of large splats.

2.2.2.2 Nanostructure

The structural condition of the coating on the nanoscale was assessed by the generalized parameters of grain size (or agglomerates of nanoparticles) and roughness. The samples of coatings with morphology parameters close to those of the bone trabeculae with mineral plates (crystal grains) of HAp were considered the best. Similar morphology of HAp coating and bone tissue is required for the activation of the geometric factor of bioactivity on the nanoscale.

Computer image processing is shown in the example image of SEM (Figure 6). Spraying distance has a significant influence on the nanostructure of HAp coatings (Figure 7). At the distance of 70 mm and without thermal activation of titanium substrate, the structure of coating splats is formed of submicron- and nano-sized crystals of HAp (Figure 7a). It reduces the actual surface area together with the geometric bioactivity of HAp coating.

Morphological heterogeneity has the maximum value when the spraying distance is 90 mm and is significantly reduced with the distance increase to 130 mm (Table 3).

The smallest nanograins of 11-12 nm and the highest surface density of their distribution are observed at the distance of 90 mm. The obtained HAp coating is characterized by considerable heterogeneity of nanostructure and formation of numerous dust particles and their agglomerates (Figure 7b and c). This structure has low mechanical characteristics, in particular microhardness, which leads to wear and coating delamination during the installation into the bone bed.

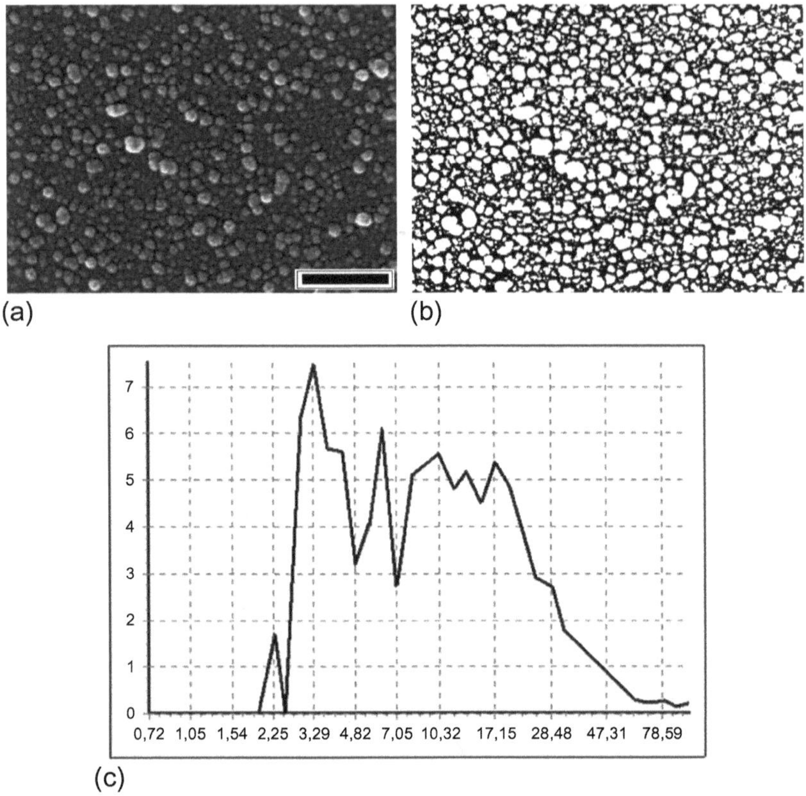

Figure 6 Morphology of the nanostructure of HAp coating obtained on regime 1-6: (a) SEM (mark 200 nm); (b) binarized image obtained by processing software; (c) distribution graph in linear dimensions of nanograins (*x*-axis, size, nm; *y*-axis, percentage, %).

The effect of IHT on the surface nanostructure of HAp coatings is explained by the previously discussed process of cooling and crystallization of splats. When the substrate is preheated to 200 °C, the structure of plasma coatings varies slightly. The average grain size increases to 26 nm; however, agglomerates of 150–300 nm are observed on the surface (Figure 8a).

With further increase of the preheating temperature to 400 °C, the structure of coatings becomes practically homogeneous (Figure 8b). Agglomerates are sporadic and consist of a small amount of nanoparticles. The average diameter reaches the maximum of about 31 nm. It should be noted that in the area of heating from 20 to 400 °C, there is a linear increase in nanograin growth (Figure 8d; line D_1).

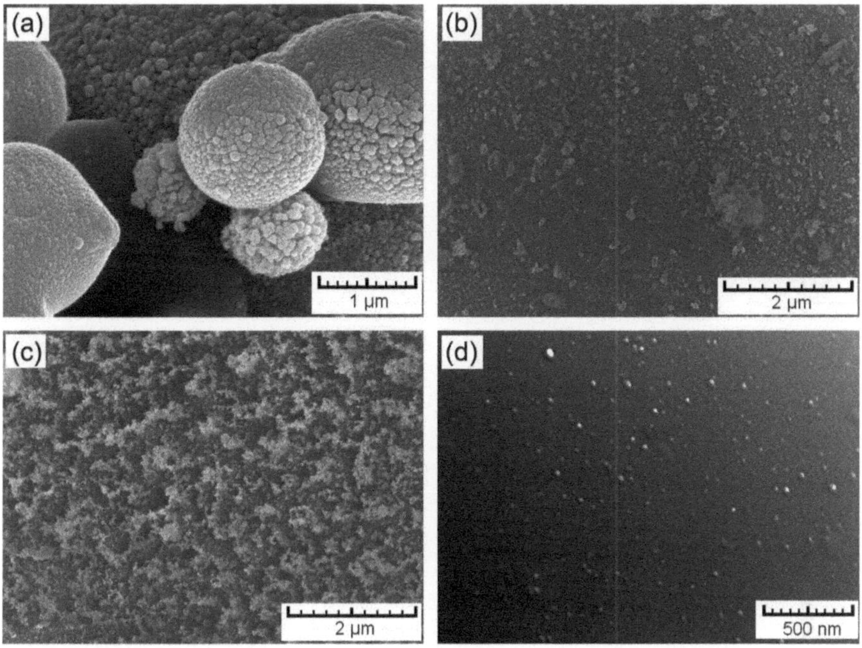

Figure 7 Nanostructure of HAp coatings obtained at the spraying distance: (a) 70 mm, without IHT; (b, c) 90 mm (regime 1-0); (d) 130 mm (regime 2-0).

Table 3 Structure parameters of HAp coatings obtained by the conventional plasma spraying

No.	Spraying distance (mm)	Nanostructure type	Density of particles (μm^{-2})	Average diameter (nm)
1	70	Nanograins of crystallization formed on the surface layer	202	50.0 ± 7.9
2	90	Dust nanoparticles after dispergating of the initial powder particles + nanograins of crystallization formed on the surface layer	123	12.0 ± 6.2
3	90	Agglomerates of dust nanoparticles	442	11.0 ± 6.1
4	130	Nanograins of crystallization formed on the surface layer	22	39.0 ± 8.1

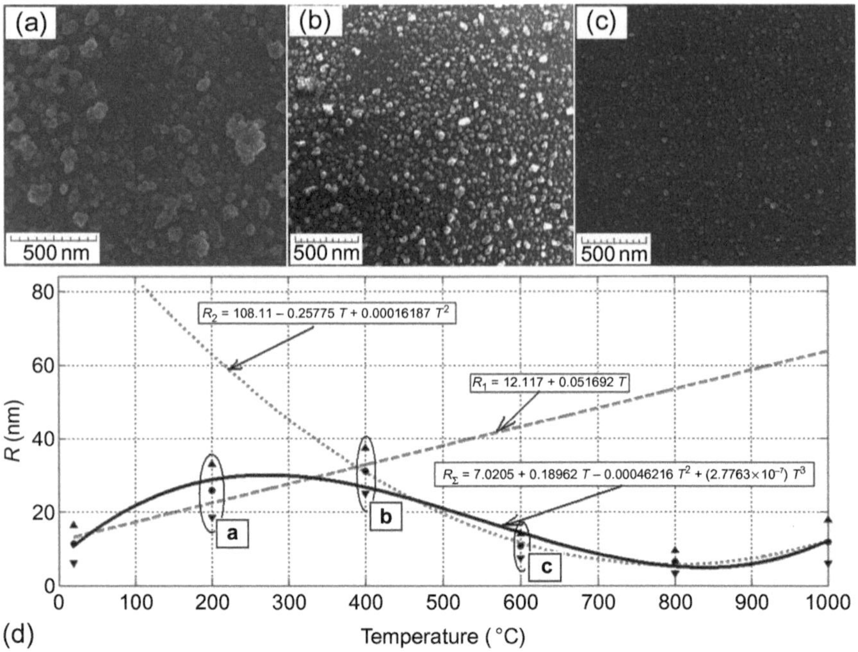

Figure 8 SEM of nanostructure of HAp coatings obtained on regimes: (a) 1-2; (b) 1-4; (c) 1-6; (d) dependency of the average grain size on the temperature (spraying distance of 90 mm).

Viscosity decrease caused by heating to 600 °C contributes to better spreading of splats, so the reduction of the average grain size D_G to the initial value of 11 nm is explicable, and it is described by the parabolic dependency (Figure 8d; curve D_2). The structure is characterized by high homogeneity; the agglomerates are sporadic. Under these conditions, the smallest nanocrystals with the size of 3-7 nm, making about 50%, are distinguished (Figure 6c).

When the preheating temperature is between 800 and 1000 °C, separate agglomerated particles of 30-90 nm are arranged on the "smooth" surface of HAp coating. There is a marked reduction in morphological heterogeneity, and the average size of nanograins equals 6-12 nm.

Analysis of the structures on the nanoscale shows that the use of preliminary IHT of titanium substrate allows to produce quite homogeneous nanocrystalline structure of the coatings. The possible convergence between morphology types of HAp coatings and bone trabeculae was established.

2.2.3 Physicomechanical Properties

To obtain HAp coatings with high physicomechanical properties, the comprehensive studies of the modulus of elasticity and hardness were carried out.

The hardness of HAp coating exceeding that of the cortical bone provides the improved biomechanical compatibility and eliminates the danger of its destruction during the installation of the implant into the bone bed. The proposed IHT of titanium substrate before the deposition of the coating enhances basic physicomechanical properties, such as adhesion and cohesion. However, when the implant is installed, its local area of the surface is exposed to the concentrated force. Thus, high hardness and modulus of elasticity of porous HAp coatings are necessary for their further reliable performance in the tissues of the human body (Figure 9).

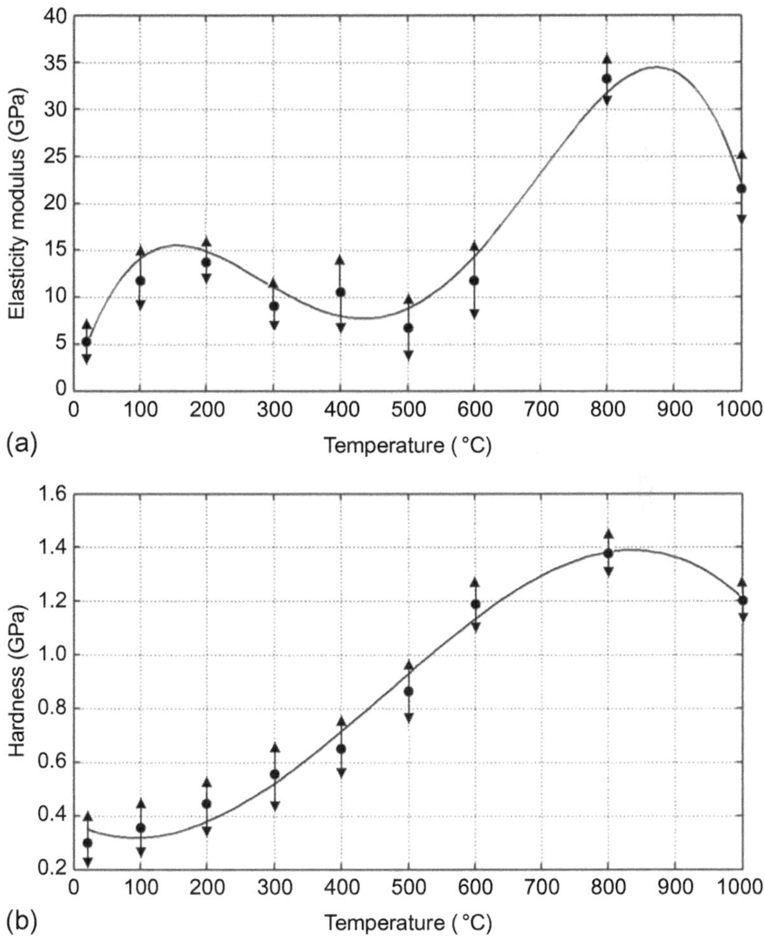

Figure 9 Dependency of modulus of elasticity (a) and hardness (b) on IHT temperature.

Low-temperature IHT of titanium substrate does not significantly increase the hardness. Modulus of elasticity for the coatings subjected to IHT at about 200 °C increased 2-2.5 times. The highest values of the modulus of elasticity for cortical bone reach 25-30 GPa with the hardness of about 0.5-0.8 GPa. When these regimes are used, there is low performance of intraosseous implants.

On the middle-temperature regime of IHT, the hardness of coatings reaches the values typical for the solid fragments of cortical bone. The modulus of elasticity remains around 8-12 GPa. Heating to 800-1000 °C provides a 2-3-fold margin of hardness at the maximum value of the modulus of elasticity of the HAp porous coating. Enhancing of these physicomechanical characteristics is also associated with a slight decrease in the average value of total porosity from 45-48% to 41-44%.

2.2.4 *Preliminary* In Vitro *Testing for Biocompatibility*

Biocompatibility of samples with different phase-structural states of the surface was studied *in vitro* using fibroblasts. Thus, it was concluded that the plasma spraying regimes of HAp coatings with preliminary IHT of the titanium substrate determine the character of their biointeraction with cells. Figure 10 shows the digital images of HAp coatings after *in vitro* testing.

In SEM images, the structural elements are observed, which are not characteristic for the surface morphology of materials subjected to thermal spraying. Surrounding space around them is filled with organic matrix and cells. The cells have an oblong or slightly curved shape (Figure 10a–c). From the cells, several large and many small filopodia begin, each of which is attached to the substrate (Figure 10d).

Fibroblasts are mainly localized on the surface of spherical microparticles and splats with submicron and nanoelements of the structure (Figure 10d). Improved interaction between the cells and HAp coating is manifested in the presence of morphologically developed surface. Cells and their filopodia "infiltrate" the porous matrix of HAp coating. It can be assumed that multiple nanopores actively contribute to this process, thereby increasing the adhesion of cells in terms of their biochemical interaction and mechanical attachment.

Expressed adhesion of cells is manifested in the samples of HAp coatings obtained on regimes 1-4, 1-6, and 1-8. The combination of micro- and nano-sized surface elements of HAp coatings also contributes to the formation of practically continuous cell layer (Figure 10e and f). Numerous

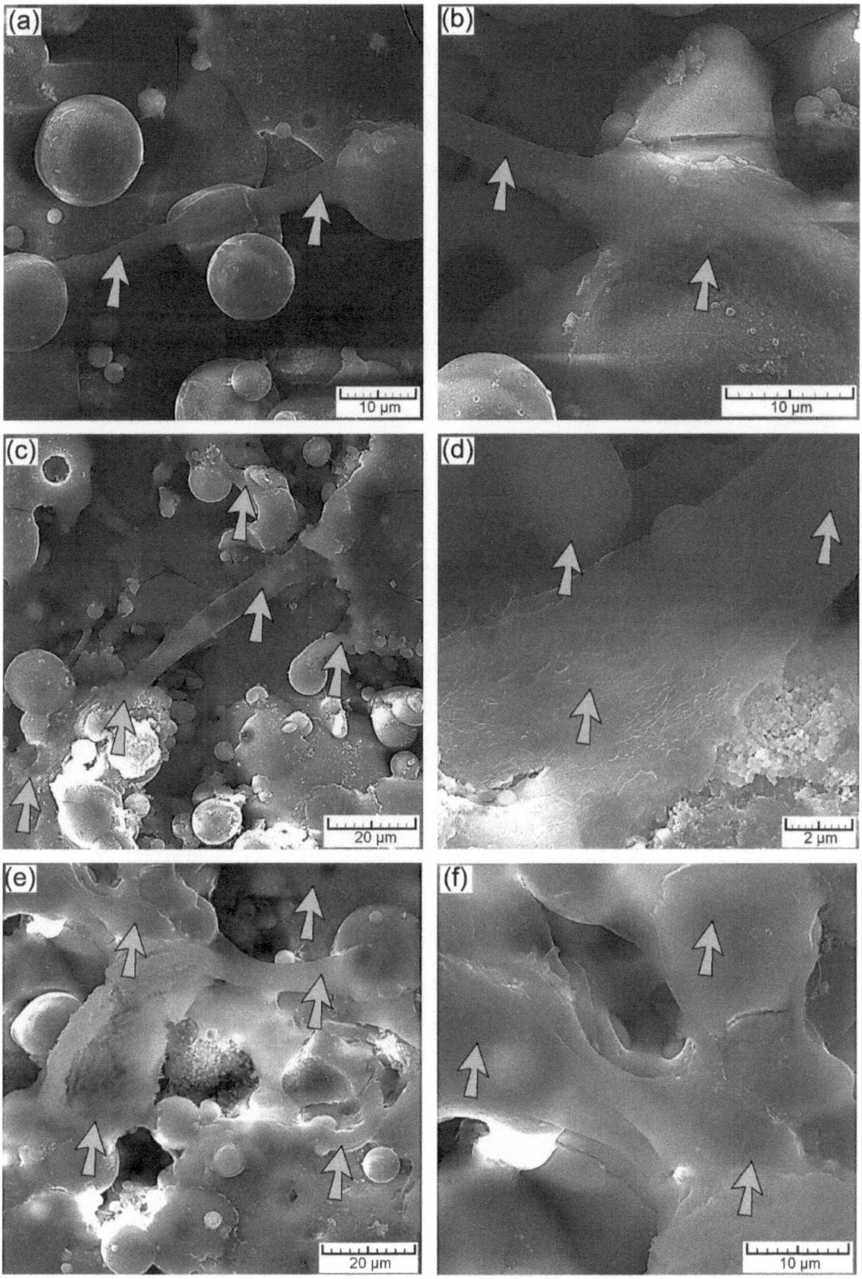

Figure 10 SEM results of *in vitro* study (cells are marked with "arrows") of the samples of HAp coatings obtained on regimes: (a) 1-0 (attachment of fibroblasts to spheroidal microparticles of relief); (b) 1-0 (fibroblast on the surface of a large splat); (c) 1-2 (fibroblasts on morphologically heterogeneous surface with numerous small splats and submicron dust particles); (d) 1-2 (interface of the type "fibroblast-nanostructured HAp coating"); (e, f) 1-6.

nanopores and protrusions serve as a sort of centers for the attachment of natural submicron and nanostructures (filopodia).

3. CONCLUSIONS

The surface of titanium VT1-00 after plasma spraying with preliminary IHT is characterized by the formation of calcium phosphate coating with the phase composition close to HAp (confirmed by XRD of phase-structural analysis), high morphological heterogeneity of microstructure, homogeneous nanostructure, high physicomechanical properties, and biocompatibility characteristics determined during preliminary *in vitro* testing showing the penetration of cell structures into the porous, rough surface of nanostructured HAp coating. HAp coatings with the best qualities are formed by micro-sized splats, the surface of which has nano-sized grains and pores. The best morphological parameters of HAp coatings combined with high hardness of about 1.2-1.4 GPa and modulus of elasticity of at least 11-12 GPa are determined. These samples of coatings correspond to the following regimes of plasma spraying with IHT 1-4 and 1-6.

Thus, it can be expected that surfaces of intraosseous implants made of titanium with hardened and porous HAp coatings will show high bioactive qualities in *in vivo* investigations.

ACKNOWLEDGMENTS

The research was carried out with financial support from RFBR grant no. 13-03-00898 "a" and scholarship of the President of the Russian Federation no. SP-1051.2012.4. The author also expresses his deep gratitude to Dr. Butovski K.G., Steinhauer A.B., and Fomina M.A. for the valuable comments during the discussion of this work.

REFERENCES

[1] S.R. Paital, N.B. Dahotre, Calcium phosphate coatings for bio-implant applications: materials, performance factors, and methodologies, Mater. Sci. Eng. R. Rep. 66 (2009) 1–70.
[2] L. Le Guéhennec, A. Soueidan, P. Layrolle, Y. Amouriq, Surface treatments of titanium dental implants for rapid osseointegration, Dent. Mater. 23 (2007) 844–854.
[3] S.A. Shabalovskaya, D. Siegismund, E. Heurich, M. Rettenmayr, Evaluation of wettability and surface energy of native Nitinol surfaces in relation to hemocompatibility, Mater. Sci. Eng. C 33 (2013) 127–132.
[4] G. Mendonça, D.B.S. Mendonça, F.J.L. Aragão, L.F. Cooper, Advancing dental implant surface technology – from micron- to nanotopography, Biomaterials 29 (2008) 3822–3835.

[5] S.K. Swain, S. Bhattacharyya, Preparation of high strength macroporous hydroxyapatite scaffold, Mater. Sci. Eng. C 33 (2013) 67–71.
[6] L. Pawlowski, Finely grained nanometric and submicrometric coatings by thermal spraying: a review, Surf. Coat. Technol. 202 (2008) 4318–4328.
[7] N. Wang, H. Li, W. Lü, J. Li, J. Wang, Z. Zhang, et al., Effects of TiO_2 nanotubes with different diameters on gene expression and osseointegration of implants in minipigs, Biomaterials 32 (2011) 6900–6911.
[8] S.D. Puckett, E. Taylor, T. Raimondo, T.J. Webster, The relationship between the nanostructure of titanium surfaces and bacterial attachment, Biomaterials 31 (2010) 706–713.
[9] X. Liu, X. Zhao, R.K.Y. Fu, J.P.Y. Ho, C. Ding, P.K. Chu, Plasma-treated nanostructured TiO_2 surface supporting biomimetic growth of apatite, Biomaterials 26 (2005) 6143–6150.
[10] P. Decuzzi, M. Ferrari, Modulating cellular adhesion through nanotopography, Biomaterials 31 (2010) 173–179.
[11] W. Pon-On, N. Charoenphandhu, I.M. Tang, J. Teerapornpuntakit, J. Thongbunchoo, N. Krishnamra, Biocomposite of hydroxyapatite-titania rods (HApTiR): physical properties and *in vitro* study, Mater. Sci. Eng. C 33 (2013) 251–258.
[12] N. Ribeiro, S.R. Sousa, F.J. Monteiro, Influence of crystallite size of nanophased hydroxyapatite on fibronectin and osteonectin adsorption and on MC3T3-E1 osteoblast adhesion and morphology, J. Colloid Interface Sci. 351 (2010) 398–406.
[13] S. Dyshlovenko, L. Pawlowski, B. Pateyron, I. Smurov, J.H. Harding, Modelling of plasma particle interactions and coating growth for plasma spraying of hydroxyapatite, Surf. Coat. Technol. 200 (2006) 3757–3769.
[14] M.F. Morks, A. Kobayashi, Influence of gas flow rate on the microstructure and mechanical properties of hydroxyapatite coatings fabricated by gas tunnel type plasma spraying, Surf. Coat. Technol. 201 (2006) 2560–2566.
[15] Y. Yan, J. Sun, Y. Han, D. Li, K. Cui, Microstructure and bioactivity of Ca, P and Sr doped TiO_2 coating formed on porous titanium by micro-arc oxidation, Surf. Coat. Technol. 205 (2010) 1702–1713.
[16] S.V. Dorozhkin, Bioceramics of calcium orthophosphates, Biomaterials 31 (2010) 1465–1485.
[17] J.M. Gomez-Vega, E. Saiz, A.P. Tomsia, G.W. Marshall, S.J. Marshall, Bioactive glass coatings with hydroxyapatite and bioglass particles on Ti-based implants. 1. Processing, Biomaterials 21 (2000) 105–111.
[18] S. González, E. Pellicer, J. Fornell, A. Blanquer, L. Barrios, E. Ibáñez, P. Solsona, S. Suriñach, M.D. Baró, C. Nogués, J. Sort, Improved mechanical performance and delayed corrosion phenomena in biodegradable Mg-Zn-Ca alloys through Pd-alloying, J. Mech. Behav. Biomed. Mater. 6 (2012) 53–62.
[19] J.P. Yuan, W. Li, C. Wang, Effect of the La alloying addition on the antibacterial capability of 316L stainless steel, Mater. Sci. Eng. C 33 (2013) 446–452.
[20] M. Gao, L. Sun, Z. Wang, Y. Zhao, Controlled synthesis of Ag nanoparticles with different morphologies and their antibacterial properties, Mater. Sci. Eng. C 33 (2013) 397–404.
[21] S. Qian, X. Liu, C. Ding, Effect of Si-incorporation on hydrophilicity and bioactivity of titania film, Surf. Coat. Technol. 229 (2013) 156–161.
[22] J. Zhou, B. Li, S. Lu, L. Zhang, Y. Han, Regulation of osteoblast proliferation and differentiation by interrod spacing of Sr-HA nanorods on microporous titania coatings, ACS Appl. Mater. Interfaces 5 (2013) 5358–5365.
[23] S. Yang, H.C. Man, W. Xing, X. Zheng, Adhesion strength of plasma-sprayed hydroxyapatite coatings on laser gas-nitrided pure titanium, Surf. Coat. Technol. 203 (2009) 3116–3122.

[24] G. Singh, S. Singh, S. Prakash, Surface characterization of plasma sprayed pure and reinforced hydroxyapatite coating on Ti6Al4V alloy, Surf. Coat. Technol. 205 (2011) 4814–4820.

[25] S. Saber-Samandari, K.A. Gross, Nanoindentation reveals mechanical properties within thermally sprayed hydroxyapatite coatings, Surf. Coat. Technol. 203 (2009) 1660–1664.

[26] B. Viswanath, R. Raghavan, U. Ramamurty, N. Ravishankar, Mechanical properties and anisotropy in hydroxyapatite single crystals, Scr. Mater. 57 (2007) 361–364.

[27] A.D. Pogrebnyak, A.A. Drobyshevskaya, V.M. Beresnev, M.K. Kylyshkanov, T. V. Kirik, S.N. Dub, F.F. Komarov, A.P. Shipilenko, Y.Z. Tuleushev, Micro- and nanocomposite Ti-Al-N/Ni-Cr-B-Si-Fe-based protective coatings: structure and properties, Tech. Phys. 56 (2011) 1023–1030.

[28] L. Lapaj, J. Markuszewski, T. Rybak, A.A. Britsko, V.V. Anosov, M. Wierusz-Kozlowska, Wear analysis of a ceramic on ceramic hip endoprosthesis, J. Frict. Wear 34 (2013) 32–37.

[29] S. Heinemann, C. Heinemann, S. Wenisch, V. Alt, H. Worch, T. Hanke, Calcium phosphate phases integrated in silica/collagen nanocomposite xerogels enhance the bioactivity and ultimately manipulate the osteoblast/osteoclast ratio in a human co-culture model, Acta Biomater. 9 (2013) 4878–4888.

[30] J. Ni, K. Noh, C.J. Frandsen, S.D. Kong, G. He, T. Tang, et al., Preparation of near micrometer-sized TiO_2 nanotube arrays by high voltage anodization, Mater. Sci. Eng. C 33 (2013) 259–264.

[31] H.-T. Chen, C.-H. Hsiao, H.-Y. Long, C.-J. Chung, C.-H. Tang, K.-C. Chen, et al., Micro-arc oxidation of β-titanium alloy: structural characterization and osteoblast compatibility, Surf. Coat. Technol. 204 (2009) 1126–1131.

[32] A. Kar, K.S. Raja, M. Misra, Electrodeposition of hydroxyapatite onto nanotubular TiO_2 for implant applications, Surf. Coat. Technol. 201 (2006) 3723–3731.

[33] S. Nag, S.R. Paital, P. Nandawana, K. Mahdak, Y.H. Ho, H.D. Vora, et al., Laser deposited biocompatible Ca-P coatings on Ti-6Al-4V: microstructural evolution and thermal modeling, Mater. Sci. Eng. C 33 (2013) 165–173.

[34] A. Siddharthan, T.S. Sampath Kumar, S.K. Seshadri, *In situ* composite coating of titania-hydroxyapatite on commercially pure titanium by microwave processing, Surf. Coat. Technol. 204 (2010) 1755–1763.

[35] S.C.G. Leeuwenburgh, J.G.C. Wolke, J. Schoonman, J.A. Jansen, Deposition of calcium phosphate coatings with defined chemical properties using the electrostatic spray deposition technique, J. Eur. Ceram. Soc. 26 (2006) 487–493.

[36] F. Maury, F.-D. Duminica, TiO_xN_y coatings grown by atmospheric pressure metal organic chemical vapor deposition, Surf. Coat. Technol. 205 (2010) 1287–1293.

[37] J. Cizek, K.A. Khor, Z. Prochazka, Influence of spraying conditions on thermal and velocity properties of plasma sprayed hydroxyapatite, Mater. Sci. Eng. C 27 (2007) 340–344.

[38] X. Liu, P.K. Chu, C. Ding, Surface modification of titanium, titanium alloys, and related materials for biomedical applications, Mater. Sci. Eng. R. Rep. 47 (2004) 49–121.

[39] C. Kim, M.R. Kendall, M.A. Miller, C.L. Long, P.R. Larson, M.B. Humphrey, et al., Comparison of titanium soaked in 5 M NaOH or 5 M KOH solutions, Mater. Sci. Eng. C 33 (2013) 327–339.

[40] P.A. Ramires, A. Romito, F. Cosentino, E. Milella, The influence of titania/hydroxyapatite composite coatings on *in vitro* osteoblasts behaviour, Biomaterials 22 (2001) 1467–1474.

[41] C. Arnould, J. Denayer, M. Planckaert, J. Delhalle, Z. Mekhalif, Bilayers coating on titanium surface: the impact on the hydroxyapatite initiation, J. Colloid Interface Sci. 341 (2010) 75–82.
[42] L. Xie, X. Ma, A. Ozturk, E.H. Jordan, N.P. Padture, B.M. Cetegen, D.T. Xiao, M. Gell, Processing parameter effects on solution precursor plasma spray process spray patterns, Surf. Coat. Technol. 183 (2004) 51–61.
[43] D. Caschera, F. Federici, L. Pandolfi, S. Kaciulis, M. Sebastiani, E. Bemporad, G. Padeletti, Effect of composition on mechanical behaviour of diamond-like carbon coatings modified with titanium, Thin Solid Films 519 (2011) 3061–3067.
[44] K. Balani, Y. Chen, S.P. Harimkar, N.B. Dahotre, A. Agarwal, Tribological behavior of plasma-sprayed carbon nanotube-reinforced hydroxyapatite coating in physiological solution, Acta Biomater. 3 (2007) 944–951.
[45] M.Y. Zhang, C. Ye, U.J. Erasquin, T. Huynh, C. Cai, G.J. Cheng, Laser engineered multilayer coating of biphasic calcium phosphate/titanium nanocomposite on metal substrates, ACS Appl. Mater. Interfaces 3 (2011) 339–350.
[46] F.-H. Lin, L. Chun-Jen, C. Ko-Shao, S. Jui-Sheng, Thermal reconstruction behavior of the quenched hydroxyapatite powder during reheating in air, Mater. Sci. Eng. C 13 (2000) 97–104.
[47] M. Bertagnolli, M. Marchese, G. Jacucci, Modelling of particles impacting on a rigid substrate under plasma spraying conditions, J. Therm. Spray Technol. 4 (1995) 41–49.
[48] R. Jaworski, C. Pierlot, L. Pawlowski, M. Bigan, M. Martel, Design of the synthesis of fine HA powder for suspension plasma spraying, Surf. Coat. Technol. 203 (2009) 2092–2097.
[49] S. Kozerski, F.-L. Toma, L. Pawlowski, B. Leupolt, L. Latka, L.-M. Berger, Suspension plasma sprayed TiO_2 coatings using different injectors and their photocatalytic properties, Surf. Coat. Technol. 205 (2010) 980–986.
[50] K.A. Khor, H. Li, P. Cheang, Processing-microstructure-property relations in HVOF sprayed calcium phosphate based bioceramic coatings, Biomaterials 24 (2003) 2233–2243.
[51] S. Dyshlovenko, B. Pateyron, L. Pawlowski, D. Murano, Numerical simulation of hydroxyapatite powder behavior in plasma jet, Surf. Coat. Technol. 179 (2004) 110–117.
[52] S.A. Catledge, M. Fries, Y.K. Vohra, Nanostructured surface modifications for biomedical implants, Encycl. Nanosci. Nanotechnol. 10 (2004) 1–22.
[53] J. Mostaghimi, S. Chandra, Splat formation in plasma-spray coating process, Pure Appl. Chem. 74 (2002) 441–445.
[54] D.L. Reviznikov, V.V. Rusakov, Heat transfer and crystallization kinetics of melted drop s under intensive cooling, Matematicheskoe Modelirovanie 11 (1999) 55–64.
[55] A.V. Evteev, A.T. Kosilov, E.V. Levchenko, O.B. Logachev, Kinetics of isothermal nucleation in a supercooled iron melt, Phys. Solid State 48 (2006) 815–820.
[56] A.A. Fomin, A.B. Steinhauer, V.N. Lyasnikov, S.B. Wenig, A.M. Zakharevich, Nanocrystalline structure of the surface layer of plasma-sprayed hydroxyapatite coatings obtained upon preliminary induction heat treatment of metal base, Tech. Phys. Lett. 38 (2012) 481–483.
[57] J.L.M. Rupp, A. Infortuna, L.J. Gauckler, Microstrain and self-limited grain growth in nanocrystalline ceria ceramics, Acta Mater. 54 (2006) 1721–1730.
[58] A.A. Fomin, A.B. Steinhauer, Coating process, RU Patent 2,430,192, 2011.
[59] J.M. Hughes, M. Cameron, K.D. Crowley, Structural variations in natural F, OH, and Cl apatites, Am. Mineral. 74 (1989) 870–876.

PART 2

Analysis and Property Profiles

CHAPTER 14

History of High-Performance Paints and Coatings of Unique Characteristics

Abdel Salam Hamdy Makhlouf
Department of Manufacturing Engineering, College of Engineering and Computer Science, University of Texas Pan-American, Edinburg, Texas, USA

Chapter Contents

1. INTRODUCTION

Today, we consider chemistry to be a science, but its roots, dating back to ancient Egyptians, lie in the art and creation of paints and colors. Ancient Egyptians expressed about their everyday life by the produced architecture, reliefs, paintings, murals, statues, decorative arts, and a variety of crafts. So many coated objects of Egyptian art have come down from ancient times to us today because they were made of durable materials like stone and clay and the hot desert air of Egypt has been ideal for preserving them [1–3].

Most of the ancient Egyptian art kept until today was oriented toward death and the quest for the afterlife. The Egyptians believed in the afterlife and the artistic renderings of images placed in tombs would become real and accompany the deceased to the afterlife.

Handbook of Nanoceramic and Nanocomposite Coatings and Materials
http://dx.doi.org/10.1016/B978-0-12-799947-0.00014-6

Visible components such as architectural and construction works were as important as invisible ones such as burial of offerings and sacrificial animals and reusing rocks carved out of other sacred places [1,2]. In fact, ancient Egyptians used funerary inscriptions and scenes of religious rituals and afterlife mainly for religious purposes rather than for mere decoration. Inscriptions and paintings were laid in endless rows on the walls of temples, courtyards, and rooms depending on their religious use. Examples included stars portrayed on ceilings, swamp lotuses on ceiling cornices, solar snakes, and celestial eagles, with rows of water and land spirits below [3–5].

2. ANCIENT EGYPTIAN PAINTING OVER WALLS AND TOMBS

The Egyptians painted on papyrus rolls, tomb walls, coffin lids, and a host of other surfaces. Hierakonpolis contains the oldest wall paintings in tombs, which were created in 3400 BC. It features glorified sticklike figures (Figure 1). Another example of the ancient colored walls at Dendera, Qena, Egypt, which still possess their brilliant color after about 2000 years.

Egyptians used a variety of materials for pigments. They made yellow and orange pigments from soil and produced blue and red from imported indigo and madder and combined them to make flesh color. They appeared to have used six colors: white, black, blue, red, yellow, and green. Egyptians mixed their colors with a gummy substance and applied them over the walls to be coated. By 1000 BC, they developed paints and varnishes using the gum of the acacia tree (gum Arabic) as their base.

Ancient Egyptian tomb painting from 2500 BC graphically illustrated everyday life along the Nile. Egyptian artists recorded people hunting, fishing, herding cattle, building boats, cultivating and irrigating their fields, harvesting crops, playing music, dancing, and making beer. They produced images of birds and animals that allow modern scientists to identify the species (Figure 2).

Tombs of kings, queens, and nobles were typically decorated with murals with images of deities and people known to the deceased. Some of the tombs have been dated back to 5000 BC. Images in tombs are often accompanied by texts from the Book of the Dead, which sometimes explain what is going on in the picture (Figures 3 and 4).

3. THREE-DIMENSIONAL PAINTING BY ANCIENT EGYPTIANS

The first ever example of a three-dimensional painting is the one discovered in the tomb of Nefertari in the Valley of the Queens (Figure 5). The first full

Figure 1 Egyptian art in the Greco-Roman period (an example of coating over the wall).

frontal representation of a pharaoh was discovered in 2004. A 3500-year-old portrait was also found sketched on a plaster-coated board in a Luxor tomb and thought to depict Thutmose III. Figure 6 shows another example of a three-dimensional painting with the artists painting a wall inside a tomb.

4. ANCIENT EGYPTIAN PAINTING OVER STONE AND MARBLE

The buildings are construction designed by ancient Egyptians with their form and decorations as stone incarnation representations of religion and rituals. Ancient Egyptians believed that such buildings had the power of giving life through the magic of simulation in this world and immortality in the afterlife.

In fact, ancient Egyptians used funerary inscriptions and scenes of religious rituals and afterlife mainly for religious purposes rather than for mere decoration. Examples are the famous Egyptian cornice over doors, temple

Figure 2 Another example of the wall painting describing the daily life of ancient Egyptians.

Figure 3 Amenhotep I (an example of a tomb painting).

gates, towers, and rooms, which were stone geometric forms, and the sets of inscriptions on the upper parts of temple walls that were replicas of arches earlier used (Figures 7 and 8). With such astounding genius of harmonizing materials, location, paintings, and ornamentation with worldly needs and

Figure 4 Another example of a tomb painting.

Figure 5 Nefertari tomb painting (an example of three-dimensional paintings).

Figure 6 Another example of a three-dimensional painting.

Figure 7 An example of painting over stone.

religious rituals, the ancient Egyptian architect could create magnificent symmetrical and highly impressive masterpieces [3–6].

5. ANCIENT EGYPTIAN PAINTING OVER PAPYRI

Ancient Egyptians used papyri mostly as a communication tool (as messages or letters) between the cities and as a record of their daily life activities.

Figure 8 An example of painting over marble.

The papyri are extremely fragmentary and fragile. Therefore, they have not been studied as a group until 1850. The Ramesseum papyri have been described as the "most precious single find of papyri" from pharaonic Egypt (Figure 9). Figures 9 and 10 show some examples of the painting and writing over papyri.

Figure 9 The Ramesseum papyri (an example of painting over papyri).

Figure 10 An example of painting over papyri.

6. ANCIENT EGYPTIAN PAINTING OVER WOODS

There are several objects such as shabties taken by ancient Egyptians for their afterlife (Figure 11). Shabty is a magical servant that would do chores for the deceased in the afterlife. Rich people had many shabties made of precious

Figure 11 Tutankhamun shabti (an example of painting over wood).

materials, including large pieces of wood, which was a rare commodity. The less fortunate had to settle for shabties made of faience, glazed earthenware.

One of the oldest artifacts discovered from ancient Egypt is a triangular 2 ft high slate carving showing King Narmer, wearing the crown of Upper Egypt, grasping the hair and humiliating an enemy from Lower Egypt to unify the nation. The carving, dating back to 3100 BC, also shows the falcon god Horus, symbolizing victory [6,7]. The other side shows him wearing the crown of Lower Egypt, leading a victory parade. King Narmer founded ancient Pharaonic Egypt by uniting Lower (northern) Egypt and Upper (southern) Egypt in 3200 BC. The funerary statue of the 11th dynasty in Figure 12 is another example of coating over wood.

Figure 13 provides a perfect example for the remarkable painting technology over wood used by ancient Egyptians. The coating shows

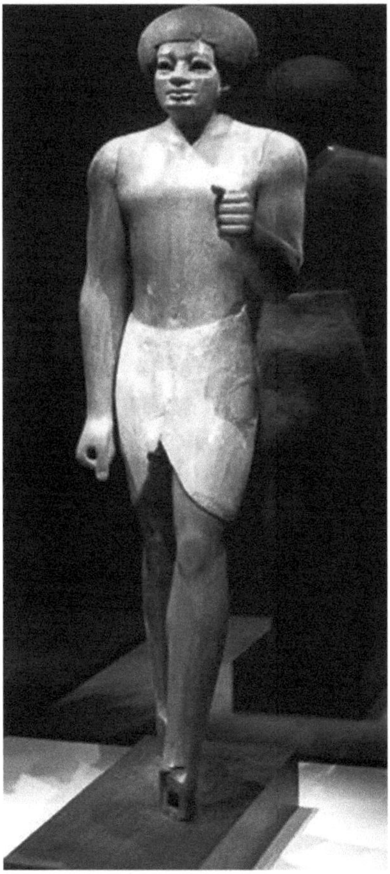

Figure 12 Eleventh dynasty funerary statue (an example of coating over wood).

Figure 13 An example of coating over wood.

outstanding brightness, smooth appearance, and stability over wood after about 4800 years of different exposure.

7. ANCIENT EGYPTIAN MUMMY PROTECTION

Due to their belief in the afterlife, ancient Egyptian discovered a unique technology to protect their body after death. The mummy was enclosed in a coffin, a sarcophagus, and four decorated and gilded wooden shrines—one inside the other. The shrines had images of the king emblazoned on them. The largest, outer golden shrine is 9 ft high, 10¾ ft wide, and 16½ ft long (Figure 14). It is inlaid with panels of brilliant blue faience with depictions of distinct symbols that protected the dead. The innermost one was covered in gold. The sarcophagus is made of yellow quartzite and has a sculpted goddess spreading protecting arms and wings over the feet area. Each shrine and the sarcophagus is displayed separately.

Many attempts have been invested over the last century by scientists and Egyptologists all over the world to discover the technology used by ancient

Figure 14 An example of mummy protection using special unknown chemicals.

Egyptians for protecting the mummies. However, the exact chemical composition of the materials used to fill the dead body, the method of application, the technique of preparation, treatment conditions, etc. have not been discovered until now.

8. CONCLUSION

Ancient Egyptians produced innovative high-performance coatings of unique characteristics to withstand corrosive environments over 4800 years. The coatings and paintings are characterized with smooth structure, brightness, adherence, stability, and durability.

Ancient Egyptians used innovative technology for the synthesis of the coating raw materials, surface preparation and pretreatment of the substrates to be coated and the coatings application method over the materials.

The smartness of these coatings and paintings lie behind their ability to maintain their outstanding appearance and stability for extremely long times and to adapt themselves to resist various severe conditions such as corrosion, erosion, abrasion, heat, cool, storms, and radiation.

This chapter provided some examples of the high-performance paints and coatings produced by ancient Egyptians over different substrates.

ACKNOWLEDGMENTS

The author would like to thank the Egyptologist team at the Egyptian Museum in Cairo for their sincere support and guidance with the required images and explanations. Most of the images included in this work were collected from the Egyptian Museum in Cairo and some of them have been collected from the publications of Louvre Museum, France; Berlin Museum; and the British Museum, the United Kingdom.

The author, in some sections, used the following media sources and collected the published comments of Egyptologists: New York Times, Washington Post, Los Angeles Times, The Times, The Guardian, National Geographic, The New Yorker, Time, Newsweek, Reuters, Lonely Planet Guides, and Compton's Encyclopedia.

REFERENCES

[1] B.G. Aston, J.A. Harrell, I. Shaw, Stones, in: P.T. Nicholson, I. Shaw (Eds.), Ancient Egyptian Materials and Technology, University of Cambridge Press, Cambridge, 2000, pp. 5–77.

[2] J. Baines, Stone and other materials in ancient Egypt—usages and values, in: C. Karlshausen, T. DePutter (Eds.), Pierres Égyptiennes—Chefs-d'Oeuvre pour l'Éternité, Faculté Polytechnique de Mons, Mons, 2000, pp. 29–41.

[3] Egyptian art characteristics, history of sculpture, painting, architecture of ancient Egypt. http://www.visual-arts-cork.com/ancient-art/egyptian.htm (accessed 02.06.13).

[4] Coatings—historical items. http://www.corrosion-club.com/historycoatings.htm (accessed 02.06.13).
[5] 11 things you may not know about ancient Egypt. http://www.history.com/news/history-lists/11-things-you-may-not-know-about-ancient-egypt (accessed 02.06.13).
[6] Ancient Egyptian art and art objects. http://factsanddetails.com/world/cat56/sub365/item1936.html (accessed 01.06.13).
[7] Z. Hawass (Ed.), Hidden Treasures of Ancient Egypt, National Geographic, Washington, DC, 2004.

CHAPTER 15

Electrical Properties of Macro-, Micro-, and Nanoceramic and Nanocomposite Coatings and Materials

Xue-Qian Fang, Jin-Xi Liu
Department of Engineering Mechanics, Shijiazhuang Tiedao University, Shijiazhuang, PR China

Chapter Contents

1. INTRODUCTION

Piezoelectricity is the important ability of some materials to convert an electric stimulus into mechanical energy, and vice versa. Nowadays, piezoelectric materials have been increasingly used in a variety of applications such as submarine hydrophones, accelerometers, microphones, ultrasonic devices, and electronic resonators [1–6].

Piezoelectric materials are typically produced by blending conductive fillers with insulating polymer matrices. A number of novel piezoelectric materials such as lead zirconate titanate (PZT) [1,2], polyvinylidene fluoride (PVDF) [3,4], and barium titanate [5,6] are available and are used as sensors and actuators in control application. At present, lead-based piezoelectric

Handbook of Nanoceramic and Nanocomposite Coatings and Materials
http://dx.doi.org/10.1016/B978-0-12-799947-0.00015-8

ceramics such as $PbTiO_3$-$PbZrO_3$ are widely used; however, it is not environmentally safe because of the toxicity of lead oxide and high vapor pressure during the sintering processing [7]. Therefore, it is necessary to develop lead-free piezoelectric ceramics with excellent piezoelectric properties comparable to those of lead-based piezoelectric ceramics such as bismuth sodium titanate ($Na_{0.5}Bi_{0.5}TiO_3$, abbreviated as NBT) [8] with R3c perovskite rhombohedral structure. Progress in a wide range of structural, functional and bio-medical applications of macro, micro and nanoceramics and piezoelectric nanocomposites coatings and materials crucially depends on the development of new fabrication and processing technologies. To this end, it is foundamental to understand the relationship between the micro-nanostructure and electrical properties of these materials [11].

Due to high performance in electrical and mechanical properties and significant processing advantages of the piezoelectric materials, there is a very fast growing interest on investigating the improvement of material properties. In general, the characterization of electrical properties can provide important information for the optimum design of materials [9,10]. Progress in a wide range of structural, functional, and biomedical applications of macro-, micro-, and nanoceramics and piezoelectric nanocomposite coatings and materials crucially depends on the development of new fabrication and processing technologies, along with the fundamental understanding of the relationship between their micro-nanostructure and electrical properties [11]. Preliminary experimental studies [6,12,13] have indicated that grain size engineering could help enhance the piezoelectric properties of materials such as $BaTiO_3$. For example, the piezoelectric coupling and charge coefficients of 3-3 composites were improved by modifying the grain size of constituents [13].

Alternatively, three-phase piezoelectric composites are emerging in recent years, because they may provide material properties superior to conventional well-known two-phase piezoelectric composites [14,15]. In these piezoelectric materials, piezoelectric fiber-reinforced composite has high piezoelectric constants, shape control (due to their weight), and high specific acoustic impedance [16].

In recent decade, the combination of nanotechnology and piezoelectric technology results in a new class of piezoelectric nanostructures (PNs) [17,18]. They exhibit enhanced piezoelectric effect, excellent resilience, and unique coupling between piezoelectric and semiconducting properties. The self-powering capability of PNs allows electronic devices to exclude energy storage components and holds great promises for directly drawing energy from ambient mechanical resources, powering small electronics, and achieving self-powered electronic devices. This distinct feature makes them attractive for a wide range of applications in nanoelectromechanical systems (NEMSs), such as nanosensors/transducers, nanogenerators, nanoresonators, diodes, and piezoelectric field-effect transistors [17–20].

In this chapter, a summary of the recent developments and trends in studying of the electrical properties of macro-, micro-, and nanoceramic and nanocomposite coatings and materials is presented. The clarification of piezoelectric composites is given in Section 2. In Section 3, the theoretical and experimental works on predicting the electrical properties of piezoelectric ceramics from macro- to nanolength scales are reviewed. The main finding and observations related to the electric properties of nanoceramics are discussed in Section 4. As an important type of piezoelectric composites, the investigations on the composites with coating layers are presented in Section 5. Some concluding remarks are given in Section 6.

2. CLARIFICATION OF PIEZOELECTRIC COMPOSITES

In the past decades, a lot of piezoelectric materials are constantly emerging and have gained wide application in engineering. Some criteria for clarification are proposed. The following two criteria are commonly accepted.

2.1 Depending on Characteristic Scales

Based on the characteristic scales of PNs, piezoelectric composites are classified into the following three main types [21]:

(1) The nanoscopic scale with a characteristic length of 10^{-9} m or a few nanometers
(2) The microscopic scale with a characteristic length of 10^{-6} m or a few micrometers
(3) The mesoscopic scale with a characteristic length of 10^{-4} m or hundreds of micrometers

2.2 Depending on Connectivity of Each Phase

Piezoelectric composites can be obtained by combining piezoelectric fibers with non-piezoelectric materials [22]. Superior mechanical and electrical properties can be achieved by tailoring most profitable properties of the constituents of the composite. To obtain the required functional properties, two or three phases are combined [23]. For two-phase piezoelectric composite

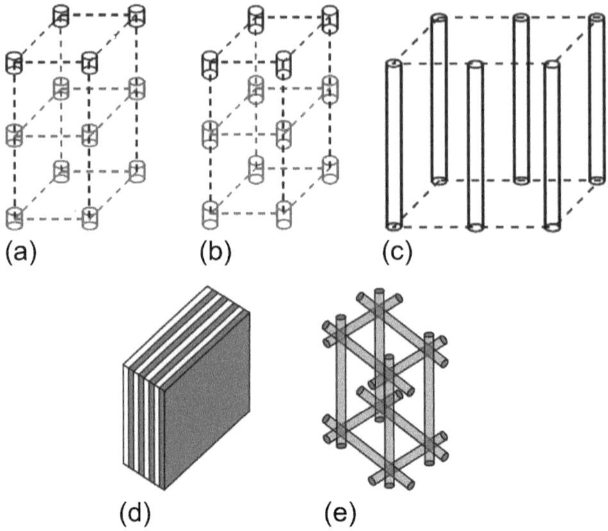

Figure 1 Schematics illustrating five types of piezoelectric composite and their spatial arrangement. (a) Particulate composite; (b) Short-fiber composite; (c) Long-fiber composite; (d) Laminate composite; (e) Interpenetrating networked composite.

materials, there are 10 distinct ways of creating a composite system, based on the connectivity of each phase, where connectivity is defined as the number of dimensions in which each component is connected to itself in the composite assemblage [12,23]. In general, for multiphase piezoelectric composites, there are several ways in which two or more components may be spatially combined to create a composite material. In particular, five principal composite structures that are frequently considered in the design of several transducers are analyzed [13].

(1) Particulate composite: 0-3 type, where the first phase (i.e., the "0" in "0-3") is enclosed in all three dimensions within a continuous matrix (Figure 1a)
(2) Short-fiber composite: 0-3 type, where the first phase is also enclosed in all three spatial dimensions within a continuous matrix (Figure 1b)
(3) Long-fiber composite: 1-3 type, where the first phase (in the form of a long fiber and with connectivity in one dimension) is embedded in a matrix (which has continuous connectivity across all three dimensions) (Figure 1c)
(4) Laminate composite: 2-2 type, where both phases exhibit connectivity in two dimensions (e.g., a layered system with the interface perpendicular to 1 direction) (Figure 1d)

(5) Interpenetrating networked composite: 3-3 type, where both phases exhibit connectivity along all three dimensions (e.g., a composite with piezoelectric fibers aligned along three mutually perpendicular directions in a non-intersecting manner) (Figure 1e)

3. ELECTRICAL PROPERTIES OF MACRO-, MICRO-, AND NANO-PIEZOELECTRIC MATERIALS

In the past decades, lead-based ceramics such as $Pb(Zr_xTi_{1-x})O_3$ (abbreviated as PZT) have been widely used for electronic and microelectronic devices due to their relatively low cost, high bandwidth, and good actuation capabilities [24,25].

The toxicity of PbO and the high vapor pressure during sintering process can cause serious environmental problems and instability of composition. Many scholars developed lead-free piezoelectric ceramics with excellent properties to replace PZT-based ceramics.

The researches on the electrical properties of macro-, micro-, and nano-piezoelectric materials can be divided into two groups: theoretical and experimental works.

3.1 Theoretical Works

For piezoelectric composites, the constitutive relations can be expressed as

$$\sigma_{ij} = C_{ijkl}\varepsilon_{kl} - e_{kij}E_k, \quad D_i = e_{ikl}\varepsilon_{kl} + \chi_{ik}E_k, \tag{1}$$

where σ_{ij}, ε_{kl}, E_k, and D_i are the stress tensor, strain tensor, electric field vector, and electric displacement vector, respectively. C_{ijkl} is the fourth-order elasticity tensor under short-circuit boundary conditions, e_{kij} is the third-order piezoelectric strain tensor, and χ_{ik} is the second-order free-body electric tensor.

In the past decades, different homogenization techniques have been proposed to predict the electromechanical properties of piezoelectric composites. These techniques include analytic mixing rules [26–28] and finite element method [29–31]. Bent and Hagood [26] used a mixing rule technique to predict the effective electrical properties. Uniform field method (UFM) was implemented into a finite element code to study the analytic formulas of the overall electromechanical properties. To apply the homogenization techniques, the representative volume element is often introduced, as depicted in Figure 2.

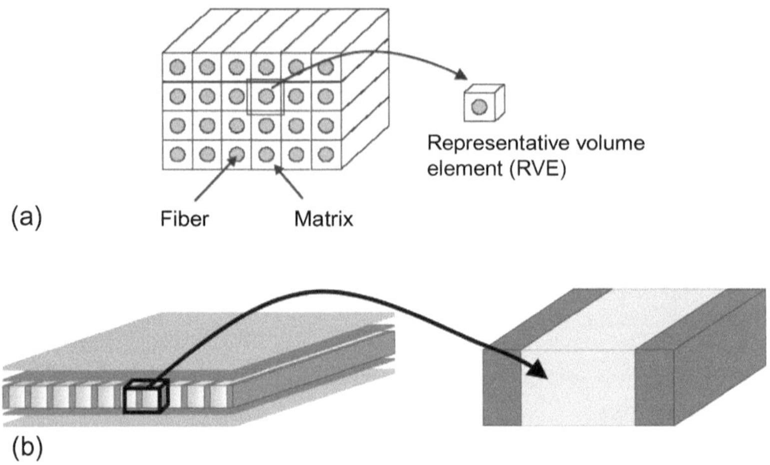

Figure 2 Examples of representative volume element. (a) Three-dimensional piezoelectric composites; (b) piezoelectric layer.

After introducing the properties of homogeneous piezoelectric active layers and finite element method, the effective electrical properties of the active layer in macro-fiber composite (MFC) transducers under plane stress were evaluated [32].

In addition, the homogenization techniques were also combined with Eshelby tensor to predict the coupled piezoelectric properties of composites [33,34]. Koutsawa et al. [35] applied the generalized self-consistent approach to estimate the electromechanical properties of piezocomposites.

In recent years, the asymptotic expansion homogenization (AEH) method is very popular in predicting the effective piezoelectric properties of composites. In this method, the generalized displacement field $u(x)$ (physical displacement and potential) is approximated, by using an asymptotic expansion, as

$$u(x) = u_i^0(x, y) + \alpha u_i^{(1)}(x, y) + \alpha^2 u_i^{(2)}(x, y) + \cdots, \quad (2)$$

where α is the scale parameter, $u_i^0(x, y)$ corresponds to the macroscopic displacement field of the homogenized material, and $u_i^{(1)}(x, y)$ are the correctors of the generalized displacement field [36].

Another important method, the asymptotic homogenization method (AHM), is also widely used to predict the electrical property. By using AHM, the original constitutive relations with rapidly oscillating material coefficients are transformed into an equivalent system with effective coefficients.

The homogenized system of differential equations is expressed as [37–39]

$$C^*_{ijkl}\widetilde{u}_{l,kj} + e^*_{kij}\widetilde{\phi}_{,kj} = 0, \quad e^*_{ikl}\widetilde{u}_{i,kl} - \chi^*_{jk}\widetilde{\phi}_{,kj} = 0, \tag{3}$$

where the asterisk denotes the overall properties. The terms $\tilde{u}$ and $\widetilde{\phi}$ represent the homogenized field.

In addition, finite element method was also applied to investigate properties of piezoelectric materials, including single crystal, PVDF, and soft and hard PZT materials [40].

In the work of Farid et al., electrical conductivities of the composites were predicted using three-point upper-bound model and two-phase self-consistent model [41].

3.2 Experimental Work

The experimental work mainly focused on the impedance analyzer method. The dielectric properties of Ti_3AlC_2/epoxy resin conductive composites were measured on a precision impedance analyzer (4294A, Agilent Technologies, CA) in the frequency range of 10^3-10^6 Hz [42]. It was found that both the electrical and dielectric properties exhibited percolation characteristics and increased with increasing Ti_3AlC_2 content. When Ti_3AlC_2 filler content was lower than 20 wt%, the dielectric constant increased slowly. However, increasing Ti_3AlC_2 filler content up to 40 wt% results in a sudden increase in the dielectric constant of conductive polymer composites and up to 50 wt% a further increase.

By measuring the resonant frequency specimens with an impedance/gain-phase analyzer, the piezoelectric constants were computed [43]. It was found that the Pt addition increases the relative dielectric constant significantly. In the work of Zhen et al., the piezoelectric constant d_{33} was measured at 110 Hz by using a quasistatic ZJ-A3 d_{33} meter [44]. Composites PE2 series were polished to different thicknesses to determine the effect of the thickness on the electrical properties. Using $Na_{0.5}K_{0.5}NbO_3$ ceramic sample, Wang et al. measured the dielectric properties in the temperature ranging from 200 to 550 °C by using an Agilent 4294A precision impedance analyzer [45]. It is found that the piezoelectric parameter (d_{33}) can reach 148 pC/N. In the work of Yang et al. [46] and Pan et al. [47], the piezoelectric constant d_{33} was also measured by a quasistatic piezoelectric d_{33} meter. Using a complex electrical admittance-curve-fitting technique, the coefficients of piezoelectric composite materials were measured in Ref. [48].

4. ELECTRICAL PROPERTIES OF NANOCERAMICS

In recent years, nanostructures with ceramic materials are the promising alternatives for the current piezoelectric composites. Owing to important nanoscopic features, such composites are referred to as ceramic nanocomposites. They can provide high-performance properties, such as higher electrical conductivity, never achieved before by conventional advanced ceramics.

The surface/interface energies essentially control the properties of a solid. Interfaces provide a mean of introducing nonhomogeneity to the material [49]. This nonhomogeneity acts as a significant modification of electrical properties of the composites. Selective mixing of materials with high percentage of interface area can lead to their enhanced properties. The properties of nanocomposites depend not only on the properties of their individual parents but also on their morphology and interfacial characteristics.

To analyze the effect of surface/interface energy on the electrical properties, the coupling surface/interface model was widely used [49]. For piezoelectric matrix embedded with piezoelectric nanofibers, the work of Chen [50], Xiao et al. [51], and Fang et al. [52] should be mentioned. They presented the electrical properties of PNs. Chen and Xiao et al. mainly focused on the static electrical properties; however, Fang et al. paid attention to the dynamic electrical properties.

Most experimental investigations on the electrical properties focused on the Fourier transform infrared (FTIR) spectroscopy measurements. In the work of David et al., the electrical properties of ZnO were measured by FTIR [53].

Li et al. measured the piezoelectric properties of ferroelectric and antiferroelectric thin layers by interferometry as a function of frequency and dc electric bias [54]. The embedded beam method and Berlincourt piezometer were used to measure the e_{31} and d_{33} piezoelectric coefficients of sputtered $PbTiO_3$ films, respectively [55].

5. ELECTRICAL PROPERTIES OF COMPOSITES: COATING LAYERS

In piezoelectric composites, the coating layer is also an important factor for the enhancement of electrical properties. Piezoelectric composites with coating layer may provide material properties largely superior to conventional well-known two-phase piezoelectric composites. Composite coatings that improve the properties of composites heavily depend on the material properties and content of coatings. The most important point is to obtain continuous, uniformly distributed, and dense-coated metal layer. Otherwise, the layer with voids or gaps may weaken and even destroy the integration between the reinforcement phase and matrix.

To evaluate the piezoelectric properties, some important investigations are reported. Ozmusul and Picu [56] studied the elastic moduli of composites containing an anisotropic coating layer and nanoparticles that are either of finite stiffness or rigid. In addition, Boutaleb et al. [57] proposed a micromechanical analytic model to predict the stiffness and yield strength of silica/polymer nanocomposites, and it was shown that the interphase was a dominant parameter for controlling the overall nanocomposite behavior. Recently, the effect of interface energy around the unidirectional coated nanofibers on the effective dynamic effective properties is explicitly addressed by effective medium method and wave function expansion method. The multiple scattering resulting from a series of coating nanofibers is reduced to the problem of one typical nanofiber in the effective medium [58].

6. CONCLUSIONS

In this chapter, the main formulation and recent developments in the researches on the electrical properties of macro-, micro-, and nanoceramic and nanocomposite coatings and materials are reviewed. The surface/interface and coating layers in piezoelectric composites can be used to enhance the piezoelectric properties of macro-, micro-, and nanoceramic and nanocomposite coatings and materials. With the development of piezoelectric materials, many other methods such as optimizing the experimental process will be introduced in the future.

The reviewed works and conclusions in this chapter can provide important references for the optimum design of piezoelectric materials.

REFERENCES

[1] R.N. Torah, S.P. Beeby, N.M. White, Improving the electroelastic properties of thick-film PZT: the influence of paste composition, powder milling process and electrode material, Sens. Actuators A Phys. 110 (2004) 378–384.
[2] H. Berger, S. Kari, U. Gabbert, R. Rodriguez-Ramos, R. Guinovart, J.A. Otero, J. Bravo-Castillero, An analytical and numerical approach for calculating effective material coefficients of piezoelectric fiber composites, Int. J. Solids Struct. 42 (2005) 5692–5714.
[3] T. Furukawa, M. Data, E. Fukuda, Hysteresis phenomena in polyvinylidene fluoride under high electric field, J. Appl. Phys. 51 (1980) 1135–1141.

[4] T. Furukawa, G.E. Johnsoni, Measurements of ferroelectric switching characteristics in polyvinylidene fluoride, Appl. Phys. Lett. 38 (12) (1981) 1027–1029.
[5] Y. Saito, H. Takao, T. Tani, et al., Lead-free piezoceramics, Nature 432 (2004) 84–87.
[6] H. Takahashi, Y. Numamoto, J. Tani, et al., Piezoelectric properties of $BaTiO_3$ ceramics with high performance fabricated by microwave sintering, Jpn. J. Appl. Phys. 45 (9B) (2006) 7405–7408.
[7] R. Sun, X.Y. Zhao, Q.H. Zhang, et al., Growth and orientation dependence of electrical properties of $0.92Na_{0.5}Bi_{0.5}TiO_3$-0.08 $K_{0.5}Bi_{0.5}TiO_3$ lead-free piezoelectric single crystal, J. Appl. Phys. 109 (2011) 124113.
[8] C.F. Buhrer, Some properties of bismuth perovskites, J. Chem. Phys. 36 (3) (1962) 798–803.
[9] J.J. Bernstein, S.L. Finberg, K. Houston, et al., Micromachined high frequency ferroelectric sonar transducers, IEEE Trans. Ultrason. Ferroelectr. Freq. Control 44 (55) (1997) 960–969.
[10] Y. Nemirovsky, A. Nemirovsky, P. Muralt, et al., Design of a novel thin film piezoelectric accelerometer, Sens. Actuators 56 (3) (1996) 239–249.
[11] F. Xu, S. Trolier-McKinstry, W. Ren, et al., Domain wall motion and its contribution to the dielectric and piezoelectric properties of lead zirconate titanate films, J. Appl. Phys. 89 (2) (2001) 1336–1348.
[12] T. Karaki, K. Yan, T. Miyamoto, et al., Lead-free piezoelectric ceramics with large dielectric and piezoelectric constants manufactured from $BaTiO_3$ nano-powder, Jpn. J. Appl. Phys. 46 (2007) L97–L98.
[13] R. Kar-Gupta, T.A. Venkatesh, Electromechanical response of piezoelectric composites: effects of geometric connectivity and grain size, Acta Mater. 56 (15) (2008) 3810–3823.
[14] C.P. Jiang, Y.K. Cheung, An exact solution for the three-phase piezoelectric cylinder model under antiplane shear and its applications to piezoelectric composites, Int. J. Solids Struct. 38 (28–29) (2001) 4777–4796.
[15] P. Yan, C.P. Jiang, F. Song, An eigenfunction expansion-variational method for the anti-plane electroelastic behavior of three-phase fiber composites, Mech. Mater. 43 (10) (2011) 586–597.
[16] P. Tan, L. Tong, Investigation of loading assumptions on the effective electroelastic constants for PFRC materials, Compos. Struct. 57 (1–4) (2002) 101–108.
[17] Z.L. Wang, J.H. Song, Piezoelectric nanogenerators based on zinc oxide nanowire arrays, Science 312 (5771) (2006) 242–246.
[18] C. Sun, J. Shi, X. Wang, Fundamental study of mechanical energy harvesting using piezoelectric nanostructures, J. Appl. Phys. 108 (3) (2010) 034309.
[19] L.C. Lew, Y. Voon, M. Willatzen, Electromechanical phenomena in semiconductor nanostructures, J. Appl. Phys. 109 (3) (2011) 031101.
[20] X.Q. Fang, Q. Yang, J.X. Liu, et al., Surface/interface effect around a piezoelectric nano-particle in a polymer matrix under compressional waves, Appl. Phys. Lett. 100 (2012) 151602.
[21] G. Lu, E. Kaxiras, Overview of multiscale simulations of materials, Handb. Theor. Comput. Nanotechnol. 10 (2005) 1–33.
[22] R.E. Newnham, D.P. Skinner, L.E. Cross, Connectivity and piezoelectric-pyroelectric composites, Mater. Res. Bull. 13 (5) (1978) 525–536.
[23] V.Y. Topolov, S.V. Glushanin, Evolution of connectivity patterns and links between interfaces and piezoelectric properties of two-component composites, J. Phys. D. Appl. Phys. 35 (16) (2002) 2008–2014.
[24] B. Jaffe, W.R. Cook, H. Jaffe, Piezoelectric Ceramics, Academic Press, New York, 1971.

[25] Z. Yang, X. Chao, L. Yang, Effect of sintering process on characteristics of multilayer piezoelectric Pb(Mg1/3Nb2/3)O-3-Pb(Zn1/3Nb2/3)O-3-Pb(Zr,Ti)O-3 ceramic transformers, Jpn. J. Appl. Phys. 46 (10A) (2007) 6746–6750.
[26] A.A. Bent, N.W. Hagood, Piezoelectric fiber composites with interdigitated electrodes, J. Intell. Mater. Syst. Struct. 8 (11) (1997) 903–919.
[27] P. Tan, L. Tong, Micro-electromechanics models for piezoelectric fiber reinforced composite materials, Compos. Sci. Technol. 61 (5) (2001) 759–769.
[28] P. Tan, L. Tong, Investigation of loading assumptions on the effective electroelastic constants for PFRC materials, Compos. Struct. 57 (1–4) (2002) 101–108.
[29] H. Berger, S. Kari, U. Gabbert, et al., A comprehensive numerical homogenisation technique for calculating effective coefficients of uniaxial piezoelectric fibre composites, Mater. Sci. Eng. A 412 (1–2) (2005) 53–60.
[30] T. Tang, W. Yu, Variational asymptotic micromechanics modeling of heterogeneous piezoelectric materials, Mech. Mater. 40 (10) (2008) 812–824.
[31] J.A. Oliveira, J. Pinho-da-Cruz, F. Teixeira-Dias, Asymptotic homogenisation in linear elasticity. Part II: finite element procedures and multiscale applications, Comput. Mater. Sci. 45 (4) (2009) 1081–1096.
[32] A. Deraemaeker, H. Nasser, Numerical evaluation of the equivalent properties of macro fiber composite (MFC) transducers using periodic homogenization, Int. J. Solids Struct. 47 (24) (2010) 3272–3285.
[33] Y. Benveniste, The determination of the elastic and electric fields in a piezoelectric inhomogeneity, J. Appl. Phys. 72 (3) (1992) 1086–1095.
[34] W. Biao, Three-dimensional analysis of an ellipsoidal inclusion in a piezoelectric material, Int. J. Solids Struct. 29 (3) (1992) 293–308.
[35] Y. Koutsawa, F. Biscani, S. Belouettar, et al., Multi-coating inhomogeneities approach for the effective thermo-electro-elastic properties of piezoelectric composite materials, Compos. Struct. 92 (4) (2010) 964–972.
[36] B. Fabio, N. Houssein, B. Salim, et al., Equivalent electro-elastic properties of macro fiber composite (MFC) transducers using asymptotic expansion approach, Compos. Part B 42 (3) (2011) 444–455.
[37] R. Guinovart-Díaz, P. Yan, R. Rodríguez-Ramos, Effective properties of piezoelectric composites with parallelogram periodic cells, Int. J. Eng. Sci. 53 (2012) 58–66.
[38] J. Bravo-Castillero, R. Guinovart-Díaz, F.J. Sabina, et al., Closed-form expressions for the effective coefficients of fibre-reinforced composite with transversely isotropic constituents-II. Piezoelectric and square symmetry, Mech. Mater. 33 (4) (2001) 237–248.
[39] F.J. Sabina, R. Rodríguez-Ramos, J. Bravo-Castillero, et al., Closed-form expressions for the effective coefficients of fibre-reinforced composite with transversely isotropic constituents-II. Piezoelectric and hexagonal symmetry, J. Mech. Phys. Solids 49 (7) (2001) 1463–1479.
[40] A. Daniels, M. Zhu, A. Tiwari, Evaluation of piezoelectric material properties for a higher power output from energy harvesters with insight into material selection using a coupled piezoelectric-circuit-finite element method, IEEE Trans. Ultrason. Ferroelectr. Freq. Control 60 (12) (2013) 2626–2633.
[41] A. Farid, J.A. Syed, A.S. Khadijah, et al., Microstructure, mechanical properties, electrical conductivity and wear behavior of high volume TiC reinforced Cu-matrix composites, Mater. Charact. 60 (4) (2009) 327–336.
[42] W.J. Wang, C.W. Li, K.P. Chen, Electrical, dielectric and mechanical properties of a novel Ti_3AlC_2/epoxy resin conductive composites, Mater. Lett. 110 (2013) 61–64.
[43] J.F. Li, K. Takagi, N. Terakubo, et al., Electrical and mechanical properties of piezoelectric ceramic/metal composites in the $Pb(Zr,Ti)O_3$/Pt system, Appl. Phys. Lett. 79 (15) (2001) 2441–2443.

[44] Y.H. Zhen, J.F. Li, H.L. Zhang, Electrical and elastic properties of 1–3 PZT/epoxy piezoelectric composites, J. Electroceram. 21 (1–4) (2008) 410–413.
[45] K. Wang, B.P. Zhang, J.F. Li, et al., Lead-free $Na_{0.5}K_{0.5}NbO_3$ piezoelectric ceramics fabricated by spark plasma sintering: Annealing effect on electrical properties, J. Electroceram. 21 (1–4) (2008) 251–254.
[46] Z.P. Yang, Y.T. Hou, H. Pan, et al., Structure, microstructure and electrical properties of $(1-x-y)Bi_{0.5}Na_{0.5}TiO_{3-x}Bi_{0.5}K_{0.5}TiO_{3-y}Bi_{0.5}Li_{0.5}TiO_3$ lead- free piezoelectric ceramics, J. Alloys Compd. 480 (2009) 246–253.
[47] H. Pan, Y.T. Hou, X.L. Chao, et al., Microstructure and electrical properties of La_2O_3-doped $Bi_{0.5}(Na_{0.68}K_{0.22}Li_{0.1})_{0.5}TiO_3$ lead-free piezoelectric ceramics, Curr. Appl. Phys. 11 (3) (2011) 888–892.
[48] M.R. Kurt, P.S. Dubbelday, Determination of the piezoelectric properties of piezoelectric composite materials by laser Doppler velocimetry, J. Acoust. Soc. Am. 88 (S1) (1990) S114–S119.
[49] X.Q. Fang, J.X. Liu, V.J. Gupta, Fundamental formulations and recent achievements in piezoelectric nano-structures: a review, Nanoscale 5 (2013) 1716–1726.
[50] T. Chen, Exact size-dependent connections between effective moduli of fibrous piezoelectric nanocomposites with interface effects, Acta Mech. 196 (3–4) (2008) 205–217.
[51] J.H. Xiao, Y.L. Xu, F.C. Zhang, Size-dependent effective electroelastic moduli of piezoelectric nanocomposites with interface effect, Acta Mech. 222 (1–2) (2011) 59–67.
[52] X.Q. Fang, J.X. Liu, M.J. Huang, Effect of interface energy on effective dynamic properties of piezoelectric medium with randomly distributed piezoelectric nano-fibers, J. Appl. Phys. 112 (9) (2012) 094311.
[53] C. David, F. Paumier, B.P. Tinkham, et al., Modification of electrical properties induced by annealing of ZnO thin films deposited by chemical vapour deposition: kinetic investigation of evolution, Mater. Sci. Eng. B 178 (9) (2013) 698–702.
[54] J.F. Li, D.D. Viehland, T. Tani, et al., Piezoelectric properties of sol-gel-derived ferroelectric and antiferroelectric thin layers, J. Appl. Phys. 75 (1) (1994) 442–448.
[55] E. Cattan, B. Jaber, P. Tronc, et al., Piezoelectric properties of sputtered $PbTiO_3$ films: growth temperature and poling treatment effects, J. Vac. Sci. Technol. A 16 (1) (1998) 169–174.
[56] M.S. Ozmusul, R.C. Picu, Elastic moduli of particulate composites with graded filler-matrix interfaces, Polym. Compos. 23 (1) (2002) 110–119.
[57] S. Boutaleb, F. Zairi, A. Mesbah, et al., Micromechanics-based modelling of stiffness and yield stress for silica/polymer nanocomposites, Int. J. Solids Struct. 46 (7–8) (2009) 1716–1726.
[58] X.Q. Fang, M.J. Huang, J.Y. Wu, et al., Effect of interface energy on size-dependent effective dynamic properties of nanocomposites with coated nano-fibers, CMC Comput. Mater. Con. 33 (2) (2013) 199–211.

CHAPTER 16

Selective Dispersion and Characterization of Fine Particle Mixture in Concentrated Suspensions for Advanced Particle Processing

Akira Otsuki
GeoRessources UMR 7359 CNRS-CREGU, ENSG, Université de Lorraine, Vandoeuvre-lés-Nancy, France

Chapter Contents

1. INTRODUCTION

This chapter aims to introduce the challenges and methods for selective dispersion and characterization of fine particle mixtures in concentrated suspensions to enhance advanced particle processing. Such process includes the sustainable mineral beneficiation of complex ores, which are composed of finely disseminated valuable elements (e.g., Au, Ni, and REE). In mineral processing, for example, proper dispersion/liberation of fine particles plays a key role in achieving selective enrichment of valuable minerals.

Handbook of Nanoceramic and Nanocomposite Coatings and Materials
http://dx.doi.org/10.1016/B978-0-12-799947-0.00016-X

Dispersion of fine particles is, however, a challenging problem. The first difficulty is to achieve a long-term stability of particle dispersions, avoiding coagulation after physical or chemical treatment. Conventional chemical/physical methods often encounter the problems of recoagulation in highly concentrated suspensions (e.g., [1,2]), which are commonly treated in plant operations. This coagulation can be caused by an increase in particle-particle interactions as suspension concentration increases. Figure 1 illustrates the effect of higher solid concentrations on the shear yield stress [3], a good indicator of increased interactions and coagulation. The relevant content of suspension rheology will be further discussed in Section 3.2. The second difficulty is to selectively disperse/liberate fine particles along grain/material boundaries. The conventional grinding/dispersion methods employ nonselective compressive forces to break/disperse particles. On the other hand, the electrical disintegration being introduced in Section 2.2 applies tensile forces that can selectively disperse/grind particles along their grain/material boundaries. This chapter will introduce a new method of applying an electric field to disperse fine particles in concentrated suspensions, based on the results of our previous studies [4,5], which showed effective particle dispersion by applied electric fields. Also, the challenge and current status associated with the characterization of concentrated suspensions will be discussed since the success of selective dispersion can only be assessed properly by the application of the appropriate methods.

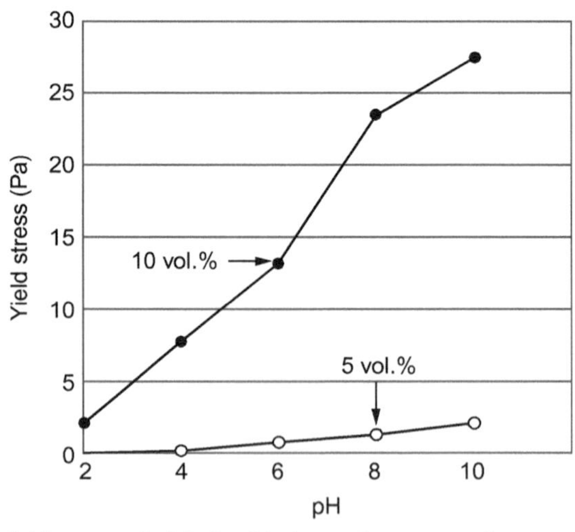

Figure 1 Shear yield stress of nickel oxide-hematite suspension as a function of solid concentration and pH. Volume % of solid = 5, 10 [modified from ref. 3].

Thus, the specific aims of this chapter are the following:

- To introduce the current status and a new method to selectively disperse/liberate fine particles
- To introduce particle characterization methods applicable to the analysis of concentrated suspensions

The details of each component will be explained more in the following sections.

2. SELECTIVE PARTICLE DISPERSION/LIBERATION

In this chapter, we are concerned with particles dispersed in a liquid (mainly water), relevant for many of the industrial particle processing operations. Selective dispersion/liberation of different components in a particle mix is a key to achieve the advanced particle processing, including selective mineral beneficiation using standard separation methods. One such method is froth flotation, where a surfactant (collector) selectively adsorbs onto a target mineral particle. Air bubbles injected in the cell attach to the hydrophobic/hydrophobized particles due to mainly hydrophobic interactions, and the particle-bubble complexes then rise to the air-water interface for collection [6]. This method relies on having good dispersion of the different mineral particles from an ore in order to have selective attachment of hydrophobic particles onto bubbles. Common particle dispersion methods can be divided into two categories: chemical (e.g., pH adjustment and surfactant addition) and physical (e.g., agitation, sonication, centrifugation, filtration (to remove fine particles), and wet milling (e.g., [7,8])). However, these dispersion methods have difficulty in achieving selective particle dispersion. For example, wet milling uses a compressive force to break the particle-particle interactions, but it is nonselective (dispersing all particles regardless of mineral type) and is also energy-inefficient (e.g., [9]). There is an urgent need for more selective dispersion techniques, such as electrical disintegration.

2.1 Electrical Disintegration

This disintegration method has been used as a tool for selective grinding of relatively large mineral particles (from 38 μm to over 2000 μm) [10–13]. As the force experienced locally is a function of the local electrical properties (and therefore the atomic spacing, charge, and crystal orientation), differential electrical forces apply at grain boundaries, allowing for selective separation of different minerals [10–13]. Upon applying high voltage to a rock sample, it can selectively disintegrate the sample depending on its physical characteristics (e.g., dielectric constant). The disintegration process starts with crack generations at the interface between different minerals with different electrical properties. The following investigations showed the applicability of the electrical disintegration method to fine particle dispersion.

Otsuki et al. applied an electric field between two electrodes filled with particle suspensions to investigate changes in the particle size distribution. The materials used were silica nanoparticles ($D_{50}=5$ nm) [5] and titanium dioxide fine particles ($D_{50}=300$ nm) [4] with high solid concentrations.

In both cases, reductions in particle size were observed with increasing supply voltages. Figure 2a shows the size distribution of TiO_2 fine particles suspended in ethanol, measured using our purpose-built equipment (called the interactive force apparatus (IFA) further introduced in Section 3.3). The results were compared with the one in dry conditions measured by scanning electron microscopy (SEM) [4]. Different supply voltages from 0.03 to 0.48 V were applied in order to investigate the effect of the electric field strength (0.03-0.48 V applied over a separation of 3000 nm, yielding average field strengths of 1×10^4-1.6×10^5 V/m). The magnitude of the electric field strength used was slightly lower than the range widely used for the investigation of electrical disintegration [10]; however, significant effects were observed. The size distribution measured by IFA decreased significantly with the increase in the voltage from 0.06 V ($D_{50}>350$ nm) to 0.12 V ($D_{50}=150$-230 nm). As the voltage increases, more and more of the larger particles are broken down. The shape of the cumulative size distribution curve, therefore,

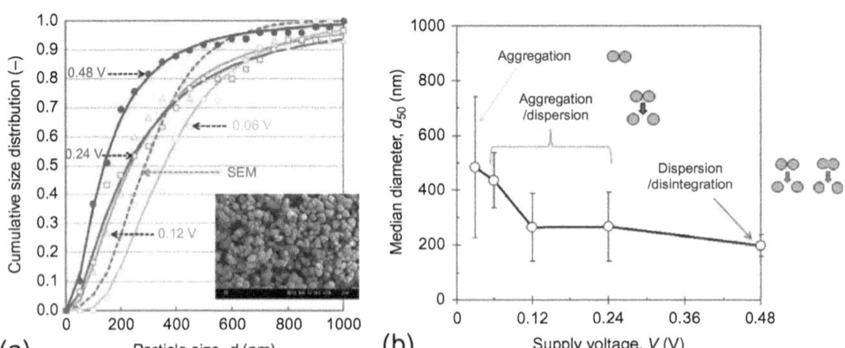

Figure 2 (a) Cumulative size distribution measured by IFA and SEM and (b) median diameters of titanium dioxide particles in ethanol at the different supply voltages from 0.06 to 0.48 V [modified from ref. 4].

changes significantly. At 0.48 V, 90% of the particles are smaller than 300 nm, whereas at 0.06 V, only 40% of the particles are smaller than 300 nm. This trend suggests that aggregates and/or doublets of particles disperse as a result of the increasing supply voltage. The SEM photograph of titanium dioxide particles (Figure 2a insert) indicates that some of the particles form aggregates and even doublets in dry conditions without applying an electric field. It explains the fact that SEM size distribution shows common features with both the low- and high-voltage suspensions. In other words, the small particle fraction is similar to the 0.06 V data, as particles have not been dispersed. At low voltage (i.e., less than 0.06 V), aggregates and doublets of particles coexisted with primary particles and they are not easily broken in the solvent, while at high voltage (i.e., 0.12-0.24 V), aggregates can disperse. With even higher voltages (i.e., 0.48 V), many of the doublets disintegrate. The observed trend is consistent with other reported measurements of silica particles in water by Otsuki et al. [5].

Median diameters as a function of supply voltage were calculated and shown in Figure 2b, in order to confirm the electrical disintegration. The width of the distribution (indicated by the error bars) decreased with increasing supply voltage. This implies that aggregates of particles dispersed at higher supply voltages and doublets may disintegrate at even higher supply voltages to produce less polydisperse size distributions. On the other hand, the distribution width is very high at the smallest supply voltage, 0.03 V. This trend is consistent with the observations in the aqueous silica system [5].

For the example shown, large aggregates exist in suspension without an electric field and require moderate voltages (0.06-0.24 V) to disperse into small aggregates (a mixture of single particles, doublets, and small aggregates). The smaller aggregates can be dispersed to predominantly single particles with even larger voltages (0.48 V). When the high voltage is applied, electrical disintegration may occur at the points where materials are weakly bound, for example, the interface between two minerals [10]. The electrical breakdown creates large currents at the interface between particles, causing expansion and/or contraction depending on the properties of the material. As the electrical breakdown generates ions that adsorb on particle surfaces, the force forming pearl chains also becomes weaker [14]. Electrical breakdown in an electrorheological fluid (ERF) has also been reported in similar experimental setups by Shibayama et al. [15] and Otsuki et al. [5].

The experiments to date have focused on model particle systems. For the future investigation, we need to explore the application of electric fields to disperse/liberate composite particles, for example, natural ore in the case of mineral separation. From such fundamental research, we will learn a great deal about the dispersion processes, and how to optimize dispersion, and will be able to apply this knowledge to scale up systems with potential for industrial use. This future investigation should also seek to understand the effects of electric fields on particle dispersion and will form the basis for the development of industry-applicable dispersion methods to enhance sustainable beneficiation of complex ores. In the medium term, this method has the potential to provide a new particle dispersion method that can be combined with traditional chemical and physical methods to yield improved outcomes.

3. CHALLENGE IN PROPER PARTICLE CHARACTERIZATION METHOD IN HIGHLY CONCENTRATED SUSPENSIONS

Many techniques are currently available for characterizing fine particles by measuring size, turbidity [16], contact angle [17], zeta potential [18], force between particle and plate [19], and a combination of these techniques [20,21]. However, these methods usually are not suitable in highly concentrated suspensions and/or nontransparent suspensions and could overlook the potential coagulation in suspensions. On the other hand, from an economic point of view, nontransparent concentrated suspensions are commonly handled in the many industrial processes, such as the separation of mineral particles by froth flotation. The conventional size distribution measurement technologies can be divided, by their applicable environments, into two categories, that is, measurements in (i) dry conditions and (ii) wet conditions [22]. Microscopic analysis methods (e.g., SEM and transmission electron microscopy (TEM)) are common in dry conditions, while light scattering methods (e.g., the dynamic light scattering (DLS) and laser diffraction) are often used in wet conditions. Each technique has advantages and drawbacks. The primary disadvantage of SEM and TEM is that they cannot be used to measure properties in suspension [4]. The primary disadvantage of light scattering methods is that they can only be carried out in relatively dilute suspensions, due to the complications of multiple scattering [23]. For DLS measurements, the concentration of micrometer-sized particles in water typically has to be below 10^{-5} vol.% [23].

In this section, potential candidates to overcome some of these problems will be introduced. They are the electroacoustic zeta potential measurement, suspension rheology, and advanced particle sizing methods.

3.1 Zeta Potential

Particle zeta potential is the experimentally measurable surface charge around the particle in a liquid. It is one of the important particle physical properties to determine their dispersion/coagulation behavior [24,25]. In the case of froth flotation, it affects the mineral-collector (e.g., [26]) and mineral-mineral interactions (e.g., [3]), which can determine mineral flotation performance. For example, a negatively charged mineral surface attracts a cationic collector. The same logic applies to particle-particle interactions where oppositely charged particles attract each other, while same-charge particles repel. Figure 3 shows an example of such interaction where the heterocoagulation of nickel oxide-hematite and nickel oxide-quartz is formed. The coagulation is proved by the zeta potential value of their mixture that is in the middle of their single systems [3]. Recently, the electroacoustic method for determining the zeta potential, which utilizes the sound waves generated from the particle movement under an AC field, was developed [27]. This method allows us to perform the above investigation in concentrated suspensions that are commonly treated in process plants. On the other hand, the conventional zeta potential measurement machine requires the diluted suspension. This dilution potentially overlooks the particle coagulation happening in a concentrated suspension. This aspect will be discussed in the next section.

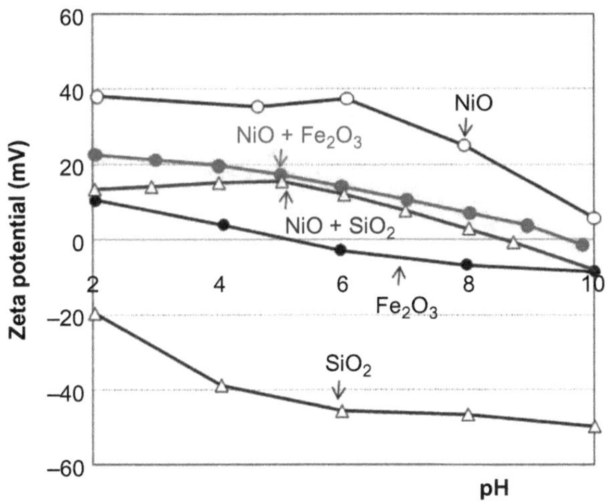

Figure 3 Zeta potential of single minerals and two-mineral mixtures in concentrated suspensions (5 vol.%) [3].

Origins of surface charge/zeta potential can be varied by many factors, such as broken bonds through the comminution of electron imbalances, adsorption of ions (e.g., H^+ and OH^- from pH adjustment) [28], substitution of ions with different valencies, ionization of compounds, and ion dissolution. Among them, pH is often a changeable parameter, while many others are predetermined at the time of the beneficiation process. Therefore, pH can be varied to investigate the separation behavior of target minerals, although a researcher may only need to carefully examine around the optimum pH reported in the past. For example, in froth flotation, often, a collector has the reported optimum solution pH range (e.g., pH 10 for dodecylamine hydrochloride-quartz system [29]). Therefore, one can play around that range to investigate the effect of other contributions, such as associated gangue minerals and slime coating.

3.2 Suspension Rheology

Rheology is a science of material transportation. It can be utilized to identify the degree of particle-particle interactions in concentrated suspensions that are processed to upgrade the target valuable minerals [24,25]. Figure 1 illustrates the effect of higher solid concentrations on the shear yield stress [3], a useful indicator of increased interactions and coagulation. The higher the shear yield stress, the higher the particle-particle interactions, that is, higher coagulation. Significantly higher particle interaction was observed in a highly concentrated mixed mineral system (28 Pa in the NiO-Fe_2O_3 system at pH 10 and 10 vol.%) than in a single-mineral system with less concentration (0.5 Pa in the NiO system at pH 10 and 5 vol.%).

As an example, a comparison study on the rheological behavior of single minerals (nickel oxide, hematite, and quartz) and their mixture system [3] is discussed here. The minerals were selected in order to provide feedback to the beneficiation of nickel ores containing hematite and quartz. Rheological properties were measured to investigate the effect of pH and solid concentration on the shear yield stress of suspensions, while zeta potential measurements of suspensions were used to explain the rheology results. The rheology results of single-mineral experiments indicated that 5 and 10 vol.% of hematite and quartz had no shear yield stress, whereas the one of nickel oxide has a positive value. The shear yield stress of the mixture increased with increasing pH, and the values were much higher than those obtained with only nickel oxide. This indicated the heterocoagulation between nickel oxide and quartz/hematite. The rheology results obtained with the two-mineral mixture system had a strong correlation with the electrokinetic properties of the minerals. The nature of particle interactions in two-mineral mixture systems was explained by the DLVO interactions. This study indicated that rheological investigation can provide significantly important insight in the particle-particle interaction in the mineral mixture system, which plays a key role in mineral separations.

3.3 Advanced Particle Sizing Methods

In this section, two different methods will be introduced. They are (1) advanced scattering methods and (2) IFA. Including the advanced scattering methods, currently, the most common method of nanoparticle sizing in a liquid is the DLS that gives the particle size distribution in an extremely diluted suspension. The use of DLS might not be suitable to study the particle-particle interactions in the actual environment of concentrated suspensions in plant operations due to multiple scattering [23]. Thus, for the concentrated suspensions, we can seek the different scattering methods, such as two-color DLS [30], differential dynamic microscopy [31], diffusing-wave spectroscopy [32], and SAXS [33]. Considering the accessibility of the equipment, SAXS can be the most appropriate option, and it does not suffer from significant multiple scattering effects and can be used for particle sizing in the most concentrated dispersions. The potential limitations of the SAXS method are that (a) it does not give the size distribution and (b) the top measurable size is normally up to several μm that is sufficient for dispersed particles, but it may limit the study on the particle coagulation.

The IFA was first designed for determining the degree of dispersion and coagulation of particles suspended in functional fluids under a magnetic or electric field [34,35]. ERF is one of the functional fluids. ERF shows an increase in suspension yield stress as a result of pearl chain formation of fine particles assembled towards an electric field between two electrodes [36]. The apparatus utilizes this assembly and employs a direct measurement system, not depending on the particle concentration and optical transparency of the suspensions. The target size range is from single nanometers to several ten micrometers. The measurement can be conducted in various solvents (e.g., aqueous solution [5] and nonaqueous solvent [4]). In 2010 and 2012, Otsuki et al. reported the size distribution measurement of inorganic nanoparticles by using IFA with an increasing number of data acquisition [4,5]. For example, titanium oxide in which its dispersion is used in the variety of industrial applications, such as pigment, was measured. In many cases, their dispersion is made with a nonaqueous solvent. It is important to study the dispersion properties of titanium oxide to enhance the applicability of this material. With the study of Otsuki et al., the applicability of IFA for the size distribution measurement in a nonaqueous solvent was also proved [4]. The current experimental setup allows us to detect the primary particle size and/or aggregate size in a sample suspension. The results were compared with the one obtained by analyzing a series of photographs taken by SEM. Their series of studies indicated the applicability of IFA to industrial uses, for example, online measurements of fine particles in their processing plants.

4. SUMMARY AND FUTURE OUTLOOK

In this chapter, the first objective was to discuss the challenges associated with the currently available methods for dispersing fine particles in highly concentrated suspensions and introduce a novel selective dispersion method. This was associated with the second objective to understand/characterize the fine particle dispersion by conventional and newly developed methods. As an example, methods applicable to particle size distribution in concentrated suspensions were discussed. Size is one of the most important and basic physical properties of particles, and size distributions are a good indicator of particle dispersion or coagulation in different environments where particle-particle interactions vary. Although the abovementioned characterization methods have been noticeably improved and provided us important information on particle-particle interactions, it is certainly impossible to cover all the aspects by only a single method. By combining the different methods with different strengths, we can have a full picture of particle-particle interactions being experienced in concentrated suspensions. A further investigation on the above subjects will develop an industry-applicable dispersion and characterization method to enhance the advanced particle processing, including sustainable beneficiation of complex ores. Such investigation has the potential to unlock new research and industry areas of particle dispersion and characterization.

REFERENCES

[1] V.M. Candelario, et al., Carbon nanotubes prevent the coagulation at high shear rates of aqueous suspensions of equiaxed ceramic nanoparticles, J. Eur. Ceram. Soc. 34 (2014) 555–563.
[2] E. Zapata-Solvas, D. Gomez-Garcia, A. Dominguez-Rodriguez, Towards physical properties tailoring of carbon nanotubes-reinforced ceramic matrix composites, J. Eur. Ceram. Soc. 32 (2012) 3001–3020.
[3] A. Otsuki, S. Barry, D. Fornasiero, Rheological studies of nickel oxide and quartz/hematite mixture systems, Adv. Powder Technol. 22 (2011) 471–475.

[4] A. Otsuki, G. Dodbiba, T. Fujita, Measurement of size distribution of titanium dioxide fine particles in a highly concentrated non-aqueous suspension by using particle self-assembly under an electric field, Adv. Powder Technol. 23 (2012) 517–522.
[5] A. Otsuki, G. Dodbiba, T. Fujita, Measurement of particle size distribution of silica nanoparticles by interactive force apparatus under an electric field, Adv. Powder Technol. 21 (2010) 419–423.
[6] A. Nguyen, Flotation, in: Encyclopedia of Separation Science, Elsevier, Amsterdam, 2007, pp. 1–27.
[7] T. Isobe, Y. Hotta, K. Watari, Dispersion of nano- and submicron-sized Al_2O_3 particles by wet-jet milling method, Mater. Sci. Eng. B 148 (2008) 192–195.
[8] M. Hussain, Y. Oku, A. Nakahira, K. Niihara, Effects of wet ball-milling on particle dispersion and mechanical properties of particulate epoxy composites, Mater. Lett. 26 (1996) 177–184.
[9] D.W. Fuerstenau, A.Z.M. Abouzeid, The energy efficiency of ball milling in comminution, Int. J. Miner. Process. 67 (2002) 161–185.
[10] U. Andres, I. Timoshkin, J. Jirestig, H. Stallknecht, Liberation of valuable inclusions in ores and slags by electric pulses, Powder Technol. 114 (2001) 40–50.
[11] U. Andres, Parameters of disintegration of rock by electrical pulses, Powder Technol. 58 (1989) 265–269.
[12] U. Andres, J. Jirestig, I. Timoshkin, Liberation of minerals by high-voltage electrical pulses, Powder Technol. 104 (1999) 37–49.
[13] U. Andres, R. Bialecki, Liberation of mineral constituents by high-voltage pulses, Powder Technol. 48 (1986) 269–277.
[14] Y. Nakajima, T. Matsuyama, Electrostatics field and force calculation for a chain of identical dielectric spheres aligned parallel to uniformly applied electric field, J. Electrostat. 55 (2002) 203–221.
[15] A. Shibayama, T. Otomo, K. Shimada, T. Fujita, Measurement of interactive surface force of suspended particles in ER and MR suspensions under electric and magnetic field, Int. J. Mod. Phys. B 19 (7–9) (2005) 1177–1183.
[16] G. Askarieh, M. Hedhammar, K. Nordling, A. Saenz, C. Casals, A. Rising, J. Johansson, S.D. Knight, Self-assembly of spider silk proteins is controlled by a pH-sensitive relay, Nature 465 (2010) 236–238.
[17] E. Kusaka, H. Tamai, Y. Nakahiro, T. Wakamatsu, Role of surface free energy in a solid surface during collectorless liquid-liquid extraction, Miner. Eng. 6 (1993) 455–464.
[18] T.K. Mitchell, A.N. Nguyen, G.M. Evans, Heterocoagulation of chalcopyrite and pyrite minerals in flotation separation, Adv. Colloid Interf. Sci. 114–115 (2005) 227–237.
[19] W. Ducker, T.J. Senden, R.M. Pashley, Direct measurement of colloidal forces using an atomic force microscope, Nature 353 (1991) 239–241.
[20] R.H. Yoon, D.H. Flinn, Y.I. Rabinovich, Hydrophobic interactions between dissimilar surfaces, J. Colloid Interface Sci. 185 (1997) 363–370.
[21] P.D. Yates, G.V. Franks, S. Biggs, G.J. Jameson, Heteroaggregation with nanoparticles: effect of particle size ratio on optimum particle dose, Colloids Surf. A 255 (2005) 85–90.
[22] T. Allen, Powder sampling and particle size measurement, in: fifth ed., Particle Size Measurement, vol. 1, Chapman and Hall, London, 1997, pp. 44–111.
[23] P.A. Kralchevsky, K.D. Danov, N.D. Denkov, Chemical physics of colloid systems and interfaces, in: K.S. Birdi (Ed.), Handbook of Surface and Colloid Chemistry, third ed., CRC Press, Boca Raton, 2008, pp. 197–377 (Chapter 7).
[24] A. Otsuki, Y. Chen, Y. Zhao, Characterisation and beneficiation of complex ores for sustainable use of mineral resources: refractory gold ore beneficiation as an example, Int. J. Soc. Mater. Eng. Resour. 20 (2) (2014).

[25] A. Otsuki, Use of microorganisms for complex ore beneficiation: bioflotation as an example, in: Encyclopedia of Biocolloid and Biointerfacial Science, Wiley STM, Hoboken, in press.
[26] A.L. Valdivieso, et al., Adsorption of isopropyl xanthate ions onto arsenopyrite and its effect on flotation, Int. J. Miner. Process. 69 (2003) 175–184.
[27] R.W. O'Brien, D.W. Cannon, W.N. Rowlands, Electroacoustic determination of particle size and zeta potential, J. Colloid Interface Sci. 173 (1995) 406–418.
[28] D. Fornasiero, J. Ralston, Iron hydroxide complexes and their influence on the interaction between ethyl xanthate and pyrite, J. Colloid Interface Sci. 151 (1992) 225–235.
[29] R.H. Yoon, J.L. Yordan, Induction time measurements for the quartz-amine flotation system, J. Colloid Interface Sci. 141 (1991) 374–383.
[30] G. Bryant, S. Martin, A. Budi, W.V. Megan, Accurate measurement of small polydispersities in colloidal suspensions, Langmuir 19 (2003) 616–621.
[31] L.G. Wilson, V.A. Martinez, J. Schwarz-Linek, J. Tailleur, G. Bryant, P.N. Pusey, W.C.K. Poon, Differential dynamic microscopy of bacterial motility, Phys. Rev. Lett. 106 (2011) 018101.
[32] D.J. Pine, D.J. Weitz, P.M. Chaikin, E. Herbolzheimer, Diffusing wave spectroscopy, Phys. Rev. Lett. 60 (1988) 1134–1137.
[33] Y. Jin, et al., Effects of ultrasound on colloidal organization at nanometer length scale during cross-flow ultrafiltration probed by *in-situ* SAXS, J. Membr. Sci. 453 (2014) 624–635.
[34] T. Miyazaki, A. Shibayama, T. Sato, T. Fujita, Measurement of interaction force between small distances sandwiched with magnetic fluid under magnetic field, J. Magn. Magn. Mater. 252 (2002) 256–258.
[35] A. Shibayama, T. Otomo, K. Shimada, T. Fujita, Measurement of interactive surface force of suspended particles in ER and MR suspensions under electric and magnetic field, Int. J. Mod. Phys. B 19 (7–9) (2005) 1177–1183.
[36] D.L. Klass, T.W. Martinek, Electroviscous fluids. I. Rheological properties, J. Appl. Phys. 38 (1967) 67–74.

CHAPTER 17

Reinforcement Effects of Carbon Nanotubes in Polypropylene: Rheology, Structure, Thermal Stability, and Nano-, Micro-, and Macromechanical Properties

Evgeni Ivanov, Rumiana Kotsilkova
Open Laboratory on Experimental Micro- and Nano Mechanics, Institute of Mechanics, Bulgarian Academy of Sciences, Sofia, Bulgaria

Chapter Contents

Handbook of Nanoceramic and Nanocomposite Coatings and Materials
http://dx.doi.org/10.1016/B978-0-12-799947-0.00017-1

1. INTRODUCTION

Isotactic polypropylene (iPP) is one of the most important commodity thermoplastics, accounting for about 20% of the world polyolefin production [1]. On the other hand, multiwall carbon nanotubes (MWCNTs) have been considered as unique reinforcements of polymer due to their exceptional physical, thermal, and mechanical properties. The tensile strength, tensile modulus, and Poisson's ratio for carbon nanotubes have been reported to be in the range of 37-100 GPa, 640 GPa to 1-2 TPa [2], and 0.14-0.28 [3], respectively. A unique feature of polymer nanocomposites is that the improved performance is reached at low filler content, resulting in lightweight materials. However, the dispersion of MWCNTs in nonpolar polymers such as polypropylene during melt processing remains a challenge.

MWCNTs are mostly dispersed in small aggregates than in single nanotubes in the polymer. The presence of aggregates decreases the surface area and disturbs the formation network structure, which is essential to improve mechanical properties; thus, the main task of processing is to disperse such aggregates as much as possible. Therefore, uniform dispersion of the nanotubes is required to realize the potentiality of the nanotubes as reinforcing fillers [4–9]. Polypropylene compatibility with fillers was improved by grafting with reactive moieties, such as acrylic acid, acrylic esters, and maleic anhydride (MA) [10–14]. Maleic anhydride-grafted polypropylene (MA-g-PP) is one of the most promising candidates that can improve the MWCNT dispersion in iPP.

Melt rheology was discussed as a method to characterize the dispersion state of polymer nanocomposites [15–17]. Pronounced shear thinning was found to be a characteristic feature of truly nanodispersed clay-containing composites. Under certain filler contents and experimental conditions, even pseudo-solid-like rheological behavior was detected, representing the network structure of the dispersed nanofiller. The authors showed that there is a qualitative relation between the extent of shear thinning and the concentration of clay nanofiller particles. To the best of our knowledge, there is no publication that discloses an approach to quantify shear thinning in order to compare debundling and nanoscale dispersion of carbon nanotubes in thermoplastic polymers.

Much research is done on rheological behavior of polypropylene/carbon nanotube composites, where a transition to solid-like response is found by increasing the filler content [18–25]. This was attributed to the network structure formation of interconnected nanotubes, associated with percolation [26], which can restrain the mobility of the polymer chains. Formation of percolating network of nanofiller in polymer (e.g., geometric percolation threshold) and its synergistic effects with electrical properties [27] and mechanical properties [28] of polymer nanocomposites is widely reported. Passing through the geometric percolation threshold, which depends on the amount of filler and on their aspect ratio, many of the properties increase suddenly. Few reports concern the rheological percolation of thermoplastic polymer nanocomposites, that is, the evolution of rheological properties from a liquid-like to pseudo-solid-like flow behavior above a critical volume fraction. The rheological percolation threshold was presumably related with the mesoscale structure formed by the percolation of nanofiller and further discussed with respect to the relationship with nanocomposite properties. A rheology-structure-property relationship for the iPP/MWCNT composites is reported [19,20], where both the flocculated structure and the percolation network are found. Such a correlation was proposed as a design approach for the processing of nanocomposites with controlled and enhanced properties. This is an important issue in terms of industrial applications, because it facilitates processing and contributes to lower final price.

Because crystallization behavior has a big influence on mechanical properties of the filled iPP, the authors [19,20,29–34] reported on crystallization behaviors of iPP in the presence of carbon nanotubes, resulting in the improvement of thermal stability [5,35–40]. However, more knowledge is needed to clarify the effects of both the crystal structure and the dispersion structure on the properties improvement.

The carbon nanotube/PP composites are recently studied as nanoreinforced materials. When PP is reinforced with pristine carbon nanotubes, its stiffness and strength are improved, but its ultimate strain and toughness are often lowered significantly [13,14,22,23,39,41–43]. More recent efforts have been aimed at measuring quantitative material properties and understanding relationships between surface properties and performance characteristics. In most of these studies, single-probe testing devices, including depth-sensing indentation and scratch systems, are used. In general, for low-modulus polymers, an increase in modulus and a decrease in friction coefficient can reduce the scratch depth and size of the plastic zone on the scratch surface [44]. However, there is no sufficient information on the joint effectiveness of compatibilizer and carbon nanotubes for the improvement of nanomechanical properties of composites at low filler contents.

In the present study, we investigate the combined effects of MWCNTs and MA-g-PP compatibilizer on the dispersion process, mechanical reinforcement, and thermal properties improvement of MWCNT/iPP composites prepared by dilution of masterbatch with iPP. The objective of this work is to understand the effect of carbon nanotubes on iPP at very low filler contents, below and around the flocculation threshold, and to improve the knowledge on the properties enhancement. In order to achieve this understanding, a wide range of characterization data on rheology, structure, crystallization behavior, and thermal properties of the neat iPP and the MWCNT/iPP composites, with and without compatibilizer, were obtained and analyzed. Then, we investigated the evolution of the structure and macro-, micro-, and nanomechanical properties around the flocculation threshold. Finally, a rheology-structure-property relation is established and the properties improvement is correlated to the changes of both the crystal structure and the dispersion structure of composites, as far as they determine the technology of products fabrication.

2. EXPERIMENTAL

2.1 Materials

The base iPP used in the formulations was supplied by Lukoil Neftochim Co, Bulgaria, as Buplen 6231 with MFI (230/2.16) of 12.2 g/10 min. The masterbatch of 20 wt% MWCNTs in polypropylene was obtained in pellet form from Hyperion Catalysis International, the United States. Typical outside diameter range of the MWCNTs was from 10 to 15 nm, the lengths between 1 and 10 microns, and density approximately 1.75 g/cm^3. The chemical compatibilizer used was MA-g-PP chains, Licomont AR504, produced by Clariant International Ltd.

2.2 Composite Preparation

The composites were produced by melt compounding of the MWCNT/PP masterbatch with neat iPP in a twin screw extruder and the extruder processing is described elsewhere [19,20]. The compositions were extruded in three runs and the MA-g-PP compatibilizer was added in concentration of 2.5, 5, and 7.5 wt% at the second extrusion run. The extruded compounds were cooled and pelletized; thus, compositions containing 0.05, 0.1, 0.5, and 1 wt% MWCNTs were prepared. The compositions were further calendared as sheets with thickness of about of 1 mm. In addition,

Table 1 Test samples with and without MA-g-PP and variation of MWCNT contents

No	Sample index	iPP content (wt%)	MWCNT content (wt%)	MA-g-PP content (wt%)
1	iPP	100	–	–
2	iPP0.05CNT	99.95	0.05	–
3	iPP0.1CNT	99.90	0.10	–
4	iPP0.5CNT	99.50	0.50	–
5	iPP1CNT	99.00	1.00	–
6	iPP5MA	95.50	–	5.0
7	iPP0.05CNT5MA	94.95	0.05	5.0
8	iPP0.1CNT5MA	94.90	0.10	5.0
9	iPP0.5CNT5MA	94.50	0.50	5.0
10	iPP1CNT5MA	94.00	1.00	5.0
11	iPP0.5CNT2.5MA	97.00	0.50	2.5
12	iPP0.5CNT7.5MA	92.00	0.50	7.5

control formulations were compounded, which contained the same amount of MWCNTs without compatibilizer.

The labeling of test samples with and without MA-g-PP, at different MWCNT contents, is given in Table 1.

2.3 Characterization Methods

The rheological measurements were carried out by means of a cone-plate Rheotron Brabender rotational viscometer (oscillatory mode) at 200 °C melt temperature. The specimens for the viscometric tests were disks with diameter of 14 mm, punched from the extruded sheets. The experimental protocol consisted of oscillatory tests in the linear range (at low strain amplitude of 0.02) and angular frequency sweep from 0.06 to 75 rad/s.

Thermal properties and crystallization processes during cooling were studied by differential scanning calorimetric (DSC) analysis performed with a Mettler Toledo 822 at the Institute for Polymers, Composites and Biomaterials, CNR, Pozzuoli, Italy. The crystallization behavior was studied by a standard procedure, as samples of about 8 mg were cooled from room temperature to −40 °C, then heated from −40 to 200 °C at 20 °C/min (1st run), and held there for 10 min to eliminate any thermal history of the material. Subsequently, the samples were cooled from 200 to −40 °C at 10 °C/min and then again heated from −40 to 200 °C at 20 °C/min (2nd run). The thermal gravimetric analysis (TGA) was performed in air atmosphere at heating rate of 10 °C/min in the temperature range from 20 to 900 °C, with sample weight of 100 mg.

Structure and morphology at micro- and nanoscales are studied by polarized optical microscopy (POM) and scanning electron microscopy (SEM). Optical measurements were carried out with an Axioskop polarizing microscope (Zeiss, Thornwood, NY) equipped with a THMS 600 hot stage (Linkam Scientific Instruments Ltd., the United Kingdom), Linkam TMS91 temperature control unit, and a video camera model JVC TK-1085E. Samples for optical microscopy were prepared by squeezing a small quantity of the material onto a glass coverslip. SEM (FEI Quanta 600F) at CfAM, University of Reading, the United Kingdom, was used for characterizing the structure and morphology. Samples are chemically etched first and then covered with vacuum evaporated gold. The hierarchy of the nanocomposite structure was estimated at different magnification within the range from 20 μm to 500 nm.

The tensile experiments were carried out with the "Tiratest 2300" universal testing machine at ambient temperature. The moving speed of the crosshead was 10 mm/min for the tensile measurements. Test specimens with size of 100 × 8 × 1 mm were used. The oriented direction of the samples during calendar processing was parallel to the tensile deformation direction. The tensile characteristic values were calculated as an average over six samples for each composition [19].

Friction behavior of the polypropylene composites was studied with module system UMT-2M (BRUKER-CETR). The experiments were performed for a distance of 10 mm by sliding test using three configurations:

1. Scratch test performed on the Tribology Tester UMT-2. In this test, polymeric materials are scratched applying the force control method with a constant force of 2 N, using a single microcutting blade of composite diamond with a tip radius of 0.8 mm at a velocity of 0.083 mm/s, over a distance of 10 mm and time of 120 s. The distance between the plaque surface and the bottom of the groove is calculated from the carriage displacement and defined as the "scratch depth."
2. Pin on plate—constant force of 2 N, test rate—2 mm/s for 5 min by using 416 stainless steel pin, 6.35 mm × 25.4 mm (dia. × length).
3. Ball on plate—increasing force from 0.5 up to 4 N, test rate—2 mm/s for 5 min using 440-C stainless steel balls, dia. 6.35 mm, correspondingly.

Thus, the coefficient of friction (COF) was determined in the bulk volume and on the surfaces, respectively, as calculated by the ratio between tangential force and normal load. At least three measurements were made for each composition.

Nanoindentation test was carried out using a Nanoindentation Tester (UNMT) with inline imaging by atomic force microscopy (AFM), produced by Bruker Co. The sample surface was polished by means of Leica RM2245 microtome with stereo microscope Leica A60S and a diamond knife, Leica Microsystems, Germany. The hardness and elastic modulus were calculated from the recorded load-displacement curves using Oliver and Pharr methods [45,46]. The indentation imprints were inline imaged using Q-Scope™ 250/400 AFM (Ambios Technology Inc.). Indenter type Berkovich Diamond with tip radius of 70 nm was used for indentations in force control mode of 5 mN. A series of 48 (4×12; spacing between indents 80 μm) nanoindentations were performed for each sample. A typical indentation experiment consists of the subsequent steps: (i) approaching the surface, (ii) loading to the peak load of 5 mN for 15 s, (iii) holding the indenter at peak load for 10 s, (iv) unloading from maximum force of 5 mN to 10% for 15 s, (v) holding at 10% of max force for 15 s, and (vi) final complete unloading for 1 s (load function 15s-10s-15s trapezoid). The hold step was included to avoid the influence of the creep on the unloading characteristics since the unloading curve was used to obtain the elastic modulus of the material.

3. RESULTS

3.1 Rheological Control on Carbon Nanotube Dispersion

The effect of compatibilizer on the degree of nanotube dispersion was investigated by testing unmodified and MA-g-PP modified samples at fixed nanotube contents. The objective of adding the coupling agent MA-g-PP was to improve both the dispersion of masterbatch in the iPP and the interfacial interactions with MWCNTs, thus creating a pathway for load transfer through the compatibilizing molecule. In this study, the effectiveness of compatibilizer was correlated to the dynamic rheological characteristics.

Figure 1 presents the dynamic viscosity η' and storage and loss moduli, G' and G'', versus angular frequency ω for an example 0.5% MWCNT/iPP composite at $T=200$ °C, with the amount of MA-g-PP compatibilizer varying from 0, 2.5, 5, to 7.5 wt%. It was found that the addition of 2.5% MA-g-PP compatibilizer to the 0.5 wt% MWCNT/PP composite has insufficient effect on the rheological characteristics, so the viscosity and dynamic modulus are obtained similar to those without compatibilizer. At low frequencies, both systems demonstrate Newtonian plateau of the viscosity and a terminal slope of dynamic moduli, $G' \sim \omega^2$ and $G'' \sim \omega$. While at 5 and 7.5 wt% MA-g-PP the rheological behavior is changed significantly, the composites show pseudo-plastic behavior of viscosity and nonterminal slopes of both moduli in the low-frequency region below $\omega < 1\ s^{-1}$,

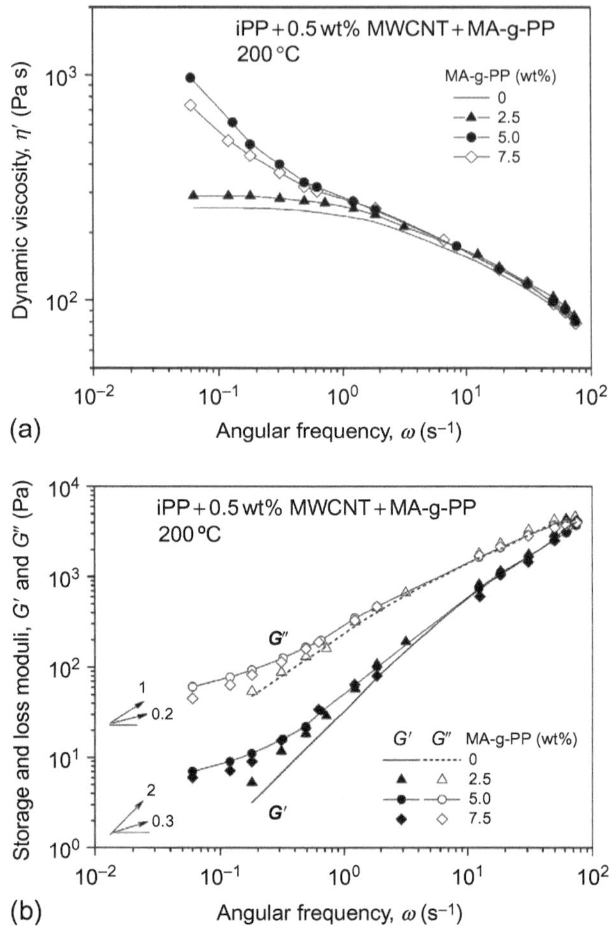

Figure 1 Dependence of the dynamic viscosity η' (a) and storage and loss moduli G' and G'' (b) on angular frequency (for the 0.5% MWCNT composite, as varying the amount of MA-g-PP compatibilizer within 0%, 2.5%, 5%, and 7.5%; $T=200$ °C).

$G' \sim \omega^{0.3}$, and $G'' \sim \omega^{0.2}$. These effects are associated with the improved dispersion of carbon nanotubes in polypropylene by addition of such amounts of compatibilizer. At high frequencies $\omega > 1\ \mathrm{s}^{-1}$, the effect of compatibilizer on viscosity and dynamic moduli is insufficient, probably due to the nanotube alignment within the flow field. Importantly, the content of 5 wt% MA-g-PP produces slightly better improvement of the rheological characteristics, compared to 7.5 wt% MA-g-PP. Hence, 5 wt% MA-g-PP content was chosen as the optimal amount of compatibilizer for further preparation of the maleinated MWCNT/iPP composites.

Rheology is one of the indirect methods that are very sensible to the changes of the structure of composites as varying the additives. In order to verify the combined effect of both nanotube content and compatibilizer, Figure 2a and b compares the complex viscosity, $|\eta^*|$, versus angular frequency, ω, as varying the MWCNT content from 0.05 to 1 wt%, with and without adding of 5% MA-g-PP compatibilizer. For the composites without compatibilizer in Figure 2a, an increase of the complex viscosity at low frequencies by increasing the nanotube content appears only at the highest amount of 1 wt% MWCNTs, while below this filler content, the viscosity of the composites remains similar to this of the neat iPP. In contrast,

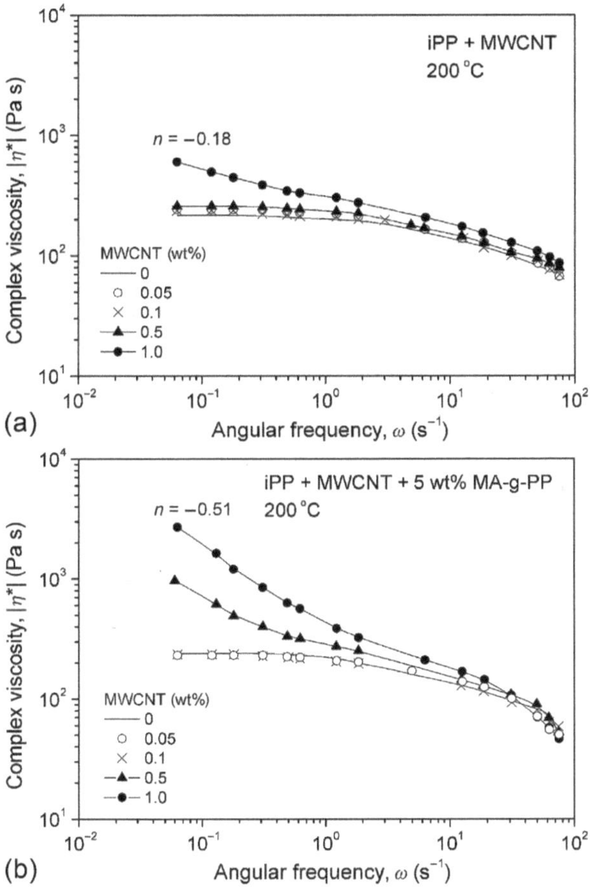

Figure 2 Dependence of the complex viscosity $|\eta^*|$ versus angular frequency (at $T=200$ °C, as varying the nanotube contents from 0 to 1 wt%): (a) nonmodified iPP/MWCNT composites; (b) 5 wt% MA-g-PP modified iPP/MWCNT composites.

the addition of MA-g-PP compatibilizer (Figure 2b), results in a significant increase of the viscosity at low frequencies, which appears at lower nanotube content of 0.5 and 1 wt% MWCNTs. For example, at $\omega \sim 0.1\ s^{-1}$, the viscosity of 1 wt% MWCNT/iPP composite with compatibilizer is a decade higher than those of neat iPP, while without compatibilizer, the viscosity is only twice higher than that of iPP. We attribute the shear thinning behavior of composites to alignment of entangled nanotubes in the flow field. The addition of MA-g-PP coupling agent obviously produces a better dispersion and debundling of nanotubes, resulting in a viscosity increase.

The rheology allows the quantity control on the degree of nanofiller dispersion in polymer matrices by using experimental data from the low-amplitude oscillatory shear and the steady-state shear flow combined with rheological models [15,17]. The shear thinning exponent (*n*), determined from the straight line fitted to the data at the lowest shear rates, is proposed as a semi-quantitative measure of the degree of exfoliation and delamination of clay-containing polymer nanocomposites. At low shear rates, the rheological response is most representative for the unoriented structure of anisotropic nanofiller in the composite. The flow curves taken under low-amplitude measurement conditions were fitted to the power law expression:

$$\eta^* = A\omega^{(n)}$$

where η^* = dynamic viscosity, ω = angular frequency, A = pre-exponential factor, and n = shear thinning exponent (flow index).

It is assumed that, when other factors are constant, the shear thinning exponent (*n*) may reflect the dispersion quality of MWCNT/iPP composite samples, prepared under intentionally varied conditions. In Figure 2a and b, the shear thinning exponent (*n*) was used to compare the degree of nanotube dispersion in both unmodified and MA-g-PP modified composite samples at a fixed filler concentration. If, for example, we compare the 1 wt% MWCNTs/PP samples, they demonstrate pronounced shear thinning. The sample with compatibilizer showed considerable shear thinning in melt rheology and the respective values of the exponent are of $n = -0.51$, representing a good dispersion of nanofiller. The sample without compatibilizer showed only moderate shear thinning of $n = -0.18$, so it is not nearly as well dispersed. Clearly, the shear thinning exponent allows a quantitative differentiation between the samples, dependent on the nano-and mesoscale composite structure formed by the debundled carbon nanotubes in the presence of MA-g-PP compatibilizer.

3.2 Rheological Thresholds of Nanodispersions

The rheological percolation threshold is mostly determined and related to the increase of some physical and mechanical properties of nanocomposites [18,21–25,47]. In our previous studies on polymer nanocomposites [15,19,20], we found that the rheological response and properties of polymer nanocomposites are very sensitive to both the short-range interconnectivity and long-range interconnectivity of nanofiller particles. Hence, we assumed two critical concentrations accounting for structural transitions of nanodispersions with increasing nanofiller content. The first rheological threshold represented the short-range connectivity of nanofiller particles, and it is related to the formation of a dispersion structure dominated by floccules of interconnected nanofiller particles incorporating matrix polymer, called "rheological flocculation threshold" (ϕ_c). This threshold is usually displayed by deviation of the concentration function of viscosity from the Einstein equation. By increasing the nanofiller contents, the number of floccules increased and they started to interact. Thus, the second rheological threshold is reached, the so-called rheological percolation threshold (ϕ_p), which represents the long-range connectivity of the nanofiller particles within the polymer matrix, resulting in the formation of a three-dimensional (network) structure. This threshold is usually displayed by the sudden rise of the viscosity versus volume fraction function.

For evaluation of the structural peculiarities of the MWCNT/iPP melts within the small nanotube contents of 0.05-1 wt% MWCNTs, Figure 3 plots the concentration dependence of the relative complex viscosity, $\eta^*_{rel} = \eta^*/\eta^*_o$, where η^* = the complex viscosity of the composite and η^*_o = those of the neat iPP. The η^*_{rel} of both MA-g-PP modified and unmodified MWCNT/iPP composites is compared at low frequency of $\omega = 0.063\ s^{-1}$, at 200 °C. As seen, the relative viscosity at very low nanotube contents is approximately 1, but above a critical filler content (ϕ_c), a deviation of the concentration function of viscosity from the Einstein equation, is observed. Moreover, the η^*_{rel} function is sharply increased for the MA-g-PP modified composites, while it is slightly increased for the unmodified composites above a characteristic concentration. We attribute this effect to the short-range junction (local interconnection) of nanotubes leading to formation of floccules of entangled nanotubes incorporating matrix polymer [19,20]. The arrows in Figure 3 show the determined rheological flocculation threshold (ϕ_c) from the relative viscosity at a low frequency; thus, for the MA-g-PP modified composites, it is determined at $\phi_c \sim 0.3$ wt%, while $\phi_c \sim 0.5$ wt% for the unmodified composites.

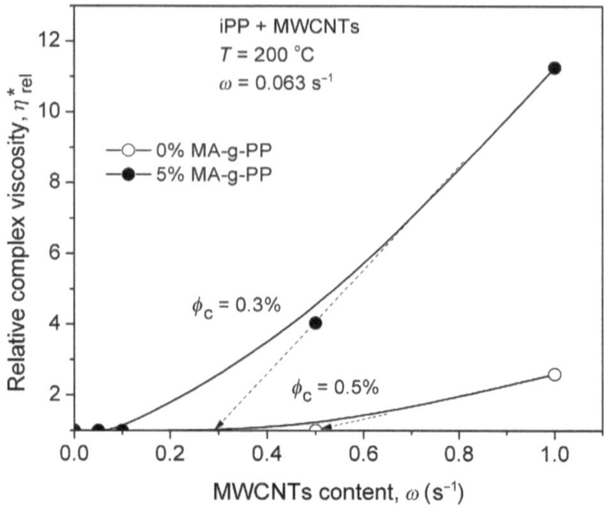

Figure 3 Relative complex viscosity, η^*_{rel} versus weight fraction of MWCNTs for the MWCNT/iPP composites without (open circles) and with MA-g-PP compatibilizer (full circles) at T=200 °C. Results are compared at $\omega \sim 0.063\ s^{-1}$. Arrows point the rheological flocculation threshold (ϕ_c).

The result confirms the positive effects of MA-g-PP compatibilizer for a better dispersion of nanotubes in iPP. This characteristic concentration threshold will be further related with the reinforcement effects of MWCNTs on iPP within the range of very small nanotube contents of 0.05-1 wt%.

Obviously, the rheological threshold of percolation is not reached for the composites studied within the low concentration range 0.05-1 wt% MWCNTs, but it is expected within a higher concentration range of nanotubes [20]. As reported by Huang and Terentjev [26], a thoroughly dispersed nanocomposite has several universal rheological features; at nanotube concentration above a characteristic value ~2-3 wt%, the effective elastic gel network is formed (so-called percolation threshold), while the low concentration composite remains a viscous liquid.

3.3 Microstructure and Morphology

Figure 4a–d presents the SEM observations of the etched 0.5% MWCNT/iPP samples as varying the amounts of MA-g-PP compatibilizer of 0, 2.5, 5, and 7.5 wt%. As seen in Figure 4a, the 0.5% MWCNT/iPP composite without compatibilizer contains large and dense agglomerates of MWCNT masterbatch having a size of about 5 μm. The addition of 2.5 wt% MA-g-PP as a

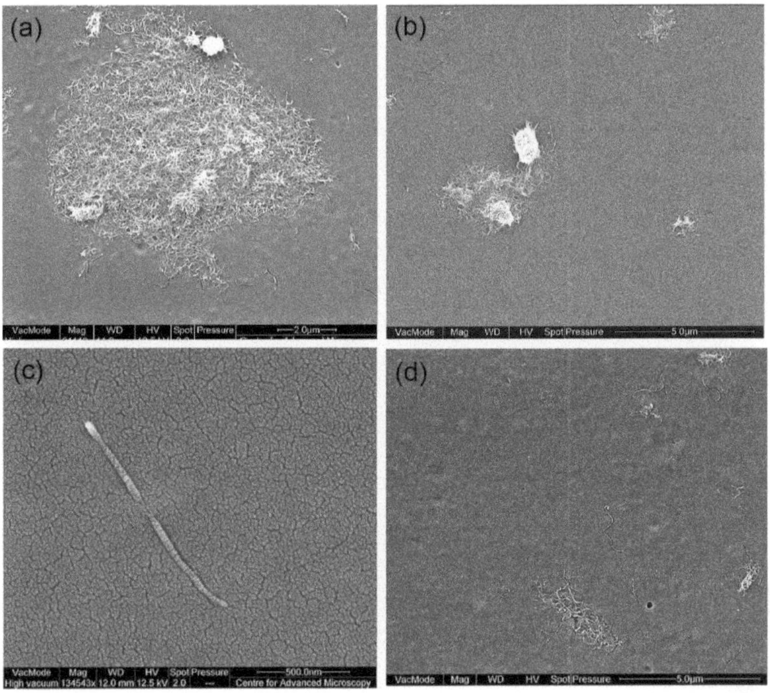

Figure 4 SEM micrographs of etched 0.5% MWCNT/iPP samples as varying the amounts of compatibilizer MA-g-PP: (a) 0%; (b) 2.5 wt%; (c) 0.5 wt% (an individual nanotube at high magnification); and (d) 7.5 wt% MA-g-PP.

coupling agent (Figure 4b), has no significant effect resulting in the presence of the same large and small aggregates. However, if the content of the compatibilizer increases to 5 and 7.5 wt% (Figure 4c and d), the agglomerates of nanotubes become smaller (below 1 μm) and look not so dense; moreover, individually dispersed nanotubes are visible at high magnification (Figure 4c).

As a conclusion, the presence of agglomerates in the MWCNT/iPP composites without compatibilizer confirms the imperfect mixing of the masterbatch with the neat iPP. Indeed, due to the high viscosity and much higher melting point of the masterbatch compared to the neat iPP, the melt blending of the masterbatch and the iPP remains difficult. Obviously, the addition of 5 and 7.5 wt% MA-g-PP to the MWCNT/iPP composites during processing leads to a better dispersion of masterbatch, and a mixed structure is formed consisting of small agglomerates and debundled individual nanotubes in the matrix polypropylene. Thus, the SEM visualization of the microstructure together with the rheological results confirms the effectiveness of the 5 wt% MA-g-PP compatibilizer for a better dispersion of MWCNTs in polypropylene.

3.4 Crystallization Behavior

Calorimetric properties of MWCNT/iPP composites are studied using DSC nonisothermal analysis, as varying nanotube content, with and without MA-g-PP compatibilizer. Figure 5 shows the thermoanalytical curves from DSC second run of iPP and its composites from 0.05 to 1 wt% MWCNTs, without and with 5 wt% MA-g-PP compatibilizer. Table 2 summarizes the melting temperature (T_m) and the glass transition (T_g) obtained from the DSC curves.

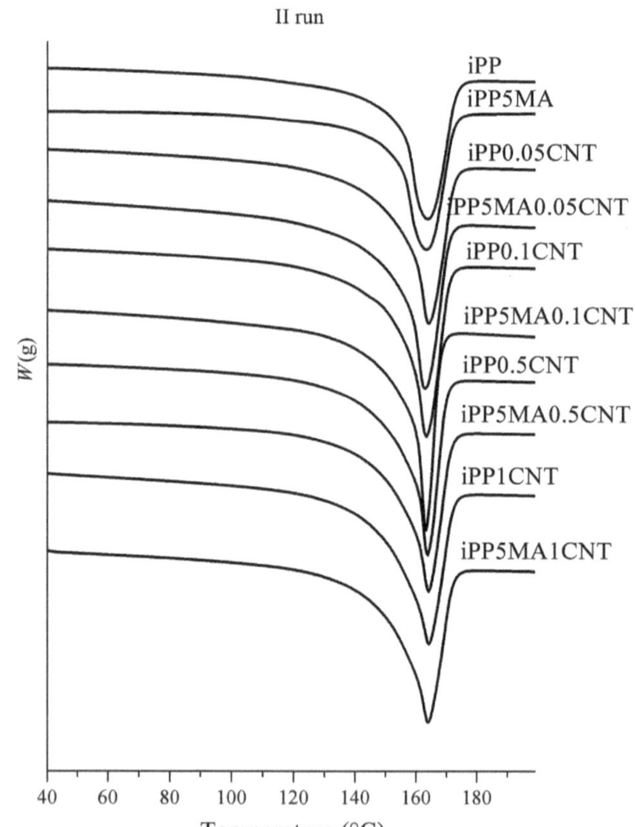

Figure 5 Thermoanalytical curves from DSC second run of iPP and its nanocomposites as varying the content of MWCNTs from 0.05 to 1 wt% in the nonmodified and the 5 wt% MA-g-PP modified composites.

Table 2 Thermal characteristics: DSC melting temperature (T_m), glass transition temperature (T_g), and crystallization temperature (T_c); TGA decomposition temperatures at weight loss of 10% ($T_{10\%}$); and DTG degradation peak temperature (T_p) in air, at heating rate of 10 °C/min

	DSC			TGA and DTG	
Sample index	T_m (°C)	T_g (°C)	T_c (°C)	$T_{10\%}$ (°C)	T_p (°C)
iPP	163.7	−2	112	300	370
iPP0.05CNT	164.3	−2	123	307	410
iPP0.1CNT	163.5	−3	124	307	405
iPP0.5CNT	163.9	−4	125	310	418
iPP1CNT	164.2	−3	126	312	418
iPP5MA	163.4	−5	110	305	405
iPP0.05CNT5MA	163.1	−4	121	310	405
iPP0.1CNT5MA	163.4	−5	123	308	410
iPP0.5CNT5MA	164.2	−6	125	311	420
iPP1CNT5MA	163.9	−6	126	311	420
iPP0.5CNT2.5MA	163.8	−5	–	310	415
iPP0.5CNT7.5MA	164.1	−6	–	310	420

The results presented in Figure 5 and Table 2 demonstrate that the two characteristic temperatures, T_m and T_g, are not changed significantly by increasing the MWCNT content and the presence of MA-g-PP compatibilizer. Therefore, the addition of MWCNTs within the concentration range studied and the use of MA-g-PP as a modifier do not influence the calorimetric properties of iPP. These results indicate for a weak interface interaction between the iPP matrix and the MWCNT surfaces in both non-modified and MA-g-PP modified compositions.

Figure 6 presents nonisothermal crystallization DSC curves at cooling rate of 10 °C/min of iPP and its composites with and without compatibilizer, 5 wt% MA-g-PP. The crystallization peak temperature (T_c) for the composites studied is reported in Table 2. As seen, the addition of very small amounts of 0.05 wt% MWCNTs to iPP produces a strong rise in the peak temperatures (~11 °C), indicating an enhanced overall crystallization rate due to the nucleation effects of carbon nanotubes. Further increase of nanotube content produces much smaller but continuous increase of the crystallization temperatures. Moreover, the crystallization peaks of the MWCNT/iPP composites are much narrow than those of the neat iPP, suggesting a narrow crystallite size distribution in the composites as compared with the neat iPP. The iPP has usually a slow crystallization rate,

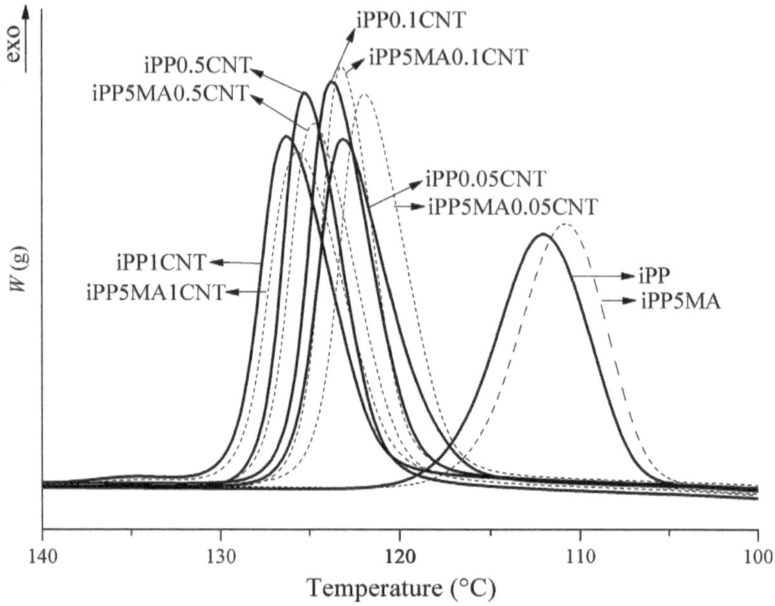

Figure 6 Nonisothermal crystallization DSC curves at 10 °C/min of iPP and its composites containing 0.05-1 wt% MWCNTs, without compatibilizer (solid line) and with 5 wt% MA-g-PP compatibilizer (dot line).

forming large spherulites, while the addition of MWCNTs causes an increase of the crystallization rate of iPP and this obviously results in a strong decrease of the spherulite size.

Hence, the MWCNTs act as a strong nucleation agent on the iPP matrix at very low nanotube contents. Apparently, the addition of 5 wt% MA-g-PP compatibilizer to both iPP and composites slightly decreases the crystallization temperature with ~2 °C. However, the narrow crystallite size distribution in the composites with compatibilizer remains similar to those without compatibilizer.

The effect of MWCNTs and compatibilizer on the crystalline morphology at the end of the dynamic crystallization process is visualized in Figure 7a–d by the optical micrographs of iPP and 0.5 wt% MWCNTs, with and without compatibilizer. As shown in Figure 7a, the typical dimension of spherulites of the neat iPP is about 80 μm. On the other hand, in Figure 7b, the addition of PP-g-MA to iPP led to the formation of spherulites of a wide size distribution polydispersity from 100 to 30 μm. Obviously, the MA in the MA-gg-PP obstructed the crystallization process of iPP, but not to so large extent, because it does not have direct nucleating effect to iPP spherulites. While Figure 7c and

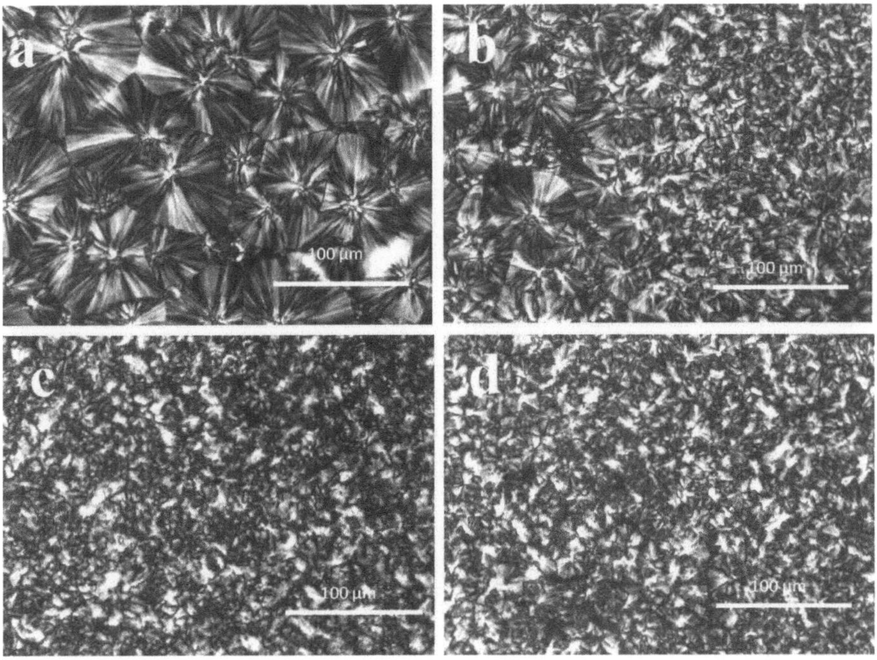

Figure 7 Optical micrographs at the end of the dynamic crystallization process of (a) iPP; (b) iPP + 5% MA-g-PP; (c) 0.05 wt% MWCNTs/iPP; and (d) 0.05 wt% MWCNTs/iPP + 5% MA-g-PP.

d shows that the addition of small amount of 0.05 wt% of MWCNTs to both neat iPP and maleinated iPP does not just reduce the crystal dimensions, it is difficult to see well-defined spherulites. The presence of nanotubes provides a tremendous amount of heterogeneous nucleation sites at crystallization process, and they act as effective nucleating agents. From a thermodynamic point of view, fillers and other kinds of impurities can never exist in a crystal [48]; therefore, the nanotubes are located in the amorphous phase between the lamellas, suppressing the crystal growth.

3.5 Thermal Properties

Our previous studies showed that MWCNTs are known to impart high thermal stability to polymeric materials, which however depends on the thermal degradation environment [19]. Here, thermogravimetric analysis was conducted in air atmosphere at a heating rate of 10 °C/min. The TG and DTG results of the MWCNT/iPP composites, as varying the MWCNTs contents within the range of 0.05-1 wt%, are shown in Figure 8a for the

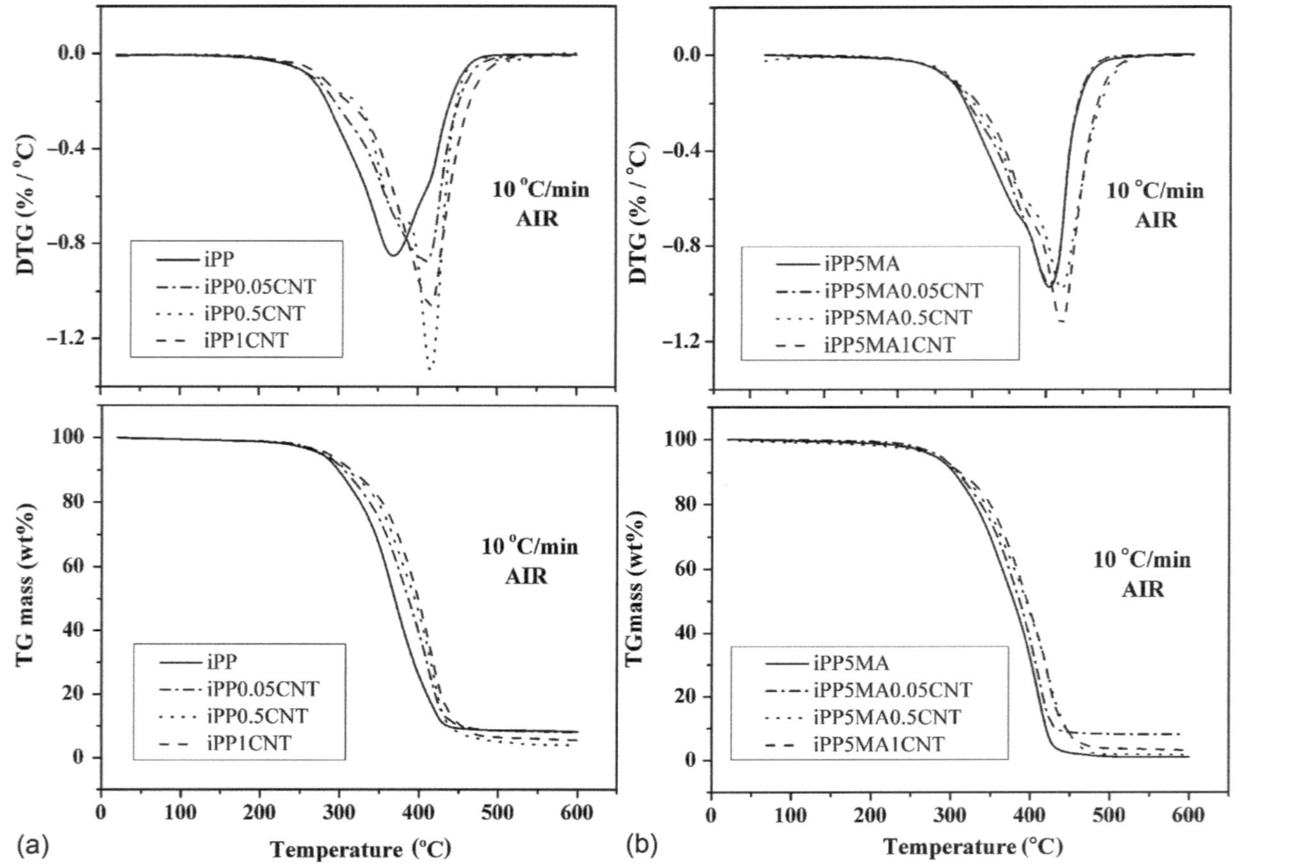

Figure 8 DTG and TG curves of MWCNT/iPP composites as varying the nanotube content of 0.05; 0.5; and 1 wt% MWCNTs: (a) without compatibilizer and (b) with 5 wt% MA-g-PP compatibilizer. Tests in air at a heating rate of 10 °C/min.

nonmodified systems and in Figure 8b for the MA-g-PP modified composites. The TGA decomposition temperatures at 10% weight loss ($T_{10\%}$) and the DTG degradation peak temperature (T_p) of the investigated compositions are shown in Table 2. More specific is that a strong effect is observed at very low nanotube contents; thus, the thermal stability ($T_{10\%}$) of composites is improved with 7-12 °C, and the degradation peak (T_p) is shifted with 40-48 °C higher temperature at 0.05 to 1 wt% MWCNTs, respectively, as compared to the neat iPP. Maleinization itself also renders better thermal stability of the neat iPP polymer, showing 35 °C higher degradation peak of the iPP with compatibilizer, compared to the nonmodified one.

Our results confirm the assumptions in Refs. [49,50] that the improvement in the thermal stability can be attributed to the much higher heat capacity of the MWCNTs than that of the polypropylene matrix. As the nanotubes are good thermal conductors, they easily take up the heat that is applied to the system. Obviously, a better dispersion of nanotubes in the composites in the presence of MA-g-PP and the formation of a relatively uniform network structure within the local flocs allow the spreading of heat uniformly, resulting in the thermal stability improvement.

3.6 Macromechanical Properties

There are three possible types of interaction in CNT-based nanocomposites (polymer-nanotube, nanotube-nanotube, and interphase-polymer interaction), but the mechanism and the magnitude of the load transfer between polymer matrices and the nanotubes are still unclear concerning mechanical properties [15,48].

Figure 9a–c presents the macroscopic mechanical parameters in tensile mode of maleinated and nonmaleinated iPP nanocomposites versus MWCNT content: (a) tensile strength, (b) tensile modulus, and (c) elongation at break. If we compare the neat iPP and the modified iPP with 5 wt% MA-g-PP, the tensile strength and the tensile modulus of both materials are similar. However, the iPP without compatibilizer demonstrates relatively higher elongation at break (~12%) compared to those of maleinated iPP with 5 wt% MA-g-PP (~9%) due to the obstructed crystallization process of iPP brought by the compatibilizer.

Obviously, the addition of MWCNTs produces a significant increase of tensile strength and the tensile modulus at very low nanotube contents, of about 0.1 wt%, Therefore, the nanotubes act as reinforcing agents below the flocculation threshold, at $\phi < \phi_c$. However, around this characteristic value ($\phi \sim \phi_c$), the tensile properties reach maximum values and tend to

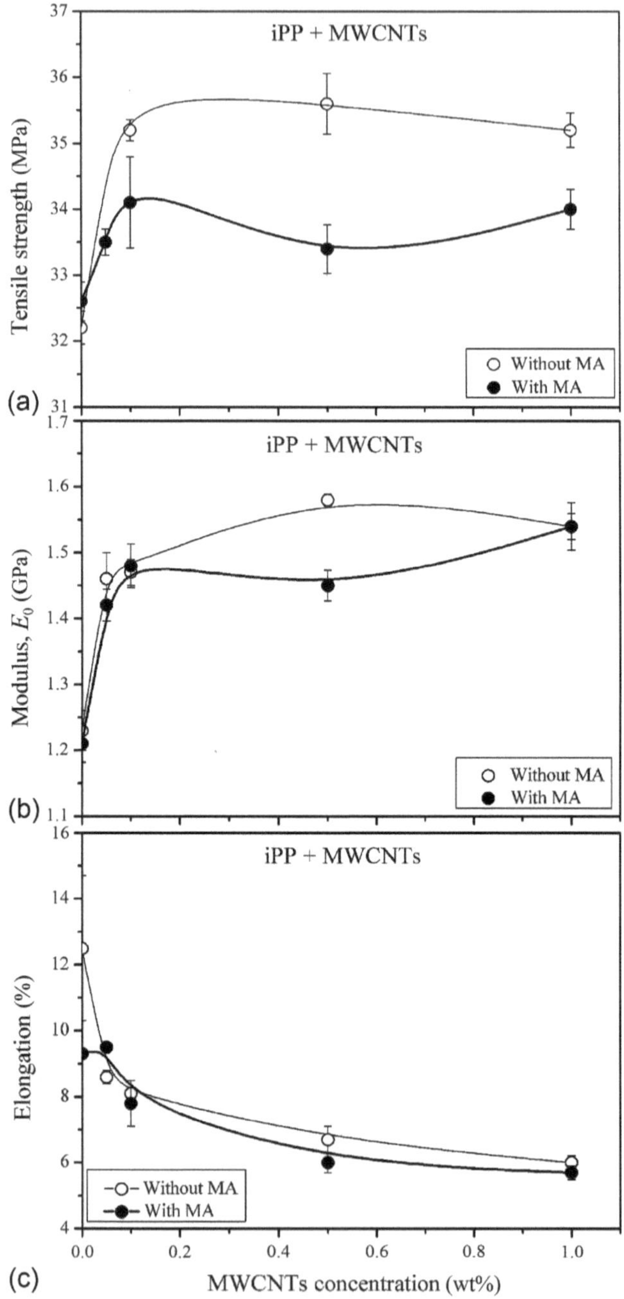

Figure 9 Mechanical parameters in tensile mode of maleinated (•) and nonmaleinated (○) iPP composites versus MWCNT content: (a) tensile strength; (b) tensile modulus; (c) elongation at break.

plateau in the concentration range studied. Thus, for example, Figure 9a and b shows that maximum enhancement of the tensile strength and the tensile Young's modulus for composites without compatibilizer is about ~11% and ~28%, respectively, at $\phi_c=0.5$ wt%. The combined addition of MA-g-PP compatibilizer and MWCNTs led to relatively lower enhancement of both functions, about ~5% for the tensile strength and ~22% for the modulus at $\phi_c \sim 0.3$ wt%, compared to nonmodified systems. When the nanotube content increases slightly above the rheological flocculation threshold (~1 wt% MWCNTs), the values of both functions tend to plateau or slightly decrease. If we consider the elongation at break (Figure 9c), it decreases with increasing the nanotube content, compared to the neat iPP. The addition of MA-g-PP strongly decreases the elongation values (~25%) of the neat iPP, but it does not affect significantly the elongation at break of the iPP/MWCNT composites, which is mostly dominated by the presence of MWCNTs.

Generally, in the MWCNT/iPP composites, partial tensile strain can be transferred to the MWCNTs embedded in iPP matrix under tensile stress, which leads to an increase of tensile strength, indicating the reinforcement effect of MWCNTs on iPP. In fact, the addition of MA-g-PP compatibilizer lowers the values of the tensile strength and modulus of both iPP and composites due to bridging of the linear PP molecules by MA compatibilizer.

3.7 Micromechanical Properties

The values of the COF strongly depend on the type of sliding test configurations used, for example, scratch, pin, or ball on flat (Table 3). Some

Table 3 Scratch and friction characteristics iPP and MWCNT/iPP composites, without and with MA-g-PP compatibilizer

Sample index	COF			Scratch depth
	Pin on flat	Ball on flat	Scratch	*Z* displacement (mm)
iPP	0.202	0.067	2.70	0.18
iPP0.05CNT	0.282	0.076	2.40	0.12
iPP0.1CNT	0.274	0.077	2.20	0.13
iPP0.5CNT	0.231	0.136	2.05	0.11
iPP1CNT	0.355	0.081	2.00	0.12
iPP5MA	0.209	0.050	1.60	0.11
iPP0.05CNT5MA	0.235	0.077	1.75	0.11
iPP0.1CNT5MA	0.244	0.083	1.70	0.06
iPP0.5CNT5MA	0.274	0.086	1.75	0.10
iPP1CNT5MA	0.305	0.095	1.75	0.09

variations at the values are observed due to the imperfect distribution and dispersion of the carbon nanotubes. In scratch on flat experiment, the tendency for the COF is to decrease with increasing the concentration of MWCNT, as compared to the neat PP. An interesting observation was made that the increase of the MWCNT filler in the iPP matrix with no MA-g-PP leads to improvement of the scratch resistance (lower COF) due to the lesser penetration into the material (lower scratch depth). Meanwhile, the combination of iPP with MA-g-PP compatibilizer gives even better results for the COF, but there is no further improvement when nanofiller is added. This effect can be explained by the observation that the MA has changed the crystal structure of the iPP, generally hardening the polypropylene. In pin on flat and ball on flat experiments, a slight increase of COF is observed by increasing the MWCNT contents, compared to the neat iPP, which presents mostly the surface properties of the nanocomposites. As a whole, the variation of the COF values is smaller in reciprocating motion test than in scratch test. This means that the sliding properties on the surface of the samples are different than in the bulk.

The computed scratch depth is the average of at least three measurements along the scratch, and the results are shown in Table 3. As can be seen, the unfilled iPP samples had a relatively deep scratch depth at a time of 120 s. The increase of the nanotube content from 0.05 to 1 wt% shifted the displacement to lower values. Interestingly, a very low nanotube content of 0.1 wt% resulted in a sharp decrease (~26%) of the scratch depth. This is associated with the significant nucleation effect of the carbon nanotubes observed for very low nanotube amount, as shown in the data for the enhanced overall crystallization rate (Figure 6) and size reduction of spherulites (Figure 7).

3.8 Nanomechanical Properties

Figure 10a shows representative load-displacement curve obtained after nanoindentation tests of example 1 wt% iPP/MWCNT composite loaded at 5 mN. The maximum load was held for 10 s for proving that the viscoelastic contribution was negligible and the Oliver-Pharr model was reliable. Figure 10b presents the calculated hardness and Young's modulus using Oliver-Pharr method [45,46]. It was found that the addition of 0.05-1 wt% MWCNTs shifted the curves to lower penetration depth compared to that of neat iPP, that is, the composite material had better resistance to penetration. This evolution evidences a hardening effect in the all tested nonmaleinated and maleinated composites. The intercept in Figure 10a presents the AFM

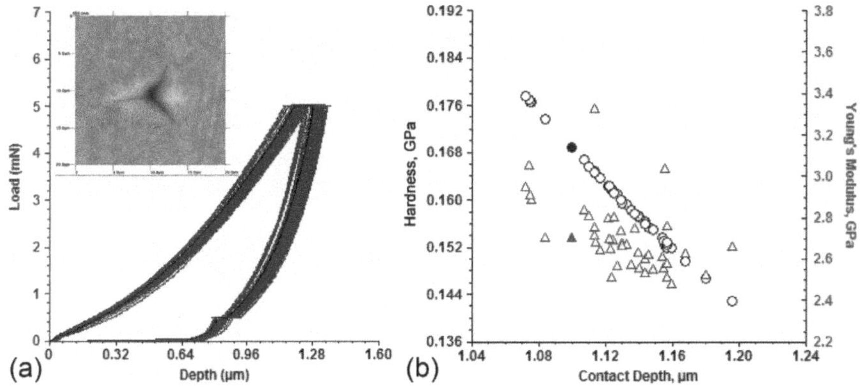

Figure 10 Representative load-displacement curves from nanoindentation tests (a) and calculated hardness (○) and Young's modulus (△) (b) for the 1 wt% MWCNT/iPP composite, loaded at 5 mN. The intercept in (a) presents AFM image of prints made on composite.

images of example imprints with Berkovich nanoindenter made on the composite sample. Such imprints provide information about the brittle-ductile behavior of polymeric materials. The neat iPP demonstrates an absence of pileup effect in the line profile across the print, which is not changed significantly by the addition of both MWCNTs and MA-g-PP compatibilizer.

The average values of hardness and apparent elastic modulus versus nanotube content from nanoindentation tests of the nonmaleinated and the maleinated MWCNT/iPP composites studied are presented in Figure 11a and b. All the values measured show small experimental dispersion, which is higher in the case of the reinforced materials than that of the neat iPP. The level of deviation of the experimental results demonstrates the sensibility of the indenter to the nonhomogeneity of the composite structure, produced by the presence of MWCNTs, when compared to the pristine polymer. The experimental errors are within the range ± 0.0009 to ± 0.001 GPa for the hardness and ± 0.008 to ± 0.01 GPa for the modulus.

It can be observed that a small addition of MWCNTs between 0.05 and 0.1 wt% results in a significant improvement of the nanomechanical properties. This is more pronounced for the maleinated sample where the improvement of the hardness and elastic modulus is of about 26% and 20%, respectively. The strong reinforcement effect of MWCNTs in iPP matrix below the flocculation threshold is followed by plateau values for hardness and modulus when the nanotube contents increase around (ϕ_c).

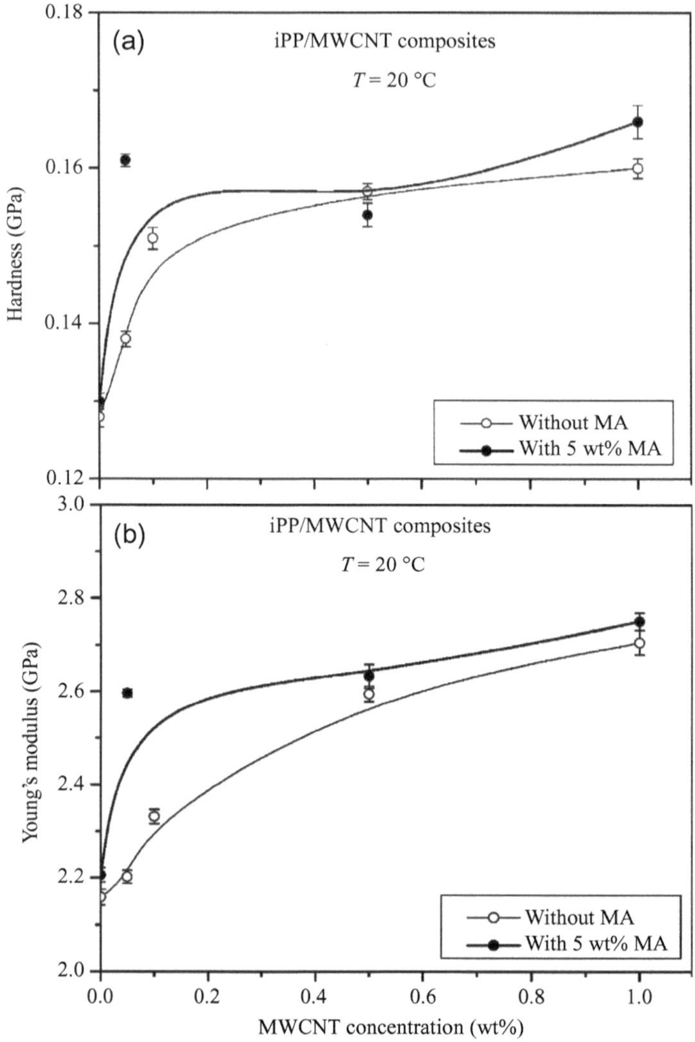

Figure 11 Average values of (a) hardness and (b) apparent elastic modulus versus nanotube content from nanoindentation tests of nonmaleinated (○) and maleinated (•) MWCNT/iPP composites at maximum load of 5 mN.

Interestingly, the composites modified with 5 wt% MA-g-PP compatibilizer showed higher values of hardness and Young's modulus compared with those of nonmodified composites. The combined effect of the MA-g-PP compatibilizer and the MWCNTs on the crystallization structure of MWCNT/iPP composites obviously contributes to the enhanced hardness of the maleinated composites.

4. DISCUSSION

4.1 Rheology-Structure-Properties Relationships

The results obtained in this study confirm several important findings about the reinforcement effect of MWCNTs and MA-g-PP compatibilizer on iPP.

First, the addition of a very small amount MWCNTs (~0.05 wt%) produces a strong nucleation effect on iPP crystallization process. The main effects of this nucleate agent on the crystallization of iPP are an increase in the crystallization temperature, which decreases the molding process cycle time, and a dramatical decrease in crystal size. While the addition of 5 wt% MA-g-PP compatibilizer produces a nonhomogeneous reduction of the crystal size of the neat iPP, thus, the size distribution ranges from very large to very small crystals. However, the joint addition of MWCNTs and compatibilizer MA-g-PP to polypropylene produces a similar dramatic decrease of the crystal size as those in the unmodified MWCNT/iPP composites. Such dramatic changes of the crystal structure by both additives strongly affect the thermal and mechanical properties of MWCNT/iPP composites at very low nanotube contents.

The second main aspect that has to be considered is the structure of the dispersed nanotubes formed in the iPP matrix and the homogeneity of the dispersion. As seen in Figure 3, in the concentration range studied, the rheological flocculation threshold is reached at relatively low nanotube contents. For the MA-g-PP modified composites, the flocculation threshold ($\phi_c \sim 0.3$ wt%) is lower than those for the nonmodified composites ($\phi_c \sim 0.5$ wt%), due to the positive effects of compatibilizer for a better dispersion of nanotubes in iPP.

As discussed in our previous investigations [19,20], the flocculated structure produces a significant growth of the process zone in front of the crack tip due to the stress-strain interaction between the nanotubes and the matrix in the floccules. We apply such approach to relate the rheology, structure, and properties when discussing the combined effect of MWCNTs and MA-g-PP compatibilizer on the performance of composites.

Thus, Figure 12a and b presents the improvement $\Delta(\%) = [(X - X_o)/X_o] \times 100$ of the example characteristic of the nanocomposite (X), compared with the corresponding characteristic of the pure iPP (X_o), versus nanotube content. Figure 12a shows the properties improvement (%) of the unmodified MWCNT/iPP composites, and Figure 12b presents these values of the systems with MA-g-PP compatibilizer. The following characteristics are compared: (i) *structural*—the crystallization temperature

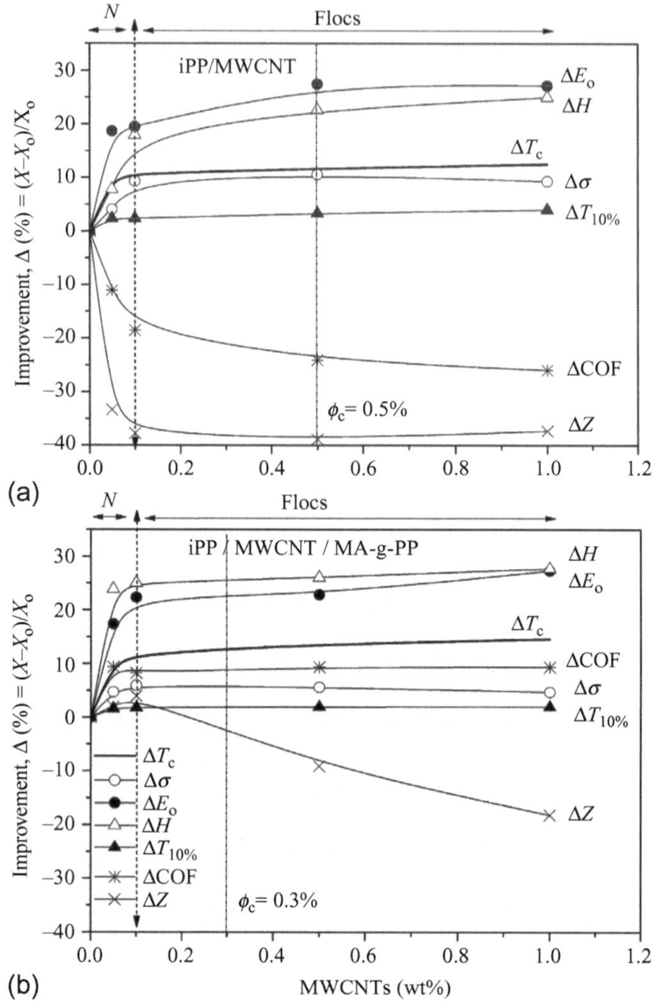

Figure 12 (a, b) Percentage of improvement of the structural, the thermal, and the mechanical characteristics versus nanotube contents. Arrows point the flocculation threshold (ϕ_c).

(ΔT_c); (ii) *thermal*—the TGA decomposition temperature ($\Delta T_{10\%}$); and (iii) *mechanical*-tensile strength ($\Delta\sigma$), tensile modulus (ΔE_o), nanohardness (ΔH), COF (ΔCOF), and scratch depth (ΔZ). The arrows point the rheological flocculation threshold of $\phi_c = 0.3$ and 0.5 wt% for the composites with and without MA-g-PP compatibilizer, respectively. The crystallization temperature (ΔT_c) versus nanotube contents is chosen as the main basis for comparison of the structure with the properties. Based on this comparison, two zones are determined: (a) the "nucleation zone" (*N*) at the lowest filler contents of 0.05-0.1 wt% MWCNTs, where the (ΔT_c) increases strongly by the nanotube addition due to the nucleation effects, and (b) the "flocculation zone" (Flocs), at relatively higher filler contents of 0.1-1.0 wt% MWCNTs, around the flocculation threshold (ϕ_c), where the (ΔT_c) curve tends to plateau values.

As seen from the correlations in Figure 12a and b, a strong enhancement of mechanical and thermal properties appears in the "nucleation zone" at very low nanotube contents, much below the rheological flocculation threshold ($\phi \ll \phi_c$). The crystallization temperature (ΔT_c) increases significantly at those nanotube contents. Moreover, the mechanical properties are improved with about 10–38%, while the thermal properties are improved slightly (~2-3%). The highest effect is observed for the increase of the modulus of elasticity (20%) and the hardness (15%), as well as the decrease of both COF (~18%) and scratch depth (38%) for the unmodified MWCNT/iPP composites. The same tendency was found for the composites with MA-g-PP compatibilizer, but some differences appear in the degree of properties improvement due to the additional effect of the MA-g-PP on the reduction of the crystal size. Obviously, the properties improvement in the "nucleation zone" is mostly governed by the nucleation effect of the MWCNTs, which appears within this concentration range (0.05-0.1 wt% MWCNTs) leading to significant changes in the crystal structure of iPP and a dramatic reduction of the spherulite size.

Further increase of MWCNT contents within the "flocculation zone" (0.1-1.0 wt% MWCNTs), around the flocculation threshold and below the percolation threshold ($\phi_c \leq \phi \leq \phi_p$), does not affect the crystallization temperature (ΔT_c) and this function tends to plateau values. The thermal and mechanical properties of the MWCNT/iPP composites are slightly affected by the addition of nanotubes. Most of the functions tend to plateau values with a slight peak around the (ϕ_c), which correlates with the formation of a flocculated structure of interconnected nanotubes incorporating matrix polymer. In this relatively broad concentration region (0.1-1.0 wt% MWCNTs), around the flocculation and below the percolation thresholds, the properties improvement is mostly dominated by the dispersion structure of MWCNTs in the matrix polymer and the nucleation is not so dominant. The formation of floccules probably not only produces a growth of the process zone in front of the crack tip from one side but also may act as mechanical failure concentrators, both resulting in plateau values of mechanical reinforcement between the two thresholds.

The addition of MA-g-PP compatibilizer has a dual effect; thus, the polydispersity of size distribution of iPP spherulites is widened and the dispersion of the nanotubes is improved. Obviously, these structural changes provoked by the addition of compatibilizer slightly decrease the tensile strength and tensile modulus but increase the hardness of the MWCNT/iPP composites. The friction properties (COF) and scratch depth of MWCNT/iPP composites are worsened by addition of compatibilizer, compared with unmodified composites.

The correlations in Figure 12 allow us to propose this rheology-structure-property relationship as a simple and novel design approach, which improves the knowledge on the joint reinforcement effects of MWCNTs and MA-g-PP compatibilizer on iPP at very low filler contents around the rheological flocculation threshold and far below the rheological percolation threshold. Initially, the structural changes of polypropylene, due to nucleation effect of nanotubes, are dominant for the properties improvement, but around the flocculation threshold (ϕ_c), the dispersion structure becomes important and governs the properties of composites. The MA-g-PP compatibilizer may add to the properties enhancement because of the effect of a better dispersion of nanotubes.

The reinforcement effect of carbon nanotubes on polypropylene at high nanotube contents around and above the rheological percolation threshold is discussed in our previous study for the unmodified MWCNT/iPP composites [20]. The strong reinforcement observed above $\phi_p \sim 2$ wt% MWCNTs was associated with the formation of a new kind of physical gel in the MWCNT/iPP composites, which originates from the network structure of interconnected nanotubes and the interactions between the nanotubes and the polymer chains.

5. CONCLUSIONS

Polypropylene nanocomposites containing 0.05-1 wt% MWCNTs, with and without MA compatibilizer (MA-g-PP), were studied, and results on rheology, structure, crystallization behavior, thermal stability, tensile, scratch, friction, and nanomechanical properties are characterized and analyzed.

Rheological study confirmed the positive effect of MA-g-PP for a better dispersion of MWCNTs in iPP, resulting in lower rheological flocculation thresholds $\phi_c \sim 0.3$ wt% of maleinated composites compared to $\phi_c \sim 0.5$ wt% for the composites without compatibilizer.

Small amount of MWCNTs below (ϕ_c) produced a strong nucleation effect on the iPP matrix and reduced significantly the size of iPP spherulites, making well-defined spherulites difficult to see, but the melting temperature and the glass transition of iPP are not influenced, while the addition of MA-g-PP compatibilizer obstructs only the crystallization process of iPP, strongly widening the polydispersity of the crystal size distribution.

Importantly, the thermal stability of iPP is strongly improved by the addition of very low nanotube contents. Thus, around the rheological flocculation threshold (ϕ_c), the MWCNTs enhanced the thermal stability of the iPP shifting the decomposition temperature ($T_{10\%}$) and the degradation peak (T_p) to higher values of ~12 and 48 °C, respectively. At nanotube contents below (ϕ_c), the maximum reinforcement of mechanical properties also appears, where the improvement is of ~11% for the tensile strength, ~28% for the tensile modulus, and 20% for the nanoindentations hardness. The friction and scratch behavior are also strongly improved by the addition of small amount of MWCNTs; thus, ~25% decrease of COF and ~48% decrease of scratch depth are observed at (ϕ_c). However, the elongation at break is generally decreased with increasing the MWCNT contents, while the addition of compatibilizer 5% MA-g-PP slightly decreases their reinforcement effect on the tensile strength and modulus of iPP and increases the hardness, making the composites more brittle. Generally, this is associated with the decrease of the spherulite size and increase of the polydispersity of the crystal size distribution by the presence of MA-g-PP.

Rheology-structure-property relationship is established for the MWCNT/iPP composites as a novel and simple design approach for the production of nanocomposites with desired properties. Two zones are identified within this low region of nanotube contents, the "nucleation zone" (up to 0.1 wt% MWCNTs) and the "flocculation zone" (0.1-1.0 wt% MWCNTs), around the rheological flocculation threshold and below the rheological percolation thresholds. It is proposed that the significant improvement of properties in the "nucleation zone" is mostly governed by the nucleation effect of the MWCNTs producing significant changes in the crystal structure of iPP and a dramatic reduction of the spherulite size. The slight enhancement of mechanical and thermal properties within the "flocculation zone" is governed mostly by the formation of floccules of entangled nanotubes incorporating matrix iPP, which enables a more efficient load transfer from the matrix to the nanotubes.

The significant improvement of the thermal, tensile, friction, and nanomechanical properties at very low nanotube contents around rheological flocculation threshold (ϕ_c) allows us to propose the 0.1 wt% MWCNT/iPP composite with high potential for application as a low viscosity nanoreinforced matrix or composition for a variety of engineering applications.

ACKNOWLEDGMENTS

The study was supported by the FP7-280987 NanoXCT project and the cofinancing project DNC7FP 01/02, supported by NSF of Bulgaria. The support from the bilateral projects BAS, Bulgaria/CNR, Italy, and DNTC/India 01/10 is also acknowledged. The authors thank Prof. Geoffrey Mitchell from CfAM, University of Reading for the SEM study, and Dr. Clara Silvestre and Dr. Sossio Cimmino from the Institute for Polymers, Composites and Biomaterials, CNR, Pozzuoli, Italy, for the thermal analysis.

REFERENCES

[1] H. Ramanathan, H. Liu, L.C. Brinson, Functionalized SWNT/polymer nanocomposites for dramatic property improvement, J. Polym. Sci. B Polym. Phys. 43 (2005) 2269–2279.
[2] M. Meo, M. Rossi, Prediction of Young's modulus of single wall carbon nanotubes by molecular-mechanics based finite element modelling, Compos. Sci. Technol. 66 (2006) 1597–1605.
[3] M.J. Barlow, H.G. Karian (Ed.), Global trends for polypropylene, Ch.1, Handbook of Polypropylene and Polypropylene Composites, CRC Press, New York, NY, 2003.
[4] A. Eitan, K. Jiang, D. Dukes, R. Andrews, L.S. Schadler, Surface modification of multi-walled carbon nanotubes: toward the tailoring of the interface in polymer composites, Chem. Mater. 15 (2003) 3198–3201.
[5] W.K. Park, J.H. Kim, Effect of carbon nanotube pre-treatment on dispersion and electrical properties of melt mixed multi-walled carbon nanotubes/poly(methylmethacrylate) composites, Macromol. Res. 13 (2005) 206–211.
[6] Y.S. Song, J.R. Youn, Influence of dispersion state of carbon nanotubes on physical properties of epoxy nanocomposites, Carbon 43 (2005) 1378–1385.
[7] Y.S. Song, J.R. Youn, Modeling of effective elastic properties for polymer based carbon nanotube composites, Polymer 47 (2006) 1741–1748.
[8] Y. Wang, J. Wu, F. Wei, A treatment method to give separated multi-walled carbon nanotubes with high purity, high crystallization and a large aspect ratio, Carbon 41 (2003) 2939–2948.
[9] S.S. Wong, E. Joselevich, A.T. Wooley, C.L. Cheung, C.M. Lieber, Covalently functionalized nanotubes as nanometer-sized probes in chemistry and biology, Nature 394 (1998) 52–58.
[10] A. Kelarakis, K. Yoon, I. Sics, R.H. Somani, X. Chen, B.S. Hsiao, et al., Shear-induced orientation and structure development in isotactic polypropylene melt containing modified carbon nanofibers, J. Macromol. Sci. B 45 (2006) 247–261.
[11] G.-W. Lee, S. Jagannathan, H.G. Chae, M.L. Minus, S. Kumar, Carbon nanotube dispersion and exfoliation in polypropylene and structure and properties of the resulting composites, Polymer 49 (2008) 1831–1840.
[12] N. Ristolainen, U. Vainio, S. Paavola, M. Torkkeli, R. Serimaa, J. Seppala, Polypropylene/organoclay nanocomposites compatibilized with hydroxyl-functional polypropylenes, J. Polym. Sci. B Polym. Phys. 43 (2005) 1892–1903.

[13] X. Zhou, X. Xie, F. Zeng, R.K.-Y. Li, Y.-W. Mai, Properties of polypropylene/carbon nanotube composites compatibilized by maleic anhydride grafted SEBS, Key Eng. Mater. 312 (2006) 223–228.
[14] Z. Zhou, S. Wang, Y. Zhang, Y.X. Zhang, Effect of different carbon fillers on the properties of PP composites: comparison of carbon black with multiwalled carbon nanotubes, J. Appl. Polym. Sci. 102 (2006) 4823–4830.
[15] R. Kotsilkova, Thermosetting Nanocomposites for Engineering Application, Rapra Smiths Group, UK, 2007.
[16] T.J. Pinnavaia, G.W. Beall, Polymer–Clay Nanocomposites, Wiley, New York, 2000 (Chapter 15).
[17] R. Wagener, T. Reisinger, A rheological method to compare the degree of exfoliation of nanocomposites, Polymer 44 (24) (2003) 7513–7518.
[18] M. Gang, B.K. Satapathy, M. Thunga, R. Weidisch, P. Potschke, D. Jehnichen, Structural interpretations of deformation and fracture behaviour of polypropylene/multi-walled carbon nanotube composites, Acta Mater. 56 (2008) 2247–2261.
[19] R. Kotsilkova, E. Ivanov, E. Krusteva, C. Silvestre, S. Cimmino, D. Duraccio, Isotactic polypropylene composites reinforced with multiwall carbon nanotubes. Part 2. Enhancement in thermal and mechanical properties related to the structure, J. Appl. Polym. Sci. 115 (2010) 3576–3585.
[20] R. Kotsilkova, E. Ivanov, E. Krusteva, C. Silvestre, S. Cimmino, D. Duraccio, Evolution of rheology, structure and properties around the rheological flocculation and percolation thresholds in polymer nanocomposites, in: C. Silvestre, S. Cimmino (Eds.), Ecosustainable Polymer Nanomaterials for Food Packaging, CRC Press, Taylor & Francis Group LLC, NW, 2013.
[21] M.K. Seo, S.J. Park, Electrical resistivity and rheological behaviors of carbon nanotubes-filled polypropylene composites, Chem. Phys. Lett. 395 (2004) 44–48.
[22] Y. Xiao, X. Zhang, W. Cao, K. Wang, H. Tan, Q. Zhang, et al., Dispersion and mechanical properties of polypropylene/multiwall carbon nanotubes composites obtained via dynamic packing injection molding, J. Appl. Polym. Sci. 104 (2007) 1880–1886.
[23] K.Q. Xiao, L.C. Zhang, I. Zarud, Mechanical and rheological properties of carbon nanotube-reinforced polyethylene composites, Compos. Sci. Technol. 67 (2007) 177–182.
[24] Q.H. Zhang, D.R. Lippits, S. Rastogi, Dispersion and rheological aspects of SWNTs in ultrahigh molecular weight polyethylene, Macromolecules 39 (2006) 658–666.
[25] Q.H. Zhang, S. Rastogi, D.J. Chen, Low percolation threshold in single-walled carbon nanotube/high density polyethylene composites prepared by melt processing technique, Carbon 44 (2006) 778–785.
[26] Y.Y. Huang, E.M. Terentjev, Dispersion and rheology of carbon nanotubes in polymers, Int. J. Mater. Form. 1 (2008) 63–74.
[27] A. Kanapitsas, P. Pissis, R. Kotsilkova, Dielectric studies of molecular mobility and phase morphology in polymer-layered silicate nanocomposites, J. Non-Cryst. Solids 305 (2002) 204–211.
[28] P.A. Cirkel, T. Okada, A comparison of mechanical and electrical percolation during the gelling of Nafion solutions, Macromolecules 33 (2000) 4921–4925.
[29] E. Assouline, A. Lustiger, A.H. Barber, C.A. Cooper, E. Klein, E. Wachtel, et al., Nucleation ability of multiwalled carbon nanotubes in polypropylene composites, J. Polym. Sci. B Polym. Phys. 41 (2003) 520–527.
[30] A.R. Bhattacharyya, T.V. Sreekumar, T. Liu, S. Kumar, L.M. Ericson, R.H. Hauge, et al., Crystallization and orientation studies in polypropylene/single wall carbon nanotube composite, Polymer 44 (2003) 2373–2377.
[31] B.P. Grady, F. Pompeo, R.L. Shambaugh, D.E. Resasco, Nucleation of polypropylene crystallization by single-walled carbon nanotubes, J. Phys. Chem. B 106 (2002) 5852–5858.

[32] W. Leelapornpisit, M. Ton-That, F. Perrin-Sarazin, K.C. Cole, J. Denault, B. Simard, Effect of carbon nanotubes on the crystallization and properties of polypropylene, J. Polym. Sci. B Polym. Phys. 43 (2005) 2445–2453.
[33] M.K. Seo, J.R. Lee, S.J. Park, Crystallization kinetics and interfacial behaviors of polypropylene compo-sites reinforced with multi-walled carbon nanotubes, Mater. Sci. Eng. A 404 (2005) 79–84.
[34] L. Valentini, J. Biagiotti, J.M. Kenny, S. Santucci, Morphological characterization of single-wall carbon nanotube/polypropylene composites, Compos. Sci. Technol. 63 (2003) 1149–1153.
[35] M.M. Hasan, Y. Zhou, S. Jeelani, Thermal and tensile properties of aligned carbon nanofiber reinforced polypropylene, Mater. Lett. 61 (2007) 1134–1136.
[36] X. Jiang, Y. Bin, N. Kikyotani, M. Matsuo, Thermal, electrical and mechanical properties of ultra-high molecular weight polypropylene and carbon filler composites, Polym. J. 38 (2006) 419–431.
[37] M.V. Jose, D. Dean, J. Tyner, G. Price, G. Nyairo, Polypropylene/carbon nanotube nanocomposite fibers: process-morphology-property relationships, J. Appl. Polym. Sci. 103 (2007) 3844–3850.
[38] S.H. Lee, E.N.R. Cho, S.H. Jeon, J.R. Youn, Rheological and electrical properties of polypropylene composites containing functionalized multi-walled carbon nanotubes and compatibilizers, Carbon 45 (2007) 2810–2822.
[39] M.A. Lopez-Manchado, L. Valentini, J. Biagiotti, J.M. Kenny, Thermal and mechanical properties of single-walled carbon nanotubes-polypropylene composites prepared by melt processing, Carbon 43 (2005) 1499–1505.
[40] S.C. Tjong, G.D. Liang, S.P. Bao, Electrical behavior of polypropylene/multiwalled carbon nanotube nanocomposites with low percolation threshold, Scr. Mater. 57 (2007) 461–464.
[41] E.M. Moore, D.L. Ortiz, V.T. Marle, R.L. Shambaugh, B.P. Grady, Enhancing the strength of polypropylene fibers with carbon nanotubes, J. Appl. Polym. Sci. 93 (2004) 2926–2933.
[42] H. Xia, Q. Wang, K. Li, G.H. Hu, Preparation of polypropylene/carbon nanotube composite powder with a solid-state mechanochemical pulverization process, J. Appl. Polym. Sci. 93 (2004) 378–386.
[43] J. Yang, Y. Lin, J. Wang, M. Lai, J. Li, J. Liu, et al., Morphology, thermal stability, and dynamic mechanical properties of atactic polypropylene/carbon nanotube composites, J. Appl. Polym. Sci. 98 (2005) 1087–1091.
[44] C. Xiang, H.-J. Sue, J. Chu, Scratch resistance and material property relationship in polymers, ANTEC 99 (1999) 3463–3469.
[45] W.C. Oliver, G.M. Pharr, An improved technique for determining hardness and elastic modulus using load and displacement sensing indentation experiments, J. Mater. Res. 7 (1992) 1564–1583.
[46] W.C. Oliver, G.M. Pharr, Measurement of hardness and elastic modulus by instrumented indentation: advances in understanding and refinements to methodology, J. Mater. Res. 19 (2004) 3–20.
[47] R. Krishnamoorti, A.S. Silva, Rheological properties of polymer-layered silicate nanocomposites, in: T.J. Pinnavaia, G.W. Beall (Eds.), Polymer-Clay Nanocomposites, John Wiley & Sons, NY, 2000, pp. 315–343.
[48] S. Abbasi, A. Derdouri, P. Carreau, Properties of microinjection molding of polymer multiwalled carbon nanotube conducting composites, Polym. Eng. Sci. 51 (2011) 992–1003.
[49] L. Chunyu, T.W. Chou, Heat capacity of multi-walled carbon nanotubes by molecular structural mechanics modeling technique, Mater. Sci. Eng. A 409 (2005) 140–144.
[50] Z. Sheng, A.R. Horrocks, A review of flame retardant polypropylene fibres, J. Polym. Sci. 28 (2003) 1517–1538.

CHAPTER 18

Effect of Nano-Additives (Al_2O_3 and NaF) on the Performance of Ceramic Coatings Formed by Microarc Oxidation on Magnesium Alloys

Abdel Salam Hamdy Makhlouf[1], Hanaa Soliman[2]
[1]Department of Manufacturing Engineering, College of Engineering and Computer Science, University of Texas Pan-American, Edinburg, Texas, USA
[2]Central Metallurgical Research and Development Institute, CMRDI, Cairo, Egypt

Chapter Contents

1. INTRODUCTION

Magnesium and its alloys possess a series of inherent superior properties, including light weight, favorable weight-to-strength ratio, strong vibrational load resistance, and efficient shielding against electromagnetic

Handbook of Nanoceramic and Nanocomposite Coatings and Materials
http://dx.doi.org/10.1016/B978-0-12-799947-0.00018-3

interference [1]. In the framework of the reduction of energy consumption, especially in the transportation industry, the lightening of the structure is necessary. These distinguishing properties are highly sought after and useful in the aerospace, aircraft, automobile, and electronic industries. However, the high chemical activity and poor corrosion resistance of magnesium and its alloys limit their application. Therefore, several surface treatment schemes have been proposed to improve the surface properties and functionality of magnesium alloys including chemical conversion [2–11], electroplating [12–16], organic coatings [17–20], physical vapor deposition [21–23], and anodic oxidization [24–26]. Anodization is one of the most effective and economical processes, because the anodic film formed has good adhesion, superior corrosion resistance, and high hardness [27–29].

Microarc oxidation (MAO) has been used for the surface modification of many metals and alloys including magnesium alloys [30–33]. The resistance of the treated samples to corrosion is not systematically improved by a thicker anodized layer; actually, the composition of the protective coatings is the key factor [34]. Yagi et al. investigated the self-repairing properties of some compounds on the protective coatings [35].

MAO was applied for the coating of Mg alloys in the presence of Na_2SiO_3 solution [36,37]. Results showed a significant improvement in the corrosion resistance after the addition of silicate. Similar results were confirmed by other researchers [38–40] where the coatings doped with fluoride or alumina showed marked improvement in the corrosion resistance and microhardness. Therefore, particular attention has been paid to the role of the additives on the electrochemical behavior of the interface during the anodizing process and the further reinforcement of the grown layer.

This chapter discusses a modified MAO process used for the formation of corrosion-resistant ceramic MgO layers over newly developed magnesium alloys. MAO coating proved to provide outstanding corrosion resistance for some magnesium alloys but has no significant effect on rare-earth-containing magnesium alloys such as AMS4429 due to the formation of multi-oxides at the surface that resulted in originating microelectrochemical cells and enhancing the galvanic corrosion. Herein, the MAO process has been modified using two kinds of nano-additives (Al_2O_3 and NaF) doped in the alkaline electrolyte system in order to improve the performance of MAO ceramic coatings formed on AMS4429. The formation of MAO coatings by anodization on the surface of rare-earth-containing magnesium alloys such as AMS4429 in an environmentally friendly electrolyte was investigated. The morphology and chemical composition of the ceramic anodic film formed were analyzed by field electron emission scanning electron microscopy (FESEM), energy-dispersive X-ray spectroscopy (EDS), and X-ray diffraction (XRD). The effects of the nano-additives on the structure and corrosion resistance of MAO ceramic coatings were discussed. The effectiveness of MAO as anticorrosion protective coatings on magnesium alloys was evaluated in aqueous 3.5% NaCl solution by means of electrochemical impedance spectroscopy (EIS) and potentiodynamic polarization techniques.

2. EXPERIMENTAL

2.1 Materials

Rare-earth-containing magnesium alloy AMS4429, provided by Magnesium Elektron, the United Kingdom, with chemical composition shown in Table 1, was employed in this study. Specimens of size 20 × 20 × 3 mm were grounded successfully with 600, 800, and 1000 grit emery sheets and then washed with acetone and dried in hot air prior to MAO treatment.

2.2 Preparation of MAO Coating

MAO treatment was carried out using AC power supply source of 200 V and 2 A capacity at room temperature. The specimens were MAO coated in a two-electrode cell with distance of 3 cm from the platinum electrode to the silicate electrolyte for 2 min. The electrolyte solution is composed of 120 g/l NaOH and 80 g/l $Na_2SiO_3 \cdot 9H_2O$ in deionized water [41]. NaF and Al_2O_3 nano-additives with an average size of about 40 nm were dispersed evenly in the solution with a concentration of 10 g/l.

2.3 Coating Characterization

2.3.1 Phase Analysis

The phases of the coatings were determined by XRD. XRD analysis was performed using a commercial X-ray diffractometer (PANalytical X'Pert Pro instrument, monochromatic Cu target radiation, 45 kV-40 mA-λ = 1.54 Å, the Netherlands). The scanning range of the diffraction angle (2γ) was set between 301 and 801 with steps of 0.021 and a step time of 40 s.

Table 1 The chemical composition of AMS4429

Alloy	Zn	Cu	Fe	Ni	Zr	Nd	Ag	Gd	Other RE elements
AMS4429	0.20-0.50	0.01	0.01	0.0020	0.4-1.0	2.6-3.1	0.05	1.0-1.7	0.4

2.3.2 Microstructure Examination

The microstructural examination was carried using the field emission scanning electron microscopy (FESEM: Quanta, FEG 250, Japan), operated at a voltage of 20 kV and a probe current of 100 pA. The filament current was 2.70 A and the detector used was a secondary electron detector. The elemental composition of the coatings was analyzed using EDS (Oxford Instruments, Quanta, Japan), attached to the scanning electron microscope mentioned above. Further, the elemental mapping was studied by EDS. To avoid surface charging, a 50-80 nm gold coating was deposited on the MAO samples by the arc deposition technique prior to insertion in the sample chamber for scanning electron microscopy.

2.3.3 Electrochemical Corrosion Behavior

2.3.3.1 Potentiodynamic Polarization

DC polarization tests were carried out using an Autolab PGSTAT302N potentiostat (Eco Chemie, the Netherlands).The cell used was a conventional three-electrode setup with the coated magnesium samples utilized as the working electrode. A platinum rod electrode was used as the counter electrode and KCl-saturated calomel electrode (SCE) as the reference electrode. The tests were performed in air at room temperature in a 3.5 wt% NaCl electrolytic solution. The edges of the samples were sealed with acrylic resin. The exposed area within the testing solution was about 1 cm^2 and the samples were immersed in the corrosive media for 15 min before the polarization test. The anodic and cathodic polarization curves were recorded using a sweep potential test in the range 700 mV with respect to the open circuit potential (OCP) with a scanning rate of 1 mV/s.

2.3.3.2 Electrochemical Impedance Spectroscopy

EIS measurements were performed using Autolab PGSTAT302N potentiostat with scan frequency ranged from 100 kHz to 10 MHz, and the perturbation amplitude was 10 mV. The tests were performed in air at room temperature in a 3.5 wt% NaCl electrolytic solution.

3. RESULTS AND DISCUSSION

The surface morphology and electrochemical characteristics of MAO-coated AMS4429 samples without additives, MAO coating with NaF, and MAO coating with Al_2O_3 nano-additives were investigated in views of the best coating condition that can offer the maximum localized corrosion resistance in NaCl solution.

3.1 XRD Analysis

Figure 1 depicts the XRD spectra of MAO-coated AMS4429 samples without additives, MAO coating with NaF, and MAO coating with Al_2O_3 nano-additives. It is shown that the coatings are mainly composed of the MgO phase (MgO is the dominant crystal phase). The main phase of the ceramic coating is not alternated with the addition of nano-additives. The diffraction peaks of magnesium substrate are detected in the ceramic coatings, which indicate that the coating is thin and porous. It seems that the insertion of suspended Al_2O_3 or NaF nanoparticles in the MAO process enhances the formation of amorphous MAO coatings with stronger intensity of MgO peaks. The addition of suspended nanoparticles results in better arrangement of MgO particles to form compact coating instead of porous ones.

3.2 Visual Inspection After Exposure to 3.5% NaCl Solution

Macroscopic images of MAO-coated AMS4429 magnesium panels without additives immersed for a week in 3.5% NaCl solution showed severe attack due to different localized corrosion forms, namely, galvanic, pitting, and crevice (Figure 2). Interestingly, MAO-coated AMS4429 panels with Al_2O_3 or NaF nano-additives showed significant improvement in the localized

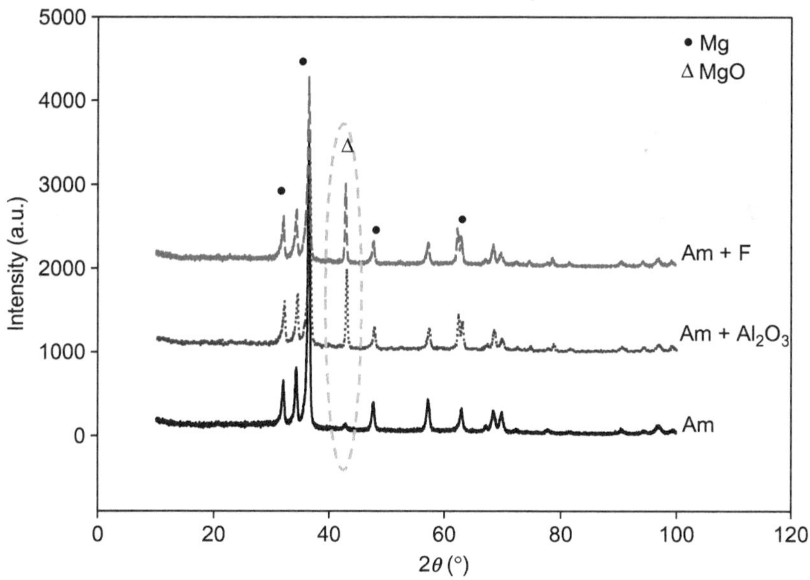

Figure 1 XRD spectra of MAO-coated AMS4429 samples without additives, with F or Al_2O_3 nano-additives.

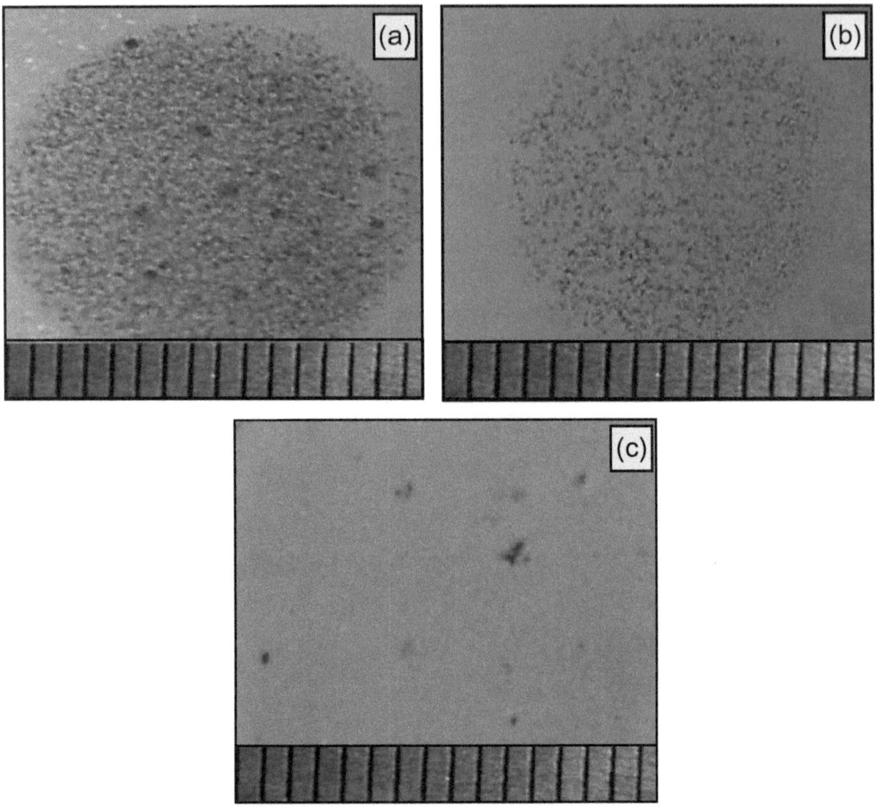

Figure 2 Visual inspection of (a) MAO-coated AMS4429 samples without additives, (b) MAO coating with F, and (c) MAO coating with Al_2O_3 nano-additives.

corrosion resistance. The number of pitting zones and crevice areas can be sorted in the following order: no additive > aluminate > fluoride additive. The pit sizes of MAO-coated samples with fluoride are smaller than aluminate and no-additive panels. Thus, the composition of the electrolytic bath seems to have a marked effect on the corrosion protection of AMS4429.

3.3 FESEM Morphology

Figure 3 shows the surface morphology of MAO ceramic coating formed on AMS4429 samples before and after corrosion. Before corrosion, MAO-coated samples without additives showed a coarser porous structure (micrometer size) than MAO-coated samples with fluoride or alumina nano-additives. In the presence of nano-additives, MAO coatings showed a more homogeneous structure with smaller pores on the surface, which

(a)
(b)
(c)

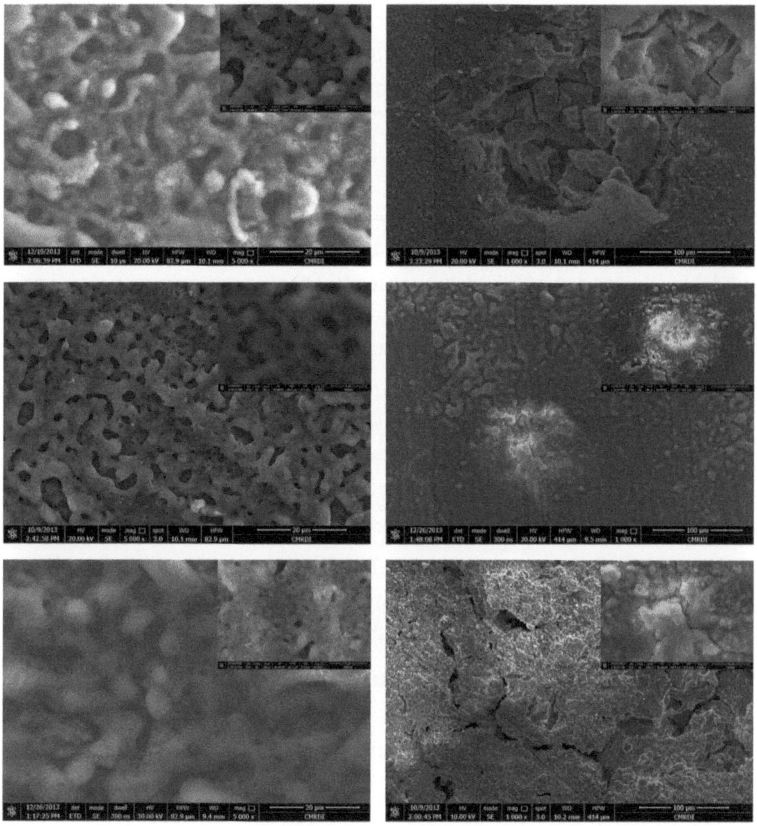

Figure 3 Surface morphology of AMS4429 sample anodized in 120 g/l NaOH and 80 g/l $Na_2SiO_3{\cdot}9H_2O$, (a) without additives, (b) MAO coating with F, and (c) MAO coating with Al_2O_3 nano-additives before corrosion (left) and after corrosion (right).

indicates that doping of nano-additives could prevent the aggressive gas evolution during the coating synthesis and enhance the formation of more homogenous coatings with very fine pores [40].

The FESEM micrographs of additives containing coats after corrosion test are illustrated in Figure 3. Fluoride additive results in the formation of some microcracks only, while aluminate decreases the pit size greatly and results in compact corrosion products. The compact layer formed from the corrosion products plays an important role in controlling the diffusion of the corrosive ions to the substrate. These results indicate that the corrosion resistance of MAO coatings with nano-additives can be deeply varied with the microstructure of the deposited particles. Thus, the cracked fluoride-anodized coating strongly explains the higher corrosion protection since it inhibits the corrosive ions from entering the MAO layer. Those fluoride-containing electrolytes produce MAO coatings with further grain refinement leading to a successful protective MAO coating.

3.4 EDS Analysis

The elemental analysis of MAO-coated AMS4429 panels with fluoride or alumina nano-additives before and after corrosion test is shown in Figure 4. The MAO-coated AMS4429 samples before corrosion test present an obvious incorporation of Al and F elements in the formed layers with increase in oxide concentration compared with no-additive MAO-coated samples. After corrosion test, the incorporated Al and F nanoparticles disappeared with lower Cl content compared with no-additive samples. It is known that the presence of Cl indicates the occurrence of corrosion, which confirms the visual inspection and microscopic examination data.

MAO-coated samples without additives and that containing nano-alumina revealed the presence of Nd element after corrosion. The detection of Nd and Gd (which are alloying elements of the alloy under investigation) seems to be due to the occurrence of deeper pits into the matrix of AMS4429 alloy. On the other hand, the F-incorporated MAO-coated sample does not reveal any Nd and Gd and presents higher O content before corrosion and lower Cl content after corrosion, indicating a higher corrosion resistance.

3.5 Potentiodynamic Measurements

Figure 5 shows the potentiodynamic curves of the MAO-coated AMS4429 alloy panels in 3.5% NaCl solution. The coating from fluoride-containing electrolytes exhibits larger corrosion resistance than that from aluminate and no-additive. Regarding the anodic side of the polarization curves, the current density of the bare alloy shows a quick increase above the corrosion potential, characteristic of an active state with dissolution of the alloy in oxidizing conditions. On the other hand, all MAO-coated samples exhibit a passive state with a reduced constant current density. The value of the passive current density seems to be independent on the composition of the coating bath, but the length of the passivation plateau is extended in the case of the fluoride-containing electrolyte.

Nevertheless, the no-additive MAO-coated sample shows no passivation plateau; the passivation behavior has been greatly improved for the sample anodized in fluoride-containing electrolytes, as an indication of a larger passivation plateau in Figure 5. The presence of fluoride as an additive in the

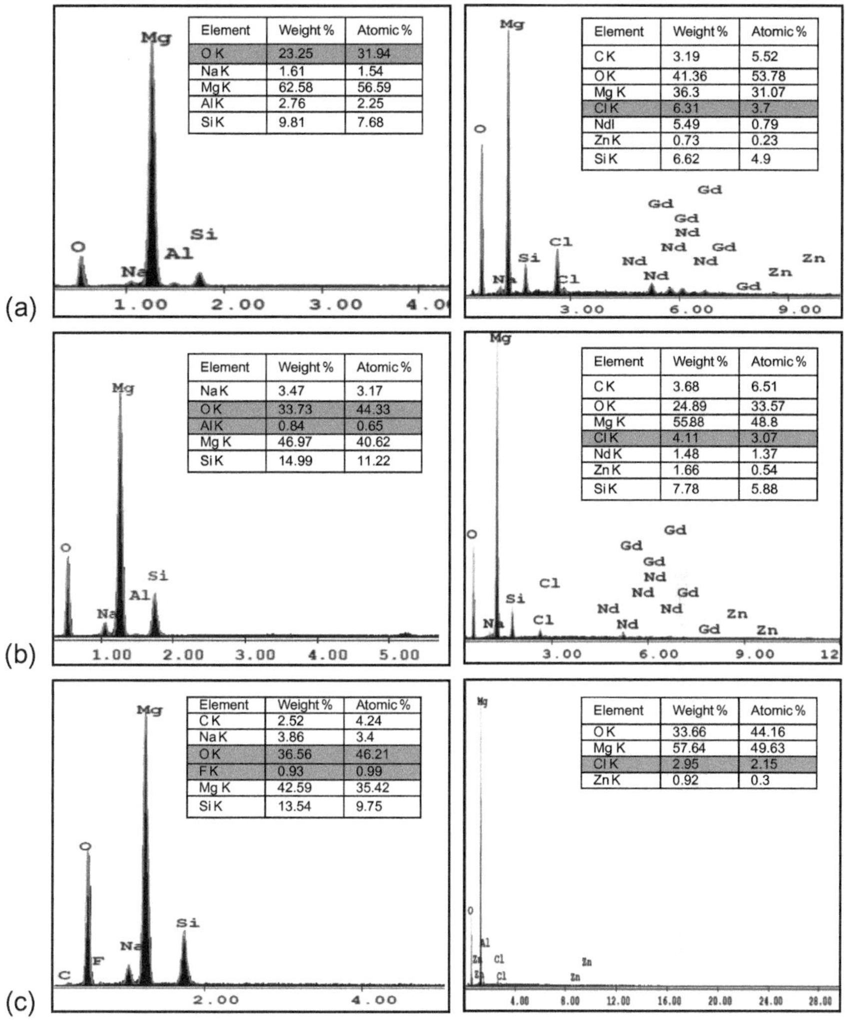

Figure 4 EDS analysis of MAO-coated AMS4429 samples with (a) no additives, (b) Al_2O_3, and (c) NaF nano-additive (left) before and (right) after corrosion testing.

anodized layer clearly enhances the corrosion resistance. Some researchers reported that fluoride-containing corrosion products could play the role of self-healing agent of porosities in corrosive conditions [34].

3.6 Electrochemical Impedance Spectroscopy

The corrosion behavior of the MgO ceramic coatings formed by the MAO process with and without alumina and fluoride nano-additives on AMS4429 alloy was investigated using EIS. Nyquist diagram (Figure 6) showed that

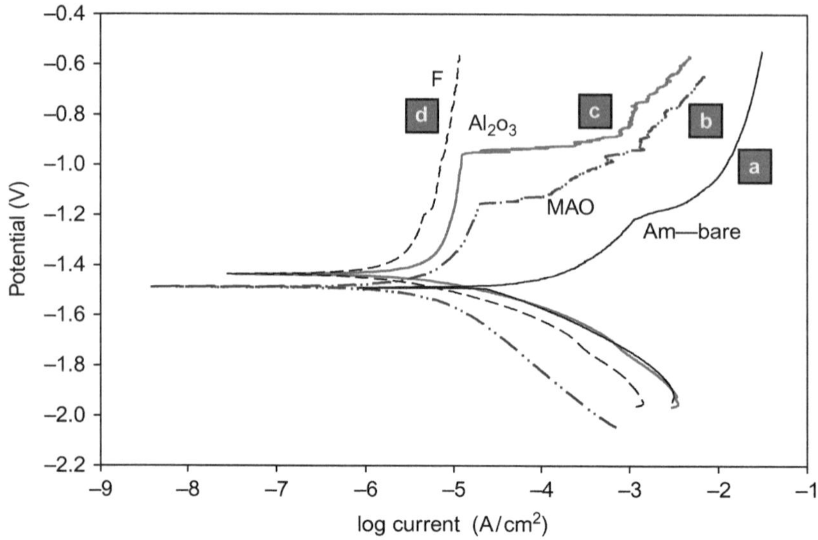

Figure 5 Potentiodynamic polarization curves of (a) bare AMS4429, (b) MAO coated without additives, (c) MAO coated with Al_2O_3, and (d) MAO coated with NaF nano-additive after immersion in 3.5% NaCl solution.

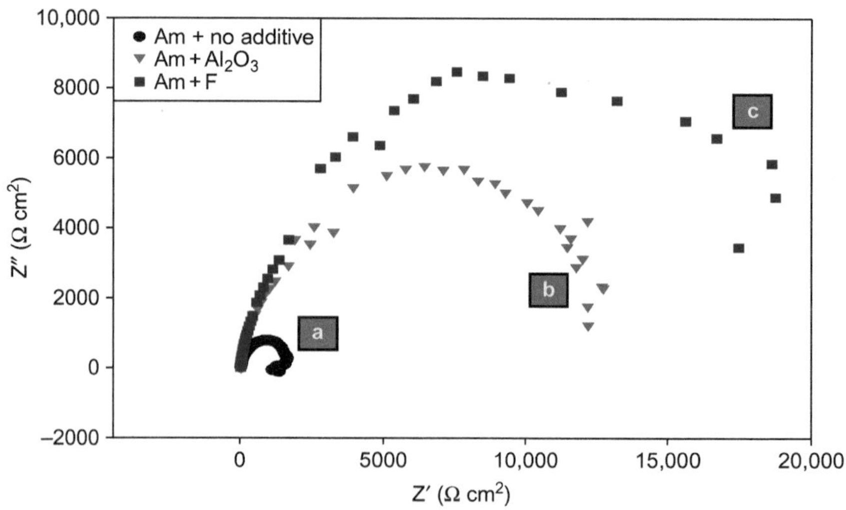

Figure 6 Nyquist plots of MAO-coated AMS4429 samples with (a) no additives, (b) Al_2O_3, and (c) NaF nano-additive after immersion in 3.5% NaCl solution at 25 °C.

MAO-coated AMS4429 alloy with fluoride additive has the highest polarization resistance compared with MAO coated without additives and with alumina nanoparticles. The polarization resistances of MAO-coated AMS4429 were measured to be $1.62 \times 10^3\ \Omega\ cm^2$, $12.8 \times 10^3\ \Omega\ cm^3$, and $18 \times 10^3\ \Omega\ cm^2$ for MAO coating without additives, MAO coating with alumina nanoparticles, and MAO coating with fluoride, respectively. Interestingly, these data point to the importance of nano-additives to the MAO process as a promising coating approach for improving the corrosion resistance of rare-earth-containing magnesium alloys.

4. CONCLUSIONS

Novel ceramic coatings were formed on rare-earth-containing magnesium alloy AMS4429 substrates using the MAO process in alkaline solution in the presence of nano-additive NaF or Al_2O_3. The coating composition is mainly a crystal phase of the MgO. The doping of the additives almost has no effect on the crystal phase of the ceramic coating. The additives alter the topography of the coatings. The doping of NaF results in MAO coatings with finer uniform nodules than that of Al_2O_3 nanoparticles, leading to a successful protective MAO coating. The corrosion resistance of the MAO coatings improves significantly by adding the nano-additives. Doping of MAO coating with fluoride provides superior corrosion resistance to the magnesium substrate better than aluminate nanoparticles.

REFERENCES

[1] J.E. Gray, B. Luan, Protective coatings on magnesium and its alloys, J. Alloys Compd. 336 (2002) 88–113.
[2] A.S. Hamdy Makhlouf (Ed.), High Performance Coatings for Automotive and Aerospace Industries, Nova Science Publishers, New York, ISBN: 978-1-60876-579-9, 2010, 415 pp.
[3] A.S. Hamdy Makhlouf, I. Tiginyanu (Eds.), Nanocoatings and Ultra Thin-Films: Technologies and Applications, Woodhead Publishing, Cambridge, UK, ISBN: 978-1-84569-812-6, 2011, 428 pp.
[4] A.S. Hamdy Makhlouf (Ed.), Handbook of Smart Coatings for Materials Protection, Woodhead Publishing, Cambridge, UK, ISBN: 9780857096807, 2014, 656 pp.
[5] A.S. Hamdy, M. Farahat, An attempt for designing economically attractive chrome-free conversion coatings for magnesium alloys, in: A.S. Hamdy (Ed.), High Performance Coatings for Automotive and Aerospace Industries, Nova Science Publishers, USA, ISBN: 978-1-60876-579-9, 2010, pp. 127–138.
[6] A.S. Hamdy, Intelligent stannate based coatings of self-healing functionality for magnesium alloys, in: A. Tiwari, J. Rawlins, L. Hihara (Eds.), Intelligent Coatings for

Corrosion Control, Elsevier Publication, USA, ISBN: 9780124114678 (Chapter 15), 2015, pp. 537–555.
[7] A.S. Hamdy, Enhancing corrosion resistance of magnesium alloy AZ91D in 3.5% NaCl solution by cerate conversion coatings, Anti-Corros. Methods Mater. 53 (6) (2006) 367–373.
[8] A.S. Hamdy, A novel approach in designing chrome-free chemical conversion coatings for automotive and aerospace materials, Eur. Coat. J. 86 (3) (2008) 43–50.
[9] A.S. Hamdy, Alkaline based surface modification prior to ceramic based cerate conversion coatings for magnesium AZ91D, Electrochem. Solid-State Lett. 10 (3) (2007) C21–C25.
[10] A.S. Hamdy, M. Farahat, Chrome-free zirconia-based protective coatings for magnesium alloys, Surf. Coat. Technol. 204 (2010) 2834–2840.
[11] A.S. Hamdy, I. Doench, H. Möhwald, Assessment of a one-step intelligent self-healing vanadia protective coatings for magnesium alloys in corrosive media, Electrochim. Acta 56 (2011) 2493.
[12] T. Saeki, Y. Seguchi, Y.H. Kourakata, Ni electroplating on AZ91D Mg alloy using alkaline citric acid bath, Electrochim. Acta 114 (2013) 827–831.
[13] T. Yin, R. Wu, Z. Leng, G. Du, X. Guo, M. Zhang, J. Zhang, The process of electroplating with Cu on the surface of Mg–Li alloy, Surf. Coat. Technol. 225 (2013) 119–125.
[14] Z. Zhang, G. Yu, Y. Ouyang, X. He, B. Hu, J. Zhang, Z. Wu, Studies on influence of zinc immersion and fluoride on nickel electroplating on magnesium alloy AZ91D, Appl. Surf. Sci. 255 (2009) 7773–7779.
[15] J. Tang, K. Azumi, Improvement of Al coating adhesive strength on the AZ91D magnesium alloy, Electrochim. Acta 56 (2011) 8776–8782.
[16] X.-B. Chen, M.A. Easton, N. Birbilis, H.-Y. Yang, T.B. Abbott, Corrosion-resistant electrochemical plating of magnesium (Mg) alloys, in: Corrosion Prevention of Magnesium Alloys, Woodhead Publishing, UK, pp. 315–346.
[17] R. Hu, S. Zhang, J. Bu, C. Lin, G. Song, Recent progress in corrosion protection of magnesium alloys by organic coatings, Prog. Org. Coat. 73 (2012) 129–141.
[18] V. Barranco, N. Carmona, J.C. Galván, M. Grobelny, L. Kwiatkowski, M.A. Villegas, Electrochemical study of tailored sol–gel thin films as pre-treatment prior to organic coating for AZ91 magnesium alloy, Prog. Org. Coat. 68 (2010) 347–355.
[19] A. Abdal-hay, M. Dewidar, J. Lim, J. Lim, Biocorrosion behavior and cell viability of adhesive polymer coated magnesium based alloys for medical implants, Ceram. Int. 40 (2014) 2237–2247.
[20] J. Hu, D. Huang, G. Song, X. Guo, The synergistic inhibition effect of organic silicate and inorganic Zn salt on corrosion of Mg-10Gd-3Y magnesium alloy, Corros. Sci. 53 (2011) 4093–4101.
[21] G. Garcés, M.C. Cristina, M. Torralba, P. Adeva, Texture of magnesium alloy films growth by physical vapour deposition (PVD), J. Alloys Compd. 309 (1–2) (2000) 229–238.
[22] M. Daroonparvar, M.A.M. Yajid, H.R. Bakhsheshi-Rad, N.M. Yusof, S. Izman, E. Hamzah, M.R. Abdul Kadir, Corrosion resistance investigation of nanostructured Si- and Si/TiO_2-coated Mg alloy in 3.5% NaCl solution, Vacuum 108 (2014) 61–65.
[23] G. Garcés, P. Pérez, P. Adeva, Oxidation behavior of a PVD-processed Mg–10.6Zr alloy, Scr. Mater. 45 (9) (2001) 1001–1007.
[24] S. Candan, M. Unal, E. Koc, Y. Turen, E. Candan, Effects of titanium addition on mechanical and corrosion behaviours of AZ91 magnesium alloy, J. Alloys Compd. 509 (2011) 1958–1963.

[25] R. Pinto, M.J. Carmezima, M.G.S. Ferreira, M.F. Montemor, Passive behavior of magnesium alloys (Mg-Zr) containing rare-earth elements in alkaline media, Electrochim. Acta 55 (2010) 2482–2489.
[26] M. Carboneras, L.S. Hernandez, J.A. del Valle, M.C. Garcia-Alonso, M.L. Escudero, Corrosion protection of different environmentally friendly coatings on powder metallurgy magnesium, J. Alloys Compd. 496 (2010) 442–448.
[27] H.Y. Hsiao, W.T. Tsai, Characterization of anodic films formed on AZ9lD magnesium alloy, Surf. Coat. Technol. 190 (2005) 299–308.
[28] H.Y. Hsiao, H.C. Tsung, W.T. Tsai, Anodization of AZ91D magnesium alloy in silicate-containing electrolytes, Surf. Coat. Technol. 199 (2005) 127–134.
[29] V. Birss, S. Xia, R. Yue, Richard G. Rateick Jr., Characterization of oxide films formed on Mg-based WE43 alloy using AC/DC anodization in silicate solutions. J. Electrochem. Soc. 151 (2004) B1–B10, http://dx.doi.org/10.1149/1.1629095.
[30] S. Durdu, A. Aytac, M. Usta, Characterization and corrosion behavior of ceramic coating on magnesium by micro-arc oxidation, J. Alloys Compd. 509 (2011) 8601–8606.
[31] I.J. Hwang, D.Y. Hwang, Y.G. Ko, D.H. Shin, Formation of black ceramic layer on aluminum alloy by plasma electrolytic oxidation in electrolyte containing Na_2WO_4, Surf. Coat. Technol. 206 (2012) 3360–3365.
[32] G. Hui-xia, M. Ying, W. Jing-song, W. Yu-shun, D. Hai-rong, H. Yuan, Corrosion behavior of micro-arc oxidation coating on AZ91D magnesium alloy in NaCl solutions with different concentrations, Trans. Nonferrous Met. Soc. China 22 (2012) 1786–1793.
[33] D. Veys-Renaux, E. Rocca, G. Henrion, Initial stages of AZ91 Mg alloy micro-arc anodizing: Growth mechanisms and effect on the corrosion resistance, Electrochem. Commun. 31 (2013) 42–45.
[34] D. Veys-Renaux, C.-E. Barchiche, E. Rocca, Corrosion behavior of AZ91 Mg alloy anodized by low-energy micro-arc oxidation: Effect of aluminates and silicates, Surf. Coat. Technol. 251 (2014) 232–238.
[35] S. Yagi, K. Kuwabara, Y. Fukuta, K. Kubota, E. Matsubara, Formation of self-repairing anodized film on ACM522 magnesium alloy by plasma electrolytic oxidation, Corros. Sci. 73 (2013) 188–199.
[36] P. Bala Srinivasan, C. Blawert, W. Dietzel, Mater. Sci. Eng. A 494 (2008) 401–406.
[37] L. Shia, Y. Xua, K. Lia, Z. Yaoa, S. Wu, Effect of additives on structure and corrosion resistance of ceramic coatings on Mg–Li alloy by micro-arc oxidation, Cur. Appl. Phys. 10 (2010) 719–723.
[38] Y. Wang, D. Wei, J. Yu, S. Di, Effects of Al_2O_3 nano-additive on performance of micro-arc oxidation coatings formed on AZ91D Mg alloy, J. Mater. Sci. Technol. 30 (2014) 984–990.
[39] B. Kazanski, A. Kossenko, M. Zinigrad, A. Lugovskoy, Fluoride ions as modifiers of the oxide layer produced by plasma electrolytic oxidation on AZ91D magnesium alloy, Appl. Surf. Sci. 287 (2013) 461–466.
[40] X. Li, B. Li Luan, A green and fast strategy for the scalable synthesis of Fe_2O_3/graphene with significantly enhanced Li-ion storage properties, Mater. Lett. 86 (2012) 88–91.
[41] W. Li, W. Li, L. Zhu, H. Liu, X. Wang, Non-sparking anodization process of AZ91D magnesium alloy under low AC voltage, Mater. Sci. Eng. B 178 (2013) 417–424.

CHAPTER 19

Chemical Composition, Structure, and Properties of the Surface of Titanium VT1-00 and Its Alloy VT16 after Induction Heat Treatment

Aleksandr Aleksandrovich Fomin, Igor Vladimirovich Rodionov
Saratov State Technical University, Saratov, Russia

Chapter Contents

1. INTRODUCTION

Biocompatible materials used in implantology are widely represented by various compact metallic [1,2] and nonmetallic [3] materials, porous blocks and granules [4–15], composites [16–20], coatings, and thin-layer membrane structures [21–24]. Different bioinert or biotolerable metal materials, for example, titanium and its alloys [1,2,25], chrome-nickel stainless steel [6,26], cobalt alloys, and alloys of zirconium and tantalum [27], often serve as constructive basis for intraosseous implants. Ceramic materials [16,17,22,28–32] such as calcium phosphate ceramics, strengthened with bioinert ceramics and carbon nanotubes, bioglasses [3,4,9–11,33–43], and

Handbook of Nanoceramic and Nanocomposite Coatings and Materials
http://dx.doi.org/10.1016/B978-0-12-799947-0.00019-5

polymers [1,8] are less frequently used for this purpose. A number of modern high-performance constructions of hip, knee, and other joints, osseofix, and dental implants are a metal structure with an interface biofunctional layer, a coating, or a thin film [1,44–47].

High functional loads and complex biomechanical interaction conditions of such implant systems with bone tissue are discussed in detail on the macro-, micro-, and nanoscale of structure [22,48–51]. Macroscale is important during installation with interference fit into the prepared bone bed. When installing an implant, its structure is subjected to dynamic axial load and in some cases (during fitting) to tangential load. In all these cases, the implant surface, including its elements, is subjected to intense friction and inevitable wear in the case of low physicomechanical and tribological characteristics [52]. These phenomena are caused by the presence of hard fragments in the cortical bone structure. There is evidence proving that hardness determined during micro- and nanoindentation of these fragments reaches 0.3-0.6 GPa, which in some cases exceeds the hardness of many commonly used porous coatings, for example, plasma sprayed calcium phosphate coatings and other types of coatings characterized by increased (over 50%) porosity and "openwork" morphology of the structure [53–56]. Hence, physicomechanical characteristics of the surface layer of metal implants, including biocompatible coatings, have increased requirements [57–62]. High values of strength (adhesion and cohesion) characteristics are achieved through the use of vacuum-condensing deposition methods [57,63–65]. However, the drawbacks of these types of coatings deposited in vacuum are low porosity, low roughness, poor morphology, and long duration. Various methods of metal oxidation are widely used, the most common of which are gas-thermal [66], microarc [67–70], steam-thermal, and anodizing [22,71–76]. Their main drawbacks are large power consumption and need for specially prepared liquid-phase or gaseous reaction oxidizing mixtures. All these technological and functional limitations of various methods for producing coatings force many implant manufacturers to abandon the use of more efficient bioactive materials. For this reason, the application of metal alloys for making implant substrates has become widely spread as they undergo only macro- and micro-texturing by mechanical processing, particularly abrasive blasting, or chemical etching [66,77]. Thus, nanoscale of interaction between the implant and bone tissue is considered according to self-assembled approach [64,78]. Some attempts to stimulate osseointegration on the nanoscale are implemented in the form of crystallization methods from simulated body fluid [16,31,53] or application of various functional biopolymers [79]. However, such micro- and nanoporous materials cannot be characterized by high strength properties due to their morphological and natural features. Strengthened porous nanostructured material or coating can be obtained by a combination of different techniques, for example, electrophysical treatment with pre- or postheating. As a result, it minimizes or eliminates undesirable phase-structural transformations, particularly amorphization, which causes accelerated resorption and further probable rejection [33,80,81]. Hence, the aim of this work is to study the influence of heat treatment with high-frequency (HF) eddy currents on the chemical composition, phase-structural state, physicomechanical properties, and biocompatibility of matrix oxide coatings (MOCs) and oxide-bioceramic coatings (OBCCs) on intraosseous titanium implants obtained by induction-heat treatment (IHT).

2. MATERIALS AND METHODS

Experimental metal samples are cylinders fabricated from commercially pure (CP) titanium VT1-00 and structural ($\alpha + \beta$)-titanium alloy VT16 (Ti-2.5Al-5Mo-5V) characterized by increased strength compared with conventional medical titanium alloy VT6 (Ti-4Al-6V). Their surface is subjected to machining and microtexturizing treatment comprising abrasive blast (sand-blast) treatment by corundum abrasive with dispersion of 200-300 μm. Etching is performed in aqueous solution of 1.5 M HF and 1.5 M HNO_3, and it is used to improve the microtexture of titanium alloy VT16. During etching, less stable phase β-Ti dissolves; thus, uniform rough morphology is formed. Obtained metal sample substrates are also subjected to ultrasonic cleaning in aqueous surfactants and alcohol. Then, the surface of samples prepared by IHT is oxidized in the air, and in case of obtaining of OBCCs, it is modified with colloidal nanoparticles of hydroxyapatite (HAp) combined with the final IHT of not less than 1 s (transition to stationary thermal treatment) and not more than 300 s. Using atomization, microdroplets of colloid solution are entered into the air stream; further, this mixture is directed onto the surface of the sample with an oxide matrix coating. Deposition length changes from 3 to 5 s. Thus, a thin layer of bioactive composition is formed on the sample surface. To produce thicker coatings, this procedure should be repeated several times. This method enables to produce quite strong and nanostructured coatings with the thickness of 0.5-5 μm. The studies determined the influence of IHT in the temperature range of 600-1200 °C on phase-structural and chemical changes occurring on the surface during the formation of coatings. Special attention is paid to the study of structure formation processes on the micro- and nanoscale, when physicomechanical and tribological properties of the resulting MOC and OBCC are defined. Double numbering was assigned to treatment regimes for experimental coating samples: The first figure corresponds to IHT temperature of metal substrate, while the second one indicates the duration of heat treatment process; for example, regime 600-120 corresponds to IHT temperature of 600 °C and exposure length of 120 s at a given temperature.

Phase-structural surface state of the samples after IHT is studied by X-ray diffraction (XRD) on Gemini/Xcalibur (CuKα-radiation, $\lambda = 1.541874$ Å, $25° < 2\Theta < 85°$). Morphology of the coatings is studied by optical and electron microscopic methods on the micro- and nanoscale to identify structure formation patterns. Morphological parameters of porous crystalline structure are defined, shape and average particle and pore size in particular. Scanning electron microscopy (SEM) and energy-dispersive X-ray (EDX) fluorescent analysis of chemical composition of the coating and surface layer of samples are performed on MIRA II LMU with INCA PentaFETx3 detector. Changes in oxygen concentration are measured in atomic fractions (at.%).

Physicomechanical properties of MOC and OBCC are evaluated by nanoindentation and scratch testing using a mechanical property tester NANOVEA Ergonomic Workstation. The basic determined quantities are hardness and modulus of elasticity, which are studied at selected load of 100 mN applied to the Berkovich indenter. Adhesion-cohesion and friction characteristics are studied by scratching with conical diamond indenter with an apex angle of 60°, tip curvature radius of 10 μm, and linear path length of 2-3 mm at a gradually increasing load from 0 to 350 mN. At the same time, points of cracking and peeling of the coating are observed, which are accompanied by sharp changes in strength, coefficient of friction, and indentation depth. Quantitative evaluation of scratch resistance and delamination is performed in accordance with the developed methodology [52].

Preliminary biocompatibility checking of studied coating samples is conducted under *in vitro* conditions (see Chapter 13).

3. RESULTS AND ANALYSIS

3.1 X-ray Analysis

On the surface of titanium having the original structure of α-Ti, oxide films and coatings are formed on the described IHT regimes. Phase composition of the surface is presented mainly by titania TiO_2, which is in the rutile phase, and titanium of various crystal structures (Figure 1a). More detailed analysis of XRD results for experimental samples of MOC and OBCC is given in the tabular form (Table 1).

In the surface layer of titanium samples, the main phases are initial phase α-Ti corresponding to CP titanium VT1-00, the main phase of oxide coating TiO_2 (rutile), and another phase β-Ti, the appearance of which is caused by rapid cooling of the titanium sample heated above 800 °C. In the low temperature range of IHT at 600 °C for all the considered interval of oxidation duration, only the initial phase α-Ti is clearly shown. The resulting oxide film is very thin (around 200-400 nm), which significantly limits the determining of crystal phase. Diffractograms show individual characteristic peaks for titanium monoxide TiO, which corresponds to the available theoretical and experimental ideas about titanium oxidation [82].

In the middle temperature range of IHT at 800 °C near-solid phase transformation α-Ti ↔ β-Ti and with limited supply of oxygen-containing medium in high temperature range of IHT at 1000-1200 °C, certain conditions appear facilitating the formation of a thin oxide sublayer TiO_2. Formed outer layer of MOC in high temperature range of IHT at 1200 °C is a fragile thick-layer coating of TiO_2. Phase composition of these samples is represented by three components: α-Ti, the amount of which decreases with the increasing duration of IHT, β-Ti, and TiO_2—rutile sublayer constituting 17-20%.

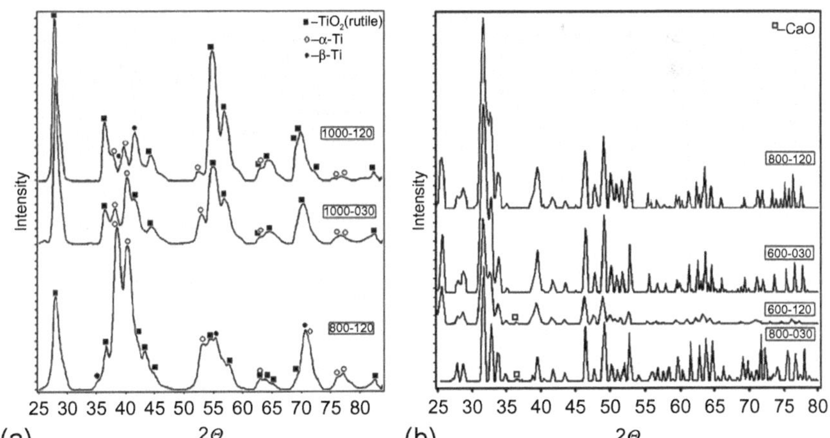

Figure 1 XRD of MOC (a) and thick-layer OBCC (b) obtained on titanium (IHT regimes are shown in rectangles).

Table 1 Phase composition of metal substrate of implants and MOC samples obtained by IHT on the surface of CP titanium VT1-00 and titanium alloy VT16

Sample	Technological regimes of IHT: Temperature (°C)	Technological regimes of IHT: Duration (s)	Phase composition (%): α-Ti	Phase composition (%): β-Ti	Phase composition (%): TiO_2
VT1-00, machining	–	–	100	0	0
800-001	800	1	100	0	0
800-030		30	100	0	0
800-120		120	36	41	23
800-300		300	44	35	21
1000-001	1000	1	22	0	78
1000-030		30	29	0	71
1000-120		120	17	0	83
1000-300		300	0	0	100
1200-001	1200	1	29	0	71
1200-001 (sublayer)		1	46	34	20
1200-300		300	0	0	100
1200-300 (sublayer)		300	28	55	17
VT16, machining	–	–	60	40	0
VT16, etching 60 s	–	–	44	56	0
VT16, etching 300 s	–	–	38	62	0
VT16, 800-120	800	120	30	0	70
VT16, 1000-030 (sublayer)	1000	30	94	0	6
VT16, 1200-030 (sublayer)	1200	30	46	0	54

A distinctive feature of phase transformations occurring on the surface of titanium alloy VT16 is reduction of the amount of α-Ti at all stages of sample preparation, including etching (Table 1). During etching (for about 5 min at room temperature), certain conditions appear, which facilitate the dissolution of α-Ti. This leads to an increase in proportion of β-Ti by 22%. During oxidation, the initial phase of β-Ti disappears almost completely and TiO_2 is formed. Oxidation is more intensive compared to CP titanium VT1-00, which is especially noticeable on IHT regime 800-120 (Table 1).

MOC formation mechanism in the considered IHT range for titanium VT1-00 and its structural alloy VT16 includes the following steps:

1. At the initial stage in the low temperature range of IHT (600 °C), slow enough transformation of α-Ti into the oxide phase occurs, the phase composition of which can be described by the general formula Ti_xO_y, according to the well-known data on oxidation of titanium and its alloys [82].

At sufficiently long IHT (300 s), Ti_xO_y transforms into a more stable phase of titania TiO_2. In this range of IHT, the most likely phase is anatase.

2. At the stage of accelerated oxidation on the middle-temperature regime of IHT (800 °C), most of the energy is consumed by the phase transformation of the initial phase α-Ti into β-Ti and its subsequent intense oxidation to the most stable (in terms of thermodynamics) phase of titania TiO_2 in the rutile phase. This phase transformation is possible in the metal substrate material of the samples only in the near-surface skin layer with the thickness of about 1 mm at a selected frequency current in the inductor. This layer has a temperature above the transition point "α-Ti → β-Ti" (at 882 °C), respectively, above the mean temperature (800 °C) of samples and has a value of about 900 °C. The difference in temperature between the "hot" skin layer and the core can reach 100 °C as it is well known in the theory of induction heating [83]. Thus, the appearance of high-temperature phase β-Ti can be explained. However, this situation occurs when test samples with a diameter exceeding 10 mm are used. These samples were selected for XRD analysis (described in Chapter 13, Figure 3b). When IHT temperature of 1000-1200 °C is used, this temperature difference becomes insignificant; that is, the mean temperature of the samples is aligned across the section due to heat transfer.

   VT16 alloy oxidation is much more intense because the alloy phase initially has a large proportion of β-Ti. In this connection, the amount of TiO_2 on the same regime of IHT on the surface of VT16 alloy is three times higher compared to the CP titanium VT1-00.

3. At the stage of intense oxidation on the high-temperature regime of IHT (1000 °C), β-Ti phase cannot be found because of its complete oxidation, which is accompanied by an increase in rutile fraction to 80-100%. The resultant TiO_2 coating on CP titanium VT1-00 is characterized by high adhesive strength and reasonable thickness (around 2-3 μm), so on IHT regime 1000-300, initial phase of α-Ti is about 0%.

   For VT16 alloy, this range of IHT leads to a sufficiently thick oxide layer, which is spontaneously separated from the surface of metal substrate. The remaining thin oxide sublayer corresponds to the rutile phase, but its proportion is less than 6% (Table 1).

4. IHT temperature increase to 1200 °C leads to extremely high, "catastrophic" oxidation rate [52,82]. Upon reaching the required temperature, the formed titania coating is separated and the sublayer formed

beneath is a porous rutile, the proportion of which is around 20%. Data analysis for the coatings also shows the presence of β-Ti phase, which is an intermediate substance for the growth of oxides.

For VT16 alloy, this range of IHT is characterized by intensive growth of sublayer, while the rutile concentration is almost two times higher than for the underlayer obtained on CP titanium. Formed sublayer thickness is characterized by high heterogeneity, which is associated with the separation of its fragments. Further increase in IHT duration in this temperature range leads to repeated separation of the outer coating, sublayer growth, and transformation of the latter into thick-layer coating.

Modification of obtained MOC samples with HAp nanoparticles can also be traced using XRD data. Colloidal modification from the solution and subsequent IHT are able to maintain the necessary phase of HAp (Figure 1b). In the presented diffractograms, all peaks correspond to HAp phase (not shown in Figure 1b); the emerging decomposition phase CaO is marked by square. Additionally, calcium oxide CaO phase is formed in the structure of OBCC, but its proportion is less than 3-5%. Thus, during the formation of sufficiently thick and continuous HAp layer, no underlying phases (oxide and metal) are found. The composition of OBCC samples should not necessarily have HAp phase, so it is advisable to obtain composite structure. In this case, MOC will be the matrix and filler—HAp nanoparticles and their proportion cannot be determined reliably by XRD analysis. Increased value of HAp fraction on the surface may indicate oversaturated oxide matrix, and excessive HAp phase will be separated in the process of brittle fracture when an implant is installed into the prepared bone bed. More detailed process of modifying and obtaining composite OBCC should be studied using SEM combined with EDX.

3.2 Morphology and Chemical Composition of Coatings

Sand blasting affects the characteristics of titanium surface, which is manifested in the formation of morphologically heterogeneous microrelief (Figure 2a). Typical microporous structure is formed on VT16 alloy by etching (Figure 2b).

Surface MOC structure is formed by intensive oxidation. Description of the crystal morphology of coatings reveals its several basic types. The shape of TiO_2 crystals depends mainly on IHT temperature, so at 600 °C, rounded elements of surface morphology are formed (Figure 2c). With the increase of temperature up to 800 °C and IHT duration, the shape of crystals changes. They elongate in one direction, while their size reaches 70 ± 10 nm (Figure 2d). When

Figure 2 Surface morphology of MOC: (a) microstructure of a titanium sample after sand blasting; (b) microstructure of a titanium alloy sample after etching; (c) low temperature range of IHT (600 °C) when exposed within 30 s; (d) middle temperature range of IHT (800 °C) when exposed within 120 s; (e) high temperature range of IHT (1000 °C) when exposed within 120 s (outer thick layer); (f) high temperature range of IHT (1000 °C) when exposed within 120 s (sublayer).

IHT duration exceeds 150-180 s, the coating has an increase in defects that contributes to the formation of crystals of split-plate shape. High temperature range of IHT (1000 and 1200 °C) is characterized by the formation of thick-layer MOC and thin sublayer of prismatic crystallites, which form porous structure with high strength characteristics (Figure 2e). The presence of pores in MOC structure contributes to local transfer of oxygen to the surface of titanium substrate, which leads to uneven growth of oxide sublayer (Figure 2f). Crystals of newly formed phase begin to grow, thereby separating the outer layer. Therefore, from a constructive point of view, during IHT in high temperature range, it is appropriate to study MOC formed by prismatic grains of sublayer.

At IHT temperature of 1000 °C and short exposure of about 1 s, the size of crystals of the oxide coating reaches 70 nm. Over the mentioned duration of IHT, intense crushing of crystals into fragments takes place. At IHT temperature of 1200 °C, rapid and spontaneous separation of the outer thick oxide layer occurs. Generalized dependency for crystal growth of rutile in the considered range of variation of IHT regimes can be represented in the graphic form (Figure 3a). A similar approach is used in the analysis of parameters for the porous structure of MOC (Figure 3b). Character of the curved surface dependency for mean pore size D_P is similar to that for grains D, but the range of maximum on high-temperature IHT regimes is less expressed. The increase in MOC pore size in the low and middle temperature range up to high temperature IHT range is characterized by 1.5-2-fold increase.

SEM results for surface samples with OBCC showed the presence of porous structure consisting of rutile matrix modified with HAp bioceramic nanoparticles (Figure 4).

Surface microstructure of titanium samples with OBCC is a relief of the initial metal substrate after preparation and oxidation (Figure 4a). Study on the nanoscale reveals fine structure represented by rounded grains, their agglomerates, and tiny pores (Figure 4e) [82]. The matrix of this composite structure is formed by an oxide matrix where relief elements (protrusions and open pores) are uniformly modified with a fine layer of HAp nanoparticles with an average size of about 40-60 nm. In the process of modification, the most important part is to ensure the conditions required to uniformly deposit a layer of nanoparticles; otherwise, micron-sized agglomerates in the shape of globules and separately arranged on the surface are formed (Figure 4a and b).

Chemical composition of OBCC includes mainly matrix components—titanium, oxygen, and natural impurities (aluminum) originally present in

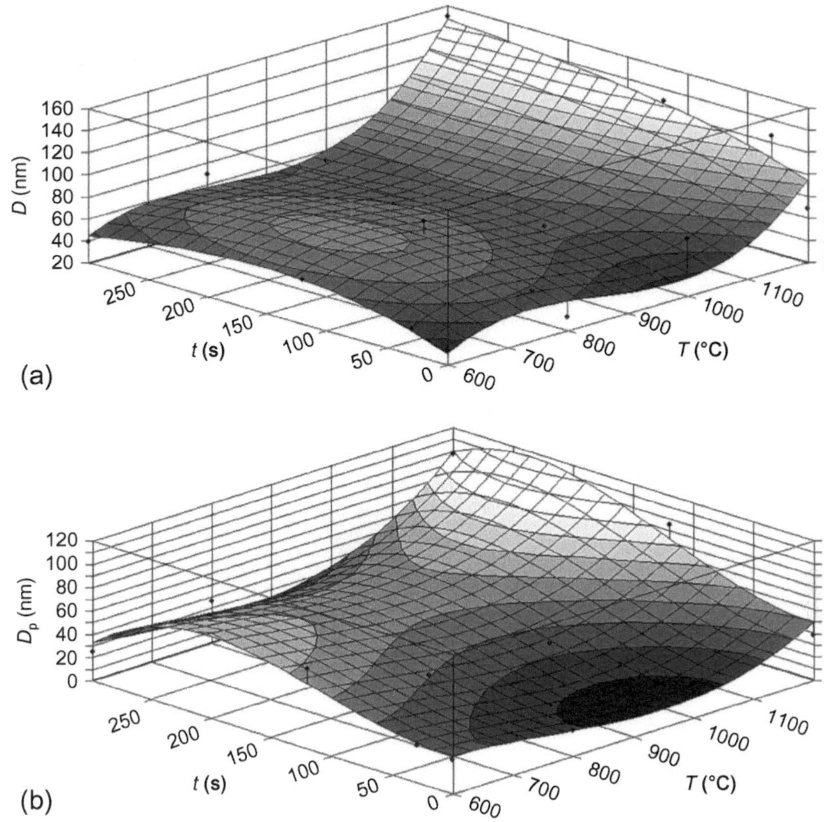

Figure 3 Summarized data of changes in the average grain size D of MOC rutile (a) and pores D_P (b) depending on IHT regimes.

CP titanium—and their proportion does not exceed 0.3 at.% (Table 2). Increased aluminum proportion of about 2 at.% can be explained by abrasive particles interspersed at early stages of metal substrate preparation, for example, during grinding or sand blasting with corundum abrasive. Characteristic ratio Ca/P for OBCC is lower than the value of 1.67 for HAp. This might be related to the small amount of deposited colloid and physical peculiarities of EDX method when the chemical composition is determined.

3.3 Physicomechanical Properties

Dependency for TiO_2 coating hardness is characterized by the highest values in high temperature range of IHT (Figure 5a). High hardness H and sufficient elasticity are ensured when oxide sublayer is formed in the MOC structure.

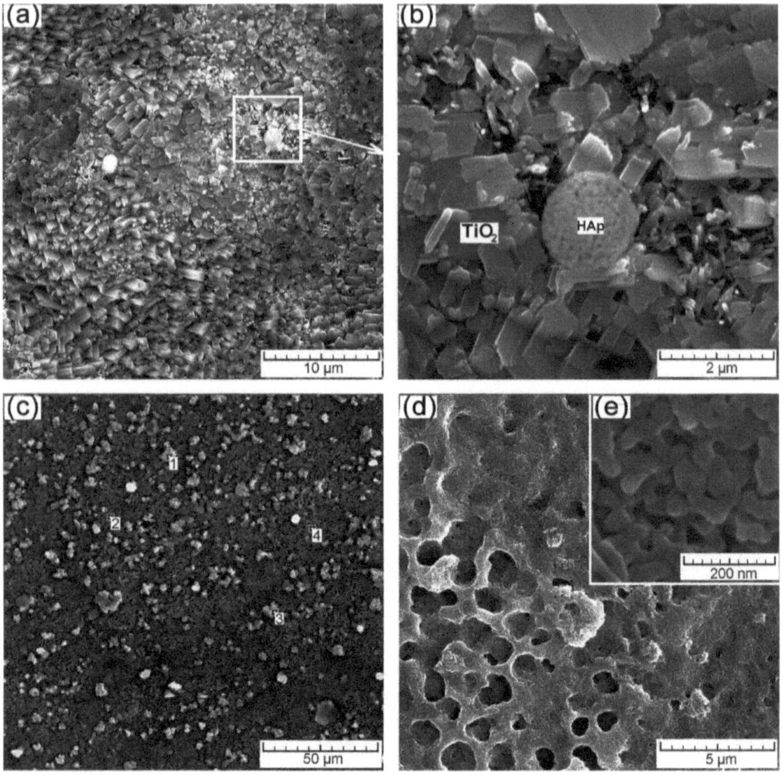

Figure 4 SEM of OBCC composite structure: (a) microstructure of MOC obtained on regime 800-300 with agglomerates of HAp bioceramics; (b) formation of globules (no wetting); (c) microstructure of OBCC combined with EDX (points 1-4 in Table 2); (d) filler from nanoparticles of HAp bioceramics that is uniformly distributed in the porous structure of MOC obtained on regime 800-120; (e) nanostructure of OBCC.

Table 2 EDX results for OBCC samples obtained on titanium VT1-00 on IHT regime 800-120 (at.%, Figure 4c)

Spectrum	O	Ti	Al	Ca	P	[Ca/P]
1	75.51	12.49	0.10	6.70	5.20	1.29
2	59.84	37.05	0.26	1.71	1.14	1.50
3	74.82	21.20	2.06	1.05	0.87	1.21
4	66.67	30.30	0.16	1.61	1.26	1.28

The initial values of modulus of elasticity E typical for CP titanium are retained on low-temperature IHT regime (regimes 600-001 and 600-300). In general, modulus of elasticity for titanium surface with MOC increases 2-4 times. There is a "peak" corresponding to 7-fold increase in modulus

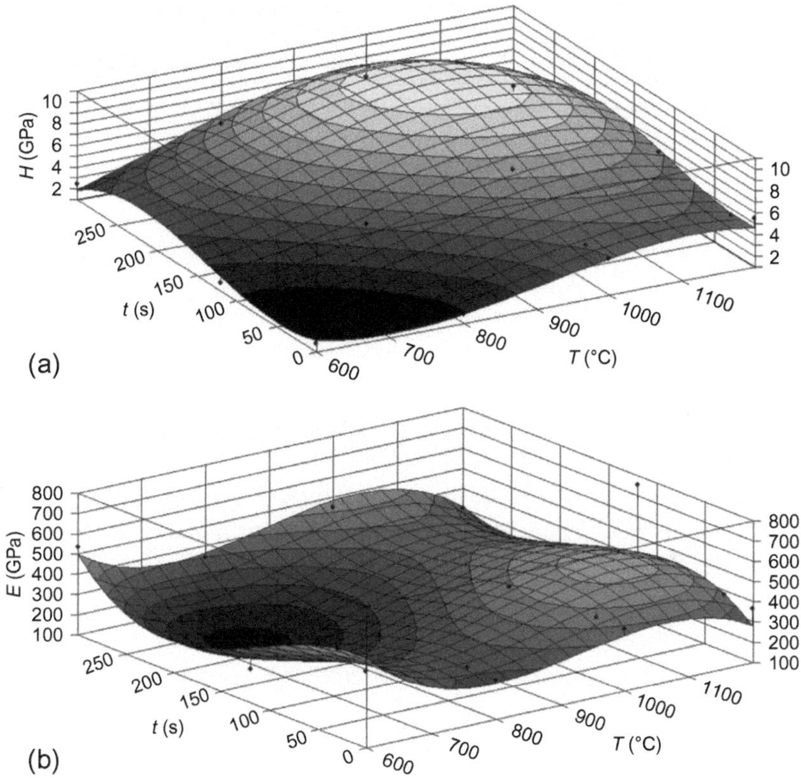

Figure 5 Hardness (a) and modulus of elasticity (b) dependencies for MOC.

of elasticity on IHT regime 1200-120, which correlates with hardness for this IHT regime (Figure 5b).

Thus, MOC crystals (sublayer), prismatic ones in particular, are characterized not only by developed surface morphology but also by high physicomechanical properties as well. The obtained values of mechanical properties characterize these MOC as high strength that is proved by 2-3.5-fold margin of hardness compared to titanium substrate and 8-15-fold to that of cortical bone (Table 3). Results on the strengthening effect of IHT for constructional titanium alloy VT16 were also obtained. On the recommended IHT regimes (800-120 and 1000-030), the hardness increases 1.4-1.5 times. Modulus of elasticity for heat-treated samples of VT16 alloy is generally high, but due to IHT, its values are slightly reduced and close to those of CP titanium (regimes 1200-120 and 800-120).

Table 3 Physicomechanical properties of bone, metal substrate, and the surface layer of MOC samples obtained on alloy VT16 at strengthening IHT

Sample	Hardness (GPa)	Elasticity module (GPa)
Cortical human bone	0.56-0.68	5.4-9.4
Spongy human bone	0.02-0.06	1.2-2.6
CP titanium VT1-00	1.86-2.68	115.3-122.8
Titanium alloy VT16	2.86-4.06	237.5-293.8
VT16 1200-120 (sublayer)	3.60-4.00	85.7-95.2
VT16 1000-030 (sublayer)	4.96-5.47	193.7-194.8
VT16 800-120	4.17-5.60	108.8-152.5

Thus, it is necessary to provide biocompatible coatings on the surface of metal constructions for implantation purposes and to ensure enhanced strength of the structure, hardness, and wear resistance when exposed to scratching with hard inclusions of cortical bone.

The data on MOC hardness correlate with their high scratch resistance, which is clearly illustrated by the data of optical microscopy and scratch test diagrams (Figure 6).

A typical diagram shows test parameters (axial load applied to the indenter and varied linearly) and measured values (friction force, friction coefficient, and depth of indentation). To calculate the stresses arising from MOC destruction (delamination), it is necessary to imagine the geometry of the indented diamond tip, which is a hemisphere. For example, for a sample

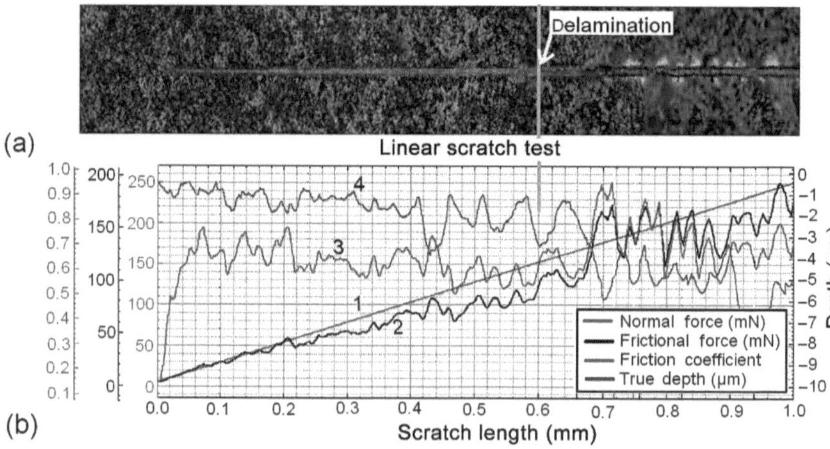

Figure 6 Scratch test data for MOC sample obtained by IHT 1000-120 (sublayer): (a) surface morphology with scratch; (b) change chart of indicators (1, base load; 2, the frictional force; 3, coefficient of friction; and 4, depth of indenter penetration).

of the coating obtained on IHT regime 1000-001, according to the microscopic analysis of scratch, increased coefficient of friction from 0.6 to 0.9 in the area of normal load of 220 mN, indenter penetration depth of 1.5 μm and load of 225 mN, and penetration depth of 4.4 μm, we can obtain the following strength characteristics: tensile strength during scratching of coating of about 4700 MPa and residual resistance after coating destruction of about 1600 MPa.

Thus, the results of scratch testing and data on the allowable load value and indentation depth not causing destruction of the functional coating and quantitative strength characteristic of scratch resistance, which is presented in graphic form, were defined (Figure 7a). Load and indentation depth can show the value of the residual strength of the coating after stress relaxation with scratch formation (Figure 7b).

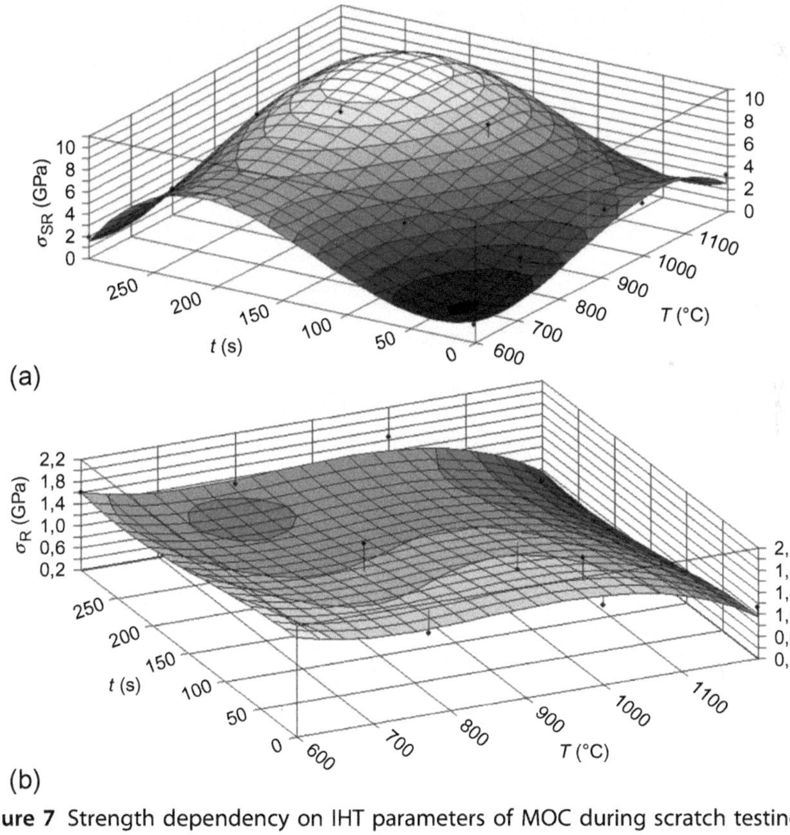

Figure 7 Strength dependency on IHT parameters of MOC during scratch testing: (a) before destruction; (b) after destruction.

3.4 Preliminary *In Vitro* Testing for Biocompatibility

In vitro testing for biocompatibility of obtained MOC confirms that high morphological heterogeneity of the surface structure contributes to stable adhesion of cells (Figure 8a and b).

It was found that increased morphological heterogeneity of needlelike structure of TiO_2 coatings seems to promote the formation of biointerface "MOC biostructure." Numerous submicron and nanopores do not prevent the penetration of biological structures (filopodia); moreover, they contribute to the process of cell adhesion (Figure 8a). Electronic images show cells as dark objects (large accumulations are marked with arrows). Analysis of MOC images after *in vitro* tests reveals that cells are fixed mainly in

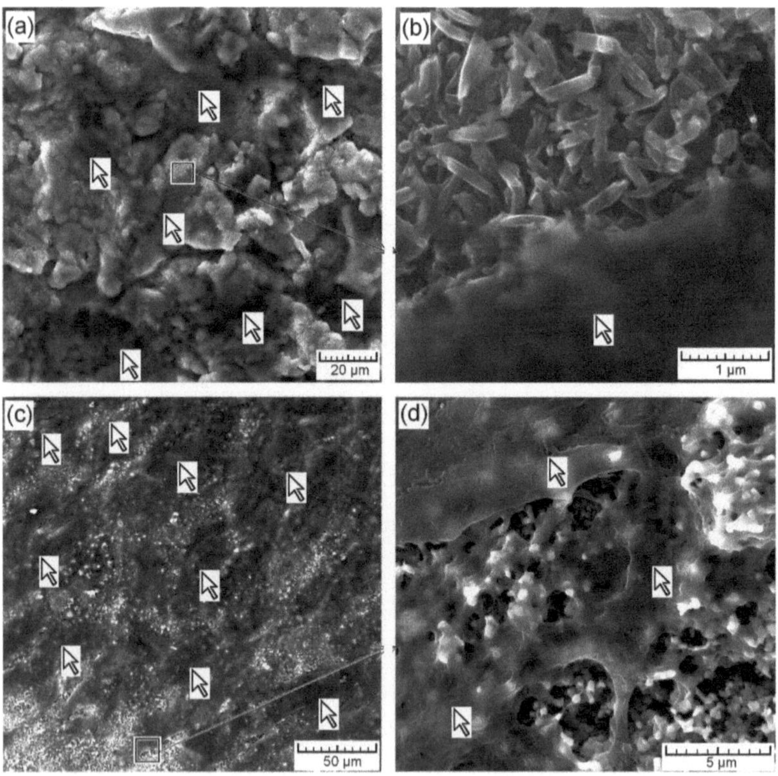

Figure 8 SEM of morphology of the coatings after *in vitro* testing: (a, b) micro- and submicron scale of studies (MOC obtained on IHT regime 800-120); (c, d) OBCC obtained on IHT regime 1000-030 (arrows indicate accumulation of cells (c) and formation of filopodia (d)).

micropores with an average size comparable with that of the cells and equaling about 5-15 μm. Cell adhesion is most active in the presence of submicron and nanograins covering rough microrelief of an implant.

Modification of TiO_2 coatings with HAp nanoparticles seems to intensify the processes of cellular adhesion and accelerate proliferation. Increased bioactivity of HAp ceramics and morphological heterogeneity of OBCC composite ensure that accelerated formation of cellular layer minimizes coating areas without fixed cells (Figure 8c and d).

4. CONCLUSIONS

Surface structure of CP titanium VT1-00 after IHT is characterized by the formation of titania coating with high morphological heterogeneity, physicomechanical properties, and biocompatibility qualities determined during *in vitro* testing. Titania coatings are formed by nano- and submicron crystals of oval, needlelike, plate, and prismatic shapes. The most suitable coating morphology parameters with increased hardness of about 4.23-9.86 GPa and scratch resistance of not less than 3.78-8.76 GPa were defined. These parameters of the structure and physicomechanical properties correspond mostly to MOC samples obtained in the following IHT regimes: 600-120, 800-120, 1000-001, 1200-001, and especially 1000-120 (during sublayer formation).

Treatment with HF eddy currents and subsequent modification and IHT of titanium medical devices provide accelerated formation of mechanically strong composite structure consisting of porous oxide matrix of rutile and filler of nanoparticles of bioactive ceramics on the surface. It has been established that porous oxide coating modified with HAp nanoparticles and formed by heating from 800 to 1200 °C and holding for at least 30 s is characterized by potential high biocompatibility according to the results of preliminary tests.

ACKNOWLEDGMENTS

The research was carried out with partial financial support from RFBR grant no 13-03-00898 "a" scholarship of the President of the Russian Federation no. SP-617.2015.4 and project number no 1189 in the framework of the basic tasks for state educational institutions of higher education, under the jurisdiction of the Ministry of Education of Russian Federation in the field of scientific activity. The authors greatly value the help of Dr. Zakharevich A.M., Dr. Skaptsov A.A., and Petrova N.V. during the laboratory experiments and for the valuable comments.

REFERENCES

[1] S.R. Paital, N.B. Dahotre, Calcium phosphate coatings for bio-implant applications: materials, performance factors, and methodologies, Mater. Sci. Eng. R Rep. 66 (2009) 1–70.
[2] X. Liu, P.K. Chu, C. Ding, Surface modification of titanium, titanium alloys, and related materials for biomedical applications, Mater. Sci. Eng. R Rep. 47 (2004) 49–121.
[3] H.W. Choi, H.J. Lee, K.J. Kim, H.-M. Kim, S.C. Lee, Surface modification of hydroxyapatite nanocrystals by grafting polymers containing phosphonic acid groups, J. Colloid Interface Sci. 304 (2006) 277–281.
[4] M. Iwasaki, Y. Miyamoto, S. Ito, T. Furuzono, W.-K. Park, Fabrication of platy apatite nanocrystals loaded with TiO_2 nanoparticles by two-step emulsion method and their photocatalytic activity, J. Colloid Interface Sci. 326 (2008) 537–540.
[5] H.-D. Jung, S.-W. Yook, T.-S. Jang, Y. Li, H.-E. Kim, Y.-H. Koh, Dynamic freeze casting for the production of porous titanium (Ti) scaffolds, Mater. Sci. Eng. C 33 (2013) 59–63.
[6] A. Dudek, R. Włodarczyk, Effect of sintering atmosphere on properties of porous stainless steel for biomedical applications, Mater. Sci. Eng. C 33 (2013) 434–439.
[7] T. Habijan, C. Haberland, H. Meier, J. Frenzel, J. Wittsiepe, C. Wuwer, et al., The biocompatibility of dense and porous nickel-titanium produced by selective laser melting, Mater. Sci. Eng. C 33 (2013) 419–426.
[8] B. Veleirinho, F.V. Berti, P.F. Dias, M. Maraschin, R.M. Ribeiro-do-Valle, J.A. Lopes-da-Silva, Manipulation of chemical composition and architecture of non-biodegradable poly(ethylene terephthalate)/chitosan fibrous scaffolds and their effects on L929 cell behavior, Mater. Sci. Eng. C 33 (2013) 37–46.
[9] C. Jelen, G. Mattei, F. Montemurro, C. De Maria, M. Mattioli-Belmonte, G. Vozzi, Bone scaffolds with homogeneous and discrete gradient mechanical properties, Mater. Sci. Eng. C 33 (2013) 28–36.
[10] S. Farhangdoust, A. Zamanian, M. Yasaei, M. Khorami, The effect of processing parameters and solid concentration on the mechanical and microstructural properties of freeze-casted macroporous hydroxyapatite scaffolds, Mater. Sci. Eng. C 33 (2013) 453–460.
[11] S.K. Swain, S. Bhattacharyya, Preparation of high strength macroporous hydroxyapatite scaffold, Mater. Sci. Eng. C 33 (2013) 67–71.
[12] A. Rezaei, M.R. Mohammadi, *In vitro* study of hydroxyapatite/polycaprolactone (HA/PCL) nanocomposite synthesized by an *in situ* sol-gel process, Mater. Sci. Eng. C 33 (2013) 390–396.
[13] P. Zhang, Z. Hong, T. Yu, X. Chen, X. Jing, *In vivo* mineralization and osteogenesis of nanocomposite scaffold of poly(lactide-co-glycolide) and hydroxyapatite surface-grafted with poly(l-lactide), Biomaterials 30 (2009) 58–70.
[14] J. Wei, F. Chen, J.-W. Shin, H. Hong, C. Dai, J. Su, et al., Preparation and characterization of bioactive mesoporous wollastonite—polycaprolactone composite scaffold, Biomaterials 30 (2009) 1080–1088.
[15] X. Liu, M.N. Rahaman, Q. Fu, Bone regeneration in strong porous bioactive glass (13-93) scaffolds with an oriented microstructure implanted in rat calvarial defects, Acta Biomater. 9 (2013) 4889–4898.
[16] C. Arnould, C. Volcke, C. Lamarque, P.A. Thiry, J. Delhalle, Z. Mekhalif, Titanium modified with layer-by-layer sol-gel tantalum oxide and an organodiphosphonic acid: a coating for hydroxyapatite growth, J. Colloid Interface Sci. 336 (2009) 497–503.
[17] C. Arnould, T.I. Korányi, J. Delhalle, Z. Mekhalif, Fabrication of tantalum oxide/carbon nanotubes thin film composite on titanium substrate, J. Colloid Interface Sci. 344 (2010) 390–394.

[18] M. Wang, Developing bioactive composite materials for tissue replacement, Biomaterials 24 (2003) 2133–2151.
[19] S. Fu, P. Ni, B. Wang, B. Chu, J. Peng, L. Zheng, et al., *In vivo* biocompatibility and osteogenesis of electrospun poly(ε-caprolactone)-poly(ethylene glycol)-poly(ε-caprolactone)/nano-hydroxyapatite composite scaffold, Biomaterials 33 (2012) 8363–8371.
[20] R. Ravichandran, J.R. Venugopal, S. Sundarrajan, S. Mukherjee, S. Ramakrishna, Precipitation of nanohydroxyapatite on PLLA/PBLG/Collagen nanofibrous structures for the differentiation of adipose derived stem cells to osteogenic lineage, Biomaterials 33 (2012) 846–855.
[21] C.Y. Flores, C. Diaz, A. Rubert, G.A. Benítez, M.S. Moreno, M.A. Fernández Lorenzo de Mele, et al., Spontaneous adsorption of silver nanoparticles on Ti/TiO_2 surfaces. Antibacterial effect on Pseudomonas aeruginosa, J. Colloid Interface Sci. 350 (2010) 402–408.
[22] J. Ni, K. Noh, C.J. Frandsen, S.D. Kong, G. He, T. Tang, et al., Preparation of near micrometer-sized TiO_2 nanotube arrays by high voltage anodization, Mater. Sci. Eng. C 33 (2013) 259–264.
[23] M.F. Morks, A. Kobayashi, Influence of gas flow rate on the microstructure and mechanical properties of hydroxyapatite coatings fabricated by gas tunnel type plasma spraying, Surf. Coat. Technol. 201 (2006) 2560–2566.
[24] P.K. Chu, Applications of plasma-based technology to microelectronics and biomedical engineering, Surf. Coat. Technol. 203 (2009) 2793–2798.
[25] H.-T. Chen, C.-H. Hsiao, H.-Y. Long, C.-J. Chung, C.-H. Tang, K.-C. Chen, et al., Micro-arc oxidation of β-titanium alloy: structural characterization and osteoblast compatibility, Surf. Coat. Technol. 204 (2009) 1126–1131.
[26] J.P. Yuan, W. Li, C. Wang, Effect of the La alloying addition on the antibacterial capability of 316L stainless steel, Mater. Sci. Eng. C 33 (2013) 446–452.
[27] J.-Y. Ha, Y. Tsutsumi, H. Doi, N. Nomura, K.-H. Kim, T. Hanawa, Enhancement of calcium phosphate formation on zirconium by micro-arc oxidation and chemical treatments, Surf. Coat. Technol. 205 (2011) 4948–4955.
[28] X. Zhao, X. Liu, C. Ding, P.K. Chu, *In vitro* bioactivity of plasma-sprayed TiO_2 coating after sodium hydroxide treatment, Surf. Coat. Technol. 200 (2006) 5487–5492.
[29] N. Wang, H. Li, W. Lü, J. Li, J. Wang, Z. Zhang, et al., Effects of TiO_2 nanotubes with different diameters on gene expression and osseointegration of implants in minipigs, Biomaterials 32 (2011) 6900–6911.
[30] M. Kazemzadeh-Narbat, B.F.L. Lai, C. Ding, J.N. Kizhakkedathu, R.E.W. Hancock, R. Wang, Multilayered coating on titanium for controlled release of antimicrobial peptides for the prevention of implant-associated infections, Biomaterials 34 (2013) 5969–5977.
[31] M. Järn, S. Areva, V. Pore, J. Peltonen, M. Linden, Topography and surface energy dependent calcium phosphate formation on sol-gel derived TiO_2 coatings, Langmuir 22 (2006) 8209–8213.
[32] F. Li, X. Yao, Z. Wang, W. Xing, W. Jin, J. Huang, Y. Wang, Highly porous metal oxide networks of interconnected nanotubes by atomic layer deposition, Nano Lett. 12 (2012) 5033–5038.
[33] S. Dyshlovenko, L. Pawlowski, B. Pateyron, I. Smurov, J.H. Harding, Modelling of plasma particle interactions and coating growth for plasma spraying of hydroxyapatite, Surf. Coat. Technol. 200 (2006) 3757–3769.
[34] R.B. Heimann, Thermal spraying of biomaterials, Surf. Coat. Technol. 201 (2006) 2012–2019.
[35] G. Singh, S. Singh, S. Prakash, Surface characterization of plasma sprayed pure and reinforced hydroxyapatite coating on Ti6Al4V alloy, Surf. Coat. Technol. 205 (2011) 4814–4820.

[36] S.V. Dorozhkin, Bioceramics of calcium orthophosphates, Biomaterials 31 (2010) 1465–1485.
[37] K. Balani, R. Anderson, T. Laha, M. Andara, J. Tercero, E. Crumpler, et al., Plasma-sprayed carbon nanotube reinforced hydroxyapatite coatings and their interaction with human osteoblasts *in vitro*, Biomaterials 28 (2007) 618–624.
[38] E. Milella, F. Cosentino, A. Licciulli, C. Massaro, Preparation and characterisation of titania/hydroxyapatite composite coatings obtained by sol-gel process, Biomaterials 22 (2001) 1425–1431.
[39] H. Li, K.A. Khor, P. Cheang, Impact formation and microstructure characterization of thermal sprayed hydroxyapatite/titania composite coatings, Biomaterials 24 (2003) 949–957.
[40] W. Pon-On, N. Charoenphandhu, I.M. Tang, J. Teerapornpuntakit, J. Thongbunchoo, N. Krishnamra, Biocomposite of hydroxyapatite-titania rods (HAp-TiR): physical properties and *in vitro* study, Mater. Sci. Eng. C 33 (2013) 251–258.
[41] S.V. Dorozhkin, Nanosized and nanocrystalline calcium orthophosphates, Acta Biomater. 6 (2010) 715–734.
[42] K. Balani, Y. Chen, S.P. Harimkar, N.B. Dahotre, A. Agarwal, Tribological behavior of plasma-sprayed carbon nanotube-reinforced hydroxyapatite coating in physiological solution, Acta Biomater. 3 (2007) 944–951.
[43] Y.C. Yang, B.Y. Chou, Bonding strength investigation of plasma-sprayed HA coatings on alumina substrate with porcelain intermediate layer, Mater. Chem. Phys. 104 (2007) 312–339.
[44] C. Arnould, J. Denayer, M. Planckaert, J. Delhalle, Z. Mekhalif, Bilayers coating on titanium surface: the impact on the hydroxyapatite initiation, J. Colloid Interface Sci. 341 (2010) 75–82.
[45] A. Maho, S. Linden, C. Arnould, S. Detriche, J. Delhalle, Z. Mekhalif, Tantalum oxide/carbon nanotubes composite coatings on titanium, and their functionalization with organophosphonic molecular films: a high quality scaffold for hydroxyapatite growth, J. Colloid Interface Sci. 371 (2012) 150–158.
[46] S. Radice, P. Kern, H. Dietsch, S. Mischler, J. Michler, Methods for functionalization of microsized polystyrene beads with titania nanoparticles for cathodic electrophoretic deposition, J. Colloid Interface Sci. 318 (2008) 264–270.
[47] C.A. Charitidis, A. Skarmoutsou, A. Tsetsekou, D. Brasinika, D. Tsiourvas, Nanomechanical properties of hydroxyapatite (HAP) with DAB dendrimers (poly-propylene imine) coatings onto titanium surfaces, Mater. Sci. Eng. B 178 (2013) 391–399.
[48] L. Le Guéhennec, A. Soueidan, P. Layrolle, Y. Amouriq, Surface treatments of titanium dental implants for rapid osseointegration, Dent. Mater. 23 (2007) 844–854.
[49] N. Ribeiro, S.R. Sousa, F.J. Monteiro, Influence of crystallite size of nanophased hydroxyapatite on fibronectin and osteonectin adsorption and on MC3T3-E1 osteoblast adhesion and morphology, J. Colloid Interface Sci. 351 (2010) 398–406.
[50] X. Liu, X. Zhao, R.K.Y. Fu, J.P.Y. Ho, C. Ding, P.K. Chu, Plasma-treated nanostructured TiO_2 surface supporting biomimetic growth of apatite, Biomaterials 26 (2005) 6143–6150.
[51] P.K. Hansma, G.E. Fantner, J.H. Kindt, P.J. Thurner, G. Schitter, P.J. Turner, S.F. Udwin, M.M. Finch, Sacrificial bonds in the interfibrillar matrix of bone, J. Musculoskelet. Neuronal Interact. 5 (4) (2005) 313–315.
[52] A.A. Fomin, A.B. Steinhauer, I.V. Rodionov, M.A. Fomina, A.M. Zakharevich, A. A. Skaptsov, A.N. Gribov, Y.D. Karsakova, Properties of titanium dioxide coatings produced by induction-thermal oxidation of VT1-00 alloy, J. Frict. Wear 35 (1) (2014) 32–39.
[53] C. Kim, M.R. Kendall, M.A. Miller, C.L. Long, P.R. Larson, M.B. Humphrey, et al., Comparison of titanium soaked in 5 M NaOH or 5 M KOH solutions, Mater. Sci. Eng. C 33 (2013) 327–339.

[54] J. Cizek, K.A. Khor, Z. Prochazka, Influence of spraying conditions on thermal and velocity properties of plasma sprayed hydroxyapatite, Mater. Sci. Eng. C 27 (2007) 340–344.
[55] N. Drnovšek, K. Rade, R. Milačič, J. Štrancar, S. Novak, The properties of bioactive TiO_2 coatings on Ti-based implants, Surf. Coat. Technol. 209 (2012) 177–183.
[56] B. Yang, M. Uchida, H.-M. Kim, X. Zhang, T. Kokubo, Preparation of bioactive titanium metal via anodic oxidation treatment, Biomaterials 25 (2004) 1003–1010.
[57] C.F. Koch, S. Johnson, D. Kumar, M. Jelinek, D.B. Chrisey, A. Doraiswamy, et al., Pulsed laser deposition of hydroxyapatite thin films, Mater. Sci. Eng. C 27 (2007) 484–494.
[58] S. Saber-Samandari, K.A. Gross, Nanoindentation reveals mechanical properties within thermally sprayed hydroxyapatite coatings, Surf. Coat. Technol. 203 (2009) 1660–1664.
[59] S. Yang, H.C. Man, W. Xing, X. Zheng, Adhesion strength of plasma-sprayed hydroxyapatite coatings on laser gas-nitrided pure titanium, Surf. Coat. Technol. 203 (2009) 3116–3122.
[60] Z.L. Dong, K.A. Khor, C.H. Quek, T.J. White, P. Cheang, TEM and STEM analysis on heat-treated and *in vitro* plasma-sprayed hydroxyapatite/Ti-6Al-4V composite coatings, Biomaterials 24 (2003) 97–105.
[61] S. Saber-Samandari, K.A. Gross, The use of thermal printing to control the properties of calcium phosphate deposits, Biomaterials 31 (2010) 6386–6693.
[62] B. Viswanath, R. Raghavan, U. Ramamurty, N. Ravishankar, Mechanical properties and anisotropy in hydroxyapatite single crystals, Scr. Mater. 57 (2007) 361–364.
[63] S. Nag, S.R. Paital, P. Nandawana, K. Mahdak, Y.H. Ho, H.D. Vora, et al., Laser deposited biocompatible Ca-P coatings on Ti-6Al-4V: microstructural evolution and thermal modeling, Mater. Sci. Eng. C 33 (2013) 165–173.
[64] R.A. Ismail, E.T. Salim, W.K. Hamoudi, Characterization of nanostructured hydroxyapatite prepared by Nd:YAG laser deposition, Mater. Sci. Eng. C 33 (2013) 47–52.
[65] M.Y. Zhang, C. Ye, U.J. Erasquin, T. Huynh, C. Cai, G.J. Cheng, Laser engineered multilayer coating of biphasic calcium phosphate/titanium nanocomposite on metal substrates, ACS Appl. Mater. Interfaces 3 (2011) 339–350.
[66] S.A. Shabalovskaya, D. Siegismund, E. Heurich, M. Rettenmayr, Evaluation of wettability and surface energy of native nitinol surfaces in relation to hemocompatibility, Mater. Sci. Eng. C 33 (2013) 127–132.
[67] A. Alsaran, G. Purcek, I. Hacisalihoglu, Y. Vangolu, Ö. Bayrak, I. Karaman, et al., Hydroxyapatite production on ultrafine-grained pure titanium by micro-arc oxidation and hydrothermal treatment, Surf. Coat. Technol. 205 (Suppl. 2) (2011) S537–S542.
[68] Y. Yan, J. Sun, Y. Han, D. Li, K. Cui, Microstructure and bioactivity of Ca, P and Sr doped TiO_2 coating formed on porous titanium by micro-arc oxidation, Surf. Coat. Technol. 205 (2010) 1702–1713.
[69] D. Wei, Y. Zhou, D. Jia, Y. Wang, Effect of heat treatment on the structure and *in vitro* bioactivity of microarc-oxidized (MAO) titania coatings containing Ca and P ions, Surf. Coat. Technol. 201 (2007) 8723–8729.
[70] L.-H. Li, H.-W. Kim, S.-H. Lee, Y.-M. Kong, H.-E. Kim, Biocompatibility of titanium implants modified by microarc oxidation and hydroxyapatite coating, J. Biomed. Mater. Res. 73A (2005) 48–54.
[71] A. Kar, K.S. Raja, M. Misra, Electrodeposition of hydroxyapatite onto nanotubular TiO_2 for implant applications, Surf. Coat. Technol. 201 (2006) 3723–3731.
[72] S.-F. Ou, C.-S. Lin, Y.-N. Pan, Microstructure and surface characteristics of hydroxyapatite coating on titanium and Ti-30Nb-1Fe-1Hf alloy by anodic oxidation and hydrothermal treatment, Surf. Coat. Technol. 205 (2011) 2899–2906.

[73] K. Huo, X. Zhang, H. Wang, L. Zhao, X. Liu, P.K. Chu, Osteogenic activity and antibacterial effects on titanium surfaces modified with Zn-incorporated nanotube arrays, Biomaterials 34 (2013) 3467–3478.
[74] I.V. Rodionov, Application of the air-thermal oxidation technology for producing biocompatible oxide coatings on periosteal osteofixation devices from stainless steel, Inorg. Mater. Appl. Res. 2 (4) (2013) 119–126.
[75] K. Vasilev, Z. Poh, K. Kant, J. Chan, A. Michelmore, D. Losic, Tailoring the surface functionalities of titania nanotube arrays, Biomaterials 31 (2010) 532–540.
[76] Y. Li, I.-S. Lee, F.-Z. Cui, S.-H. Choi, The biocompatibility of nanostructured calcium phosphate coated on micro-arc oxidized titanium, Biomaterials 29 (2008) 2025–2032.
[77] G. Mendonça, D.B.S. Mendonça, F.J.L. Aragão, L.F. Cooper, Advancing dental implant surface technology—from micron- to nanotopography, Biomaterials 29 (2008) 3822–3835.
[78] H. Cao, X. Liu, Plasma-sprayed ceramic coatings for osseointegration, Int. J. Appl. Ceram. Technol. 10 (1) (2013) 1–10.
[79] J.A. Inzana, D. Olvera, S.M. Fuller, J.P. Kelly, O.A. Graeve, E.M. Schwarz, et al., 3D printing of composite calcium phosphate and collagen scaffolds for bone regeneration, Biomaterials 35 (2014) 4026–4034.
[80] S. Dyshlovenko, C. Pierlot, L. Pawlowski, R. Tomaszek, P. Chagnon, Experimental design of plasma spraying and laser treatment of hydroxyapatite coatings, Surf. Coat. Technol. 201 (2006) 2054–2060.
[81] A.A. Fomin, I.V. Rodionov, A.B. Steinhauer, M.A. Fomina, N.V. Petrova, A. M. Zakharevich, A.A. Skaptsov, A.N. Gribov, V.S. Atkin, Nanostructure of biocompatible titania/hydroxyapatite coatings, Proc. SPIE 9031 (2014) 90310H.
[82] P. Kofstad, High-Temperature Oxidation of Metals, Wiley, New York, 1966, 340 pp.
[83] V. Rudnev, D. Loveless, R. Cook, M. Black, Handbook of Induction Heating, Marcel Dekker, New York, 2003, 800 pp.

PART 3

Application Fields

CHAPTER 20

Bone Apposition on Nanoporous Titanium Implants

Laëtitia Salou[1,2,3], Alain Hoornaert[4], Guy Louarn[3], Pierre Layrolle[1]
[1]Inserm U957, Lab. Pathophysiology of bone resorption, University of Nantes, Nantes, France
[2]Biomedical Tissues SAS, IRSUN, Nantes, France
[3]Institute of Materials Jean Rouxel, CNRS, University of Nantes, Nantes, France
[4]CHU Nantes, Nantes, France

Chapter Contents

1. INTRODUCTION

Titanium and its alloys are widely used for manufacturing dental and orthopedic implants because of their good biocompatibility. Surface properties of titanium implants are key factors for rapid and stable bone tissue integration [1]. Upon implantation in the body, titanium implant surfaces interact with proteins, cells, and tissues in the nanometer range. Nanotechnologies may produce surfaces with controlled topography and chemistry that would help in understanding biological interactions and developing novel implant

Handbook of Nanoceramic and Nanocomposite Coatings and Materials
http://dx.doi.org/10.1016/B978-0-12-799947-0.00020-1

surfaces with predictable tissue-integrative properties [2–4]. Various surface treatments such as templating, hydrothermal processing, and electrochemical anodization may be applied to the surfaces of dental and orthopedic implants to produce controlled features at the nanometer scale. These nanostructured surfaces exhibit titania nanotubes of various sizes and distributions.

Basic research on the self-organization of nanostructures made by Zwilling and coworkers showed that the porous surface of titania could be formed electrochemically in fluorinated electrolytes by an anodization process [5]. Two years later, Gong and coworkers [6] first reported the observation of the growth of highly ordered self-assembled TiO_2 nanotube arrays via anodic oxidation of titanium in a hydrofluoric (HF) electrolyte. This new titania structure was shown to improve properties compared with any other form of titania making some interest for different applications such as photocatalysor [7,8], sensors [9,10], and photovoltaic applications [11,12] and in the biomedical field [13–15] including drug delivery [16–19].

Titania nanotubes may be highly relevant to biomedical applications, particularly in dental and orthopedic implants. The optimal surface may be screened by using high-throughput biological assays *in vitro*. For instance, specific protein adsorption, cell adhesion, and differentiation of stem cells should be studied in relation to the surface properties. Following *in vitro* screening, nanostructured surfaces may then be verified in animal models to validate hypotheses in a complex *in vivo* environment. This approach may define the ideal surface for a specific biological response such as bone tissue integration.

In this chapter, the different methods to produce titania nanotubes and their application in the biomedical field will be reviewed.

2. METHODS OF FABRICATION OF TITANIA NANOTUBES

Titanium metal is naturally covered by a thin layer of titania or titanium oxide (TiO_2). This layer is a few nanometers thick and generally amorphous. This passivation film forms spontaneously when titanium is exposed to air. In the biological environment, this titanium oxide layer is very stable and prevents biocorrosion and the release of metal ions ensuring the biocompatibility of implants. However, the oxide layer surface is often contaminated with hydrocarbon components from the environment. Indeed, Ogawa and colleagues have shown that the bioactivity and osteoconductivity of titanium degrade over time after surface processing [20]. In general, manufacturers of implants do not control the formation and properties of the titanium oxide layer in the nanometer range. Several groups have demonstrated that nanostructured titanium surfaces have a great impact on the fate of cells and may ultimately control the integration of implants in bone tissue. Among other nanostructures, titania nanotubes have received much attention. Titania nanotubes and nanotube arrays can be produced using various routes. One of the important ways is a template-assisted method that was used by Hoyer [21]. It consists of forming TiO_2 nanotubes by using alumina templates [21–24]. Other methods for fabricating TiO_2 nanotubes using hydrothermal [25,26] or anodization processes have been described [27–29]. These different methods to produce titania nanotubes are presented hereafter. The advantages and drawbacks of the different methods are reported in Table 1.

Table 1 Advantages and disadvantages of the different fabrication methods for TiO_2 nanotubes

Technique	Advantage	Disadvantage
Positive template-assisted method	Dimension control of TiO_2 nanotubes by the template	Not uniform in length and open ends Only large tubes and small wall Complicated fabrication process Not applicable to complex-shaped implants
Negative template—sol-gel technique	Ordered array of TiO_2 nanotubes with uniform diameters	Large tubes Risk of contamination during dissolution Time of pre- and postprocessing, difficult to scale up in industry
Hydrothermal treatment	Good crystallinity (anatase) Titania fiber/tube 100 nm to micron	Nonreproducibility of nanostructures Long reaction time ($\approx$20 h) Highly concentrated NaOH used Individual free nanotubes Numerous critical parameters
Anodization	Simple and cost-effective process applicable to different implants Well-ordered and vertically aligned nanotubes Wide range of nanotube dimensions Directly attached on substrate	Amorphous nanotubes Requires a postannealing step Highly expensive fabrication apparatus for large batches (Pt electrode)

2.1 Template-Assisted Methods

The template-assisted method utilizes a pattern transfer made of porous alumina. The alumina template is generally made of anodic aluminum oxide (AAO). The length or diameter of the TiO_2 nanotubes directly depends on template dimensions. Two types of template, positive and negative, are described in the literature.

In the positive template method, an anodic alumina mask is placed on the titanium surface. The AAO film serves as a mold for a polymer on which TiO_2 is electrochemically deposited. Then, the polymer mold is dissolved and reveals TiO_2 nanostructures corresponding to the AAO pattern. This technique was explored by some authors [22,30] but it does not produce uniform length nanotubes.

The negative template method consists of a sol-gel technique. An anodic alumina mask is placed on the titanium substrate and then dipped into a Ti $(OR)_4$ solution that penetrates through the AAO film. After calcination of over 400 °C, the template is removed by chemical dissolution and reveals an ordered array of TiO_2 nanotubes with uniform diameter [23,24].

However, these template-assisted methods are not easily applicable to complex-shaped medical devices such as dental and orthopedic implants. Furthermore, their scalability for industrial production is also difficult.

2.2 Hydro-/Solvothermal Processes

This method is based on the self-assembly of TiO_2 nanoparticles or powder treated into a NaOH solution that deposits on the surface of titanium implants. Following the hydrothermal process at temperatures ranging from 100 to 150 °C, the Ti-O-Ti bond breaks out to Ti-O-Na or Ti-OH, leading to the formation of a nanostructured layer. Chemical acid treatment in HCl converts this layer into nanotubes by removing electrostatic charges of the sodium salt [25,26]. This procedure may be applicable to medical devices. However, the obtained nanostructures are not easily controlled as they depend on multiple parameters (e.g., size of titanium oxide, concentration of NaOH, temperature, and neutralization).

2.3 Electrochemical Anodization Method

Anodization of titanium implants is widely used in the medical device industry for color coding of products. It is a relatively simple and cost-effective process that does not change the tolerances of implants. However, the anodization is usually performed without any control of the nanostructure of the titanium oxide layer.

A self-assembly or "bottom-up" technique has been described to produce nanomodifications of the titanium oxide surface by the electrochemical process of anodization with or without fluoride ions. By using specific electrochemical parameters, an array of oxide nanotubes aligned perpendicularly to the surface can be formed. Some authors have described self-standing nanostructured membranes by detaching the titanium oxide layer from the surface [31]. Dimensions of the nanotubes can be precisely controlled. A large variety of sizes from 20 up to 100 nm in pore diameter, length from 50 to 6000 nm, and a wall thickness ranging from 7 to 34 nm can be produced. [27,32].

In the following paragraph, the mechanism and the influence of the different electrochemical parameters on the properties of this self-assembled TiO_2 nanotube array are discussed. Their application in the biomedical field as a new surface treatment for implants with or without biologically active molecules is then presented.

3. TiO_2 NANOTUBE FORMATION UNDER ANODIC POLARIZATION

As shown in Figure 1, the electrochemical process of anodization consists of applying a potential (potentiostatic method) or a current (galvanostatic method) between the titanium implant and a counter electrode in an electrolyte solution. The titanium implant acts as an anode and a platinum inert electrode as a cathode. When applying a current, an oxidation reaction occurs at the titanium surface. Depending on the anodizing conditions, four behaviors are possible. At first, there is a continuous solvation of metal ions due to electrocorrosion and the process is known as electropolishing. Metal ions can also react with oxygen and form a compact metal oxide (MO) layer. This process is used as a color-coding surface treatment, by phenomenon of constructive and destructive interference with light. The colored surface can be used for decorating or for the identification of medical instrumentation. Under high electrochemical conditions (i.e., 100-300 V), the thickness of the oxide layer is greatly increased, leading to dielectric breaks with sparks that cause the melting of the oxide layer [33]. In the latter conditions, two layers are formed: a compact oxide layer and a microporous outer oxide layer. The first layer has shown to increase the fatigue strength and to reduce wear of friction. During anodization, the competition between solvation and growth of the oxide layer forms a nanoporous array of dioxide layer. By using more specific parameters, these nanopores self-organize into oxide

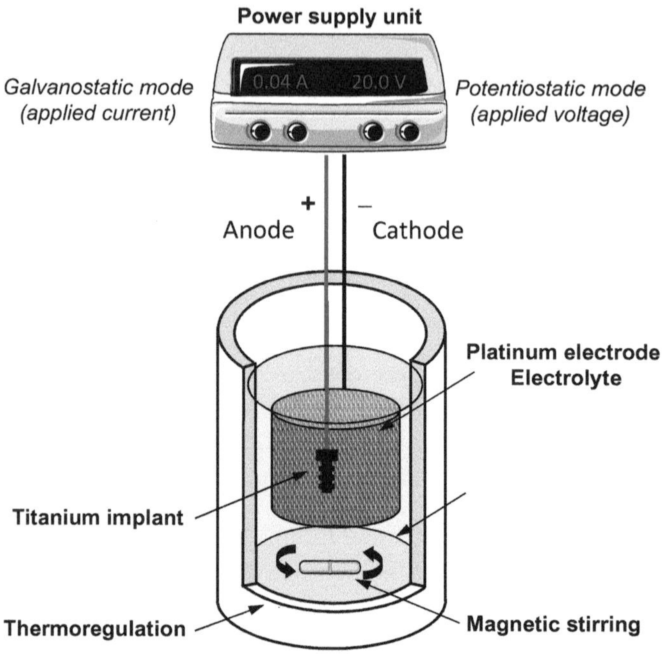

Figure 1 Experimental setup of the electrochemical anodization process for the formation of nanoporous TiO_2 on a titanium implant surface.

nanotubes, perpendicularly aligned on the metal surface. This nanostructure formation has been demonstrated and mainly developed at first not only on aluminum substrate but also on titanium, niobium, tantalum, zirconium, and a large range of other transition metals and alloys. These materials are called "valve metals" because of their dense, stable, well-attached electrical insulating oxide layer. In the following paragraphs, the mechanism of the formation of a nanotubular titanium dioxide film, the influence of the electrochemical condition, and the mean thermodynamic stabilization are discussed.

3.1 Mechanism of TiO_2 Nanotube Formation

The mechanism of TiO_2 nanotube formation is illustrated in Figure 2. During the anodization process, a chemical reaction corresponding to water decomposition appears at the cathode and produces dihydrogen gas according to Equation (1):

$$2H^{+} + 2e^{-} \rightarrow H_2 \tag{1}$$

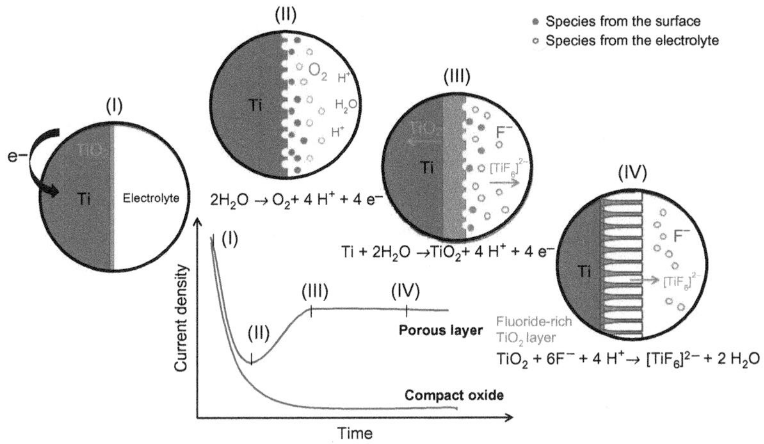

Figure 2 Mechanism of the electrochemical anodization process of titanium implant surfaces into TiO_2 nanotubes with ring used for species from the electrolyte and full circle for species from the surface.

At the anode interface, metal oxidation (Eq. 2) and decomposition of water take place and generate dioxygen gas (Eq. 3):

$$M \rightarrow Mn^{n+} + ne^{-} \tag{2}$$

$$2H_2O \rightarrow O_2 + 4H^{+} + 4e^{-} \tag{3}$$

In the case of titanium substrate, anodization follows the same scheme, titanium ion (Ti^{4+}) formations that react with O^{2-} originating from water deprotonation to form the oxide layer (Eq. 4). The thickness of the oxide layer increases with the applied current, leading to current resistance.

In the presence of sufficient fluoride in the bath, a water-soluble species $[TiF6]^{2-}$ is formed on the substrate surface (Eq. 5). Chemical dissolution of the oxide layer by the electrolyte creates pitting corrosion on the surface. Difference in the oxide thickness and thus electrical conductivity leads to the preferential passage of current. Therefore, the competition between oxide formation and Ti^{4+} solvatization creates etching at some point forming a nanopit array. The resistance of the oxide layer decreases the current density in other locations. A good balance between specific electric power, oxide formation, and chemical dissolution can form a nanoporous or nanotube array:

$$Ti + 2H_2O \rightarrow TiO_2 + 4H^{+} + 4e^{-} \tag{4}$$

$$TiO_2 + 6F^{-} + 4H^{+} \rightarrow [TiF_6]^{2-} + 2H_2O \tag{5}$$

Fluoride rapidly migrates into the oxide layer and accumulates at the oxide-metal interface and at grain boundaries [29]. This fluoride-rich layer seems to play a role in the transition from a porous hexagonal structure to a tubular structure due to a preferential chemical dissolution between pores forming individual nanotubes. Examples of TiO_2 nanopores and nanotubes are shown in Figure 3. These nanostructures uniformly cover the surface of titanium. Depending on experimental conditions, the size of the pores may vary from 20 to 110 nm in diameter. The thickness of the TiO_2 nanotube layer may be adjusted from few tens to several hundred nanometers by increasing the processing time but is generally around 150 nm in thickness. As shown in the cross section, columnar nanotubes are oriented perpendicular to the titanium surface.

Nanoanodization is applicable to complex-shaped medical implants, and either nanopores or nanotubes have been uniformly produced on dental implants, orthopedic implants, or spine fusion devices. The main disadvantage of nanoanodization is that the formed nanotubes are amorphous and so thermodynamically instable. Annealing from 280 to 580 °C transforms the amorphous structure into crystalline phases. The tetragonal anatase phase is metastable and formed at lower temperatures (around 280 °C) and transforms irreversibly to rutile at higher temperatures (Figure 4). This nucleation-growth type of the phase transition is not instantaneous but is time-dependent [34]. Depending on the nanotubular structure, crystallized rutile phase grows up from the surface, destroying the nanotubular morphology at higher temperatures than 600 °C [28,35].

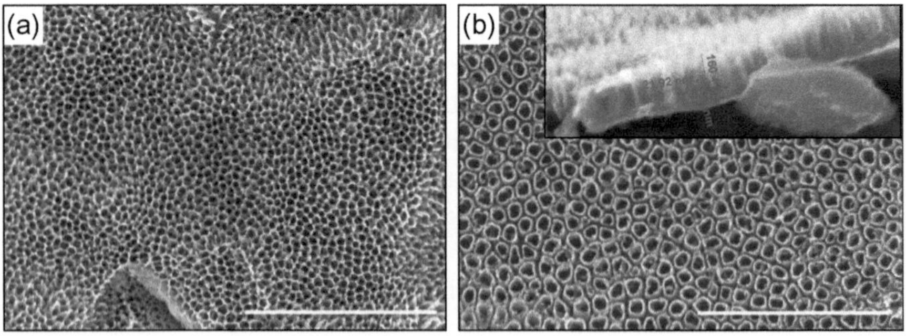

Figure 3 Scanning electron microscopy images of TiO_2 porous structures formed by anodization in 1 wt% HF and 1 M CH_3COOH at (a) at 10 V nanopores and (b) at 20 V nanotubes are formed. Inset shows the thickness of the layer of nanotubes (≅160 nm).

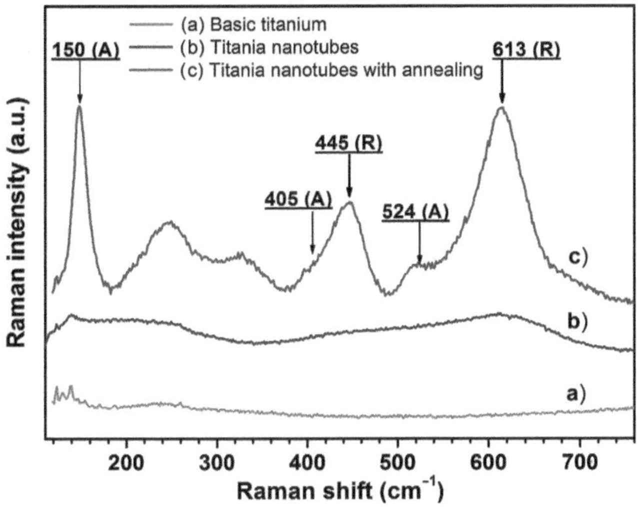

Figure 4 FT-Raman spectra of the titanium surfaces before and after the anodization process with or without annealing at 500 °C (A: anatase; R: rutile).

3.2 Importance of the Electrolyte in the Nanoanodization Process

Since the TiO_2 nanotubes have been discovered, a large number of articles have studied the electrochemical parameters that can modulate nanotube morphologies such as length, pore size, and ordering. However, the electrolytes govern both the formation and dissolution rates of the oxide layers and subsequently may be the most important parameter for nanopore formation and control. Four different generations of electrolytes have been reported in the literature. The first generation was used at the beginning with Gong and his coworkers [6] and is based on an aqueous dilute HF electrolyte. Because of the low pH value (pH < 5), the length of nanotubes is limited to approximately 500 nm. Therefore, a second generation of electrolytes was used in buffered aqueous solution at higher pH. By using these solutions, the nanotube lengths could be increased to microns. For example, Sreekantan et al. obtained nanotubes of 3 μm in length by using an electrolyte solution at pH 7 [36]. Other authors have investigated the use of a the third generation electrolyte containing a polar organic electrolytes, such as dimethyl sulfoxide (DMSO) and ethylene glycol (EG), together with or without water (1-3%). These organic electrolytes enabled the production of long nanotubes from 18-220 mm [37,38] to 1000 mm [39,37,36,35]. Finally, some studies have proposed a fourth generation of electrolytes to prepare titania nanotube arrays fabricated in a non-fluoride-based electrolyte that contains an acidic sulfate solution. Other electrolyte parameters that have been found to influence the titania nanotube arrays are the viscosity, the water content and the fluoride concentration, the bath temperature, etc. [29].

3.3 Influence of the Electrochemical Parameter on Titania Nanotube Arrays

3.3.1 Electrophysical Parameters

The literature mainly reports about experiments carried out under constant voltage (potentiostatic) conditions, whereas only few studies have used a constant current (galvanostatic) condition that should be easily transposed to the industrial surface treatment of medical devices. The electrochemical characteristic curve depends on the mode used (Figure 2). However, in both cases, the reaction time and dissolution rate define nanotube lengths. Under the potentiostatic condition, the diameter of nanotubes grows proportionally with voltage. Constant voltage produces straight nanotubes, while a sweeping voltage forms conical structures. By increasing the magnetic stirring and current density, the production of hydrogen gas bubbles increases. Unfortunately, the formation of hydrogen gas could cause current fluctuations [40]. In the characteristic curve of the galvanostatic condition, the voltage increases linearly with reaction time until stationary regime at which a porous oxide layer is formed. Current densities from 10 to 30 mA cm^{-2} in HF aqueous solutions promote titania nanotubes from 50 to 200 nm in diameter.

3.3.2 Electrode Surface

In electrochemistry, both the anode and cathode influence the anodization process. The main material used as cathode is a conductive material, platinum, due to its high stability. Nevertheless, some authors [41] have used different cathode systems to develop alternatives to the platinum-based electrode as it is quite expensive. Therefore, various cathode materials such as Ni, Pd, Fe, Co, Cu, Ta, W, C, Al, and Sn were used for producing TiO_2 nanotubes. Only the Ni, Pd, and C cathodes have produced similar homogen nanotubular structures as Pt. However, nickel shows lower stability than the two other materials. Instable materials could lead to element contamination, which should be problematic in biomedical application.

In our case, titanium implants or alloys are used as anode. The chemical composition of the surface could vary from alpha phase or both alpha and beta phases. Those two titanium phases have found to form nanostructure under different conditions and thus give microheterogeneity into the nanotubular array as observed in [42,43]. It has also been shown that a more uniform and regular nanotube array is obtained after a second anodization [44]. Nanoanodization with a short treatment time does not modify the initial surface roughness of the titanium surface. Therefore, the nanostructuration of implants is compatible with microroughened surfaces obtained by grit blasting or polished surface. As the developed surface of the implant increases, the current rises proportionally. This increase of surface area and current could modify bath temperature and thus the oxide layer characteristics [28]. In general, this electrochemical processing enables to treat complex geometries (Tullio Monetta 2013). However, in the case of masked area from the cathode, the equilibrium between the formation and dissolution of the oxide layer can be broken, which can lead to heterogeneity of structures.

4. BIOMEDICAL APPLICATIONS OF NANOSTRUCTURED TITANIUM DIOXIDE COATING

Surface properties play a determinant role in biological interactions. In particular, the nanometer-sized roughness and the chemistry have a key role in the interactions of surfaces with proteins and cells. These early interactions will in turn condition the late tissue integration. In this prospect, nanostructuration of the TiO_2 layer may be beneficial for enhancing bone healing around titanium implants.

The nanoanodization is applicable to complex-shaped medical implants, and either nanopores or nanotubes have been uniformly produced on dental implants, orthopedic implants, or spine fusion devices. Anodization is a method commonly used to obtain nanoscale oxides on metals including titanium. By adjusting the anodization condition such as voltage, time, and shaking, nanoscale properties can be controlled. Shankar et al. [38] had reported that the diameters of the nanotubes could be adjusted to a range from 20 nm up to 100 nm by modifying voltage conditions. On the other hand, Kang et al. [45] found that TiO_2 nanotube arrays were more uniform on electropolished titanium than machined titanium.

4.1 Medical Device Applications and Biocompatibility

During surgery, blood vessels are injured and thus, implant surfaces interact with blood components. Various plasma proteins get adsorbed on the material surface within a minute. Platelets from the blood interact also with the implant surface. Plasma proteins modify the surface, while activated platelets are responsible for thrombus formation and blood clotting. Subsequently, various cell types will after migration to the surface interact with the modified surface through membrane integrin receptors. These early events occur prior to peri-implant tissue healing.

Following blood clotting around implants, several cells interact with surfaces for tissue healing. Mesenchymal stem cells (MSCs) attracted to the injured site by chemotactic factors have a determinant role in peri-implant tissue healing. Numerous studies have shown that nanometer-sized features on surfaces direct stem cell fate through protein adsorption and cell signaling pathways [2,46,47]. MSCs are multipotent cells that are present in the bone marrow, peripheral blood, and other tissues in very low levels. These cells are attracted to the peri-implant healing site by cytokines and are able to migrate through the blood clot to colonize the implant surface. Depending on their microenvironment, MSCs can differentiate into different lineages such as osteoblasts (bone), chondrocytes (cartilage), adipocytes (fat), or fibroblasts (skin).

The surface properties of implants may control the differentiation of MSCs and thus the type of peri-implant tissue. Nevertheless, this hypothesis has mainly been validated *in vitro*. For instance, it has been shown that cytoskeleton tensions modulate gene expression through the Ras homologue gene family, member A (RhoA) molecular pathway and regulate stem cell lineage commitment [48]. Dalby et al. have shown that random nanostructured surfaces induce the osteoblastic differentiation of MSCs, even without osteogenic supplements [49]. Titania nanotubes obtained by anodization have been shown to control the adhesion and osteoblastic differentiation of MSCs [50,51]. These studies have demonstrated that nanotubes measuring 30 nm in diameter favor cell adhesion, whereas larger nanopores of 70–100 nm induce cell elongation, cytoskeletal stress, and selective differentiation into osteoblastic cells. It has been shown that TiO_2 nanotubes on Ti improved the activity of alkaline phosphatase (ALP) by MSCs. Since ALP is a marker of osteogenic differentiation, these surfaces may demonstrate enhanced bone tissue-integrative properties. We have indeed demonstrated a correlation between the osteoblastic cell differentiation and bone tissue integration of nanostructured titanium implants [14].

In a more recent study, we have compared the bone tissue integration of machined (MA), standard microroughened (MICRO), and nanostructured (NANO) titanium implants by implantation into the femoral epiphysis of rabbits. After 4 weeks of healing, all types of implants appeared well

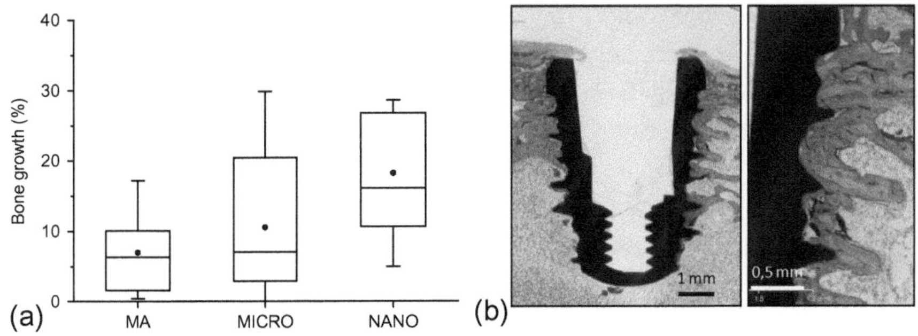

Figure 5 (a) Bone growth into titanium implants after 4-weeks implantation in femoral epiphyses of rabbits with three different surface treatments: machined (MA), grit-blasted and acid-etched creating micro-roughness (MICRO) and anodizing in HF electrolyte creating nanotubes array on the titanium surface (NANO) (* if $p<0.05$). (b) Nondecalcified histology sections of dental implant inserted in bone tissue at 4 weeks (basic fuchsine and methylene blue staining; original magnifications ×1 and ×10).

integrated into the bone without macroscopic signs of osteolysis. As shown in Figure 5, the quantitative assessment of the bone growth was higher for the NANO implants than for the MICRO and MA implants. The histological sections corroborated bone tissue healing around the implants (Figure 5). Newly formed bone apposition was visible at the cortical and trabecular levels. The quantity of bone around the implants appeared comparable for the three types of implant surfaces. At high magnification, a thin, fibrous tissue gap between the bone and the MA implants was observed, whereas direct bone-to-implant contact was seen on both NANO and MICRO implants.

4.2 Immobilization of Biologically Active Molecules in the Coatings

Titania nanotubes have an ideal size for many biomolecules, and therefore, these nanotubes are particularly attractive for local drug delivery. For instance, nanotubes may be loaded with growth factors and act as reservoirs for delivering molecules that would trigger tissue repair in the peri-implant region. In this perspective, Neupane et al. have successfully incorporated gelatin-stabilized gold nanoparticles into titania nanotubes by simple vacuum drying. Since gelatin is a thermally denatured form of collagen, it represents a good substrate for cell adhesion [52]. These authors have consequently studied the adhesion and proliferation of osteoblastic cells on standard and modified Ti plates with polished Ti, nanotubes, or

AuNPs-gelatin. After only 1 day of culture, an elongated flattened cell morphology was observed, suggesting a rapid cell spreading on the nanotubes loaded with AuNPs-gelatin surface. Furthermore, they observed a significantly higher osteoblastic cell adhesion and proliferation on the AuNPs-gelatin surface as compared with other groups.

Titanium implants are widely used clinically but postsurgical infections lead to serious complications. During surgery, and particularly in oral implantology, microorganisms may colonize the surface of implants and hamper the tissue integration. It becomes especially difficult for the immune system to clear robustly adherent antibiotic-resistant biofilm infections on medical devices. In addition, the expansion of bacterial antibiotic resistance is a growing problem today. For many years, it has been known that silver ions exhibit strong inhibitory effects towards a broad spectrum of bacterial strains. However, Ag^+ is also toxic for mammalian cells at high concentrations. In a recent paper, Zhao et al. have loaded titania (TiO_2) nanotubes with silver nanoparticles. After anodization, titanium implants were simply soaked in $AgNO_3$ solution, followed by ultraviolet light irradiation. Due to the outstanding photocatalytic properties of TiO_2, reduction of Ag^+ occurred and nanoparticles of Ag adhered tightly to the wall of the TiO_2 nanotubes. The amount of Ag introduced to the nanotubes could be easily varied by changing processing parameters such as the $AgNO_3$ concentration and immersion time. Depending on loading conditions, the release of Ag^+ was in the range of 0.05-0.5 ppm for up to 2 weeks, a relatively low concentration compatible to cell viability. They showed that the TiO_2 nanotubes loaded with Ag nanoparticles were effective against bacteria growth and in preventing bacterial adhesion on the implant surface without obvious decline for 30 days. This time span is long enough to prevent postsurgical infection in the early and intermediate stages and perhaps even late infection around dental implants. The authors have also studied the cytotoxicity of silver-loaded nanotubes using osteoblasts. They found limited cytotoxicity of leached samples that could be reduced by controlling the Ag^+ release rate [53]. In conclusion, using TiO_2 nanotubes as release systems to fight together with controlled biological response and accelerated tissue integration may rapidly become an attractive solution for dental and orthopedic implants. This simple method used by Zhao et al. [53] may be an interesting strategy in order to prevent peri-implantitis, one of the major drawbacks in the growing market of dental implants. Future work should examine preclinical models for studying the osseointegration of implants in infected sites.

5. CONCLUSION

Nanotechnology may produce surfaces with controlled topography and chemistry that help to improve our understanding of biological interactions and help to develop novel implant surfaces with predictable tissue-integrative properties. In this perspective, nanotopographies were created, particularly titania nanotubes, by a simple electrochemical anodization of Ti implants. TiO_2 nanotubes grow perpendicularly to the surface, forming a regular array of nanopores with controllable dimensions. It has been shown by several groups that these nanotubes control the fate of MSCs and promote differentiation to osteoblastic lineage. Consequently, many studies have shown that nanostructured Ti implants were similar or even faster and better integrated into bone tissue than untreated ones or with other surface modifications [15,43,54].

6. FUTURE PERSPECTIVES

TiO_2 nanotubes applied on orthopedic or dental implants are potent carriers of proteins that can trigger adhesion, proliferation, or differentiation of cells. The control of cell responses by nanotechnology is a fast-growing field of research as it potentially opens new directions in peri-implant tissue healing. The technology of nanotubes may be introduced to other medical devices such as cardiovascular stents that locally release antithrombotic drugs or growth factors.

REFERENCES

[1] L. Le Guéhennec, A. Soueidan, P. Layrolle, Y. Amouriq, Surface treatments of titanium dental implants for rapid osseointegration, Dent. Mater. 23 (7) (2007) 844–854.
[2] L. Bacakova, E. Filova, M. Parizek, T. Ruml, V. Svorcik, Modulation of cell adhesion, proliferation and differentiation on materials designed for body implants, Biotechnol. Adv. 29 (6) (2011) 739–767.
[3] M.J. Dalby, Topographically induced direct cell mechanotransduction, Med. Eng. Phys. 27 (9) (2005) 730–742.
[4] H.W. Liwen Lin, Enhanced osteointegration of medical titanium implant with surface modifications in micro/nanoscale structures, J. Orthop. Transl. 2 (1) (2014) 35–42.
[5] V. Zwilling, M. Aucouturier, E. Darque-Ceretti, Anodic oxidation of titanium and TA6V alloy in chromic media. An electrochemical approach, Electrochim. Acta 45 (6) (1999) 921–929.
[6] D. Gong, C.A. Grimes, O.K. Varghese, W. Hu, R.S. Singh, Z. Chen, et al., Titanium oxide nanotube arrays prepared by anodic oxidation, J. Mater. Res. 16 (12) (2001) 3331–3334.

[7] D. Guan, P.J. Hymel, C. Zhou, Y. Wang, Engineering bamboo-type TiO_2 nanotube arrays to enhance their photocatalytic property, J. Nanosci. Nanotechnol. 14 (6) (2014) 4541–4550.
[8] R. Yuan, B. Zhou, D. Hua, C. Shi, L. Ma, Effect of metal-ion doping on the characteristics and photocatalytic activity of TiO_2 nanotubes for the removal of toluene from water, Water Sci. Technol. J. Int. Assoc. Water Pollut. Res. 69 (8) (2014) 1697–1704.
[9] P.X. Yunhuai Zhang, Carbon monoxide annealed TiO_2 nanotube array electrodes for efficient biosensor applications, J. Mater. Chem. 19 (7) (2009) 948–953.
[10] Z.-D. Gao, J. Guo, N.K. Shrestha, R. Hahn, Y.-Y. Song, P. Schmuki, Nickel hydroxide nanoparticle activated semi-metallic TiO_2 nanotube arrays for non-enzymatic glucose sensing, Chem. Eur. J. 19 (46) (2013) 15530–15534.
[11] H.M.A. Javed, W. Que, Z. He, Anatase TiO_2 nanotubes as photoanode for dye-sensitized solar cells, J. Nanosci. Nanotechnol. 14 (2) (2014) 1085–1098.
[12] X. Zhen, D. Yu, F.-Q. Xiong, M. Li, Z. Yang, J. Zhu, et al., Controlled growth of semiconductor nanofilms within TiO_2 nanotubes for nanofilm sensitized solar cells, Chem. Commun. (Camb.) 50 (33) (2014) 4364–4367.
[13] K.S. Brammer, C.J. Frandsen, S. Jin, TiO_2 nanotubes for bone regeneration, Trends Biotechnol. 30 (6) (2012) 315–322.
[14] S. Lavenus, V. Trichet, S. Le Chevalier, A. Hoornaert, G. Louarn, P. Layrolle, Cell differentiation and osseointegration influenced by nanoscale anodized titanium surfaces, Nanomedicine 7 (7) (2012) 967–980.
[15] Y.-T. Sul, Electrochemical growth behavior, surface properties, and enhanced *in vivo* bone response of TiO_2 nanotubes on microstructured surfaces of blasted, screw-shaped titanium implants, Int. J. Nanomedicine 5 (2010) 87–100.
[16] P. Chennell, E. Feschet-Chassot, T. Devers, K.O. Awitor, S. Descamps, V. Sautou, *In vitro* evaluation of TiO_2 nanotubes as cefuroxime carriers on orthopaedic implants for the prevention of periprosthetic joint infections, Int. J. Pharm. 455 (1–2) (2013) 298–305.
[17] Y. Hu, K. Cai, Z. Luo, D. Xu, D. Xie, Y. Huang, et al., TiO_2 nanotubes as drug nanoreservoirs for the regulation of mobility and differentiation of mesenchymal stem cells, Acta Biomater. 8 (1) (2012) 439–448.
[18] T. Shokuhfar, S. Sinha-Ray, C. Sukotjo, A.L. Yarin, Intercalation of anti-inflammatory drug molecules within TiO_2 nanotubes, RSC Adv. 3 (38) (2013) 17380–17386.
[19] Y.-Y. Song, F. Schmidt-Stein, S. Bauer, P. Schmuki, Amphiphilic TiO_2 nanotube arrays: an actively controllable drug delivery system, J. Am. Chem. Soc. 131 (12) (2009) 4230–4232.
[20] N.H. Wael Att, Time-dependent degradation of titanium osteoconductivity: an implication of biological aging of implant materials, Biomaterials 30 (29) (2009) 5352–5363.
[21] P. Hoyer, Formation of a titanium dioxide nanotube array, Langmuir 12 (6) (1996) 1411–1413.
[22] C. Bae, H. Yoo, S. Kim, K. Lee, J. Kim, M.M. Sung, et al., Template-directed synthesis of oxide nanotubes: fabrication, characterization, and applications, Chem. Mater. 20 (3) (2008) 756–767.
[23] J. Qiu, W. Yu, X. Gao, X. Li, Fabrication and characterization of TiO_2 nanotube arrays having nanopores in their walls by double-template-assisted sol-gel, Nanotechnology 18 (29) (2007) 295604.
[24] M. Zhang, Y. Bando, K. Wada, Sol-gel template preparation of TiO_2 nanotubes and nanorods, J. Mater. Sci. Lett. 20 (2) (2001) 167–170.
[25] N. Liu, X. Chen, J. Zhang, J.W. Schwank, A review on TiO_2-based nanotubes synthesized via hydrothermal method: Formation mechanism, structure modification, and photocatalytic applications, Catal. Today 225 (2014) 34–51.

[26] C.L. Wong, Y.N. Tan, A.R. Mohamed, A review on the formation of titania nanotube photocatalysts by hydrothermal treatment, J. Environ. Manage. 92 (7) (2011) 1669–1680.

[27] A.W. Tan, B. Pingguan-Murphy, Review of titania nanotubes: fabrication and cellular response, Ceram. Int. 38 (6) (2012) 4421–4435.

[28] G.K. Mor, O.K. Varghese, M. Paulose, K. Shankar, C.A. Grimes, A review on highly ordered, vertically oriented TiO_2 nanotube arrays: fabrication, material properties, and solar energy applications, Sol. Energy Mater. Sol. Cells 90 (14) (2006) 2011–2075.

[29] P. Roy, S. Berger, P. Schmuki, TiO_2 nanotubes: synthesis and applications, Angew. Chem. Int. Ed. 50 (13) (2011) 2904–2939.

[30] J.-H. Lee, I.-C. Leu, M.-C. Hsu, Y.-W. Chung, M.-H. Hon, Fabrication of aligned TiO_2 one-dimensional nanostructured arrays using a one-step templating solution approach, J. Phys. Chem. B 109 (27) (2005) 13056–13059.

[31] G. Liu, K. Wang, N. Hoivik, H. Jakobsen, Progress on free-standing and flow-through TiO_2 nanotube membranes, Sol. Energy Mater. Sol. Cells 98 (2012) 24–38.

[32] S. Minagar, C.C. Berndt, J. Wang, E. Ivanova, C. Wen, A review of the application of anodization for the fabrication of nanotubes on metal implant surfaces, Acta Biomater. 8 (8) (2012) 2875–2888.

[33] M.V. Diamanti, M. Ormellese, Anodic titanium oxide as immobilized photocatalyst in UV or visible light devices, J. Hazard. Mater. 186 (2–3) (2011) 2103–2109.

[34] D.A.H. Hanaor, C.C. Sorrell, Review of the anatase to rutile phase transformation, J. Mater. Sci. 46 (4) (2011) 855–874.

[35] P. Xiao, D. Liu, B.B. Garcia, S. Sepehri, Y. Zhang, G. Cao, Electrochemical and photoelectrical properties of titania nanotube arrays annealed in different gases, Sens. Actuators B Chem. 134 (2) (2008) 367–372.

[36] S. Sreekantan, Z. Lockman, R. Hazan, M. Tasbihi, L.K. Tong, A.R. Mohamed, Influence of electrolyte pH on TiO_2 nanotube formation by Ti anodization, J. Alloys Compd. 485 (1–2) (2009) 478–483.

[37] S. Sreekantan, K.A. Saharudin, Z. Lockman, T.W. Tzu, Fast-rate formation of TiO_2 nanotube arrays in an organic bath and their applications in photocatalysis, Nanotechnology 21 (36) (2010) 365603.

[38] K. Shankar, G.K. Mor, H.E. Prakasam, S. Yoriya, M. Paulose, O.K. Varghese, et al., Highly-ordered TiO_2 nanotube arrays up to 220 μm in length: use in water photoelectrolysis and dye-sensitized solar cells, Nanotechnology 18 (6) (2007) 065707.

[39] M. Paulose, H.E. Prakasam, O.K. Varghese, L. Peng, K.C. Popat, G.K. Mor, et al., TiO_2 nanotube arrays of 1000 μm length by anodization of titanium foil: phenol red diffusion, J. Phys. Chem. C 111 (2007) 14992–14997.(cité 5 déc 2014). Disponible sur: http://wenku.baidu.com/view/5dd6824a852458fb770b5603.

[40] Y.C. Satoshi Kaneco, Fabrication of uniform size titanium oxide nanotubes: impact of current density and solution conditions, Scr. Mater. 5 (2007) 373–376.

[41] N.K. Allam, C.A. Grimes, Effect of cathode material on the morphology and photoelectrochemical properties of vertically oriented TiO_2 nanotube arrays, Sol. Energy Mater. Sol. Cells 92 (11) (2008) 1468–1475.

[42] C. Pérez-Jorge, A. Conde, M.A. Arenas, R. Pérez-Tanoira, E. Matykina, J.J. de Damborenea, et al., *In vitro* assessment of Staphylococcus epidermidis and Staphylococcus aureus adhesion on TiO_2 nanotubes on Ti-6Al-4V alloy, J. Biomed. Mater. Res. A 100A (7) (2012) 1696–1705.

[43] L. Salou, A. Hoornaert, G. Louarn, P. Layrolle, Enhanced osseointegration of titanium implants with nanostructured surfaces: an experimental study in rabbits, Acta Biomater. 11 (2015) 494–502.

[44] Y. Lai, J. Gong, C. Lin, Self-organized TiO_2 nanotube arrays with uniform platinum nanoparticles for highly efficient water splitting, Int. J. Hydrog. Energy 37 (8) (2012) 6438–6446.

[45] S.H. Kang, H.S. Kim, J.-Y. Kim, Y.-E. Sung, An investigation on electron behavior employing vertically-aligned TiO_2 nanotube electrodes for dye-sensitized solar cells, Nanotechnology 20 (2009) 355307.
[46] K. Cai, J. Bossert, K.D. Jandt, Does the nanometre scale topography of titanium influence protein adsorption and cell proliferation? Colloids Surf. B: Biointerfaces 49 (2) (2006) 136–144.
[47] S. Lavenus, J.-C. Ricquier, G. Louarn, P. Layrolle, Cell interaction with nanopatterned surface of implants, Nanomedicine 5 (6) (2010) 937–947.
[48] R. McBeath, D.M. Pirone, C.M. Nelson, K. Bhadriraju, C.S. Chen, Cell shape, cytoskeletal tension, and RhoA regulate stem cell lineage commitment, Dev. Cell 6 (4) (2004) 483–495.
[49] M.J. Dalby, N. Gadegaard, R. Tare, A. Andar, M.O. Riehle, P. Herzyk, et al., The control of human mesenchymal cell differentiation using nanoscale symmetry and disorder, Nat. Mater. 6 (12) (2007) 997–1003.
[50] J. Park, S. Bauer, K. von der Mark, P. Schmuki, Nanosize and vitality: TiO_2 nanotube diameter directs cell fate, Nano Lett. 7 (6) (2007) 1686–1691.
[51] S. Oh, K.S. Brammer, Y.S.J. Li, D. Teng, A.J. Engler, S. Chien, et al., Stem cell fate dictated solely by altered nanotube dimension. Proc. Natl. Acad. Sci. 106 (7) (2009) 2130–2135, http://dx.doi.org/10.1073/pnas.0813200106.
[52] M.P. Neupane, I.S. Park, T.S. Bae, H.K. Yi, M. Uo, F. Watari, et al., Titania nanotubes supported gelatin stabilized gold nanoparticles for medical implants, J. Mater. Chem. 21 (32) (2011) 12078–12082.
[53] L. Zhao, H. Wang, K. Huo, L. Cui, W. Zhang, H. Ni, et al., Antibacterial nanostructured titania coating incorporated with silver nanoparticles, Biomaterials 32 (24) (2011) 5706–5716.
[54] L.M. Bjursten, L. Rasmusson, S. Oh, G.C. Smith, K.S. Brammer, S. Jin, Titanium dioxide nanotubes enhance bone bonding *in vivo*, J. Biomed. Mater. Res. A 92 (3) (2010) 1218–1224.

CHAPTER 21

Biocomposite and Bioceramic Coatings and Materials

Cornelia Wolf-Brandstetter, Dieter Scharnweber
Technische Universität Dresden, Institute of Materials Science, Max Bergmann Center of Biomaterials, Dresden, Germany

Chapter Contents

1. INTRODUCTION

In 1926, de Jong et al. identified the chemistry and structure of bone mineral such as that of hydroxyapatite (HAp) [1]. This formed the basis for using HAp-either synthetically or biologically derived-as a bone substitute material and as coatings of implants with bone contact [2,3]. Healing time frames for bone are usually over 2-3 months, and it is a limitation of Hap that its remodeling time *in vivo* does not correspond to this. To improve the remodeling time, other calcium phosphate phases (CPPs) such as tricalcium

Handbook of Nanoceramic and Nanocomposite Coatings and Materials
http://dx.doi.org/10.1016/B978-0-12-799947-0.00021-3

phosphate (TCP) and brushite and mixtures of different CPPs have been introduced for the treatment of bone defects. Figure 1 gives an overview of currently used CPP-based bone substitute materials [4]. Besides ceramic materials of a given shape, either granules or cements are in wide use as they can fill even irregularly shaped defects almost perfectly.

A variety of methods were developed to deposit CPP (mainly HAp) onto metal surfaces. The available techniques can be divided into high-energy physical methods and low-energy wet-chemical techniques. The physical methods comprise atmospheric plasma spraying (APS), vacuum plasma spraying, suspension plasma spraying, liquid plasma spraying (LPS), high-velocity suspension flame spraying, high-velocity oxygen fuel, gas tunnel-type plasma spraying, pulsed laser deposition (PLD), matrix-assisted pulsed laser evaporation (MAPLE), ion beam-assisted deposition (IBAD), and radio-frequency (RF) magnetron sputtering [5]. All have the major disadvantage that organic additives for specific osteogenic induction would be damaged during the fabrication process. Furthermore, for most methods based on spray processes, it would be impossible to coat irregular geometries homogeneously, particularly in the case of undercuts. For both reasons, new technologies based on wet-chemical techniques were developed. The three most important methods are the biomimetic deposition from saturated

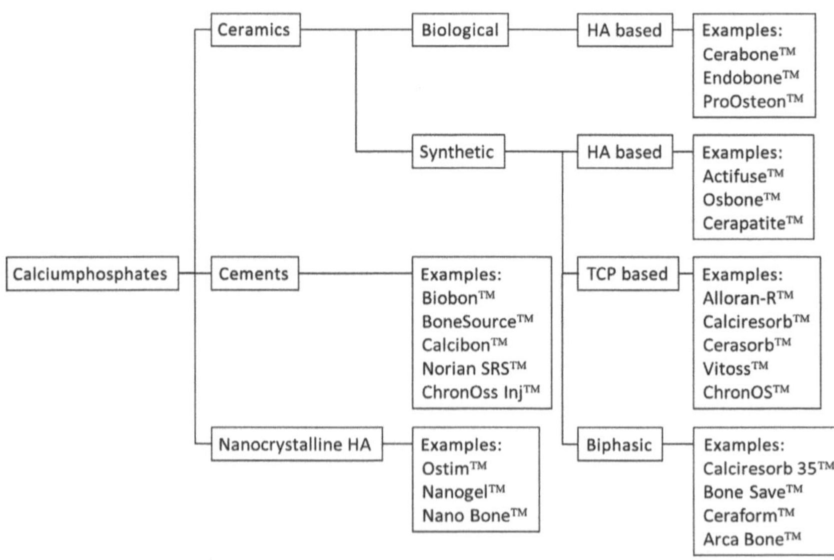

Figure 1 CPP-based bone replacement materials based on chemical compositions and application technique [4].

calcium phosphate solutions, sol-gel-based coatings (described in more detail in Chapter 4), and the electrochemically assisted deposition (ECAD). Although the most recent publications focus on the initially mentioned biomimetic deposition, this process requires comparatively long process times (up to days). Also, it is difficult to transfer to commercialized processes because of difficulties with the reproducibility of the process.

2. BIOCOMPOSITES

2.1 Purely Inorganic Materials

From the previously mentioned mixtures of CPP, especially, ceramics based on HAp with β-TCP additions are used because of their tunable degradation behavior [6]. Cho et al. [7] had developed a flame spray pyrolysis pathway for the synthesis of nanosized biphasic CPP of a wide range of Ca/P molar ratios.

This is in line with a recent tendency toward using nanocrystalline CCP for biomedical applications because of their improved degradation behavior and mechanical properties [6]. Among nanocrystalline samples, the highest resorption was found for nano-HAp with a low carbonate content [8]. CPP resorption by osteoclast-like cells did not show a strong effect of grain size.

Additionally, Zn and carbonate substitution have been used to tune the degradation behavior of nano-HAp [9]. One approach to improve the mechanical properties of nano-β-TCP used a reinforcement of the porous structures with bioglass [10]. Comparable effects are reported for doping bioglass with additional cations such as Zn [11] and for using Sr-doped bioglass to improve the mechanical properties of HAp bioceramics without negative effects on the cell's biological behavior [12].

To improve the mechanical properties, grain growth of nanocrystalline CPP during sintering processes should be avoided. Adding carbon black as a source of carbonate has shown excellent efficiency in this regard [13]. Similar effects have been reported for the addition of zirconia via a continuous hydrothermal synthesis, resulting in the formation of HAp-zirconia nanocoprecipitates [14].

Pore size and geometry of CPP can also be influenced by specific sintering regimes [15]. This has effects on protein adsorption and desorption phenomena [16,17] as well as the cellular response [18] and bone healing capacity [19], with nanocrystalline materials giving improved performance. One possible explanation for this advantageous biological behavior of nano-HAp can potentially be traced back to nonstoichiometric (010) surfaces coexisting on HAp nanocrystals. These were shown by Ospina et al. [20] using high-resolution transmission electron microscopy and density functional theory.

To improve the mechanical properties of highly porous HAp scaffolds for tissue engineering applications, Zhang et al. [21] developed a double slip-casting method using powders of different particle size similar to Lin et al. [22].

Additionally, there is ongoing work to combine the osteoconductive properties of CPP with antibacterial ones through the incorporation of silver nanoparticles [23].

2.2 Inorganic Organic Compound Materials

To prepare highly porous scaffolds with sufficient mechanical properties, CPPs have been combined with organic components such as chitosan [24,25], gelatin [26], and poly(lactide-co-glycolide) (PLGA) [27,28]. Materials with both good mechanical and biological property profiles typically show CPP contents in the range of 30-40% [24,27,28].

Kweon et al. [29] had developed a nano-HAp graft with silk fibroin bone grafting materials by transforming eggshell $CaCO_3$ into HAp in a three-step process that includes the treatment of calcinated and crushed eggshell with phosphoric acid. Bone regeneration, however, was best in the nano-HAp group without silk fibroin. As another biological source for nano-CPP, dried starfish bone has been used [30].

Contrary to granular or preshaped ceramics, self-setting CPP cements have the great advantage of being able to fill complex bone cavities. The addition of type I collagen to such cements has been shown to improve the work of fracture more than tenfold in combination with a twofold increase in cell attachment [31]. Similar results have been reported for using chitosan instead of collagen [32].

3. CPP COATINGS PREPARED BY ECAD

3.1 Basic Principle of the ECAD Process

The term *electrochemically assisted* reflects the fact that, even though electrochemical reactions play a major role in the overall process, in the actual deposition of CPP, there is no transfer of charge carriers in the metallic substrate/electrolyte interface in the sense of an electrochemical reaction. Instead, the charge transfer gives rise to the formation of OH^- ions. This results locally in (i) an increase of the pH value, (ii) a change of the equilibrium of the phosphate dissociation [33], and (iii) an increase of the relative supersaturation σ of the electrolyte solution (containing Ca^{2+} and $H_xPO_4{}^{(3-x)-}$ ions) with respect to CPP. It thus *induces* the formation of particulate CPP and its deposition on the substrate surface. In air-saturated solutions, the two following overall reactions at the substrate will have to be taken into account:

$$2H_2O + 2e^- \rightarrow H_2 + 2OH^- \quad (1)$$

$$O_2 + 4e^- + 2H_2O \rightarrow 4OH^- \quad (2)$$

In the current density range of several milliamperes, which is typically used for the ECAD process, the predominant electrochemical reaction consists in the formation of hydrogen according to Equation (2).

For the electrochemical polarization, three different methods can be used [34]: constant cell voltage, constant potential of the working electrode (potentiostatic), and constant current (galvanostatic). Due to changes in the resistance of the surface by the deposited mineral phases, only in the latter case, the turnover of the electrochemical reactions at the substrate is independent of time.

3.2 Parameters Influencing the Coating Properties

The formation of a specific CPP and the morphology of the resulting coating generated during the ECAD process are highly determined by a variety of parameters such as electrochemical conditions, electrolyte composition, temperature, and deposition time. The individual influence of each of these parameters strongly depends on the actual settings of the other parameters. Hence, in the following, the general trends of the impact of each parameter will be explained and some typical examples will be given.

3.2.1 Influence of pH

Based on the principle of the ECAD process and the pH-dependent solubility of different CPPs, the outstanding role of pH is clearly understandable. From Figure 2, it can be seen that the solubility product decreases with increasing pH up to pH 8 for all investigated CPPs. With further pH increase, the decrease in solubility continues only for the alkaline CPPs, namely, octacalcium phosphate (OCP), TCP, and HAp.

For the main components reported in literature as deposited by means of ECAD, the brutto reaction equations are given below. The consumption of OH ions is a measure of the influence of the pH on the formation of the respective CPP. As indicated by the indexes and factors x in the equations,

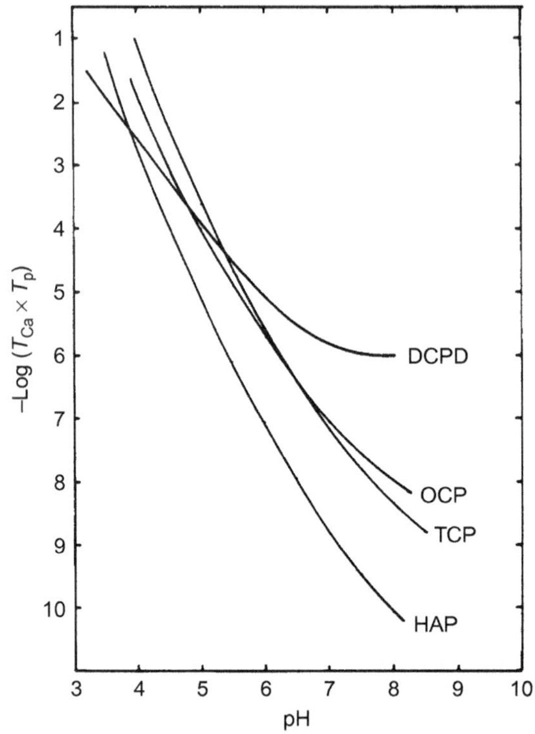

Figure 2 Solubility isotherms of CPPs at 37 °C and 0.1 mol/L ionic strength. *(From [35]).*

the consumption is clearly a function of the pH of the used electrolyte. Furthermore, because there is a relation between composition and the buffer capacity, x also depends on the concentration of the electrolyte:

- Formation of brushite (BR) $M_{BR}=172$ g/mol

$$Ca^{2+} + H_xPO_4^{(3-x)-} + (x-1)OH^- + (3-x)H_2O \rightarrow CaHPO_4{\cdot}2H_2O \tag{3}$$

- Formation of OCP $M_{OCP}=892$ g/mol

$$8Ca^{2+} + 6H_xPO_4^{(3-x)-} + (6x-2)OH^- \rightarrow Ca_8H_2(PO_4)_6 + (6x-2)H_2O \tag{4}$$

- Formation of amorphous calcium phosphate (ACP) $M_{ACP}=310$ g/mol

$$3Ca^{2+} + 2H_xPO_4^{(3-x)-} + 2xOH^- \rightarrow Ca_3(PO_4)_2 + 2xH_2O \tag{5}$$

- Formation of HAp $M_{HAp}=1004$ g/mol

$$10Ca^{2+} + 6H_xPO_4{}^{(3-x)-} + (6x+2)OH^- \rightarrow Ca_{10}(PO_4)_6(OH)_2 + 6_xH_2O \quad (6)$$

For pH values of the electrolyte between 3 and 6 (which is valid for most investigations in the literature), it follows from the pK_s values of phosphoric acid [33] that x in the equations given above is close to 2. Apart from the pH of bulk solution, the actual local pH occurring at the site of deposition depends on all parameters that affect the number of OH^- generated by the process and the consumption by precipitation and transport processes.

3.2.2 Current Density and Deposition Time

As stated above, the deposition principle consists in the generation of OH^- at the substrate surface. The two process parameters that directly influence the amount of OH^- generation caused by the electrochemical reaction are the current density and the polarization time. However, the degree of changes provoked by these two parameters can be strongly modulated by the actual settings of other solution parameters (bulk pH, concentration of Ca and PO_4, temperature, and convective conditions).

To elucidate the individual contribution of each of the two parameters, a design of experiments would be required keeping the total charge applied constant and varying all important process parameters for at least two different settings to investigate the interaction with other process parameters. Unfortunately, such comprehensive data are not available from published literature. As most authors focusing on the impact of current density investigated several combinations of current density and process time, this section will deal with the combined impact of both parameters.

Redepenning and McIsaac [36] varied the current density (between 0.1 and 10 mA/cm^2) and the deposition time (8-133 min) under galvanostatic conditions working at room temperature with acidic (pH 3.5) solutions saturated with $Ca(H_2PO_4)_2$. Under these conditions, the current density and polarization time mainly affected the crystal size of the CCP, which in this case was brushite.

Using a similar but fluoride-containing electrolyte and working at elevated temperature, Monma et al. [37] observed the transition of apatitic crystals with polarization time from ellipsoidal rods with conical heads over hexagonal pyramidal rods to hexagonal prismatic rods for current densities of 2 and 6 mA/cm^2, respectively. Ban and Maruno [38] varied the current density between 1.3 and 12.9 mA/cm^2 for deposition times between 20 and 720 min, additionally investigating the impact of temperature (52-92 °C).

For the low current densities, the coating transformed from OCP to HAp with increasing deposition time and process temperature. For the high current densities, only HAp was detected irrespective of the temperature and time of deposition. For the change from brushite to HAp, similar trends were observed by Kuo and Yen [39] under the conditions of at least 10 mA/cm^2 and electrolytes with higher Ca and PO_4 concentrations.

Rossler et al. [40] and Sewing et al. [41] studied several phases of the CPP deposition. The formation of an initial CPP base layer has been observed for the whole investigated range of current densities (0.5-10 mA/cm^2), being faster for higher ones. Atop this initial layer, spherical amorphous CPPs were formed. The formation of the nanocrystalline HAp started directly on the surfaces of ACP spheres and on the initially formed CPP islets. With continuing polarization, the substrate surface is finally completely covered with needlelike HAp crystals with typical dimensions of <500 nm in length and <60 nm in diameter for current densities above 7 mA/cm^2 as shown in a pseudo phase diagram in Figure 3 [40].

According to Nancollas [35], the formation of HAp from ACP can be described as a transformation into a crystalline product accompanied by a dehydration:

$$3Ca_3(PO_4)_2 \cdot nH_2O + Ca^{2+} + 2OH^- \rightarrow Ca_{10}(PO_4)_6(OH)_2 + 3nH_2O \tag{7}$$

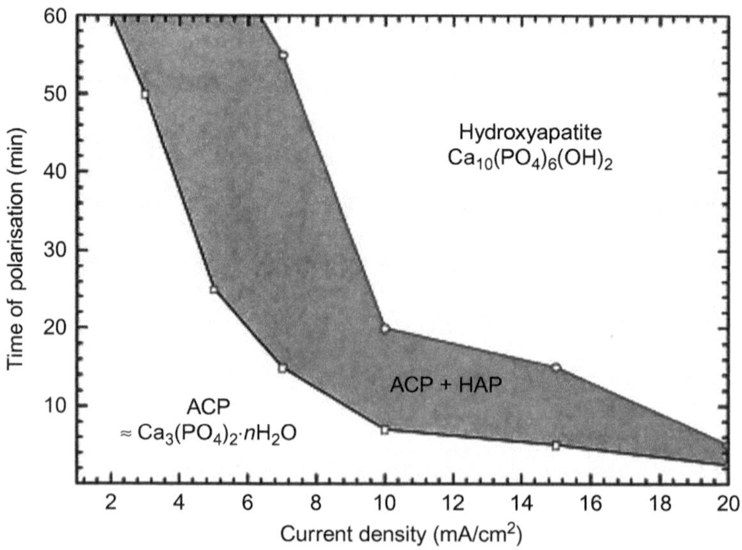

Figure 3 Phase diagrams of the composition of ECAD-CPP layers as identified by SEM and FTIR as a function of current density and deposition time [40].

Irrespective of slight process differences caused by the mode of polarization, the same trends were observed for potentiostatic polarizations and experiments with defined cell voltages when increasing current densities were applied due to increasing potentials. For cell voltages of up to 6 V, Wang et al. [42] detected deposition of brushite phases with increasing crystal sizes, whereas for voltages of >7 V, HAp started to appear, becoming the major phase at 10 V. Varying potentials in a study by Montero-Ocampo et al. [43] between 1.6 and 2.0 V resulted in steady-state current densities of 3 and 10 mA/cm^2. The respective coatings were identified as $Ca_3(PO_4)_2$ (TCP) and hexagonal HAp as the main phase for the more negative potential. For high current densities, prolonging deposition times resulted in coarsening of the crystal morphology [44].

Ban and Maruno [45] and Montero-Ocampo et al. [43] demonstrated a square root relation between the deposited mass and the deposition time for potentiostatic conditions, which indicates that, irrespective of the electrochemical conditions, deposition can be limited by diffusion processes.

3.2.3 *Temperature*

Another important parameter influencing the formation of CCP coatings is the process temperature. It is of special importance for the deposition of highly crystalline coatings but additionally serves as a limiting factor in the development of coatings containing organic components like proteins in codeposition procedures.

From SEM images and XRD studies, it was observed by several authors that in general, higher temperatures lead to the formation of crystalline phases with increasing crystal size [38,44,45] and to the formation of higher percentages of HAp [38,42]. This is exemplarily shown in Figure 4 for XRD studies.

Additionally, rising temperature also increased the deposited amount of CPP [38,44–46]. Except for short deposition times, it was concluded that an increase in temperature always provokes a higher rate of crystal growth compared to the rate of nucleation [46] explaining the abovementioned results.

3.2.4 *Concentration*

The formation of CPP close to the surface depends on the degree of supersaturation, which is mainly affected by the used concentrations of Ca and

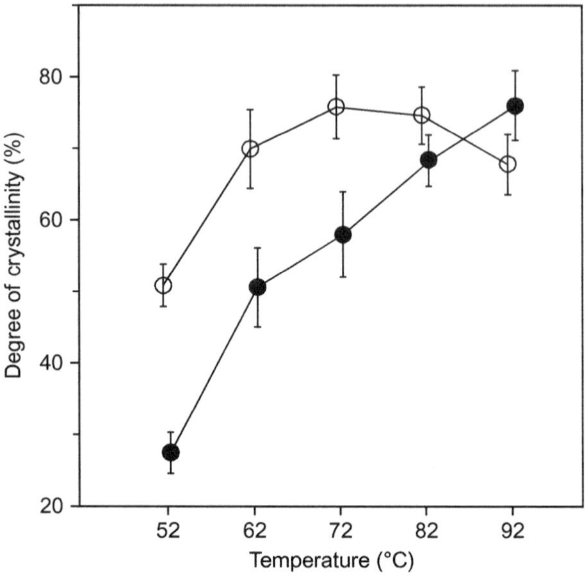

Figure 4 Intensities of signals from X-ray diffraction assigned to OCP and HAp for coatings prepared for 1 h at current densities of 1.3 (○) and 12.9 (•) mA/cm^2 as a function of the electrolyte temperature. *(From [38]).*

PO_4 (expressed as the ionic product of the solution in Eq. 8), besides the solubility product at the actual pH and temperature:

$$i_{\text{eff}} = i_{\text{total}} \frac{1}{(1-\theta)} \tag{8}$$

with i_{eff} as the current density with respect to the substrate surface available for electrochemistry (not coated by CPP), i_{total} as the current density with respect to the total substrate surface, and θ as the fraction of surface covered by CPP.

In the different published experiments, calcium concentrations varied over a broad range from 0.5 to 21 mM, sometimes also varying the Ca/PO_4 ratio but often keeping it constant at 1.67. Solutions with high calcium concentration require acidic pH of the solution to avoid precipitation in the bulk electrolyte.

Shirkhanzadeh [47] observed more resolved and intense FTIR bands corresponding to HAp in the case of low calcium concentrations, while higher Ca concentrations resulted in mixtures of OCP and HAp (with no details given about the used pH).

If depositions were performed at a neutral pH of 7.2 and comparatively low concentrations varying between 2.4 and 4 mM Ca and 1.0 and 2.5 mM PO_4 (resulting in Ca/PO_4 ratios between 1.0 and 4.0), the morphology of the obtained coatings (amorphous CPP) was not influenced by the Ca/PO_4 ratio of the electrolyte [48].

A slightly different impact of Ca/PO_4 ratio was concluded from a more comprehensive study performed at pH 6.4 by Sewing et al. [41]. This study showed that a decrease in the Ca/PO_4 ratio of the electrolyte did not significantly influence the onset of HAp crystal formation from an ACP coating, but delayed the transformation process of ACP to HAp. This resulted in a much larger parameter field where ACP and HAp coexist (green area in the lower plane of Figure 5).

3.3 Composites

Apart from variations in deposited amounts or type of CPP, degree of crystallinity, and crystal size, the opportunities to further increase the biological

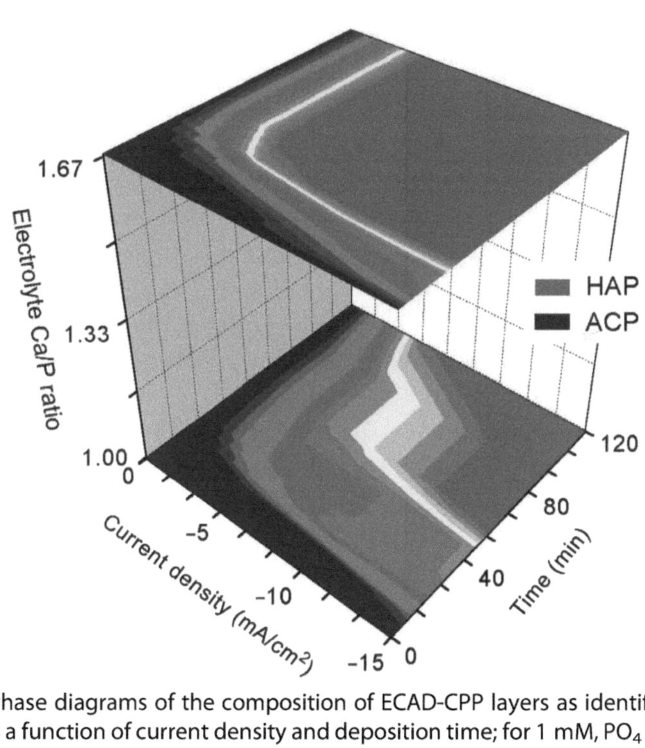

Figure 5 Phase diagrams of the composition of ECAD-CPP layers as identified by SEM and FTIR as a function of current density and deposition time; for 1 mM, PO_4 and Ca/PO_4 ratios between 1 and 1.67. *(From [41]).*

performance were limited. Hence, recent research focused on the incorporation of specific inorganic additives or organic molecules aiming for different physicochemical properties [49] and very specific aspects like antimicrobial effects, improved cellular adhesion, proliferation and differentiation, and inhibition of osteoclastogenesis.

3.3.1 Composite Coatings with Inorganic Additives

Several modifications of the original ECAD method based on calcium- and phosphate-containing solutions were performed with the aim of CPP deposition with ion substitution of anions and cations. Early attempts concentrated on the naturally occurring mineral phases like fluorinated or carbonated apatites. While the first was achieved by the presence of fluoride ions in the solution [50], the latter was often the result of processing modes without special control of the gas atmosphere, but was also actively modulated by Haddow et al. [51].

More recently, a series of cations, in particular Mg, Zn, and Si and Sr ions, were included in CPP coatings because of their ability to promote specific osteoblast response and-in the case of Sr-inhibit osteoclasts.

Li et al. [52] presented the incorporation of silicon into the apatite lattice by changing the molar ratio of $SiO_3/(SiO_3+PO_4)$ in the range of 0.01 to 0.3. Ca and PO_4 concentrations were kept constant at 1.2 and 0.72 mM, respectively. Uniform Si-HAp phase coatings were obtained with 0.7-1.4 μm in thickness. Percentages of silicon in the coatings increased linearly to about 0.48 wt% up to 0.03 molar ratio, and a plateau level of 0.55 wt% was reached at a molar ratio of 0.1. The presence of increasing silicon in the electrolyte provoked an inhibition of the growth of Si-HAp crystals, leading to the decrease of coating thickness and the formation of treelike Si-HAp crystals for the high molar ratios.

As a potential treatment for osteoporosis [53], strontium became a focus of research due to its ability to stimulate collagenous and noncollagenous protein synthesis and effectively block osteoclastic bone resorption [54]. Sr-doped CPP coatings with different microstructures and morphologies have been produced by Li et al. [55]. Using a high current density of 20 mA/cm^2 resulted dominantly in needlelike HAp or Sr-HAp, while at low current densities from 5 to 10 mA/cm^2, deposits with a faceted crystallite morphology, possibly monetite, were more prevalent. Doping with strontium in general introduced more faceted crystallites of the monetite phase compared to the pure HAp coatings when deposited under the same process parameters.

In another recent study by Huang et al. [56], Sr was incorporated simultaneously with gelatin into CPP layers from acidic solutions with a $(Ca+Sr)/PO_4$ ratio of 1.67 and a molar ratio of Sr:Ca of 1:9. Obtained coatings were composed of floral crystals or flake agglomerates that mainly consisted of carbonated CPP, with small amounts of incorporated Sr (confirmed by EDX). Neither the amount of deposited gelatin nor the identity of deposited CPP was analyzed. From the $(Ca+Sr)/PO_4$ ratio of 1.1 in the coatings, a mixture of brushite and HAp was concluded.

Other elements like silver or copper were incorporated to impart antibacterial properties while maintaining good biocompatibility or even improving it by proangiogenic effects [57,58]. The first attempts to immobilize Ag on CPP coatings concentrated on ion exchange in the preformed coatings of Ca for Ag ions by exposure to $AgNO_3$ solutions [59]. Although such coatings showed a burst release of nearly 50% after incubation in SBF for 1 day, they resulted in larger initial inhibition zones in agar diffusion tests than surfaces generated by codeposition of Ag and CPP. The best results have been shown for a combination of codeposition with ion exchange incorporation of Ag.

By the application of pulsed electrochemical deposition for electrolytes containing Ca, PO_4, and Ag ions and L-cysteine as coordination agent, Lu et al. [60] were able to deposit nano-Ag-loaded HAp coatings on titanium. The surfaces demonstrated initial antimicrobial activity against Gram-positive *Staphylococci* and Gram-negative *Escherichia coli*, but long-term behavior or release of Ag ions over time has not been investigated.

Only recently the application of copper came into the focus of research. Apart from its long known bacteriostatic properties [61], additional proangiogenic effects were described *in vitro* and *in vivo* [62,63]. The antibacterial effect is reached at concentrations below those that cause cell damage [64]. Furthermore, bacteria are not able to develop resistance in the case of decreasing copper concentration, the same being valid for Ag ions.

The successful homogeneous incorporation of copper into HAp layers has been demonstrated in our own investigations studying the impact of pulsed electrochemical deposition on resultant copper phases [57]. The presence of copper did not change the crystalline phase of deposited calcium phosphate (HAp), but provoked a decrease by a factor of 3 or 4 in the deposited amounts. The copper was deposited mainly as Cu(I) species with a minor fraction of basic copper (II) phosphates. Reduction of copper occurred not only at the surface of titanium but also within the HAp coating due to the reaction with hydrogen produced by the electrolysis of water during the cathodic polarization of the substrate.

Neumann and Prinz developed a process (24) that assumes a 3-stage biological interaction: (i) burst-like release of Cu to prevent microbial attachment, (ii) osteoconductive effect due to the dissolution of the brushite phase, and (iii) slow dissolution of the remaining phase with the lowest solubility ($CuPO_4$) to assist in mineralization and angiogenesis of the new bone.

Cobalt-another trace element-was incorporated into CPP coatings in earlier studies mainly to achieve a better bonding strength of deposited layers. In a sequential manner, first, a coating consisting of several CPP phases was converted to HAp by heat treatment with 0.1 M NaOH, and then, cobalt was deposited from an electrolyte with 0.5 M $CoSO_4$ and boric acid (pH 5.2). Homogeneous coating with Co contents between 12 and 33 wt% and bonding strength values between 6.7 ($\pm$24.4) and 20.4 ($\pm$3.1) MPa were achieved [65].

3.3.2 Composite Coatings with Organic Additives

Calcium phosphates are known for their adsorption of proteins due to their high specific surface area and specific interaction with acidic proteins such as osteocalcin and osteopontin that play dominant roles in biomineralization [66,67]. Hence, CPP coatings prepared by ECAD process were subject to several postdeposition adsorption treatments [59,68,69]. This processing based on adsorptive binding has the advantage that it can be performed from small volumes, allowing for higher concentrations even of expensive organic compounds. However, as these coating approaches do not use any electrochemical techniques, they will not be discussed in this part of the chapter.

One of the most studied proteins in combination with the ECAD process is type I collagen. The interest in this protein can be attributed to the fact that it is the main organic component of the bone and can thus potentially create implant surfaces with similar composition and morphology to the bone. It is also well known for its ability to improve osteoblast adhesion, proliferation, and differentiation [70,71]. Other advantages lie in the compatibility with the basic principle that allows an enrichment of positively charged molecules at the cathode surface and the decreased solubility with increasing pH and, last but not least, in the lower price and better availability compared with other proteins like growth factors.

In principle, two different approaches were documented: (i) the adsorption of preformed fibrillar collagen onto either metallic [72] or CPP precoated [73] surfaces followed by further CPP deposition resulting in the mineralization of deposited collagen (Figure 6) or (ii) the simultaneous

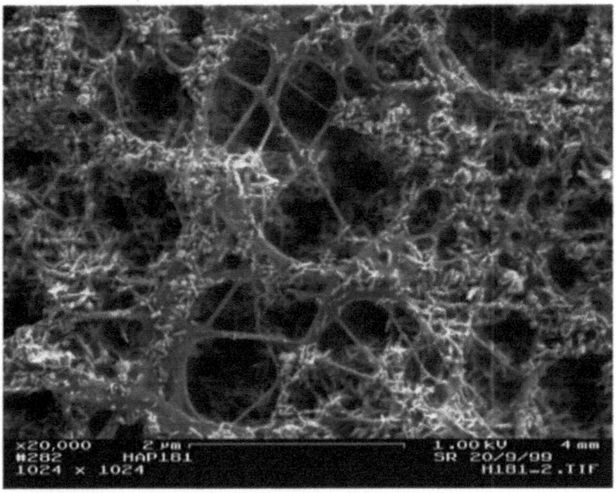

Figure 6 Mineralized fibrillar type I collagen as derived by ECAD formation of HAp onto collagen adsorbed to a preformed HAp base coating. *From [73].*

deposition of CPP and fibrillar collagen based on acidic solution of tropocollagen units and subsequent collagen fibrillogenesis.

High-resolution SEM investigations following the mineralization process showed that the characteristic banding pattern of collagen fibrils (63-67 nm) remained visible. The broadening and shift to lower wave numbers in FTIR spectra for the amide I band in the mineralized coating corresponded well with the broad amide I band of the bone, indicating an interaction between the two components of this coating in analogy to the situation *in vivo*.

The first studies using the second approach that aims to combine collagen fibrillogenesis with the deposition of CPP were published in a very short technical paper by Okamura et al. [74]. From electron spin resonance (ESR) studies, the authors concluded the occurrence of hydrogen bonding between H^+ of HPO_4^{2-} groups of the defect apatite and the carbonyl groups of collagen.

Fan et al. [75] varied acid-soluble type I collagen in concentrations between 0.1 and 0.5 mg/ml in slightly acidic electrolytes (pH 4.8-5.3). The ECAD process was performed at 30 °C in a potential range between -1 and $-3\ V_{SCE}$. Starting with a 100 nm thin basic layer of CPP, a three-dimensional porous network of mineralized collagen fibers up to a 100 μm thick was formed at longer polarization times. Mineralization was claimed to nucleate on the collagen fibers with crystal clusters of about 300-500 nm in size. Also, size and number increased with deposition time and ion concentration of the electrolyte. The final CPP was identified as OCP by electron diffraction. However, in SEM and TEM investigations of the mineralized samples, no banding of the collagen fibers could be shown. That indicates that the self-assembly of tropocollagen to supramolecular structures under the chosen conditions does not result in the formation of fibrils with native substructure. Similar coatings composed of a porous network of mineralized collagen consisting mainly of HAp/OCP attached to a thin CPP layer at the substrate surface were developed by Ling et al. [76]. Porosity and degree of mineralization was modulated by the presence of H_2O_2 (resulting in a higher local OH^- concentration due to electrolysis).

Another application of the ECAD process consists in the codeposition of potential therapeutic agents with CPP for a sustained release (68). The codeposition of the model protein bovine serum albumin (BSA) and HAp was performed by the immersion of preformed HAp coatings into BSA solutions containing calcium and phosphate ions. After a three-step process with increasing current densities from -0.09 to -0.47 mA/cm^2, the coatings consisted of 500 μg/cm^2 BSA within a total mass of 3.3 mg/cm^2. In comparison, pure adsorption of BSA resulted only in a load of about 7 μg/cm^2, which desorbed in a burst release within 2 h.

However, it has to be taken into account that the ECAD process significantly increases the pH values at the interface between substrate and electrolyte. This might have consequences for the conformation and thus biological activity of labile compounds immobilized in this manner.

A number of studies dealt with the incorporation of chitosan, a natural cationic polysaccharide, for different reasons such as increasing biocompatibility [77], increasing coating stability [78], or use as a drug carrier [79]. Redepenning et al. [78] investigated the combination of electrolytes based on different Ca- and PO_4-concentrations (0.18-0.6 mM Ca and 0.43-1.46 mM PO_4) with varying chitosan molalities from 0.008 to 0.047. The resulting chitosan/brushite composite coatings were converted to chitosan/HAp coatings by immersion in 0.1 M NaOH for 7 days at room temperature, resulting in chitosan contents between 1.2 and 16 wt%. Wang et al. [77,80] worked with slightly higher concentrated solutions of 2.5-4 mM Ca, 1-2 mM PO_4, and 0-0.2 g/L chitosan. He additionally varied the current densities and polarization times from 30 min to 15 h. In the SEM images, CPP and chitosan are clearly visible as two different phases for all deposition times. With intermediate phases of OCP showing maximum peak intensity in XRD spectra after 6 h process time, the final composition after 12 h was carbonated HAp with minor amounts of OCP containing 2-5% chitosan.

Zhou et al. [79] investigated the binding and drug release behavior of gentamicin to either pure CPP layers, dense chitosan coatings, and microporous or porous chitosan/CPP composites. The combination of chitosan with calcium and phosphate concentrations of the same molarity (10 mM) resulted in microporous composites with brushite as mineral phase. Coatings with higher porosity and additional HAp phase were obtained for increasing the calcium concentration up to 30 mM. Although no analysis of chitosan content was performed, gentamicin binding was supposed to depend both on available chitosan for direct adsorptive binding and on increased physical loading due to the higher number of pores for porous coatings.

Antimicrobial agents like chlorhexidine (CHD) or antimicrobial peptides (AMPs) were also included in the ECAD process. For the deposition of CHD, its pH-dependent solubility was utilized for coimmobilization with CPP [81]. A wide range of deposited amounts of CHD (150 ng/cm^2 to 65 μg/cm^2) was achieved by these authors. SEM and FTIR analysis demonstrated the codeposition of both phases and the impact of deposited CHD on the resulting CPP phases.

In another study, the broad spectrum AMP Tet213 (KRWWKWWRRC) was selected and loaded by adsorption onto a preformed OCP coating [82]. A 7 μm thick porous CPP coating could bind up to 9 μg of AMP/cm^2. The drug-loaded CPP coating was not cytotoxic for MG-63 osteoblast-like cells and demonstrated antimicrobial activity against both Gram-positive (*Staphylococcus aureus*) and Gram-negative (*Pseudomonas aeruginosa*) bacteria with 10^6-fold reductions.

3.3.3 In Vitro *Cell Culture Studies on ECAD-CPP Coatings*

The scope of *in vitro* studies analyzing ECAD-generated CPP layers ranged from simple cell attachment [83] to proliferation [77,84] and cell differentiation (mainly evaluated via ALP activity) [85]. As the investigations were performed under different experimental conditions and using a variety of cells (primary cells and cell lines), the experimental results are in many cases not directly comparable.

The published studies reported well-spread cells and higher proliferation on brushite and HAp-coated samples compared to the respective reference (collagen coating) [83]. Further, better proliferation was found for SAOS-2 on HAp-coated surfaces compared to tissue culture polystyrene (TCPS) control [84] and for primary human osteoblasts compared to machined and Al_2O_3-blasted reference surfaces [85]. Contrary to this, in another study, lower adhesion and proliferation of cells on HAp coatings compared to TCPS [86] were observed.

A disadvantage of such cell experiments lies in the lack of more complex reactions that occur when implants are brought in contact with the body, in particular with aspects of the blood system (immune system, complement system, and fibrinolytic system, to name only a few). Furthermore, in mammals, the calcium level in the extracellular space is maintained within a normal range by regulatory mechanisms that coordinate the metabolic activities of the kidneys, intestines, parathyroid glands, and bones [87]. In contrast, the dissolution and reprecipitation behavior of different CPP coatings in cell culture media will probably be quite different from the *in vivo* situation and will certainly differ from study to study. Minor changes in the medium concentration of calcium and phosphate can provoke tremendous differences in the differentiation behavior of osteoblast-like cells [88]. *In vitro* studies on CPP generated by other wet-chemical procedures and by physical methods reviewed by Surmenev et al. [89] revealed the important influence of the surface morphology besides the chemical composition. Hence, differences in several *in vitro* studies might be attributed to differences in the mineral phase composition and to the actual surface topography.

Other studies focused on the effect of inorganic Sr with or without further organic additives. In the first case, the cells preferred to spread on more flattened CPP structures, which were more dominant in case of Sr doping [55]. As it is not clear if the effect relies on the specific chemical modification by the presence of Sr or the changed morphology, no conclusion can be drawn on the specific effect of the Sr. For the triple composite of CPP (HAp and brushite) with Sr and gelatin, an improved *in vitro* biocompatibility in comparison to the pure CPP coating was claimed based on a higher proliferation rate in the case of the additives [56].

In vitro studies on CPP coatings with antimicrobial additives were mainly restricted to the demonstration of the antimicrobial effect using bacteria suspensions or agar diffusion tests. Results were already cited in the respective paragraph above.

If organic additives were codeposited or adsorbed onto CPP coatings, more specific effects could be demonstrated *in vitro* such as improved adhesion for deposited peptides with adhesion motifs like RGD or laminin-derived peptide [73]. MC3T3-E1 cells gave the best adhesion on mineralized collagen. Additionally, adsorbed RGD gave a further positive effect.

Wang et al. [77] could demonstrate a positive influence of codeposited chitosan on cell proliferation and differentiation (determined by gene and protein level of osteogenic markers) when using bone marrow stromal cells. However, the authors also argued that the observed changes could be attributed either to a specific impact of chitosan or to changes in the physicochemical properties of the coatings.

3.3.4 Animal Experimental Investigations

As for the very first ECAD coatings, mainly brushite phases were produced, the first published animal study investigated the impact of brushite and brushite converted to HAp [90]. In this preliminary study using a rabbit model, cylindrical samples were inserted press fit into femoral shafts. The significantly highest bone ingrowth was found for the HAp coating after 6 weeks.

Further studies on a later developed commercial brushite coating (BONIT®) were performed with screw-shaped implants in rabbits. A biomechanical removal torque test showed significantly increased values for coated implants only for the later time point of 12 weeks ($p < 0.05$), while higher bone-implant contact was found for the coated implants in the tibia after 6 weeks and for both tibial and femoral screws after 12 weeks ($p < 0.05$) [91]. The authors concluded that the BONIT®-coated implants resulted in a stronger biomechanical implant fixation compared to the controls of uncoated Al_2O_3-blasted Ti6Al4V and based this on the rougher surfaces of the BONIT®-coated screws. A later study of the same group compared the BONIT® coating with thin plasma-sprayed coatings [92]. Higher removal torque values (almost twice that of BONIT® after 6 weeks) at 6 and 12 weeks were observed for plasma-sprayed HAp (pHAp) and claimed to be based on their stronger bioactive properties. In contrast, the BONIT®-coated implants exhibited increasing fixation from 6 to 12 weeks and from 12 to 52 weeks. Also, the coating was resorbed within 6 weeks with minimal activity of multinucleated giant cells or bone resorption, while coat flaking and delamination of the plasma-sprayed screw led to multinucleated giant cell activity and bone resorption. The results suggested BONIT® to represent a truly resorbable calcium phosphate coating, transforming the interface into direct bone-implant contact resembling the fixation of titanium-alloy implants. No indication of any side effect of the coating has been observed.

Cylindrical implants with HAp coatings produced directly in the ECAD process were implanted into femurs of Japanese white rabbits by Ban et al. [93]. From the beginning, the values of the Weibull modulus *m* for the

ECAD-coated samples showed a higher level compared to the uncoated references, and this increased with time from week 3 to week 6. In contrast, uncoated samples had changed little after 6 weeks but increased after 9 weeks. Together with the higher levels for the bonding strength, the predicted failure level, and additional FE-SEM and micro-FTIR data, the authors concluded that the ECAD process undoubtedly contributes to early-stage fixation between implant and bone.

Composites of CPP layers with organic compounds were investigated by Schliephake et al. [94] using a model with cylindrical samples inserted press fit into the mandible of beagle. Two mineralized collagen layers with different thicknesses (500 nm and 2-3 μm), a pure HAp layer, fibrillar collagen partially integrated into an anodic oxide layer, and the polished reference were investigated. After 1 month healing time, a significantly higher percentage of mean bone-implant contact was found for the HAp-coated implants compared to polished and collagen-coated ones. After a healing period of 3 months, these differences had disappeared. Bone apposition next to the implants after 3 months was found to be significantly higher for the 2-3 μm collagen/HAp coating compared to the 500 nm coating and the polished surfaces.

In summary, it was demonstrated that HAp coatings produced by the ECAD process show a clear tendency to increase bone-implant contact in the early ingrowth period, while the biomechanical fixation for brushite coatings develops over time. The inclusion of collagen may hold some promise when used as a sequential HAp/collagen coating of sufficient thickness.

3.3.5 Clinical Applications

Clinical experience for coatings prepared by the ECAD process is available only for BONIT®-based surfaces and the BoneMaster® (BM) coating manufactured by Biomet.

A study by Wedemeyer et al. [95] reports on the midterm results of patients suffering from arthritis of the knee that were treated with a cementless second-generation CPP-coated (BONIT®) tibial component. Seventy-six knees were evaluated according to the Knee Society clinical, functional, and radiological score; the Hospital for Special Surgery rating system; and the patella score. The clinical and radiological parameters were assessed preoperatively and after a mean follow-up of 8.7 years. All three score systems revealed excellent clinical outcomes after the follow-up period. The mean preoperative Knee Society clinical score was 124.41 ± 12.99 and the mean postoperative score was 187.07 ± 14.59 at the time of the final consultation ($p = 0.0008$). The survival rate was 97.5%. Radiolucency of <1 mm around the uncemented CPP-coated tibial component and accompanying pain symptoms were seen in fewer than 6% of cases. In conclusion, these findings indicate that the uncemented second-generation CPP-coated tibial component performed well at midterm follow-up and provides sufficiently stable bone ingrowth fixation.

BONIT® coatings used on conical, self-cutting femoral neck prostheses did not show any specific complications within 1 year, and the rehabilitation time showed a tendency to be shorter than after stem arthroplasty [96]. Radiological follow-up investigations after 3 months showed the development of increasing trabecular reinforcement of the femoral neck and pertrochanteric regions.

A short-term study compared pHAp coatings with BM coatings prepared via the ECAD process. The stems had subsided 0.25 (pHAp) and 0.28 (BM) mm with no statistically significant differences between the groups regarding both migration and rotation. The BM group retained significantly more bone than the pHAp group in Gruen zone 1 during the first 2 years [97].

In summary, the ECAD method combines a variety of advantages:

- Deposition of thin, nonmonolithic layers with a composition, crystal size, and chemical history as close as possible to that of bone mineral.
- Composition, mineral phase, and crystallinity of CPP can be accurately tuned by process parameters.
- Thickness and chemical composition of coatings can be controlled down to the sub-μm level.
- Generation of coatings with a high specific surface area is possible.
- Low-energy and low-cost process (with respect to production costs and waste management).
- Compared to incubation in SBF, a more defined and higher relative supersaturation at the interface, resulting in shorter processing times.
- Due to the possible use of physiological processing parameters (pH, temperature, and aqueous solution), the deposition of CPP can be combined with the immobilization of organic components like proteins or peptides.
- Excellent homogeneity of coatings on structured and porous surfaces and on irregularly formed structures as it is no line of sight process.
- Positive effects of coatings prepared by ECAD *in vitro* and *in vivo* and no adverse effects related to delamination and/or inflammatory effects.

Nevertheless, going from the laboratory to the industrial scale, a few points have to be considered:

- Most discussed is a potentially low adhesive strength of the coatings. However, this is potentially not a problem for rough or structured surfaces that are common for implants in contact with bone and can be overcome by combining them with anodic growth of oxide layers [73].
- Electrically conductive substrates or substrate top layers are necessary.
- For extreme electrochemical conditions (high current densities and long duration of polarization), there is the possible danger of hydrogen embrittlement of the metallic substrate material due to an uptake of atomic hydrogen as an intermediate product of the cathodic electrode reaction.

REFERENCES

[1] W.F. de Jong, La Substance Minérale Dans les Os, Recl. Trav. Chim. Pays-Bas 45 (6) (1926) 445–448.
[2] S.V. Dorozhkin, A detailed history of calcium orthophosphates from 1770s till 1950, Mater. Sci. Eng. C 33 (6) (2013) 3085–3110.
[3] S. Dorozhkin, History of calcium phosphates in regenerative medicine, in: B. Ben-Nissan (Ed.), Advances in Calcium Phosphate Biomaterials, Springer, Berlin, Heidelberg, 2014, pp. 435–483.
[4] C. Rentsch, et al., Knochenersatz, Unfallchirurg 115 (10) (2012) 938–949.
[5] J. Jansen, B. León, Thin Calcium Phosphate Coatings for Medical Implants, Springer Science + Business Media, New York, 2009.
[6] C. Shuai, et al., Nano-hydroxyapatite improves the properties of β-tricalcium phosphate bone scaffolds, Int. J. Appl. Ceram. Technol. 10 (6) (2013) 1003–1013.
[7] J. Cho, et al., Synthesis of nano-sized biphasic calcium phosphate ceramics with spherical shape by flame spray pyrolysis, J. Mater. Sci. Mater. Med. 21 (4) (2010) 1143–1149.
[8] R. Detsch, et al., The resorption of nanocrystalline calcium phosphates by osteoclast-like cells, Acta Biomater. 6 (8) (2010) 3223–3233.
[9] G.S. Kumar, et al., Synthesis, characterization and *in vitro* studies of zinc and carbonate co-substituted nano-hydroxyapatite for biomedical applications, Mater. Chem. Phys. 134 (2–3) (2012) 1127–1135.
[10] D. Tang, et al., Biphasic calcium phosphate nano-composite scaffolds reinforced with bioglass provide a synthetic alternative to autografts in a canine tibiofibula defect model, Chin. Med. J. 127 (7) (2014) 5.
[11] M.-R. Badr-Mohammadi, S. Hesaraki, A. Zamanian, Mechanical properties and *in vitro* cellular behavior of zinc-containing nano-bioactive glass doped biphasic calcium phosphate bone substitutes, J. Mater. Sci. Mater. Med. 25 (1) (2014) 185–197.
[12] S. Hesaraki, et al., Mechanical and *in vitro* biological properties of hydroxyapatite bioceramics reinforced with strontium-containing nano-bioactive glass, Curr. Nanosci. 8 (4) (2012) 612–622.
[13] W.-C. Chen, et al., Carbon black sintering effects on the composition of multiphase calcium phosphate bioceramics, J. Ceram. Process. Res. 14 (3) (2013) 279–283.
[14] A.A. Chaudhry, et al., Phase stability and rapid consolidation of hydroxyapatite-zirconia nano-coprecipitates made using continuous hydrothermal flow synthesis, J. Biomater. Appl. 27 (1) (2012) 79–90.

[15] K.-T. Chu, et al., Research of phase transformation induced biodegradable properties on hydroxyapatite and tricalcium phosphate based bioceramic, Ceram. Int. 39 (2) (2013) 1455–1462.
[16] Y. Chen, et al., Adsorption and release behaviors of vascular endothelial growth factor on porous hydroxyapatite ceramic under competitive conditions, J. Biomater. Tissue Eng. 4 (2) (2014) 155–161.
[17] S.P. Victor, C.P. Sharma, Calcium phosphates as drug delivery systems, J. Biomater. Tissue Eng. 2 (4) (2012) 269–279.
[18] S. Reddy, et al., Evaluation of nano-biphasic calcium phosphate ceramics for bone tissue engineering applications: *in vitro* and preliminary *in vivo* studies, J. Biomater. Appl. 27 (5) (2013) 565–575.
[19] Y. Tokudome, A. Ito, M. Otsuka, Effect of zinc-containing β-tricalcium phosphate nano particles injection on jawbone mineral density and mechanical strength of osteoporosis model rats, Biol. Pharm. Bull. 34 (8) (2011) 1215–1218.
[20] C.A. Ospina, et al., Experimental evidence and structural modeling of nonstoichiometric (010) surfaces coexisting in hydroxyapatite nano-crystals, Colloids Surf. B: Biointerfaces 89 (2012) 15–22.
[21] Y. Zhang, et al., Fabrication of porous hydroxyapatite ceramic scaffolds with high flexural strength through the double slip-casting method using fine powders, J. Am. Ceram. Soc. 95 (1) (2012) 147–152.
[22] K. Lin, et al., Improvement of mechanical properties of macroporous β-tricalcium phosphate bioceramic scaffolds with uniform and interconnected pore structures, Ceram. Int. 37 (7) (2011) 2397–2403.
[23] C. Bergmann, et al., Temperature-dependent morphology changes of noble metal tricalcium phosphate-nanocomposites, Ceram. Int. 40 (6) (2014) 7931–7939.
[24] M. Jafarkhani, et al., Fabrication and characterization of PLLA/chitosan/nano calcium phosphate scaffolds by freeze-casting technique, Ind. Eng. Chem. Res. 51 (27) (2012) 9241–9249.
[25] H.H.K. Xu, L. Zhao, M.D. Weir, Stem cell-calcium phosphate constructs for bone engineering, J. Dent. Res. 89 (12) (2010) 1482–1488.
[26] L. Bakhtiari, et al., Preparation of porous biphasic calcium phosphate-gelatin nanocomposite for bone tissue engineering, J. Nano Res. 11 (2010) 6.
[27] M. Ebrahimian-Hosseinabadi, et al., Evaluating and modeling the mechanical properties of the prepared PLGA/nano-BCP composite scaffolds for bone tissue engineering, J. Mater. Sci. Technol. 27 (12) (2011) 1105–1112.
[28] D. Barbieri, et al., Heterotopic bone formation by nano-apatite containing poly(D,L-lactide) composites, Eur. Cells Mater. 19 (2010) 10.
[29] H. Kweon, et al., Development of nano-hydroxyapatite graft with silk fibroin scaffold as a new bone substitute, J. Oral Maxillofac. Surg. 69 (6) (2011) 1578–1586.
[30] S.-J. Lee, H.-S. Choi, M.-H. Lee, Highly sinterable calcium phosphate fabricated by using starfish bone, J. Nanosci. Nanotechnol. 11 (2) (2011) 1815–1817.
[31] J.L. Moreau, M.D. Weir, H.H.K. Xu, Self-setting collagen-calcium phosphate bone cement: mechanical and cellular properties, J. Biomed. Mater. Res. A 91A (2) (2009) 605–613.
[32] J.L. Moreau, H.H.K. Xu, Mesenchymal stem cell proliferation and differentiation on an injectable calcium phosphate—chitosan composite scaffold, Biomaterials 30 (14) (2009) 2675–2682.
[33] Q. Zhang, Y. Leng, Electrochemical activation of titanium for biomimetic coating of calcium phosphate, Biomaterials 26 (18) (2005) 3853–3859.
[34] D. Scharnweber, S. Bierbaum, Electrochemically assisted deposition of thin CaP coatings, in: Thin Calcium Phosphate Coatings for Medical Implants, Springer, New York, 2009, pp. 215–261.

[35] M.S.-A. Johnsson, G.H. Nancollas, The role of brushite and octacalcium phosphate in apatite formation, Crit. Rev. Oral Biol. Med. 3 (1) (1992) 61–82.
[36] J. Redepenning, J.P. McIsaac, Electrocrystallization of brushite coatings on prosthetic alloys, Chem. Mater. 2 (6) (1990) 625–627.
[37] H. Monma, et al., Electrolytic formation and morphology of biomimetic apatite coatings, J. Electroceram. 4 (1999) 135–140.
[38] S. Ban, S. Maruno, Morphology and microstructure of electrochemically deposited calcium phosphates in a modified simulated body fluid, Biomaterials 19 (14) (1998) 1245–1253.
[39] M.C. Kuo, S.K. Yen, The process of electrochemical deposited hydroxyapatite coatings on biomedical titanium at room temperature, Mater. Sci. Eng. C Biomim. Supramol. Syst. 20 (1–2) (2002) 153–160.
[40] S. Rossler, et al., Electrochemically assisted deposition of thin calcium phosphate coatings at near-physiological pH and temperature, J. Biomed. Mater. Res. A 64A (4) (2003) 655–663.
[41] A. Sewing, et al., Influence of Ca/P ratio on electrochemical assisted deposition of hydroxyapatite on titanium. Key Eng. Mater. 254–256 (2004) 419–422, http://dx.doi.org/10.4028/www.scientific.net/KEM.254-256.419.
[42] S.H. Wang, et al., Morphology of calcium phosphate coatings deposited on a Ti-6Al-4V substrate by an electrolytic method under 80 Torr, J. Eur. Ceram. Soc. 25 (14) (2005) 3287–3292.
[43] C. Montero-Ocampo, D. Villegas, L. Veleva, Controlled potential electrodeposition of calcium phosphate on Ti6Al4V, J. Electrochem. Soc. 152 (10) (2005) C692–C696.
[44] S. Ban, S. Maruno, Hydrothermal-electrochemical deposition of hydroxyapatite, J. Biomed. Mater. Res. 42 (3) (1998) 387–395.
[45] S.J. Ban, S. Maruno, Effect of temperature on electrochemical deposition of calcium-phosphate coatings in a simulated body-fluid, Biomaterials 16 (13) (1995) 977–981.
[46] M. Yousefpour, et al., Nano-crystalline growth of electrochemically deposited apatite coating on pure titanium, J. Electroanal. Chem. 589 (1) (2006) 96–105.
[47] M. Shirkhanzadeh, Microneedles coated with porous calcium phosphate ceramics: effective vehicles for transdermal delivery of solid trehalose, J. Mater. Sci. Mater. Med. 16 (1) (2005) 37–45.
[48] S. Ban, S. Maruno, Effect of Ph buffer on electrochemical deposition of calcium-phosphate, Jpn. J. Appl. Phys. 2 Lett. 33 (11A) (1994) L1545–L1548.
[49] R. Bosco, et al., Instructive coatings for biological guidance of bone implants, Surf. Coat. Technol. 233 (2013) 91–98.
[50] M. Shirkhanzadeh, Calcium-phosphate coatings prepared by electrocrystallization from aqueous-electrolytes, J. Mater. Sci. Mater. Med. 6 (2) (1995) 90–93.
[51] D.B. Haddow, et al., Compositional and structural control in bone regenerative coatings, J. Mater. Sci. Mater. Med. 10 (4) (1999) 219–222.
[52] D.H. Li, et al., Synthesized silicon-substituted hydroxyapatite coating on titanium substrate by electrochemical deposition, J. Mater. Sci. Mater. Med. 22 (5) (2011) 1205–1211.
[53] P.J. Meunier, et al., Effects of long-term strontium ranelate treatment on vertebral fracture risk in postmenopausal women with osteoporosis, Osteoporos. Int. 20 (10) (2009) 1663–1673.
[54] E. Bonnelye, et al., Dual effect of strontium ranelate: stimulation of osteoblast differentiation and inhibition of osteoclast formation and resorption *in vitro*, Bone 42 (1) (2008) 129–138.
[55] L. Li, et al., Comparison study of biomimetic strontium-doped calcium phosphate coatings by electrochemical deposition and air plasma spray: morphology, composition and bioactive performance, J. Mater. Sci. Mater. Med. 23 (10) (2012) 2359–2368.

[56] Y. Huang, et al., Bioactivity and corrosion properties of gelatin-containing and strontium-doped calcium phosphate composite coating, Appl. Surf. Sci. 282 (2013) 583–589.
[57] C. Wolf-Brandstetter, et al., Influence of pulse ratio on codeposition of copper species with calcium phosphate coatings on titanium by means of electrochemically assisted deposition, J. Biomed. Mater. Res. B Appl. Biomater. 102 (1) (2014) 160–172.
[58] H.G. Neumann, C. Prinz, Antibacterial and osteoinductive implant coating, method of producing such coating, and implant coated with same, d. GmbH, Editor 2013, Google Patents.
[59] M. Shirkhanzadeh, M. Azadegan, G.Q. Liu, Bioactive delivery systems for the slow release of antibiotics: incorporation of Ag+ ions into micro-porous hydroxyapatite coatings, Mater. Lett. 24 (1–3) (1995) 7–12.
[60] X. Lu, et al., Nano-Ag-loaded hydroxyapatite coatings on titanium surfaces by electrochemical deposition, J. R. Soc. Interface 8 (57) (2011) 529–539.
[61] D. Hubacher, et al., Use of copper intrauterine devices and the risk of tubal infertility among nulligravid women, N. Engl. J. Med. 345 (8) (2001) 561–567.
[62] J. Barralet, et al., Angiogenesis in calcium phosphate scaffolds by inorganic copper ion release, Tissue Eng. A 15 (7) (2009) 1601–1609.
[63] C.K. Sen, et al., Copper-induced vascular endothelial growth factor expression and wound healing, Am. J. Physiol. Heart Circ. Physiol. 282 (5) (2002) H1821–H1827.
[64] M. Egler, Rolle von RpoE-homologen Sigmafaktoren in der Schwermetall-Homöostase von *Escherichia coli* und Cupriavidus metallidurans, (PhD Thesis), in Biology, Martin-Luther-Universität Halle-Wittenberg, 2005. http://sundoc.bibliothek.uni-halle.de/diss-online/05/05H204/index.htm or: http://nbn-resolving.de/urn/resolver.pl?urn=nbn%3Ade%3Agbv%3A3-000009507.
[65] J. Zhang, Z. Tian, Hydroxyapatite/metal composite coatings prepared by multi-step electrodeposition method, J. Mater. Sci. Lett. 17 (13) (1998) 1077–1079.
[66] M. Mazzali, et al., Osteopontin—a molecule for all seasons, QJM 95 (1) (2002) 3–13.
[67] K. Flade, et al., Osteocalcin-controlled dissolution-reprecipitation of calcium phosphate under biomimetic conditions, Chem. Mater. 13 (10) (2001) 3596–3602.
[68] K.K. Pereira, et al., Progression of osteogenic cell cultures grown on microtopographic titanium coated with calcium phosphate and functionalized with a type I collagen-derived peptide, J. Periodontol. 84 (8) (2013) 1199–1210.
[69] M. Shirkhanzadeh, Interaction between self-assembled protein vesicles and microporous apatite surface, J. Mater. Sci. Mater. Med. 9 (9) (1998) 503–508.
[70] U. Geissler, et al., Collagen type I-coating of Ti6Al4V promotes adhesion of osteoblasts, J. Biomed. Mater. Res. 51 (4) (2000) 752–760.
[71] C. Roehlecke, et al., Synergistic effect of titanium alloy and collagen type I on cell adhesion, proliferation and differentiation of osteoblast-like cells, Cells Tissues Organs 168 (3) (2001) 178–187.
[72] A. Roguska, et al., Collagen immobilization on 316L stainless steel surface with cathodic deposition of calcium phosphate, Appl. Surf. Sci. 257 (11) (2011) 5037–5045.
[73] S. Roessler, et al., Biomimetic coatings functionalized with adhesion peptides for dental implants, J. Mater. Sci. Mater. Med. 12 (10–12) (2001) 871–877.
[74] H. Okamura, M. Yasuda, M. Ohta, Synthesis of calcium-deficient hydroxy apatite-collagen composite, Electrochemistry 68 (6) (2000) 486–488.
[75] Y. Fan, K. Duan, R. Wang, A composite coating by electrolysis-induced collagen self-assembly and calcium phosphate mineralization, Biomaterials 26 (14) (2005) 1623–1632.
[76] T. Ling, et al., Mineralized collagen coatings formed by electrochemical deposition, J. Mater. Sci. Mater. Med. 24 (12) (2013) 2709–2718.
[77] J. Wang, J. de Boer, K. de Groot, Proliferation and differentiation of MC3T3-E1 cells on calcium phosphate/chitosan coatings, J. Dent. Res. 87 (7) (2008) 650–654.

[78] J. Redepenning, G. Venkataraman, J. Chen, N. Stafford, Electrochemical preparation of chitosan/hydroxyapatite composite coatings on titanium substrates, J. of Biomed. Mater. Res. A 66 (2) (2003) 411–416.

[79] J. Zhou, et al., Release behaviors of drug loaded chitosan/calcium phosphate coatings on titanium, Thin Solid Films 519 (15) (2011) 4658–4662.

[80] J. Wang, A. van Apeldoorn, K. de Groot, Electrolytic deposition of calcium phosphate/chitosan coating on titanium alloy: growth kinetics and influence of current density, acetic acid, and chitosan, J. Biomed. Mater. Res. A 76A (3) (2006) 503–511.

[81] D. Scharnweber, et al., Adjusting the chlorhexidine content of calcium phosphate coatings by electrochemically assisted co-deposition from aqueous solutions, J. Mater. Sci. Mater. Med. 18 (2) (2007) 391–397.

[82] M. Kazemzadeh-Narbat, et al., Antimicrobial peptides on calcium phosphate-coated titanium for the prevention of implant-associated infections, Biomaterials 31 (36) (2010) 9519–9526.

[83] P. Becker, et al., Cellular investigations on electrochemically deposited calcium phosphate composites, J. Mater. Sci. Mater. Med. 15 (4) (2004) 437–440.

[84] H. Hu, et al., Electrochemical deposition of hydroxyapatite with vinyl acetate on titanium implants, J. Biomed. Mater. Res. A 65A (1) (2003) 24–29.

[85] M. Prado da Silva, et al., *In vitro* cellular response to titanium electrochemically coated with hydroxyapatite compared to titanium with three different levels of surface roughness, J. Mater. Sci. Mater. Med. 14 (6) (2003) 511–519.

[86] J. Wang, et al., Biomimetic and electrolytic calcium phosphate coatings on titanium alloy: physicochemical characteristics and cell attachment, Biomaterials 25 (4) (2004) 583–592.

[87] V. Kartsogiannis, K.W. Ng, Cell lines and primary cell cultures in the study of bone cell biology, Mol. Cell. Endocrinol. 228 (1–2) (2004) 79–102.

[88] Y.K. Liu, et al., The effect of extracellular calcium and inorganic phosphate on the growth and osteogenic differentiation of mesenchymal stem cells *in vitro*: implication for bone tissue engineering, Biomed. Mater. 4 (2) (2009) 025004.

[89] R.A. Surmenev, M.A. Surmeneva, A.A. Ivanova, Significance of calcium phosphate coatings for the enhancement of new bone osteogenesis—a review, Acta Biomater. 10 (2) (2014) 557–579.

[90] J. Redepenning, et al., Characterization of electrolytically prepared brushite and hydroxyapatite coatings on orthopedic alloys, J. Biomed. Mater. Res. 30 (3) (1996) 287–294.

[91] O. Reigstad, et al., Improved bone ingrowth and fixation with a thin calcium phosphate coating intended for complete resorption, J. Biomed. Mater. Res. B Appl. Biomater. 83 (1) (2007) 9–15.

[92] O. Reigstad, et al., Different patterns of bone fixation with hydroxyapatite and resorbable CaP coatings in the rabbit tibia at 6, 12, and 52 weeks, J. Biomed. Mater. Res. B Appl. Biomater. 99B (1) (2011) 14–20.

[93] S. Ban, et al., Effect of electrochemically deposited apatite coating on bonding of bone to the HA-G-Ti composite and titanium, J. Biomed. Mater. Res. 36 (1) (1997) 9–15.

[94] H. Schliephake, et al., Biological performance of biomimetic calcium phosphate coating of titanium implants in the dog mandible, J. Biomed. Mater. Res. A 64A (2) (2003) 225–234.

[95] C. Wedemeyer, et al., Cementless second-generation hydroxyapatite CaP-coated tibial component: an 8.7-year follow-up, Arch. Orthop. Trauma Surg. 132 (12) (2012) 1759–1764.

[96] B. Birkenhauer, H. Kistmacher, J. Ries, Conception and first results of the Spiron cementless femoral neck screw prosthesis, Orthopade 33 (11) (2004) 1259–1266.

[97] B.G. Bøe, et al., A prospective randomized study comparing electrochemically deposited hydroxyapatite and plasma-sprayed hydroxyapatite on titanium stems: 55 hips followed for 2 years with RSA and DXA, Acta Orthop. 82 (1) (2011) 13–19.

CHAPTER 22

Biosafety of Nanomaterials Used in Nanoceramics and Nanocomposites

Sandhya Sanand, Ahmad Hussain, Gautam Kaul
Department of Biochemistry, National Dairy Research Institute (Deemed University), Karnal, India

Chapter Contents

1. INTRODUCTION

In recent years, nanotechnology has been one of the most wanted areas in research because of the ability of nanomaterials to be manipulated at nano-range with a high surface area/volume ratio, which in itself is unique in a biological system. According to a new market report [1], there is a huge surge in the demand for nanomaterials in the last couple of years, which is expected to rise manyfold in the foreseeable future. The demand for nanocomposites that was over USD 1.91 billion in 2011 is expected to increase to USD 5.91 billion by 2018, growing at a CAGR (compounded annual growth rate) of 18.0% from 2012 to 2018. In terms of volume, the global market for nanocomposites is expected to grow at a CAGR of 17.5% between 2012 and 2018.

Several types of nanomaterials are being produced, of which nanoceramics/nanocomposites have proved to be of great use in the orthopedic

Handbook of Nanoceramic and Nanocomposite Coatings and Materials
http://dx.doi.org/10.1016/B978-0-12-799947-0.00022-5

system and dentistry because of their inert nature and their high and resilient bioactivity. Their applications have been growing at a rapid rate from orthopedics, to drug delivery and tissue engineering, to food packaging. They are also addressing global challenges in the field of energy and environment. Nanocomposites are polymer mixtures of distinct size, geometry, and chemistry and possess advantages like biodegradability and biocompatibility both *in vitro* and *in vivo* [2]. Organic-inorganic nanocomposite coatings, which combine biopolymers and bioactive ceramics that mimic the bone structure to induce biomineralization, with the addition of biomolecules, represent alternative systems and ideal materials for smart implants. But there are limited studies that report the nanocomposite- and nanoceramic-mediated biological effects inside the living system. So the common questions that come to our mind while using such "nano" materials with several uses are as follows: How do these materials stand up with transplantation materials? Up to what extent are they safe? How is their toxicity, if any, related to their size, chemical composition, and aspect ratio?

2. DEFINITION AND CLASSIFICATION

Nanoceramics and nanocomposites are a product of nanotechnology with at least one of the dimensions in the nanoscale range (<100 nm) [3]. These encompass a wide range of materials including nanostructured particles, fibers, rods, tubes, sheets, and nanofilms. Several materials and complex systems of different compositions contain nanocomposites and nanoceramics as their integral component.

Nanoceramics and nanocomposites have been classified on the basis of nanofiller dimensionality: zero-dimensional (e.g., nanoparticles (NPs) that include oxides, metals, semiconductors, and fullerenes), one-dimensional (e.g., nanofibers that include nanowires, nanorods, and nanotubes) two-dimensional (e.g., nanolayers that include multilayers, monolayer, self-assembled, and mesoporous thin films), and three-dimensional (interpenetrating network that includes nanocomposites, hybrid nanograins, micro- and mesoporous hybrids, and organic-inorganic hybrids) systems [4]. On basis of their architectural differences, they have been placed into two categories: NPs and nanoscaffolds.

3. THEIR IMPORTANCE AND USE

One of the most important facts regarding nanomaterials is that they can be manipulated in order to have better and desirable characteristics. This virtue applies to nanocomposites and nanoceramics as well. Nanoceramics have been a part and parcel of many present and emerging technologies. Due to their insulating and/or conducting properties, nanoceramics have uses in the construction of the next-generation high-speed computer chips and in other aspects of electronics. For the same reasons, they also have applications in power generation. In medicine, nanoceramics are starting to be used to develop new bone implants and artificial organs, since working at the nanoscale allows device manufacturers to fine-tune how the implant will interact with the human body, thus improving the chance of the body accepting it. Apart from these, nanoceramics have applications in space exploration [5], weapons manufacturing [6], construction [7], consumer goods [8] and transport [9]. Composites (e.g., Pt, Au, Pd, $Zr(HPO_4)_2$, MoO_3, MnO_2, Mo_3Se_3, c-Fe_2O_3, Fe_3O_4, and IrO_2) have been applied in a number of fields ranging from electrocatalysis to energy storage devices and electronics.

Many of the features that make ceramic NPs unique are longer biodegradation times, least affected by changes in temperature and pH, and most importantly their inherent capacity to mimic the structural aspects of the targeted tissue. Slow degradation kinetics is an excellent advantage in the case of diffusion-controlled drug release studies [10]. Their stability at a long range of physiological variations adds to their multitude of applications. This feature allows a slow release of drug rather than a burst commonly observed with hydrogels such as with poly(2-hydroxyethyl methacrylate) (pHEMA) systems. If synthesized properly, these materials show the highest bioactivity and biocompatibility. In recent times, these materials have been further engineered to produce materials with desirable electrical and mechanical properties [11]. These manipulations have rendered them with high porosity, high surface area/volume ratios, high surface area, high structural stability, and long degradation times, and these properties make them ideal model systems for the storage and controlled release of drugs, especially drugs for *in situ*. Like nanoceramics, nanocomposites possess the following properties:

- Lightweight
- Corrosion resistance (long life)
- Design and formulation flexibility
- Fatigue resistance
- Good damping characteristics
- Low thermal expansion
- Unitized structure
- High strength and stiffness

3.1 Properties and Exposure Route of Nanomaterials

Physical properties	Chemical properties
• Specific surface area, aspect ratio	• Molecular structure
• Particle size	• Chemical Composition, homo or hetro
• State, agglomeration/aggregation	• Degree of purity
• Surface morphology/tonography	• Hydrophobicity or hydrophilicity
• Nature of solubility, aqueous or non aqueous	• Surface chemistry
	• Reactivity
	• Photocatalytic property

So now, we will try to explore the interaction of cells with nanoceramic/nanocomposite materials, their biocompatibility, chemical durability, their adverse effects in the simulated biological system, and biosafety assessment *in vitro* and *in vivo* with various cell lines and animal models.

It has been observed that the behavior of these nanomaterials is influenced by the surrounding environment, for example, media in which they are suspended. To assess their effect, it is compulsory to study the composition in which they are manufactured or in the form that they are delivered. Moreover, prior to the biosafety evaluation of the nanomaterial, it is very important to characterize the physical and chemical properties of the nanomaterial (Table 1).

Despite their agglomerating effects, nanomaterials tend to remain intact in terms of bioactivity due to the least change in their surface area, which is compulsory for interaction with cellular structures [12]. Different studies have suggested ways on how to best disperse the NPs and the most apt protocols may vary between the different nanomaterials.

4. HOW THESE NANOCERAMICS/NANOCOMPOSITES INTERACT WITH BIOLOGICAL CELLS?

Nanomaterials are used in tissue replacement, bone regeneration, dental restoration, and drug delivery due to their excellent mechanical, electrical, and conductive properties and their ability to mimic the organic component of the extracellular matrix of the bone [13]. The use of nanoceramics and nanocomposites is not limited to the medical sector; they are also being used in the food sector as food preservative and food packaging materials. These materials enter into the body either accidentally as in the case of industries

Table 1 Characterization of nanoceramic/nanocomposite nanomaterials [37]

Nanoceramics/ nanocomposite materials	Cell line	Parameter assessed	Major outcome or biocompatibility	Applicable field	References
AgNP soft nanocomposite	HepG2 cells	Cell viability	Increased cell viability	As bactericidal agent	[38]
Selenium-ferroferric oxide nanocomposites	Human osteoblast-like MG-63 cells	MTT assay	Induced cell apoptosis, inhibited cell proliferation	Bone tissue engineering	[39]
Poly(3-hydroxybutyrate) (PHB) composite scaffolds	MG63 cell line	Cell proliferation and alkaline phosphatase (ALP) assays	Significantly improved cell proliferation, inducing better osteoconductivity, and increased cell attachment	Bone tissue engineering	[40]
Nanohydroxyapatite/ gelatin/carboxymethyl chitin (n-HA/gel/ CMC)	MG63 osteoblast cells	MTT assay, cell morphology, and hemocompatibility (hemolysis, platelet adhesion, and protein adsorption)	Enhancement of cell adhesion, proliferation, and network formation	Bone tissue engineering	[41]
Halloysite nanotube (HNT)-doped poly (lactic-co-glycolic acid) (PLGA) composite nanofibers	Mouse fibroblast cells	MTT assay and SEM	Promotion of cell attachment and proliferation of HNT-doped PLGA scaffolds allow more protein adsorption	Tissue engineering and drug delivery	[42]

Continued

Table 1 Characterization of nanoceramic/nanocomposite nanomaterials [37]—cont'd

Nanoceramics/ nanocomposite materials	Cell line	Parameter assessed	Major outcome or biocompatibility	Applicable field	References
Nanohydroxyapatite/ gelatin	Rat somatic stem cells	Alizarin red test, histomorphometry	Higher osteoblast attachment and proliferation, more osteoblast differentiation on 2-D cell culture; did not change morphology	Bone regeneration	[43]
Thermoresponsive nanocomposites generated by poly (N-isopropylacrylamide)	Human umbilical vein endothelial cells	Cytotoxicity test	Excellent biocompatibility	As an embolization agent in cancer treatment	[44]
Hydrogel nanocomposites	Lens epithelial cells (LECs)	*In vitro* cell adhesion test	No deleterious effect on the biocompatibility	A potential UV/ blue light filtering material for ophthalmic lenses	[45]
Mg-Zn-Zr nanocomposites	Osteoblasts	*In vitro* corrosion resistance and cytocompatibility	Better adhesion and proliferation of cells	Biodegradable bone fixation material	[46]
Biogenic silica-metal phosphate nanocomposites (BSMPNs)	Human mesenchymal stem cells	Cell viability assay	Nontoxic and showed excellent biocompatibility	Bone tissue engineering	[47]
Titania-graphene nanocomposites	Simulated body fluid and human cell lines (AGS and MG-63)	MTS assay, confocal microscopy, SEM, FITR	More favorable materials to biomimic bone action	As biomimicking agent	[48]

Liquid crystalline epoxy nanocomposites	–	MTT and thermocycling	Showed good biocompatibility and greater microhardness	As fabricated aesthetic brackets for orthodontic treatment	[49]
Epoxy/clay nanocomposite	*Staphylococcus aureus* (ATCC11632), *Escherichia coli* (ATCC 10536), *Mycobacterium smegmatis* (ATCC14468), and *Candida albicans* (ATCC 10231) strains	MTT and RBC hemolytic assays	Excellent cytocompatibility	As antimicrobial agent	[50]
	Wistar rat	Histopathologic study of the brain, heart, liver, and skin and proliferation of dermatocytes without induction of any sign of toxicity to the above organs	Adherence and proliferation of cells, nontoxic	Tissue regeneration	

and workplace exposure or deliberately as in the case of the medical sector (for delivering of drug, peptide, hormone, vaccine, etc.), agricultural sector, and veterinary sector (Figure 1). Rafeeqi and Kaul [15] experimentally proved that carbon nanotubes act as a scaffold for spermatogonial cell maintenance. However, during their use in the abovementioned areas, they have been observed to exhibit some pathological changes or adverse effects in various cell lines and also in animal models [16]. When administered, the nanomaterials associate with biological fluid and subsequently with the proteins present in it, which is determined by the size and hydrophobicity of these particles. This coating of proteins affects their desirable response.

But still, nanotechnology, or the use of nanomaterials, may be the best solutions since these can mimic surface properties (including topography and energy) of natural tissues (Figure 2). They have a bright future in biomedical engineering due to their superior properties as compared to conventional materials [17].

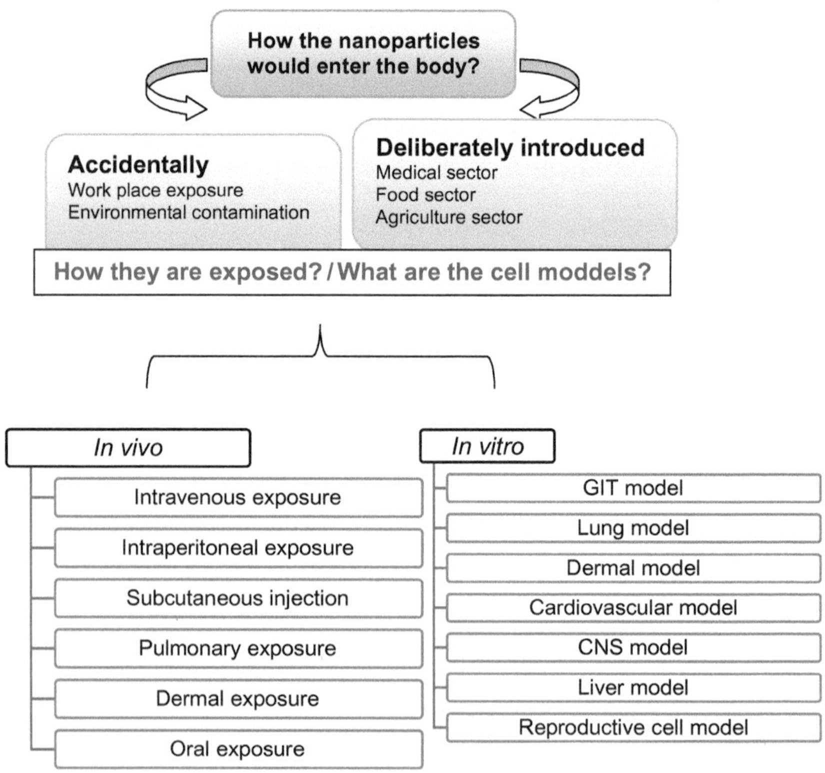

Figure 1 Possible ways of exposure of nanomaterials in *in vivo* and *in vitro* system [14].

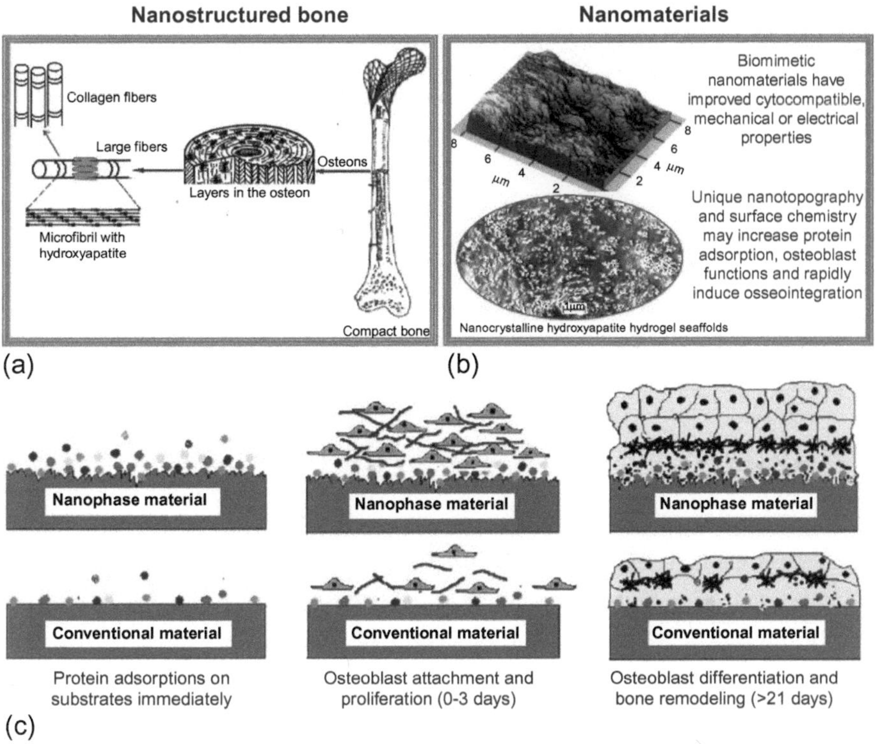

Figure 2 The biomimetic advantages of nanomaterials. (a) The nanostructured hierarchal self-assembly of the bone. (b) Nanophase titanium (top, the atomic force microscopy image) and nanocrystalline HA/HRN hydrogel scaffold (bottom, the SEM image). (c) Schematic illustration of the mechanism by which nanomaterials may be superior to conventional materials for bone regeneration. The bioactive surfaces of nanomaterials mimic those of natural bones to promote greater amounts of protein adsorption and efficiently stimulate more new bone formation than conventional materials. *(Image adapted from Ref. [18]).*

4.1 Adverse Effect of Nanoceramics/Nanocomposites

Despite having a long list of usage and applications in several areas, these nanomaterials are not free from some undesirable effects, which have been observed in several cell lines and animal models (Figure 2). Toxicity of titanium oxide NPs causes functionality and DNA damage in buffalo (*Bubalus bubalis*) sperm *in vitro* [19]. There are various health hazards related with nanomaterials [20]. To elucidate nanotoxicity, a detailed characterization of nanomaterials including shape, size, and composition along with the fate of these nanomaterials inside the body is required. Very little information on these aspects is presently available and this implies that there is an urgent need for generating toxicokinetic data for nanomaterials.

4.2 *In Vitro* Testing of Nanoceramics/Nanocomposites

Quite a few nanocomposites have shown no demonstrable cytotoxic effects on cells albeit at low concentrations, but there was reduction in cell viability with increasing concentrations. Cell viability assays like MTT have revealed a drastic decrease in fibroblast cell viability with higher concentrations of these nanocomposites. On the contrary, not only is cell viability unaffected at lower concentrations, but also there has been a proliferative action on such cells; for example, in a study, it was found that at a lower concentration of the CPC/MWCNT-OH/BSA composite (6.25 μg/ml) immersed in SBF for 7 days, the percentage of cell viability was 102%, which was observed to be due to higher metabolic activity of fibroblasts, but when concentration was increased to 200 μg/ml, a significant reduction in cell viability was observed [21]. In this regard, these nanocomposites show a differential dose dependent cytotoxicity. Zhang et al. [22] reported that ceramic zirconia and silicon nitride NPs interfere with MG-63 cell function and remarkably stimulate the secretion of TNF-α in RAW264.7 cells. Magnetically enhanced cytotoxicity of paramagnetic selenium-ferroferric oxide nanocomposites on human osteoblast-like MG-63 cells was studied by Zhou et al. [23]. The biocompatibility of poly(vinylidene fluoride-trifluoroethylene)/barium titanate composite using cultures of human periodontal ligament fibroblasts and keratinocytes was studied by Teixeira et al. [24]. *In vitro* biological study of gelatin/poly(D,L-lactide-co-glycolide) (PLG) nanocomposite using MCF-7 breast cancer cells was done by Haroun et al. [25]. Cytotoxicity studies with other nanomaterials like apatite/cisplatin systems have shown a similar trend of toxic effects on K8 cells as observed previously for CPC/MWCNT-OH/BSA composites. Studies with biopolymers like PLG-based nanocomposite materials are also more or less similar in their toxicodynamics. Ziaa et al. [26] had investigated the cytotoxicity of chitin-based polyurethane bionanocomposites (PUBNCs), most commonly used in food packaging. These are prepared with bentonite nanoclay enriched in montmorillonite (MMT). This study also concluded with the fact that increasing doses of nanomaterials become increasingly toxic to cells in terms of their viability.

Polymers incorporating clay NPs have also been investigated for their cytotoxic effects. These nanomaterials were one of the first polymer composites to emerge as improved food packaging materials [27]. Examples include nonmodified clay (Cloisite Na^+) and an organoclay (Cloisite 30B). Studies have been ambiguous regarding their cytotoxicity, with researchers differing based on their data. Further, Maisanaba et al. reported that Cloisite 30B showed cytotoxicity mainly via DNA damage in hepatic cell line HepG2 and with increasing concentrations up to 88 μg/mL after 48 h exposure. Lordan et al. [28] also investigated the cytotoxicity of Cloisite Na^+ in HepG2 cells and found a reduction of viability from 1 μg/mL with the MTT assay. Similar adverse effects have also been observed in Caco-2 cells with organo-modified clay [29]. Besides, Sharma et al. [29] and Gutiérrez-Praena et al. [30,31], while studying the cytotoxicity of the same clay in Caco-2 cells (a human colon cancer cell line), observed no cytotoxic effects. This fact has been reiterated by other authors who found no significant differences in the cytotoxic effects of Cloisite Na^+ in U937 monocytic cells with respect to control [32].

4.3 *In Vivo* Testing

In vivo evaluation has always been found to be a more effective means and approach for toxicity studies as they present a complex system in which several physiological mechanisms are working in tandem to reduce the foreign influence. Besides, it is also useful in reducing time as *in vitro* studies need to be complemented by *in vivo* studies. *In vitro* studies are good at providing a clue or an idea that could be implemented when subjecting live animals to toxicity studies. Of the three nanomaterials, BG0, BG0.5, and BG24, that have been tested *in vivo*, BG0 and BG0.5 can be defined as *in vivo* biocompatible, whereas BG24 might be classed as toxic. Severe hepatic cytopathy has been observed indicated by cellular necrosis, malignancy, and drastic increases in alanine aminotransferase (ALT), aspartate aminotransferase (AST), and blood urea nitrogen (BUN). A significant aspect of this toxicity has been the pore structure of BG24, which is ultimately linked to the size of the nanomaterial. Rapid biodegradability produces large amounts of debris, which results in organ and cellular toxicity. Yun et al. [33], this mechanism does not result in the formation of a new bone. Similarly, synthetic GZNC has been observed to be cytotoxic on midgut cells of *Drosophila melanogaster* larvae evidenced by hsp70 expression and comet assay [34]. The toxicity of GZNC was observed only at 0.199 and 3.999 μg/μL for 24 and 48 h of duration of exposure. Wound healing was reported by Archanaa et al. [35] after the administration of chitosan-PVP-titanium dioxide nanocomposite as wound dressing material in wounded albino rat models. Li et al. [36] evaluated antibacterial and remineralizing nanocomposite and adhesive in rat tooth cavity model. The abovementioned evidences clearly indicate the potential health hazard of these nanomaterials when used without regulation and a comprehensive and complete evaluation. Hence, the full implementation of such nanomaterials in biological applications needs to be investigated in depth and by prioritizing not only humans but also animals and surrounding ecology. Nanocomposites and nanoceramics and their associated toxicity could limit a number of biomedical applications. Thus, it could be said that nanomaterial toxicity is very poorly studied and understood. Therefore, any future action in this area must be focused on the investigation of the toxicity and improvement of biocompatibility of multifunctional composites and ceramics along with due consideration to environment (Table 1).

5. RISK ASSESSMENT AND REGULATION/LEGISLATION

In the United States, the risk assessment and regulation/legislation related to nanotechnology is monitored jointly by the Environmental Protection Agency (EPA) and the Food and Drug Administration (FDA). The EPA [51] has changed its approach towards the regulation of nanomaterials. Previously, it encouraged nanomaterial manufacturers to provide information voluntarily through the Nanoscale Materials Stewardship Program (NMSP), but now, it gathers information and imposes standards on the manufacture, use, and disposal of nanomaterials. These changes are made by EPA often in cooperation with agencies in the United States and around the world. EPA follows the Toxic Substances Control Act (TSCA), which encompasses certain new uses of chemicals before they enter the market, imposes restrictions, and then adds those new chemicals to the toxic list. Reporting, recordkeeping, and testing of chemicals that may pose an unreasonable risk to human health or the environment or that may reach certain production or exposure levels are also necessary under TSCA norms. Besides these, relationship between nanotechnology and EPA's statutory mandates includes providing recommendations relating potential environmental benefits or research needs associated with nanotechnology and risk assessment. Under EPA norms, it is mandatory for nanomaterial manufacturers to give EPA information about the production volume, methods of manufacture, exposure, and health safety data. The main focus of EPA is on single-walled and multiwalled CNTs and on nanoscale clays and silica. EPA makes certain test rule for nanomaterials in commerce, particularly those not already being tested by other international organizations. Nanomaterial users are not required to supply information to EPA; they may be asked for information given by their suppliers especially related to use exposure and release data.

The FDA [52] does not categorize all products that involve the application of nanotechnology as harmful. They regulate nanotechnology products under existing statutory authorities, in accordance with the specific legal standards applicable to each type of product under its jurisdiction. FDA's product-specific guidance documents address issues such as the regulatory status, safety, effectiveness, performance, quality, and public health impact of nanotechnology products. Besides EPA and FDA, Consumer Product Safety Commission (CPSC), Occupational Safety and Health Administration (OSHA), and National Institute for Occupational Safety and Health (NIOSH) look after the nanosafety of nanomaterials in the United States.

In Europe, Registration, Evaluation, Authorisation and Restriction of Chemicals (REACH) and Classification, Labelling and Packaging (CLP) control nanosafety measures. Asia-Pacific has its own agencies for controlling the safety of nanomaterials such as, in Japan, the Ministry of Economy, Trade and Industry (METI) and National Institute of Advanced Industrial Science and Technology (AIST) and, in Thailand, the National Nanotechnology Center (NANOTEC) that puts "Nano Q" standard nanomark for selected Thai nanoproducts to identify types, sizes, and properties of NPs. In Australia, the National Industrial Chemicals Notification and Assessment Scheme (NICNAS) controls nanosafety measures.

International agencies and organizations also contribute towards the biosafety of nanomaterials. The Organisation for Economic Co-operation and Development (OECD) formed the Working Party on Manufactured Nanomaterials (WPMN) to check nanomaterial biosafety. The International Agency for Research on Cancer (IARC) operates under the auspices of the World Health Organization and American Society for Testing and Materials (ASTM).

6. SUMMARY

Nanocomposites and nanoceramics have been found to be incorporated within a large variety of systems and materials composed of strikingly dissimilar substances mixed together at a range from 1 to 100 nm. Their applications have been growing at a rapid rate from orthopedics, to drug delivery and tissue engineering, to food packaging. They are also addressing global challenges in the field of energy and environment. Nanocomposites are polymer mixtures of distinct size, geometry, and chemistry and possess advantages like biodegradability and biocompatibility both *in vitro* and *in vivo*. Nanocomposites have been known to combine biopolymers that mimic the organic component of the extracellular matrix of the bone and bioactive nanoceramics to induce biomineralization. In a study between nanosized biphasic calcium phosphate (BCP) ceramics and conventional microsized BCP ceramics, the former was found to be a better substrate for osteoblast proliferation. Considering their immense use in several biological systems, it becomes pertinent to address their biosafety issues. This aspect has been studied by several authors where ceramic NPs have been observed to be bioactive to osteoblast and macrophage-like cells with no significant toxicity. New bioceramics such as hydroxyapatite (HA), calcium phosphates, and bioactive glasses (BGs) are some of the most promising biomaterials for hard tissue regeneration. These materials have excellent cell proliferation response and differentiation behavior *in vitro* and *in vivo* without serious toxicity issues. Not only this, the nanocomposites are also being employed in drug delivery systems, which makes them potent candidates for cytotoxicity investigation studies. Recently, nanocomposites have been shown to exhibit low systematic toxicity and high *in vivo* antitumor therapy efficacy. A thorough knowledge of biocompatibility and bioefficacy issues about these nanomaterials needs to be revisited and is being comprehensively addressed in this chapter.

REFERENCES

[1] Market Research Reports. Nanocomposites (carbon nanotubes, polymer metal fiber, nanofibers, graphene, nanoplatelet and others) market for automotive, aviation, electronics, energy, construction, healthcare, plastics, military, consumer goods and other applications—global industry analysis, size, share, growth, trends and forecast, 2012-2018, in: Transparency Market Research, 2013, pp. 1–118.
[2] S.T. Dutta, T. Kar, D. Mandal, et al., Structure and properties of cholesterol-based hydrogelators with varying hydrophilic terminals: biocompatibility and development of antibacterial soft nanocomposites, Int. J. Artif. Organs 35 (11) (2013) 1015–1024.
[3] B. Basu, A brief report on international workshop on nanoceramics and nanocomposite, Trends Biomater. Artif. Organs 21 (2) (2008) 121–123.
[4] EuroNanoForum: Nanotechnology in Industrial Applications, European and international forum on nanotechnology, Düsseldorf (Germany), 19-21 June 2007, in: EUR, 22833, Office for Official Publications of the European Communities, 2007, EuroNanoForum, 3 (Düsseldorf), 2007.
[5] L. Madison, New "Smart" Clothes, Superfilter Named to Space Tech Hall of Fame, NASA News, Houston, 2005.
[6] J. Altmann, Potential military applications of nanotechnology, in: Military Nanotechnology: Potential Applications and Preventive Arms Control, Routledge, New York, 2006, pp. 71–118.
[7] W. Zhu, P.J.M. Bartos, A. Porro, Application of nanotechnology in construction, Mater. Struct. 37 (9) (2004) 649–658.
[8] Q. Chaudhry, L. Castle, R. Watkins, Nanotechnologies in the food arena: new opportunities, new questions, new concerns, in: Nanotechnologies in Food, The Food and Environment Research Agency, Sand Hutton, 2010, pp. 1–17.

[9] R.T. Enrique, S.S. Jaime, V.M. Ruben, et al., Microstructure and electrical transport in nano-grain sized $Ce_{0.9}Gd_{0.1}O_{2-\delta}$ ceramics, J. Solid State Chem. 180 (11) (2007) 3093–3100.
[10] Q. He, J. Shi, Mesoporous silica nanoparticle based nano drug delivery systems: synthesis, controlled drug release and delivery, pharmacokinetics and biocompatibility, J. Mater. Chem. 21 (2011) 5845–5855.
[11] G.D. Zhan, A.K. Mukherjee, Processing and characterization of nanoceramic composites with interesting structural and functional properties, Rev. Adv. Mater. Sci. 10 (2005) 185–196.
[12] P. Broz, Strategies for transmembrane passage of polymer-based nanostructures, in: Polymer-Based Nanostructures—Medical Applications, RSC Nanoscience and Nanotechnology, University of Basel, Switzerland, 2010, pp. 16–70.
[13] S.H.R. Ali, M.M.A. Almaatoq, A.S.A. Mohamed, Classifications, surface characterization and standardization of nanobiomaterials, Int. J. Eng. Technol. 2 (3) (2013) 187–199.
[14] R.A. Yoke, R.C. MacPhail, Engineered nanomaterials: exposures, hazards, and risk prevention, J. Occup. Med. Toxicol. 6 (7) (2011) 1–27.
[15] T. Rafeeqi, G. Kaul, Carbon nanotubes as a scaffold for spermatogonial cell maintenance, J. Biomed. Nanotechnol. 6 (6) (2010) 710–717.
[16] S. Maisanaba, S. Pichardo, J.B. María, et al., Cytotoxicity and mutagenicity studies on migration extracts from nanocomposites with potential use in food packaging, Food Chem. Toxicol. 66 (2014) 366–372.
[17] S. Oh, N. Oh, M. Appleford, et al., Bioceramics for tissue engineering applications—a review, Am. J. Biochem. Biotechnol. 2 (2006) 49–56.
[18] L. Zhang, T.J. Webster, Nanotechnology and nanomaterials: promises for improved tissue regeneration, NanoToday 4 (2009) 66–80.
[19] K. Pawar, G. Kaul, Toxicity of titanium oxide nanoparticles causes functionality and DNA damage in buffalo (*Bubalus bubalis*) sperm *in vitro*, Toxicol. Ind. Health 30 (6) (2012) 520–533.
[20] G. Pattan, G. Kaul, Health hazards associated with nanomaterials, Toxicol. Ind. Health 30 (6) (2012) 499–519.
[21] S.H.S. Zein, Cytotoxic properties of calcium phosphate/multi-walled carbon nanotubes/bovine serum albumin on the CCD-18Co cell line, J. Tissue Sci. Eng. 2 (5) (2011) 1–3.
[22] Y.F. Zhang, Y.F. Zheng, L. Qin, A comprehensive biological evaluation of ceramic nanoparticles as wear debris, Nanomed. Nanotechnol. Biol. Med. 7 (6) (2011) 975–982.
[23] Y. Zhou, X. Jia, L. Tan, et al., Magnetically enhanced cytotoxicity of paramagnetic selenium-ferroferric oxide nanocomposites on human osteoblast-like MG-63 cells, Biosens. Bioelectron. 25 (5) (2010) 1116–1121.
[24] L.N. Teixeira, G.E. Crippa, A.C. Trabuc, et al., *In vitro* biocompatibility of poly(vinylidene fluoride-trifluoroethylene)/barium titanate composite using cultures of human periodontal ligament fibroblasts and keratinocytes, Acta Biomater. 6 (2010) 979–989.
[25] A.A. Haroun, M.A. Abo-Zeid, A.M. Youssef, et al., *In vitro* biological study of gelatin/PLG nanocomposite using MCF-7 breast cancer cells, J. Biomed. Mater. Res. A 101 (5) (2013) 1388–1396.
[26] K.M. Ziaa, M. Zubera, M. Barikanib, et al., Cytotoxicity and mechanical behavior of chitin-bentonite clay based polyurethane bio-nanocomposites, Int. J. Biol. Macromol. 49 (2011) 1131–1136.
[27] S. Maisanaba, M. Puerto, S. Pichardo, et al., *In vitro* toxicological assessment of clays for their use in food packaging applications, Food Chem. Toxicol. 57 (2013) 266–275.
[28] S. Lordan, J.E. Kennedy, C.L. Higginbothamb, Cytotoxic effects induced by unmodified and organically modified nanoclays in the human hepatic HepG2 cell line, J. Appl. Toxicol. 31 (2011) 27–35.
[29] A.K. Sharma, B. Schmidt, H. Frandsen, et al., Genotoxicity of unmodified and organo-modified montmorillonite, Mutat. Res. 700 (1–2) (2010) 18–25.

[30] D. Gutiérrez-Praena, S. Pichardo, M. Jordá, et al., Cytotoxicity study of microarcilla Nanofil116 cellular line in human intestinal Caco-2, Rev. Toxicol. 28 (2011) 87.
[31] D. Gutiérrez-Praena, S. Maisanaba, S. Pichardo, et al., Toxic effects of a modified montmorillonite clay on the human intestinal cell line Caco-2, J. Appl. Toxicol. 34 (6) (2014) 714–725.
[32] S. Lordan, C.L. Higginbotham, Effect of serum concentration on the cytotoxicity of clay particles, Cell Biol. Int. 36 (2012) 57–61.
[33] H.S. Yun, J.W. Park, S.H. Kim, et al., Effect of pore structure of bioactive glass ball on biocompatibility *in vitro* and *in vivo*, Acta Biomater. 7 (6) (2011) 2651–2660.
[34] Y.H. Siddique, W. Khan, S. Khanam, et al., Toxic potential of synthesized graphene zinc oxide nanocomposite in the third instar larvae of transgenic, BioMed Res. Int. 2014 (2014), Article ID 382124, 10 pp.
[35] D. Archanaa, B.K. Singha, J. Duttab, et al., *In vivo* evaluation of chitosan-PVP-titanium dioxide nanocomposite as wound dressing material, Carbohydr. Polym. 95 (2013) 530–539.
[36] F. Li, P. Wang, M.D. Weir, et al., Evaluation of antibacterial and remineralizing nanocomposite and adhesive in rat tooth cavity model, Acta Biomater. 10 (2014) 2804–2813.
[37] P.C. Lin, S. Lin, P.C. Wang, R. Sridhar, Techniques for physicochemical characterization of nanomaterials, Biotechnol. Adv. 32 (2014) 711–726.
[38] S. Dutta, A. Shome, S. Maiti, et al., *In situ* synthesized Ag nanoparticle in self-assemblies of amino acid based amphiphilic hydrogelators: development of antibacterial soft nanocomposites, Soft Matter 7 (2011) 3011–3022.
[39] Y. Zhou, X. Jia, L. Tan, et al., Magnetically enhanced cytotoxicity of paramagnetic selenium-ferroferric oxide nanocomposites on human osteoblast-like MG-63 cells, Biosens. Bioelectron. 25 (5) (2010) 1116–1121.
[40] H. Hajiali, M. Hosseinalipour, S. Karbasi, et al., The influence of bioglass nanoparticles on the biodegradation and biocompatibility of poly (3-hydroxybutyrate) scaffolds, J. Biomed. Mater. Res. B Appl. Biomater. 100 (3) (2012) 624–636.
[41] N. Sagar, V.P. Soni, J.R. Bellare, Influence of carboxymethyl chitin on stability and biocompatibility of 3D nanohydroxyapatite/gelatin/carboxymethyl chitin composite for bone tissue engineering, PLoS ONE 8 (10) (2013) 1–24.
[42] R. Qi, X. Cao, M. Shen, et al., Biocompatibility of electrospun halloysite nanotube-doped poly(lactic-co-glycolic acid) composite nanofibers, J. Biomater. Sci. Polym. Ed. 23 (1–4) (2011) 299–313.
[43] S. Tavakol, M. Azami, A. Khoshzaban, et al., Effect of laminated hydroxyapatite/gelatin nanocomposite scaffold structure on osteogenesis using unrestricted somatic stem cells in rat, Cell Biol. Int. 37 (11) (2013) 1181–1189.
[44] X. Chen, L. Huang, H.J. Sun, et al., Stimuli-responsive nanocomposite: potential injectable embolization agent, Macromol. Rapid Commun. 35 (5) (2014) 579–584.
[45] D. Bozukova, C. Pagnoulle, M.C. De Pauw-Gillet, et al., Hydrogel nanocomposites: a potential UV/blue light filtering material for ophthalmic lenses, J. Biomater. Sci. Polym. Ed. 22 (14) (2011) 1947–1961.
[46] X. Ye, M. Chen, M. Yang, et al., *In vitro* corrosion resistance and cytocompatibility of nano-hydroxyapatite reinforced Mg-Zn-Zr composites, J. Mater. Sci. Mater. Med. 21 (4) (2010) 1321–1328.
[47] J. Athinarayanan, V.S. Periasamy, A.A. Alshatwi, Biogenic silica-metal phosphate (metal = Ca, Fe or Zn) nanocomposites: fabrication from rice husk and their biomedical applications, J. Mater. Sci. Mater. Med. 25 (7) (2014) 1637–1644.
[48] K. Kandiah, P. Muthusamy, S. Mohan, et al., TiO_2-graphene nanocomposites for enhanced osteocalcin induction, Mater. Sci. Eng. C Mater. Biol. Appl. 38 (2014) 252–262.

[49] Y.Y. Tai, S.H. Hsu, R.S. Chen, et al., Liquid crystalline epoxy nanocomposite material for dental application, J. Formos. Med. Assoc. 114 (1) (2014) 46–51, pii: S0929-6646 (14)00041-2.
[50] S. Barua, N. Dutta, S. Karmakar, et al., Biocompatible high performance hyper-branched epoxy/clay nanocomposite as an implantable material, Biomed. Mater. 9 (2) (2014) 025006.
[51] M.N. Duvall, A.M. Wyatt, Regulation of Nanotechnology and Nanomaterials at EPA and Around the World: Recent Developments and Context, Beveridge & Diamond, P.C., Washington, DC (USA), 2011, pp. 1–25.
[52] Food and Drug Administration, Guidance for Industry Considering Whether an FDA-Regulated Product Involves the Application of Nanotechnology, Food and Drug Administration, U.S. Department of Health and Human Services, Silver Spring, USA, 2014. http://www.fda.gov/RegulatoryInformation/Guidances/ucm257698.htm, pp. 1–14.

CHAPTER 23

Nanocomposite Coating for Antibacterial Purposes

Hideyuki Kanematsu[1], **Michiko Yoshitake**[2]
[1]Department of Materials Science and Engineering, Suzuka National College of Technology (SNCT), Suzuka, Japan
[2]MANA, National Institute for Materials Science (NIMS), Tsukuba, Japan

Contents

1. INTRODUCTION

Bacteria and materials have coexisted with each other in the globe history, since bacteria appeared firstly on this globe more than 4 billion years ago. The affinity is very strong and stronger than we anticipated before. While we had been ignorant about it, bacteria have continued to survive and

Handbook of Nanoceramic and Nanocomposite Coatings and Materials
http://dx.doi.org/10.1016/B978-0-12-799947-0.00023-7

to hand their characteristics down to their posterity in its long history. We wrote "materials" in the description above. However, we might as well include metallic materials or minerals here. Metals and bacteria have been correlated with each other in these 4 billion years or more.

Some metals have high affinities with bacteria. And the others do not at all. Some metals are incorporated into the inside of bacteria. And the others kill them. From this viewpoint, the correlation varies so much with the combination and case. In this chapter, we would like to focus on the antibacterial effect of metals among those various correlations. In many advanced countries such as the United States, some European countries, and Japan, people are surrounded by high-function materials, devices, and apparatuses and enjoy amenities of life. In those many countries, the increase of population hit the ceiling. The birth rate is declining and the society is being aged, usually. In such a mature society, materials based on the infrastructure must have not only high function but also user-friendliness. The antibacterial effect by metallic materials is also one of the characteristics. Technically, it will be needed in food industries, medical field, plumbing products in households, etc. Since it relates to our daily lives directly, the antibacterial effect provided by materials would bring us the safety and secure feeling. There are some bulk matters showing antibacterial effects directly. However, the commercial products usually need versatile properties such as mechanical, chemical, and physical ones. Therefore, materials could not be designed only from the viewpoint of antibacterial effect, since it would be very hard for a material to have all of those beneficial properties at the same time. To make a material get those beneficial properties simultaneously, surface coating is the most attractive, useful, accessible, and reasonable process. While the surface could show antibacterial effect, the bulk could keep other characteristics for the application. From this point of view, we would say that surface coating would be generally a kind of composite material.

In this chapter, the antibacterial effects of metals will be described first, and it will be compared with photocatalytic oxides and organics. Then, antibacterial coatings that have been proposed in practical industries will be introduced. The merits and demerits will be discussed, and the concept of composite materials for antibacterial effect will be proposed followed by a detail discussion. Some possibilities will be mentioned. And finally, the two concrete examples by the authors will be introduced and explained.

2. ANTIBACTERIAL EFFECTS AND METALS

2.1 What is Antibacterial Effect?

We knew that some materials would have antibacterial effects so far [1–4], even though we still do not know the mechanism correctly. Particularly, some metallic materials, such as silver and copper, have had been well known for the characteristics, and they had been considered as magic or miracles. Conventionally, silver has been utilized for eating utensils. It means that people have known the antibacterial effect of the metal since the ancient time. So why and how do some materials including metals show antibacterial effect?

Before the explanation of antibacterial effect shown by materials, we have to clarify the definition of antibacterial effect of materials more clearly. In English, almost all cases to suppress the growth of bacteria could be expressed as antibacterial effect. However, Japanese generally subdivide the concept into smaller ones, depending on the extent and the effect.

Usually, we have five technical terms for antibacterial effects [5,6]. The first one is microbiocidal effect. In this case, all bacteria are killed perfectly. However, the concept does not contain any objects nor extents. For example, if only several bacteria would be killed and the rest would be still alive, one could say that the microbiocidal effect would still work in that case. The second is the sterilization that has broader meaning. Not only to kill but also to remove bacteria or viruses from the target objectives could be included, even though they might be harmful or harmless. From the viewpoint of probability, it is impossible to make the bacterial number completely zero. Therefore, SAL (sterility assurance level) is often used to judge the extent of sterilization. Internationally, sterilization is defined as the situation where SAL is smaller than 10^{-6}. The third is disinfection by which the number of pathogenic bacteria would decrease down to the nonharmful extent. This could be achieved by microbiocidal effect, the first concept. In this point, one may say that disinfection would be almost the same with microbiocidal effect. However, disinfection could be achieved only by causing the pathogenicity to disappear. Therefore, the two concepts are different with each other, strictly speaking. The fourth is Jokin. This Japanese technical term should be translated into "decreasing the bacterial number." The concept does not include targets nor extents. It means that the concept doesn't matter; the target would be how much bacteria might be removed. And the fifth is "Kokin." Kokin can be defined as an action that would suppress the growth of bacteria in the target object. Usually, we need to confirm the decrease of bacteria in the order of 10^2. As for these five antibacterial effects, the authors used some technical terms in English tentatively. However, they are not correct, strictly speaking. All of them in Japanese might be included in the concept of antibacterial effect used in the western world. Kokin is the most desirable among these five categories, when we come to think about the environmental friendliness, since it doesn't require any harmful or strong chemicals to kill them completely. Regarding the onset of antibacterial effects by materials, they are not so strong to kill bacteria completely in usual cases. Therefore, we deal with Kokin as antibacterial effect in this chapter.

2.2 Metallic Antibacterial Groups

The materials are classified into three groups from the viewpoint of antibacterial effect on the target. These classifications include metallic materials, some photocatalytic materials such as titanium oxide, and organic materials [5,7]. And each has its own characteristic mechanism of action for the antibacterial effect. The first group is metallic materials. Some metals having obvious antibacterial effects are dissolved to bind with some proteins, leading to changes of the higher-order structure. The 3-D conformation of proteins [8] is generally composed of relatively weak bonding among amino acids, such as hydrogen bonds, ionic bonds, hydrophobic bonds, and S–S bonds. The bond between metallic ions and proteins could weaken the bonds between amino acids and sometimes break them. The high-order structure generally has a close relation to the function of protein. Thus, the following high-order structural change would disturb the microbial activity and the metabolism to develop antibacterial effects. Usually, there are many pores (channels) on membrane cells, where ions and various matters on cell membranes are moved into the inside cells. Ikegai et al. [9] pointed out in the past that several divalent metallic ions tend to bind with the "channel" to decrease its function as the moving passage of matters. They also confirmed that zinc ion, nickel ion, and copper ion inhibited the function of channel proteins on the cell membranes of *Thiobacillus ferrooxidans*. The mechanism is shown in Figure 1 schematically.

Conventionally, silver, as well as some heavy metals, has also been known for its antibacterial effect. Regarding the antibacterial effect of silver, there have been lots of discussions about the mechanism. Generally, silver ion tends to react with function groups containing S, N, O, etc. very easily whose electron densities are generally high. Particularly, the silver ion binds

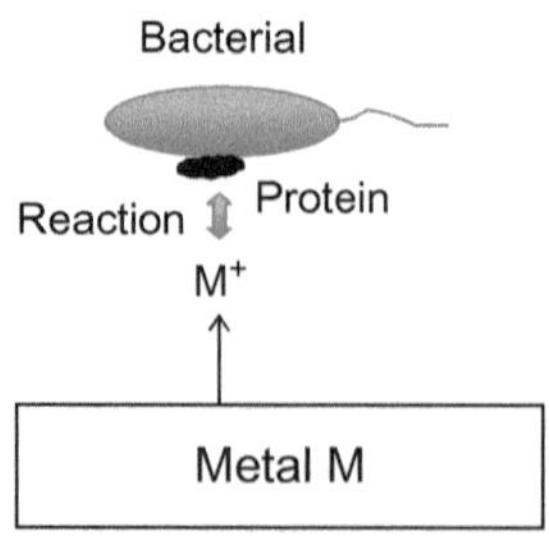

Figure 1 The mechanism of antibacterial effect by metals (hypothesis).

with sulfhydryl group (–SH). As a result, the enzyme's function is inhibited. And in addition, silver ion can react with amino acid residue, DNA, RNA, and NH group to form insoluble salts, which would lead to inactivation of enzymes on the cell membranes.

We presume that antibacterial metals have the same effects with those described above. Therefore, the antibacterial effects by metallic ions should be considered from the viewpoint of inherent affinity with metallic ions and degree of dissociation for them.

As for some commercial coating metals, the authors et al. also confirmed the extent of antibacterial effect in the past [10]. Those metals were zinc, manganese, nickel, copper, cobalt, chromium, titanium, aluminum, silver, iron, and magnesium. These metallic powders were added into a usual agar media. Then, the 0.1 ml solution containing 10^6 ml^{-1} bacteria (*Escherichia coli*, *Klebsiella pneumoniae*, and *Staphylococcus aureus*) was added into the agar media. After the culture in 18 h at 37 °C, the formed colonies were measured. The relation between the colony number and metal content evaluated the antibacterial effect, shown in Figure 2 schematically. The colony formation units for these metals could be shown with the metal contents, respectively. Generally, the colony formation unit decreased with the increase of content. The extent of decrease and the behavior could show their antibacterial effects. As a result, these commercially available metals were classified into three types. The first group was composed of metals that show strong antibacterial effects. Zinc, manganese, nickel, copper, and cobalt constitute this strong antibacterial group. On the other hand, the second group was composed of chromium, titanium, aluminum, and silver whose antibacterial effect was medium. And the third group was composed of iron and magnesium. Since this evaluation method did not take the specimen's shape, diameter of particles, and the inherent dissolution characteristics into consideration, the results might not be appropriate for detailed quantitative

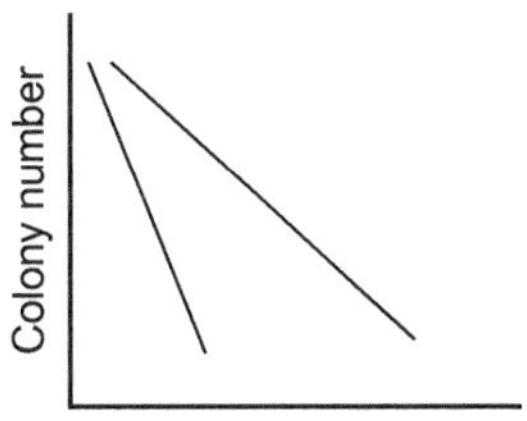

Figure 2 Evaluation for tentative antibacterial effect of metal powder.

analysis. However, the differences among these elements were convictive for us, when we compared the results with those in the past.

2.3 Antibacterial Test and the Standard

As quantitative evaluation method of antibacterial effect, an international standard is already established as ISO 22196 [11]. The test is usually called the "film covering method." The standard could provide us a useful way to evaluate the antibacterial effect for practical materials and products. The procedure can be described as follows.

The specimens were put in a plastic Petri dish, while the bacterial solution was prepared as follows: the bacteria were incubated in 10 mL of a nutrient broth for 24 h at 35 °C and then diluted 2000-fold with sterilized water and established as a bacterial solution. The diluted bacterial solution was applied to the specimen (16 μL/cm), and then, a polymer film was laid over the solution. The sample was kept in an incubator for 24 h at 35 °C. After the incubation, a solution of 10 mL of sterilized water containing 200 μL of Tween® 80 (a nonionic surfactant and emulsifier) was introduced into the plastic Petri dish, and the bacteria attached to the specimen and polyethylene film were washed into the aqueous solution. To determine the number of viable cells, serial decimal dilutions of the cell suspension were made, a 0.1 mL portion of which was uniformly spread on an agar medium. The plate was incubated at 35 °C for 24 h and the colonies formed were counted. The viable cell count was represented as colony-forming unit per milliliter (CFU/mL). The final colony formation number was measured to evaluate the antibacterial properties.

This method is very useful for us to check the antibacterial effect not only for metallic materials but also for most industrial solid materials. However, the concept for the antibacterial effect might be too static. It means that the antibacterial effect evaluated by the method would not show us any time-dependent characteristics. Figure 3 shows a problem for an example, which

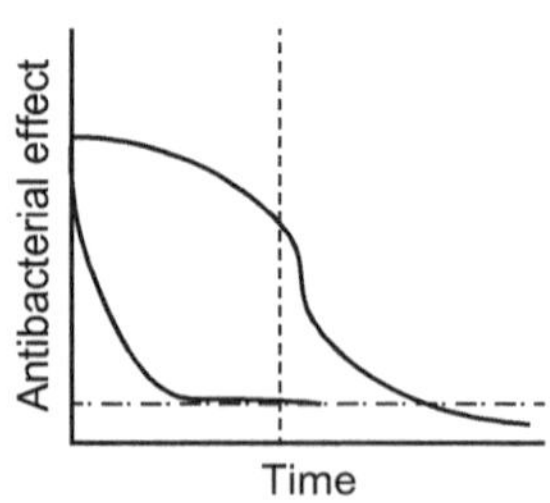

Figure 3 A problem of film covering method for antibacterial effect.

has been often discussed as personal communication between authors. The vertical axes correspond to an indication. The vertical solid lines in the figures indicate for antibacterial effect such as colony formation unit or bacterial numbers, while the horizontal ones to time. In any case for antibacterial effects, they would decrease with the increase of time, more or less. The dotted lines would correspond to the points when the film covering method fixed antibacterial effects. However, who could tell which case would have higher antibacterial effect, when the final values would be compared. At this point, there are no alternative standards to show the time dependence of antibacterial effect. To predict the effect more properly, such a test method is needed.

2.4 Other Antibacterial Group [5]

Other than metallic materials, there are still two groups of materials showing antibacterial effect. The second group is composed of photocatalytic substances, such as titanium oxide. As shown in Figure 4 schematically, photocatalytic oxides such as titanium oxide have the energy gap between valency and conduction band, since they are basically semiconductors. When they are irradiated by ultraviolet light, electrons in valency band become energized to conduction band. Then, positive holes form in valency band. Generally, the positive hole has the strong oxidation power to compensate the lack of electrons and to disappear themselves by depriving of electrons. As a result, hydroxyl ions near the surface of oxide are changed to hydroxyl radical. They could kill bacteria and inhibit the growth. Titanium oxide is the most well known, even though there are some photocatalytic substances such as tungsten oxide.

The third group for antibacterial substances is organic matters. As shown in Figure 5 schematically, organic antibacterial agents generally damage cell membranes. When they would be damaged, denaturation of membrane

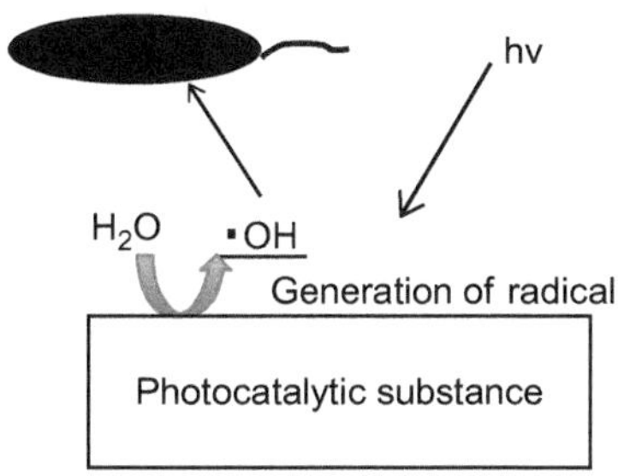

Figure 4 The schematic principle of antibacterial effect by photocatalytic substance.

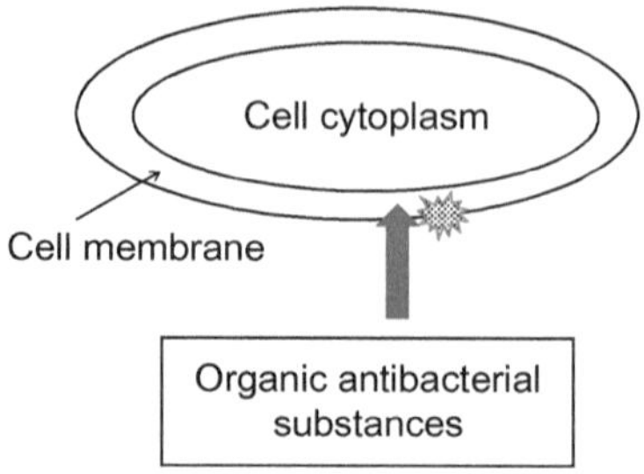

Figure 5 The schematic principle of antibacterial effect by organic substance.

proteins, the change of membranous structures, etc. might occur. Antibacterial effect by organic agents would appear in such a way.

3. ANTIBACTERIAL SURFACE COATINGS

3.1 Application to Surface Coating

There are many types of antibacterial agents. They are usually classified and organized systematically in the authors' country (Japan) and other countries. In Japan, antibacterial matters have been generally classified into six types from the viewpoint of patent application—photocatalyst, silver, heterocyclic compound, copper, vegetable matter, and amine/quaternized ammonium salts [12]. And all of them are applied and used in the following classified fields—water treatment, domestic housewares, basic goods (fibers, leather, paper encasement, container, clothes, etc.), machinery/appliances, construction/pain, separation, and others. The most utilized substances for antibacterial purposes have been photocatalyst (titanium oxides) and silver in Japan, followed by heterocyclic compounds and copper. Therefore, we should think that antibacterial surface coatings would be the proper application of these antibacterial substances to surface coating. Of course, only surface coatings could realize the effective outcomes from the viewpoint of some inevitable restrictions.

3.2 Polymer Coating: Antifouling Coating

Most of polymer coating aims to make the surface hydrophilic, since the hydrophilic surface very often could repel proteins, bacteria themselves, etc. From the viewpoint, the coating does not correspond to antibacterial effect based on killing or controlling bacteria. The basic concept for such a polymer coating would be antibiofouling effect. Biofouling can be defined as the whole process composed of attaching of organisms on materials and the following phenomena. The biofouling brought about by bacteria is called "microfouling." The antibacterial effect is related to the microfouling. Figure 6 [13] shows the general case for microfouling occurring on materials surfaces. Generally, the biofilm formation process is explained in the following way [14–18].

Carbon compounds form on materials surface in advance. It is called "conditioning film." That should be nutrition for bacterial. Floating bacteria called "planktonic bacteria" move toward the materials surface to get the nutrition. Through some complicated phenomena, bacteria attach to the surface and the number of bacteria attached to the materials surface increases. When the number exceeds a certain threshold value, polysaccharides are excreted from bacteria and the sticky matters surround materials surface including bacteria on the surface and also water in the vicinity of surface. Then, the sticky and inhomogeneous filmlike matters form on the materials surface. It is called biofilm. However, it is not the homogenous film, strictly speaking. The film is composed of consecutive towerlike thin matters covering materials surfaces. Biofilm is composed of more than 80% water, bacteria, and extracellular polymeric substance (EPS). In the past, researchers believed that EPS is almost equal to polysaccharide. However, we got to know recently that EPS contains a lot of proteins and polysaccharides [18]. Hydrophilic polymer coatings could repel the proteins and bacteria mainly. Usually, such a coating is called "brush polymer coating." The outline of the mechanism is shown in Figure 7 [19]. The nonhydrated polymers

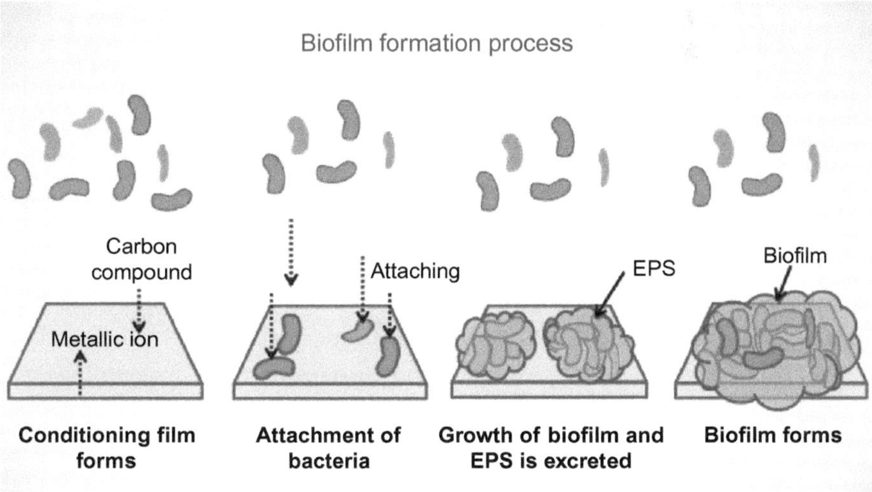

Figure 6 Mechanism of microfouling process on material surface.

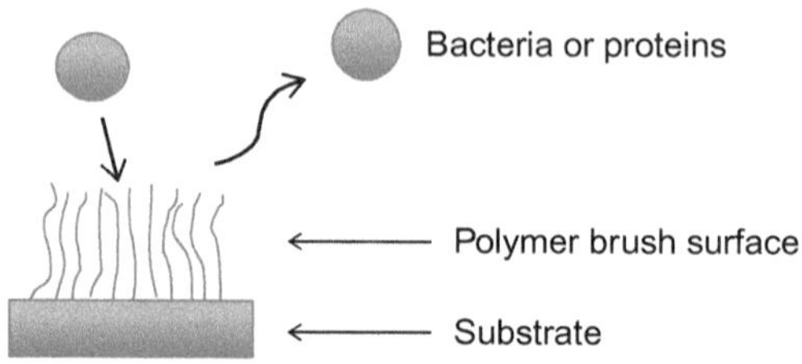

Figure 7 Polymer brush surfaces repelling bacteria and protein.

are randomly coated on the surface. When the coating is too difficult, some pretreatment may be required to enhance the bonding of polymers with the substrate. Then, tightly packed highly hydrated brush is created in an aqueous environment. As shown in the figure, proteins and bacteria would be repelled by the action of bound water in the brush and the elasticity of the polymer chains. According to this concept, Khoo and coresearchers achieved polymer coating through a peptide-linked PEG. Since the polymer film would have longer polymer chains, the repellent effect would be stronger. This is just an example for medical applications, and the reader can refer to many good reviews about this topic. However, such a polymer coating does not relate to "Kokin" where the bacterial growth and metabolism are controlled by the direct reaction of antibacterial substances. It controls biofilm formation and biofouling, the wider and more comprehensive concepts relating to bacterial activity.

3.3 Metallic Coating

As described above, silver and copper have been antibacterial metals. Figure 8 shows the percentage of dominated application by silver and tin in Japanese patents. When the cases would be restricted to coating technique, 177 cases for silver-related coating and 167 cases for copper-related coating were applied to "Japan Patent Office" as antibacterial coating from 1993 to 2013 [20]. Even though the mechanism would be not fixed, many circumstances, phenomena, and results show clearly that dissolved metallic ions must play an important role to reveal antibacterial effects. As described above, there have been so many intentions also for metallic coatings; we are afraid that we could not mention all of them. For the detailed information, the reader could refer to one of our papers in the past [20]. We have to mention their application to surface coating first of all as well as the dominated patents for bulk materials.

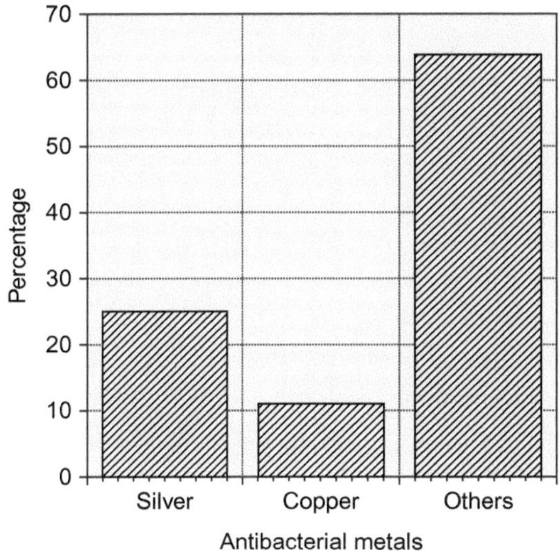

Figure 8 Application percentage of antibacterial metals, silver and copper, in Japan.

Koura and Sakado proposed silver-zeolite coating [21]. Zeolite could contain silver ion in a stable condition and control the reactivity of silver ion well. They devised a solution composed of aluminum alkoxide, alkoxysilane, alkoxide of alkali metal or alkaline earth metal, amines, amino-based organoalkoxysilane, and a silver salt. The specimen was coated with silver-zeolite film by dipping and spraying after calcined at 200-800 °C.

Miyasaka et al. devised the stacked films composed of n-type semiconductor oxide and silver compounds [22]. N-type semiconductor film would show photocatalytic antibacterial effect, while silver compounds would produce silver ion. They aimed to achieve combined effects by photocatalytic and silver ion effects simultaneously.

Otani et al. devised laminated silicate carrying some antibacterial metals such as silver, copper, and zinc [23]. They achieved the coating by spray coating, dipping or bushing, etc. Hirose et al [24] devised dissolutive glass containing silver and copper ions. The antibacterial metal ions gradually dissolve to work as antibacterial agent. They crushed the glass down to tiny particles in micrometer orders and successfully attached on clothes, etc.

Toda et al. devised a new agent to achieve antibacterial and antifungal ceramic coating [25]. The coating films are composed of titanium oxides and one of the antibacterial metals. They also aimed to show combined effects for antibacterial effects deriving from different groups.

Mase et al. devised the coating method for some antibacterial metals by plating, sputtering, vacuum coating, etc [26].

Miyauchi et al. also devised sputtering mask composed of silver, zinc, copper, titanium oxide, and zinc oxide [27]. These are concrete examples that physical vapor deposition processes were applied to antibacterial coating concretely.

Recently, nanoparticles of antibacterial metals have come under the spotlight. And the nanopowder has been applied to various coatings, particularly for medical fields [19]. Kowsett and Strom-Versloot et al. found that dressings could be impregnated with silver sulfadiazine and other silver compounds [28,29]. The antibacterial effect of the wound dressing with silver and silver salt for medical applications has been still discussed [30–33], since the experimental conditions were versatile and complicated. External ventricular drains (EVDs) that have been used for hydrocephalus treatment are also one of the medical tools where silver coating was applied. Fichtner et al. applied the nanosilver coating to EVD and confirmed that the colonization of drains decreased by a factor of 2-4 [34]. However, the negative results were also obtained [35]. The nanopowder coating was also applied to catheters. However, the positive result was still obtained only for animals [36,37]. The results for human bodies are still investigated.

Medical applications always have such difficulty. Generally, it takes a long time to pass the safety regulations and to get approved from government and authorities. And in addition, the environments are too complicated to simulate and analyze on laboratory scales. The more accurate and appropriate evaluation tèchniques should be improved and highly developed.

4. NEW POSSIBILITIES AND PREDICTIONS FOR COMPOSITE COATING

4.1 The Basic Principle for Antibacterial Composite Coating

As described in previous section, metallic coating of silver, copper, etc. might kill bacteria. However, metals on material surfaces must dissolve to exhibit antibacterial effects. It might lead to the two defects, as shown in Figure 9. One of them is the change and deterioration of surfaces. Since the dissolution means corrosion, the surface would be lost more or less. When the extent would be serious, the surface would be damaged and deteriorated. They might not be able to keep the other purposes such as wear resistance and color tone. And the coating thickness would be decreased due to the dissolution. That

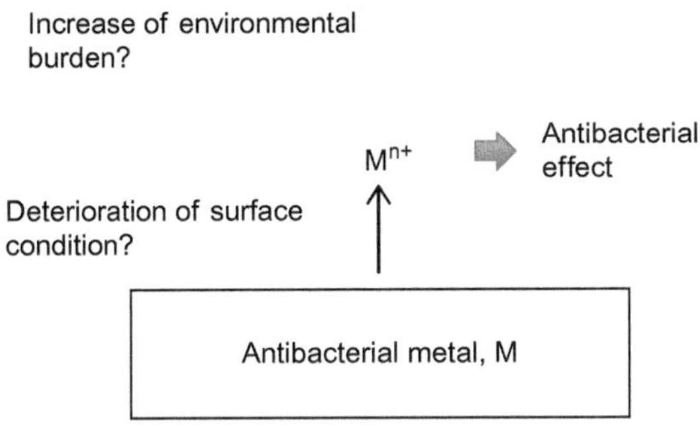

Figure 9 Problems for antibacterial metal coating.

might lead to the decrease of life cycles for the coating. Secondly, the dissolution might lead to the infringement of some environmental regulations such as "REACH" in European Community. Even if some metallic components would not be regulated at this point, they might be listed as hazardous components in the future. Due to those reasons and some possible others, the direct coating of antibacterial heavy metals would not be always good ideas to solve the solution for antibacterial effects. So what type of coating would be desirable? The answer should be composite coating. Concretely, the problem solution should contain the following points:

1. Antibacterial metals should be contained in the coating more or less.
2. The antibacterial effect should appear by the action of metallic ions.
3. The dissolution rate should be controlled properly.

To achieve those points simultaneously, composite coatings are the most beneficial. The problem is how to combine antibacterial metals into the other basic component(s). According to the concept, there are already some investigations and proposals as follows in these recent years, even though there are still not so many investigations at this point.

4.2 Recent Research and Proposals

Voccia et al. proposed a new method for the electrochemical deposition of a metal/polymer composite layer on a conducting substrate [38]. The electrochemical solution was a mixture of an acrylate (ethyl acrylate, EA; 2-phenyl-2-(2,2,6,6-tetramethylpiperidin-1-yloxy) ethyl acrylate, PTEA; and 8-quinolinyl acrylate, 8QA), a metallic salt (silver (I) acetate), and a conducting salt in dimethylformamide.

Sunkara et al. investigated and showed a distinct enhancement of antimicrobial activity of $W4^{+}$-doped titania that was coated on nickel ferrite nanoparticles in comparison with undoped titania [39]. The composite nanoparticles were synthesized by uniquely combining reverse micelle and chemical hydrolysis synthesis methods.

Ahn et al. proposed experimental composite adhesives (ECAs) containing silica nanofillers and silver nanoparticles and compared with two conventional adhesives (composite and resin-modified glass ionomer, RMGI) to analyze surface characteristics, physical properties, and antibacterial activities against cariogenic *streptococci* [40]. And they confirmed that their new ECAs were effective to control the bacterial growth.

Necula et al. carried out the synthesis of a porous TiO (2)-Ag composite coating and assessment of its *in vitro* bactericidal activity against methicillin-resistant *S. aureus* [41]. The coating was produced by plasma electrolytic oxidation of Ti-6Al-7Nb medical alloy in a calcium acetate/calcium glycerophosphate electrolyte bearing Ag nanoparticles. Following oxidation, the surface of the titanium substrate was converted into the corresponding oxide (TiO_2) bearing Ca and P species from the electrolyte. In addition, Ag was detected to be associated with particles present in the oxide layers. They confirmed successfully that the coating could kill *S. aureus* effectively.

Shirokova et al. proposed a new polymer composite film based on carboxymethyl chitin and silver nanoparticles [42]. The composite films revealed a pronounced concentration-dependent antibacterial activity toward strains *Salmonella typhimurium* and *S. aureus*.

Dastjerdi et al. proposed a novel method to overcome the problem that the treatment of textiles with Ag/TiO_2 nanoparticles caused a brownish color and limited the application of this otherwise good composite [43]. To achieve the purpose, they investigated the effect of various concentrations of cross-linkable polysiloxane (XPs) and Ag mixed with XPs on TiO_2-treated fabrics.

Qu et al. proposed a new hydroxyapatite (HA) coating loaded with nanosilver particles [44]. They produced porous Ti scaffolds with high porosity and interconnected structures by polymer impregnating method. And then, they successfully made uniform Ag/HA composite coatings on the surfaces of porous Ti substrates by a sol-gel process. They used Ca$(NO_3)_2 \cdot 4H_2O$ and P_2O_5 in an ethyl alcohol-based system to prepare the sol, which ensured the homogeneous distribution of Ag in the sol. They successfully achieved the good balance between the biocompatibility and antibacterial properties of the coatings.

Yu et al. prepared neat TiO_2 and Ag-TiO_2 composite nanofilms on silicon wafer via the sol-gel method by the spin-coating technique successfully [45].

The synthesized Ag-TiO_2 thin films showed enhanced bactericidal activities compared to the neat TiO_2 nanofilm both in the dark and under UV illumination. They achieved antibacterial effects by both metallic ion and photocatalytic oxides.

Basheer et al. synthesized W-TiO_2 composite and incorporated them into the zinc bath during the hot-dipping process [46]. They found that the composite was effective in controlling the growth of bacteria and formation of biofilm thereafter.

Apart from all of the composite films, the authors proposed some composite films for the antibacterial effect. In the next section, some of them are introduced.

5. EXAMPLES OF COMPOSITE COATINGS FOR ANTIBACTERIAL PURPOSES BY THE AUTHORS

5.1 Heating Stacked Single Layers Process [47–49]

We, the authors, proposed intermetallic compounds between an antibacterial element and non-antibacterial element to show antibacterial effect, while the dissolution of antibacterial metallic ion could be controlled properly. Figure 10 shows the concept of this method. In the conventional alloy plating from an aqueous solution, alloying process takes the path shown in Figure 10a. However, this process would take the double steps path shown in Figure 10b. The latter requires a more complicated process clearly. However, the degree of freedom is larger in the latter than that in the former. We usually call it HSSL process standing for *h*eating *s*tacked *s*ingle *l*ayers process. Both for copper and silver systems, the intermetallic compound films could be produced by the processes. The concrete example cases were introduced as follows.

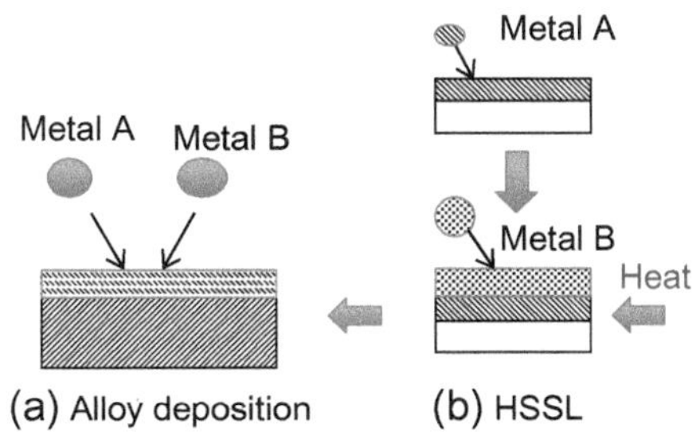

Figure 10 Conventional alloy plating (a) and HSSL process (b).

5.1.1 *Tin-Copper Alloy Film* [50,51]

A carbon steel (JIS SS400) was immersed in copper sulfate solution composed of 150 g/L copper sulfate and 40 g/L sulfuric acid at 25 °C. The steel was used as cathode and a platinum wire was used as anode. The constant current source provided the cathode 1 A/dm^2 for three minutes to form 10 μm copper layer on the cathode specimen. Then, the steel was immersed into tin sulfate (30 g/L)/sulfuric acid (180 g/L) and electrolyzed with 1 A/dm^2 for a certain time to form tin layer as top layer. The stacked single layer specimens were heat-treated at around melting point of tin. Table 1 shows the produced phases by HSSL process, being identified by X-ray diffraction analyses. When the specimens were heated to the temperatures below the melting point of tin (200 °C), the produced phase was only tin and the alloying did not occur. When the specimens were heated to the temperatures above the melting point (250 °C), various intermetallic compounds formed. For an hour treatment, Cu_6Sn formed, while Cu_3Sn also formed for 2 h treatment. It suggests that the alloying process made progress with time. When the time was prolonged to 3 h and more, tin oxide, as well as Cu_3Sn, was found. It suggests that the specimen should be heat treated for an appropriate heat treatment time.

Table 2 shows the results by Film Covering Method (ISO 22196). The bacteria used were *E. coli* (ATCC25032). Table 2a shows the result for the stacked

Table 1 Surface phases produced by HSSL process for tin-copper system

Heat treatment temperature	1 h	3 h	5 h
200 °C	Sn, Cu	Sn, Cu	Sn, Cu
250 °C	Cu, Sn, Cu_6Sn_5	Cu_6Sn_5, Cu_3Sn, SnO_2	Cu_3Sn, SnO_2

Table 2 Evaluation of specimens' antibacterial effects for tin-copper by film covering method

Time (h)	Specimen	*S. aureus*
(a) Before heat treatment of stacked single layers specimens		
0	Control	9.5×10^4/a plate
24	Control	2.04×10^6/a plate
	Plating specimen	7.51×10^5/a plate
(b) After heat treatment of stacked single layers specimens		
0	Control	9.5×10^4/a plate
24	Control	2.04×10^6/a plate
	Plating specimen	0/a plate

single layers without heat treatment. On the other hand, Table 2b shows the result for the heat-treated specimen at 250 °C for 5 h. For non-heat-treated specimen, the number of colonies increased with time, and it indicates that the specimen did not show any antibacterial effects. For heat-treated specimen, the numbers of colonies were zero in any cases. All of these results suggest that intermetallic compounds between copper and tin showed antibacterial effects clearly. According to our hypothesis that metallic ions would be the main cause for antibacterial effect, the dissolution of intermetallic copper compounds on materials surfaces would work to control the bacterial growth. Since the intermetallic compounds are generally very stable, the infinitesimal quantity of copper ion on materials surfaces seems to work as antibacterial substance. In this point, three principles for antibacterial composites are satisfied by the intermetallic compounds. The schematic principle is shown in Figure 10.

5.1.2 *Tin-Silver Alloy Film* [52,53]

In the same way, silver intermetallic compounds could be produced. For example, the authors prepared the stacked single layers as follows. The tin under layer and silver top layer were prepared by high-frequency magnetron sputtering to produce stacked single layers of tin and silver. Then, the specimens were heated at various temperatures and for various times. Table 3 shows some of the experimental results. The evaluation of their antibacterial effect was carried out also by film covering method and the bacteria used for the evaluation were *E. coli*. The produced phase after heat treatments was Ag_3Sn. As the table shows clearly, the colony number decreased down to zero with the existence of the intermetallic compound. As in the case of

Table 3 The results of tin-silver surface coating (Sn-Ag alloy)

Heat treatment temperature	Treatment time	Intermetallic compounds?	Ag thickness (μm)	Colony number after heat treatment
200	60	Yes	0.01	0
200	60	Yes	0.1	0
200	60	Yes	1	0
200	120	Yes	0.01	0
200	120	Yes	0.1	0
200	120	Yes	1	0
300	120	Yes	0.01	0
300	120	Yes	0.1	0
300	120	Yes	1	0

copper, the stable intermetallic compound could control the bacterial growth due to their infinitesimal dissolution. However, the existence of intermetallic phase is very important for antibacterial effect. The formation depends on the time and temperature for the heat treatment. The authors proposed the following experimental equation to get the intermetallic compounds of silver by HSSL process:

$$Y = 64\exp\left(\frac{-X}{46}\right) + 1681\left(\frac{-X}{1.4}\right) \tag{1}$$

where Y is the heat treatment temperature and X is the heat treatment time kept at a certain temperature.

5.2 Theory of Segregation Prediction and Its Application [54–59]

The authors proposed other possibility by HSSL process. We have not carried out the experiments concretely. Instead, we predicted a certain type of antibacterial composite films, using simulation software. It is called Surf-Seg, and anyone can use the simulation software on the Internet. The software on a web is based on a theory of segregation prediction by one of the authors. Yoshitake established the theory to predict if the substrate atom might move through the surface film to reach the top of the surface or not, when metallic atom B would exist on the other metallic substrate A. The theory made it possible to evaluate the energy of atoms, when the excited atom by heating would adsorb onto grain boundaries of surface materials with a probability. If the energy would be small (stable), then diffusion process would continue, while it might stop for higher energy conditions. Concretely speaking, the calculation and evaluation require the following equations:

$$\begin{aligned}\Delta H_{ad}(\mathrm{A\ on\ B}) = & -F \times \gamma_B \times S_A \\ & + (1-F) \times \gamma_A \times S_A + F \times \Delta H_{sol}(\mathrm{A\ in\ B}) - \Delta H_{vap}(\mathrm{A})\end{aligned} \tag{2}$$

where $\Delta H_{vap}(\mathrm{A})$ is the sublimation enthalpy of metal A, $\Delta H_{sol}(\mathrm{A\ in\ B})$ is the heat of dissolution when A dissolves into B, γ_A and γ_B are the surface free energy of metals A and B at absolute zero point, S_A is the average surface area per mole of metal A, and F is the ratio of surface area for metal A setting in touch with metal B. (When the cover ration of the film is low, the ratio is usually from 0.3 to 0.4. In this case, 0.4 was used.)

ΔH_{sol}(A in B) of Equation (2) is calculated by Equation (3):

$$\Delta H_{\mathrm{sol}}(\mathrm{A\,in\,B}) = \left(\frac{2V(\mathrm{A})^{2/3}}{n(\mathrm{A})^{-1/3} + n(\mathrm{B})^{-1/3}}\right) \times N_0 \times P\left\{-\exp\left(\Delta\phi^2\right) + \frac{Q}{P(\Delta n^{1/3})^2} - \frac{R}{P}\right\} \quad (3)$$

where V(A) is the molar volume of metal A, n(A) and n(B) are the electron densities of metals A and B at the boundary of Wigner-Seitz cell, and $\Delta\phi$ is the difference between the work functions of metals A and B.

When $\Delta H_{\mathrm{ad}}(\mathrm{A\,on\,B})$ would be compared with $\Delta H_{\mathrm{ad}}(\mathrm{A\,on\,A})$ from the viewpoint of energy value, one can estimate the possibility of segregation.

Using the theory, we predicted that the following case might be realized. Figure 11 shows the schematic illustration for the new proposal. The upper layer should be the base metal, such as nickel and zinc chromium. Antibacterial metals such as silver and copper are set as under layer. When the heat would be applied to the system, we could predict if the under layer nickel could diffuse into the upper layer or not. If the antibacterial atom could reach the top of the surface by diffusion, the surface would have antibacterial effect. When the top layer would be nickel and antibacterial element would be silver, $\Delta H_{\mathrm{ad}}(\mathrm{Ag\,on\,Ni})$ was calculated as 288 kJ/mol and $\Delta H_{\mathrm{ad}}(\mathrm{Ag\,on\,Ag})$ was done as 222 kJ. In this case, we could predict that the silver atoms would appear gradually, while the surface would be heated. This case could be applied to antibacterial composite coating. If some devices to heat the surface area would be set to the system, the infinitesimal amount of antibacterial silver could continue to appear on the top of the surface gradually and consecutively. Then, the three principles would be satisfied.

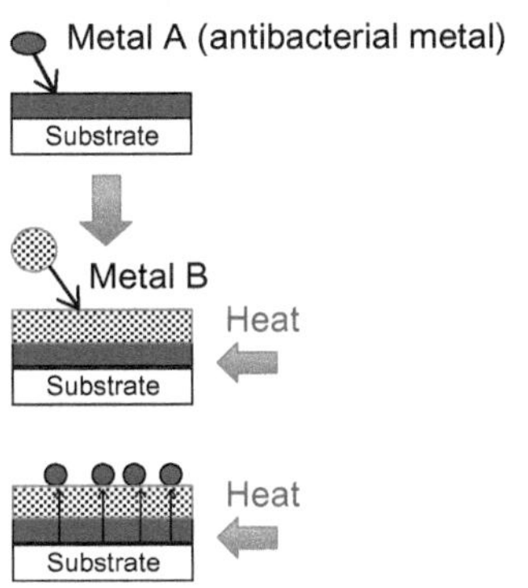

Figure 11 New proposal by segregation prediction theory.

5.3 Dispersion Film of Inorganic Metals [60]

The authors have devised the new composite coating composed of silane base coating and the dispersed nanopowders of antibacterial metal elements (Ag, Cu, and Ti) or those organic substances. The concept for the composite coating was shown in Figure 12 [13].

Three kinds of raw materials for the coating were used. (A) Alkoxysilane oligomer including methyl and phenyl (Permeate, D&D Cooperation, Japan), (B) N-2-(aminoethyl)-3-aminopropyltrimethoxysilane (KBM-603, Shin-Etsu Chemical Co., Ltd.), (C) tetra-*n*-butoxytitanium (B-1, Nippon Soda Co., Ltd.), tin acetylacetonato (Sigma-Aldrich Co.), or nickel (II) acetylacetonato dehydrate (Wako Pure Chemical Industries, Ltd.). The following process was an example, when float glasses were used as substrate. When the other substrates would be chosen, some pretreatment processes could be slightly changed to other appropriate ones. Float glasses were cut down to the tiny sheets whose sizes were 10×5 cm. And they were cleaned in an ultrasonic bath filled with clean water containing various surfactants to remove tiny glass residues and oil components. Then, the specimen surfaces were cleaned by distilled water to remove tiny glass residues, oil, etc. Then, the surfactant was removed by pure water completely. After cleaning, glass specimens were immersed into a solution composed of

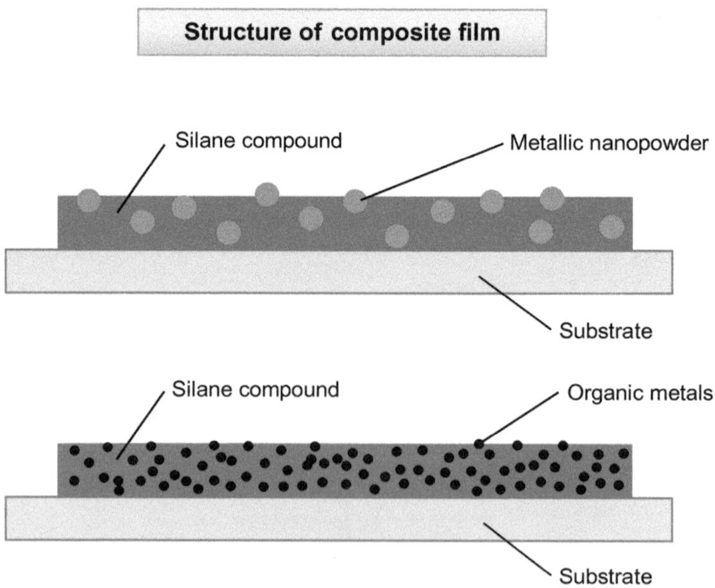

Figure 12 The schematic illustration of composite film's structure.

hydrogen peroxide solution and ammonia water (equivalent volume ratio) until ammonia water was vaporized and the pH could reach neutral values. During this process, the temperature was kept constant at 110 °C. This process made the glass specimens surfaces hydrophilic.

The raw materials, (A), (B), and (C), were mixed for an hour by a disperser (Paint Shaker, Toyo Seiki Seisaku-sho, Ltd.). And the solution as product was used as coating material. The material was coated on those hydrophilic glasses by an air spray. The thickness of coating was 10 μm. All of coated glasses were cured for a week.

The structure of the silane-based antibacterial coating is shown in Figure 13 [13]. At this point, the nanopowder dispersion coating can't achieve the initial goal except for copper nanopowder, since their coatings have high roughness. The rough surface generally attracts lots of bacteria, and the negative effect seemed to be dominant over all of other factors. However, inorganic chemical dispersion coatings worked well to control or inhibit bacterial growth. Since the coating has high transparency, we could evaluate the bacterial growth and the following biofilm formation as the decrease of transparency for the specimens quantitatively, using UV-VIS. Specimens were immersed into a newly develop laboratory biofilm reactor (LBR) for 14 days, and the decrease of transparency for each specimen was measured. According to the experimental results, organic titanium, organic silver, and organic copper specimens showed only 2-4%

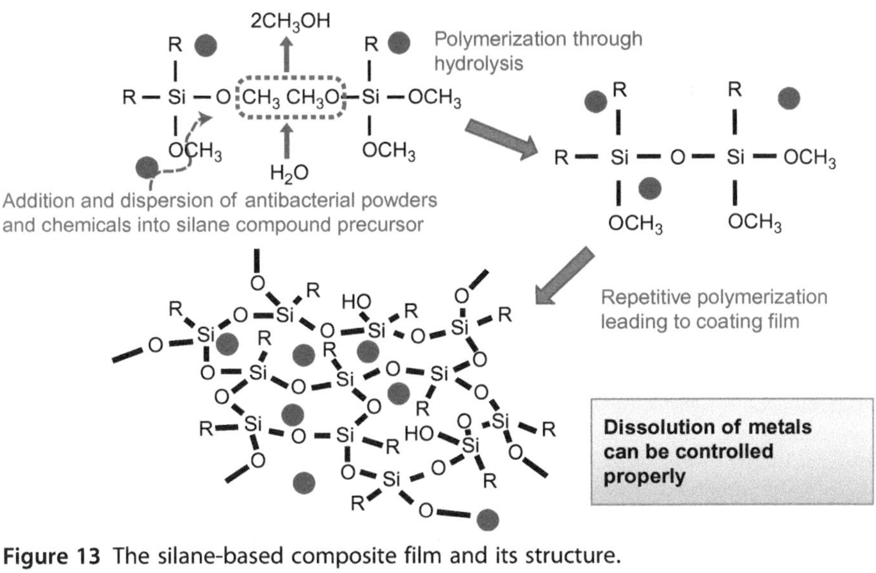

Figure 13 The silane-based composite film and its structure.

transparency drops after 14 days of immersion. The LBR is a kind of acceleration test, and we confirmed in advance that 10 days of immersion test usually corresponds to 10 years of natural exposure in a Japanese area, where the annual amount of rain fall would be 1316 mm and the average temperature 14.9 °C. As for the new composite coating, the authors investigated further from various viewpoints.

6. FUTURE SCOPE

In this chapter, the composite coatings utilizing metallic components were introduced. We human beings have already known the antibacterial effects of silver, copper, and some other metals. However, the appropriate dissolution would be required, when we come to think about the mechanism and practical applications. From the viewpoint, composite coatings would be desirable. The investigation in this area is now going to be accelerated. Photocatalytic oxides and organics are rivals for the antibacterial substances. However, the former needs daylight or ultraviolet light specially. And the latter is often environmentally harmful. From both points of view, the metallic antibacterial substance should be investigated much more. If this chapter would be an opportunity for the acceleration of the study, we would be very happy.

REFERENCES

[1] J.W. Alexander, History of the medical use of silver, Surg. Infect. 10 (2009) 289–292.
[2] M. Spear, Silver: an age-old treatment modality in modern times, Plast. Surg. Nurs. 30 (2010) 90–93.
[3] B.S. Atiyeh, M. Costagliola, S.N. Hayek, S.A. Dibo, Effect of silver on burn would infection and healing: review of the literature, Burns 33 (2007) 139–148.
[4] S. Silver, L.T. Phung, G. Silver, Silver as biocides in burn and would dressings and bacterial resistance to silver compounds, J. Ind. Microbiol. Biotechnol. 33 (2006) 627–634.
[5] H. Ikigai, H. Kanematsu, Mechanism for antibacterial effects by metallic elements and their inhibition of bacterial infections, J. Jpn. Soc. Biomater. 29 (4) (2011) 232–239.
[6] H. Kohrai, How-nual Visual Guide Book to Understand Kohkin and Sakkin Easily, Shuwa System Co, Tokyo, Japan, 2012, 206 pp.
[7] H. Kanematsu, Antibacterial Materials for Safety, Security and Reliability, in: H. Kanematsu (Ed.), Yoneda Shuppan, Chiba, 2010, p. 154.
[8] D.L. Nelson, M.M. Cox, Lehninger Principles of Biochemistry, fifth ed., W.H. Freeman, New York, USA, 2008, p. 92.
[9] H. Ikigai, T. Ono, M. Iwata, T. Nakae, T. Shimamura, El Tor hemolysin of Vibrio cholera OI forms channels in planar lipid bilayer membranes, FEMS Microbiol. Lett. 150 (1997) 249–254.

[10] H. Kanematsu, H. Ikigai, M. Yoshitake, Antibacterial eco-plating based on HACCP and biofilm, Bull. Iron Steel Inst. Jpn. 13 (1) (2008) 27–34.
[11] H. Kanematsu, H. Ikigai, M. Yoshitake, Evaluation of various metallic coatings on steel to mitigate biofilm formation, Int. J. Mol. Sci. 10 (2) (2009) 559–571.
[12] H. Kanematsu, H. Ikegai, M. Yoshitake, Patents for antibacterial metallic coating and its future trend in Japan, Res. Inventy: Int. J. Eng. Sci. 3 (6) (2013) 48, 53.
[13] K. Sano, H. Kanematsu, N. Hirai, A. Ogawa, T. Kogo, T. Tanaka, Silane coating for antifouling and the evaluation method, in: The 168 Annual Meeting of ISIJ (Iron and Steel Institute of Japan), Nagoya University, 2014.
[14] H.M. Lappin-Scott, J.W. Costerton (Eds.), Microbial Biofilms, Plant and Microbial Biotechnology Research Series, Cambridge University Press, Cambridge, 1995.
[15] W.G. Characklis, Fouling biofilm development: a process analysis, Biotechnol. Bioeng. 23 (1981) 1923–1960.
[16] J.W. Costerton, The Biofilm Primer, Springer, New York, 2007, 199 pp.
[17] H.-C. Flemming, P. Sriyutha Murthy, R. Venkatesan, K.E. Cooksey, Marine and Industrial Biofouling, Springer, Berlin/Heidelberg, Germany, 2009, 333 pp.
[18] Z. Lewandowski, H. Beyenal, Fundamentals of Biofilm Research, second ed., CRC Press, Boca Raton, London, New York, 2014.
[19] M.L.W. Knetsh, L.H. Koole, New strategies in the development of antimicrobial coatings: the example of increasing usage of silver and silver nanoparticles, Polymers 3 (2011) 340–366.
[20] H. Kanematsu, H. Ikegai, M. Yoshitake, Patents for antibacterial metallic coating and its future trend in Japan, Res. Inventy: Int. J. Eng. Sci. 3 (6) (2013) 49, 54.
[21] S. Koura, K. Sakoda, Silver-zeolite coating solution and coating method, JPO, Japan, JP10-265959A, 1998.
[22] A. Miyasaka et al., Metallic material excellent in antibacterial property, JPO, Japan, JP09-195061A, 1997.
[23] A. Otani et al., Antimicrobial silicate having film-forming property, JPO, Japan, JP02-019308A, 1990.
[24] K. Hirose et al., Antifungal and deodorant agent, JPO, Japan, JP03-007201A, 1989.
[25] M. Toda et al., Antibacterial and antifungal ceramics and their production, JPO, Japan, JP06-065012A, 1994.
[26] H. Mase et al., Antibacterial film, JPO, Japan, JP2010-247450A, 2009.
[27] A. Miyauchi et al., Antibacterial metal sputtering mask, JPO, Japan, JP06-065012A, 1994.
[28] C. Dowsett, The use of silver-based dressing in wound care, Nurs. Stand. 19 (2004) 56–60.
[29] M.N. Strom-Versloot, C.G. Vos, D.T. Ubbink, H. Vermeulen, Topical silver for preventing wound infection, Cochrane Database Syst. Rev. 3 (2010) CD006478.
[30] M.J. Carter, K. Tingley-Kelley, R.A. Warriner III, Silver treatments and silver-impregnated dressings for the healing of leg wounds and ulcers: a systematic review and meta-analysis, J. Am. Acad. Dermatol. 63 (2010) 668–679.
[31] S. Lo, C. Chang, W. Hu, M. Hayter, Y. Change, The effectiveness of silver-releasing dressing in the management of non-healing wounds: a meta-analysis, J. Clin. Nurs. 18 (2009) 716–728.
[32] K.Y. Woo, Review: silver-releasing dressings improve healing and quality of life in patients with non-healing chronic wounds, Evid. Based Nurs. 12 (2009) 117. Polymers 3 (2011) 363.
[33] G. Gravante, R. Caruso, R. Sorge, F. Nicoli, P. Gentile, V. Cervelli, Nanocrystalline silver, Reconstr. Surg. Burns 63 (2009) 201–205.
[34] J. Fichtner, E. Güresir, V. Seifert, A. Raabe, Efficacy of silver-bearing external ventricular drainage catheters: a retrospective analysis, J. Neurosurg. 112 (2010) 840.

[35] P. Lackner, R. Beer, G. Broessner, R. Helbok, K. Galiano, C. Pleifer, B. Pfausler, C. Brenneis, C. Huck, K. Engelhardt, A.A. Obwegeser, E. Schmutzhard, Efficacy of silver-nanoparticles-impregnated external ventricular drain catheters in patients with acute occlusive hydrocephalus, Neurocrit. Care. 8 (2008) 360–365.
[36] D. Roe, B. Karandikar, N. Bonn-Savage, B. Gibbins, J.B. Roullet, Antimicrobial surface functionalization of plastic catheters by silver nanoparticles, J. Antimicrob. Chemother. 61 (2008) 869–876.
[37] S.H. Hsu, H.J. Tseng, Y.C. Lin, The biocompatibility and antibacterial properties of waterborne polyurethane-silver nanocomposites, Biomaterials 31 (2010) 6796–6808.
[38] S. Voccia, M. Ignatova, R. Jérôme, C. Jérôme, Design of antibacterial surfaces by a combination of electrochemistry and controlled radical polymerization, Langmuir 22 (20) (2006) 8607–8613.
[39] B.K. Sunkara, R.D. Misra, Enhanced antibactericidal function of W4+-doped titania-coated nickel ferrite composite nanoparticles: a biomaterial system, Acta Biomater. 4 (2) (2008) 273–283.
[40] S.J. Ahn, S.J. Lee, J.K. Kook, B.S. Lim, Experimental antimicrobial orthodontic adhesives using nanofillers and silver nanoparticles, Dent. Mater. 25 (2) (2009) 206–213.
[41] B.S. Necula, L.E. Fratila-Apachitei, S.A. Zaat, I. Apachitei, J. Duszczyk, *In vitro* antibacterial activity of porous TiO_2-Ag composite layers against methicillin-resistant *Staphylococcus aureus*, Acta Biomater. 5 (9) (2009) 3573–3580.
[42] L.N. Shirokova, V.A. Aleksandrova, E.M. Egorova, G.A. Vikhoreva, Macromolecular systems and bactericidal films based on chitin derivatives and silver nanoparticles, Prikl. Biokhim. Mikrobiol. 45 (4) (2009) 422–426.
[43] R. Dastjerdi, M. Montazer, S. Shahsavan, A novel technique for producing durable multifunctional textiles using nanocomposite coating, Colloids Surf. B: Biointerfaces 81 (1) (2010) 32–41.
[44] J. Qu, X. Lu, D. Li, Y. Ding, Y. Leng, J. Weng, S. Qu, B. Feng, F. Watari, Silver/hydroxyapatite composite coatings on porous titanium surfaces by sol-gel method, J. Biomed. Mater. Res. B Appl. Biomater. 97 (1) (2011) 40–48.
[45] B. Yu, K.M. Leung, Q. Guo, W.M. Lau, J. Yang, Synthesis of Ag-TiO_2 composite nano thin film for antimicrobial application, Nanotechnology 22 (11) (2011) 115603.
[46] R. Basheer, G. Ganga, R.K. Chandran, G.M. Nair, M.B. Nair, S.M. Shibli, Effect of W-TiO_2 composite to control microbiologically influenced corrosion on galvanized steel, Appl. Microbiol. Biotechnol. 97 (12) (2013) 5615–5625.
[47] H. Kanematsu, T. Kobayashi, N. Wada, T. Oki, Environmental friendly alloy film formation through heating of stacked single layers, Materia 41 (10) (2002) 713–719.
[48] H. Kanematsu, T. Oki, Heat treatment-alloying, plating process, J. High Temp. Soc. Jpn. 30 (4) (2004) 191–196.
[49] H. Kanematsu, Alloy plating by heating stacked single layers and the possibility of its application in the future, Prod. Finish. 78 (3) (2013) 1–14.
[50] H. Kanematsu, H. Ikigai, M. Yoshitake, Various properties of tin-copper alloy film produced by heat treatment, in: SFIC Sur/Fin 2006, SFIC, Milwaukee, WI, 2006, pp. 864–870.
[51] H. Kanematsu, H. Ezaki, H. Ikigai, T. Oki, Sn-Cu alloy films for antibacterial activities, Sn-Cu alloy film products for antibacterial activities & a production method for antibacterial Sn-Cu alloy film products, in: J.P. Office (Ed.), Suzuka National College of Technology, Japan, 2005.
[52] H. Kanematsu, H. Ikigai, M. Yoshitake, Antibacterial tin-silver plating by the combination of multistage plating and heat treatment, J. Appl. Surf. Finish. 2008 (3) (2008) 114–118.

[53] M. Yoshitake, H. Kanematsu, H. Ikigai, Anti-bacterial products and the production process of anti-bacterial films, in: J.P. Office (Ed.), National Institute of Materials Science, Japan, 2007.
[54] M. Yoshitake, Y.-R. Aparna, K. Yoshihara, General rule for predicting surface segregation of substrate metal on film surface, J. Vac. Sci. Technol. A 19 (4) (2001) 1432–1437.
[55] M. Yoshitake, Y.-R. Aparna, K. Yoshihara, Novel interface modification technique applied from the top of a coated layer, J. Vac. Sci. Technol. A 19 (5) (2001) 2612–2616.
[56] M. Yoshitake, Y.-R. Aparna, K. Yoshihara, Modification of coating layer/base material interface, J. Vac. Soc. Jpn. 44 (3) (2001) 151–154.
[57] M. Yoshitake, Metal segregation prediction system, SurfSeg, Metal (Kinzoku) 81 (12) (2011) 73.
[58] NIMS web page, SurfSeg. http://surfseg.nims.go.jp/SurfSeg/menyu.html.
[59] H. Kanematsu, M. Yoshitake, D. Kuroda, Anti-bacterial alloy film formation by heat treatment and the design possibility based on segregation prediction system, Netsushori (J. Jpn. Soc. Heat Treat.) 52 (3) (2012) 143–149.
[60] H. Kanematsu, S. Sasaki, Y. Miura, T. Kogo, K. Sano, N. Wada, M. Yoshitake, T. Tanaka, Composite coating to control biofilm formation and the effect of alternate electromagnetic field, Mater. Technol. 2 (1) (2015) B21–B26.

CHAPTER 24

Recent Advances in Nanocomposite Coatings for Corrosion Protection Applications

Nedal Y. Abu-Thabit[1], Abdel Salam Hamdy Makhlouf[2]
[1]Department of Chemical and Process Engineering Technology, Jubail Industrial College, Jubail Industrial City, Saudi Arabia
[2]Department of Manufacturing Engineering, College of Engineering and Computer Science, University of Texas Pan-American, Edinburg, Texas, USA

Chapter Contents

Handbook of Nanoceramic and Nanocomposite Coatings and Materials
http://dx.doi.org/10.1016/B978-0-12-799947-0.00024-9

1. INTRODUCTION

The word "nano" comes from the Greek language meaning "dwarf." It is related to a dimension scale within 10^{-9} m. Nanotechnology and nanoscience is an interdisciplinary research area where the concepts of physics, chemistry, biology, and engineering are utilized cooperatively to study and manipulate matter at the atomic and molecular scale.

Compared to bulk materials, nanomaterials and nanoparticles typically exhibit high surface area to volume ratio, which has many interesting effects on the subsequent behavior of these materials. Recently, nanocomposite-based materials have been extensively investigated as novel and alternative materials for various applications. Nanocomposites offer an exciting and practical approach for designing and fabrication of new technological products and materials with superior mechanical, electrical, optical, antimicrobial, catalytic, and reactive properties.

Nanocomposite coatings are one of the active and rapidly developing areas that utilize various approaches and aspects of nanotechnology and nanoscience for designing and fabrication of different types of surface coatings with superior performance and enhanced durability at lower cost. Moreover, nanocomposite coatings provide room for implementing new and diverse functionalities that enable the possibility for achieving innovative characteristics through engineering smart multifunctional coatings [1–9].

Anticorrosion coatings provide barrier functionality for the metal surface against the corrosive species present in the environment. External factors such as temperature, humidity, and the presence of aggressive chemicals can accelerate the formation of defects and micropores that deteriorate the barrier effect of the organic coating and initiate corrosion. This deterioration effect is more pronounced with traditional organic coatings that have macrosized particles that lead to large phase separation between the solid inorganic macroparticles and the organic/polymeric liquid phase in the formulated coating dispersions. On the other hand, nanocomposite protective coatings that employ nanosized inorganic fillers/particles offer superior, robust, and durable corrosion protection performance due to the strong interactions between the inorganic nanoparticles and the host organic matrix, resulting in lower porosity and lower cracking potential of the formulated nanocomposite coatings.

Several approaches have been reported for developing nanocomposite anticorrosive coatings with different complexity ranging from the simple barrier protection functionality to smart autoresponsive and self-healing functionalities [10–18]. In this chapter, most recent findings in this area will be highlighted and discussed.

2. HYBRID NANOCOMPOSITE COATINGS

2.1 Nanocomposite Coatings Containing Nanoparticles and Nanofillers

Organic polymeric coatings usually provide corrosion protection functionality by forming a barrier to isolate the metal from the surrounding environment. However, all polymeric coatings are permeable to corrosive species such as oxygen, water, and chloride ions. This permeability facilitates corrosion initiation process and decreases the service time for the polymeric coating. Incorporation of inorganic fillers into the polymeric coating formulation can reduce the permeability and prolong the lifetime of the composite coating. In particular, inorganic nanofillers, with very fine gain size and high boundary volume, offer enhanced barrier properties compared with their conventional filler counterpart [19–23].

Behzadnasab et al. [24] prepared epoxy-based nanocomposite coatings containing various amounts of nanoclay and 3-aminopropyltrimethoxysilane (3-APTMOS)-treated zirconia nanoparticles by slurry method. The incorporation of nanoclay and zirconia nanoparticles (1 wt.% each) provided the best film-forming conditions and improved the corrosion protection performance. The enhanced corrosion protection performance in the presence of spherical zirconia nanoparticles was attributed to a better exfoliation of clay silicate layers in the formulated nanocomposite coating.

Zhou et al. [25] incorporated silica nanoparticles into functional polybenzoxazine-trimethoxysilane (PBS-TMOS) coating for corrosion protection of mild steel. Silica nanoparticles were employed as nanofillers and provided cross-linking sites for the PBS-TMOS coating. The formed cross-linked covalent bonds, between SiO_2 nanoparticles and PBS-TMOS coating, greatly improved the interfacial interactions at the polymer/filler interfaces and led to a better corrosion resistance. The enhanced corrosion protection performance in the presence of (1-5 wt.%) silica nanoparticles was attributed to the synergetic effect of improved barrier properties of the hybrid coating and the formation of passive layer of Fe-silicate compounds.

Elhalawany et al. prepared novel water-based coatings containing polyaniline (PANI) conductive polymer nanoparticles (CPNs) as corrosion inhibitor [26]. The hybrid coating was formulated by using poly(vinyl acetate) PVAc as a binder and nanohybrid particles (nHPs) of CPNs and nanosilica with 1:1 ratio. The formulated coating was utilized for the preparation of painted films on six different substrates. The chemical, mechanical, and corrosion protection properties of the nanocomposite paint films (PVAc/nHP) were found to be superior to the blank paint film (only PVAc), which was attributed to the improved barrier properties that resulted from the presence of silica nanoparticles. Similar results were obtained for water-based coatings containing CNPs of polytoluidine (PTol) and polyanisidine (PAns) with styrene/butyl acrylate emulsion binder [27].

Hamdy et al. [28–33] designed high-performance anticorrosion nanocomposite and ternary alloy Ni-P coatings via electroless deposition technique on low carbon steel. The optimum conditions under which such coatings can provide the highest anticorrosion characteristics were determined. The coating performance was investigated using electrochemical impedance spectroscopy (EIS) and polarization measurements. Ni-P-W alloy coatings showed the highest hardness and anticorrosion properties compared with Ni-P-Al_2O_3 and Ni-P. Surface examination of Ni-P-W and Ni-P-Al_2O_3 using XRD, SEM-EDS, and macroscopic images revealed grain refining of Ni grains. A protective layer of nickel phosphide and tungsten phosphide was formed over steel substrate due to Ni-P-W, while a protective layer of nickel phosphide enriched with aluminum oxide was formed due to Ni-P-Al_2O_3.

2.2 Nanocomposite Coatings Containing One-Dimensional Nanowires

One-dimensional nanomaterials have attracted considerable interest in both fundamental and applied science [34,35]. For example, amorphous cerium molybdate (Ce/Mo) nanowires were suggested to be utilized as corrosion-inhibiting pigments due to their improved solubility in comparison with the low-solubility conventional crystalline cerium molybdate oxides [36].

Yasakau et al. reported the utilization of amorphous cerium molybdate nanowires (CMNs) into a hybrid sol-gel coating for corrosion protection of AA2024 alloys [37]. The hybrid sol-gel coating was prepared by controlled hydrolysis of metal organic (zirconium(IV) propoxide (TPOZ)) and organosiloxane (3-glycidoxypropyltrimethoxysilane (GPTMS)) precursors in 2-propanol solution with aqueous nitric acid solution [38]. A slurry of CMN material in ethanol containing 14 wt.% of solid content was introduced into hybrid sol-gel solution at the step of mixing both TPOZ and GPTMS-containing solutions. EIS results for etched AA2024 samples coated by sol-gel containing CMN revealed higher mixed oxide layer resistance and polarization resistance demonstrating an improvement of active corrosion protection during immersion in 0.5 M NaCl solution. The inhibition mechanism was attributed to the combined action of cerium and molybdate ions. Release studies showed a higher amount of cerium ions leached out from the sol-gel matrix compared to molybdate anions [39].

2.3 Nanocomposite Coatings Containing Conductive Components

This section highlights the nanocomposite coatings embedded with electrically conductive components including electrically conductive polymers (ECPs) and conductive carbon nanostructures such as conductive carbon nanotubes CNTs and conductive graphene sheets. The mechanisms for corrosion protection using such conductive components will be discussed.

2.3.1 Nanocomposite Coatings Containing ECPs

ECPs belong to those classes of polymers that have π conjugation along the polymer backbone such as PANI, polypyrrole (PPy), and polythiophene (PTh), as shown in Figure 1. ECPs are known for their ability to undergo oxidation-reduction reactions, by gaining or loosing electrons from the surrounding environment. Due to this interesting property, ECPs have been utilized as novel corrosion protective coatings for a variety of metals and alloys [40–42].

ECPs function as corrosion protective coatings by one or more of the following three mechanisms [43]: First, ECPs can provide barrier protection by isolating the metal surface from the surrounding environment. Second, ECPs can provide anodic protection through the formation of passive layer of metal oxide (ennobling mechanism) [44,45]. In this case, ECPs act as oxidizing agent for the leached metal ions, hence maintaining the metal in the

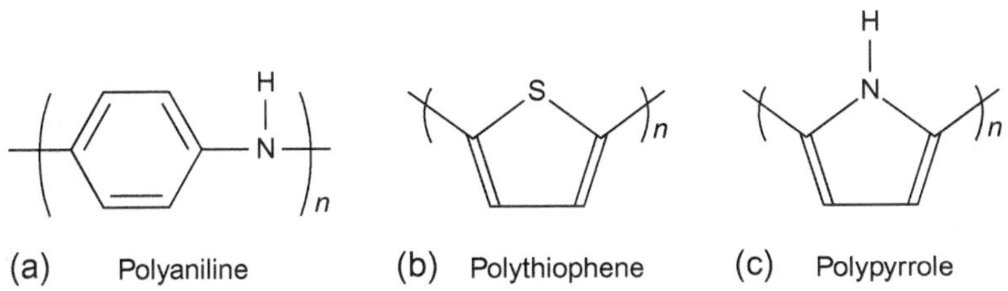

Figure 1 Chemical structures of common ECPs.

passive domain by providing a thin layer of metal oxide. In the third protection mechanism, which was proposed by Kendig et al. [15], ECPs provide self-healing functionality for the coated metal substrates. In this case, ECPs act as reservoir for immobilization of the doped corrosion inhibitor anions [14]. The release of the stored inhibitor anions to the corroding defect is triggered by the electrochemical reduction of the ECP.

Steel substrates coated with PPy immobilized by inorganic inhibitors such as molybdate [46] and phosphomolybdate $[PMo_{12}O_{40}]^{3-}$ [47,48] anions showed significant self-healing corrosion protection performance. This was attributed to the migration of molybdate anions through the PPy film in a short time and subsequent formation of passive layer at the defected sites.

Kowalski et al. [49] investigated bilayered PPy coating composed of molybdate-doped internal layer and dodecyl sulfate-doped external layer. The bilayered coating provided corrosion protection *via* two parallel mechanisms. The penetration of small aggressive anions was prevented by the presence of immobile and bulky counteranion (dodecyl sulfate) through anion exchange reaction, whereas the internal layer provided the self-healing functionality through the controlled release of the molybdate anions. When a dodecyl sulfate-doped PPy monolayer coating was applied to steel substrate, the coating was able to provide corrosion protection for only 10 h. Steel substrate coated with similar thickness of the duplex coating system provided corrosion protection for 190 h in chloride-containing solution. Hence, it was concluded that the molybdate-doped internal layer was able to maintain the metal substrate in the passive domain for longer period, which is attributed to the formation of salt layer of ferric molybdate complex at the defect site [50].

Liu [51] prepared PANI-containing waterborne coating for corrosion protection applications by using oxidative dispersion polymerization of aniline monomer in the presence of partially phosphorylated polyvinyl alcohol (P-PVA). P-PVA was able to stabilize the PANI dispersion and provided PANI nanoparticles with uniform diameters in the range of 60-100 nm. Waterborne epoxy resin was used as a matrix for the preparation of nanocomposite coating, with a dry film thickness of $\approx$90 μm. Mild steel samples were dip-coated with the waterborne PANI/P-PVA-containing coating and exposed to 3.0 wt.% NaCl solution. The corrosion protection performance was evaluated by open circuit potential versus time of exposure and EIS. After 30 days of salt spray tests, samples coated with 2.5 wt.% PANI/P-PVA content showed impedance values greater than $1 \times 10^7\ \Omega\ cm^2$, which proved that the PANI/P-PVA coating was able to provide superior corrosion protection for long period.

Hosseini et al. [52] investigated the hybrid epoxy/polypyrrole-montmorillonite (E/Ppy-MMT) nanocomposite coating for corrosion protection of aluminum substrates. The corrosion inhibition performance of epoxy/MMT (EM), epoxy/Ppy (EP), and (E/Ppy-MMT) was evaluated in 3.5 wt.% NaCl solutions. EIS revealed that the hybrid E/Ppy-MMT coating provided the best corrosion performance over long time. This performance was attributed to the morphology of the nanocomposite coatings that improves the barrier properties of the formulated coating.

Peng et al. reported a novel nanocasting methodology for the preparation of biomimetic superhydrophobic polyaniline (SH-PANI) surface for corrosion protection applications [53]. The first step utilized the superhydrophobic natural leaves (*Xanthosoma sagittifolium)* for the preparation of polydimethylsiloxane (PDMS) imprinting substrate. This step was followed by casting of the soluble PANI emeraldine base (EB) in the PDMS negative substrate.

The coating was prepared as follows: The EB form of PANI powder was dissolved in NMP ($\approx$2 wt.%). Drops of the PANI solution were spread on the cold rolled steel (CRS) surface, and then the PDMS template was pressed against the precursor solution surfaces. The solvent was evaporated at 90-100 °C under the hood for one day. Finally, the PDMS template was peeled off to obtain coating with superhydrophobic imprinted substrates.

The contact angle of water droplet on sample surface was increased from $\sim$90° of flat PANI surface (F-PANI) to $\sim$156° of SH-PANI. SH-PANI coated on CRS rods provided superior corrosion protection over the F-PANI coating. The improved corrosion protection was attributed to the outstanding water-repellant property of the biomimetic SH-PANI film and the consequent improvement in the barrier properties of the superhydrophobic coating.

2.3.2 Nanocomposite Coatings Containing Electrically Conductive Carbon Nanostructures

Yu et al. [54] prepared high-performance polystyrene (PS)-/graphene-based nanocomposite coating for corrosion protection application. Well-dispersed PS/modified GO nanocomposites were successfully prepared using *in situ* miniemulsion polymerization. The preparation steps are illustrated in Figure 2. The prepared nanocomposite material was uniformly dispersed in a solution of NMP and THF having equal mass ratio. The prepared dispersion was used to coat a steel electrode with a final coating thickness of $\approx$16 μm. According to their experimental results, the corrosion potential and polarization resistance

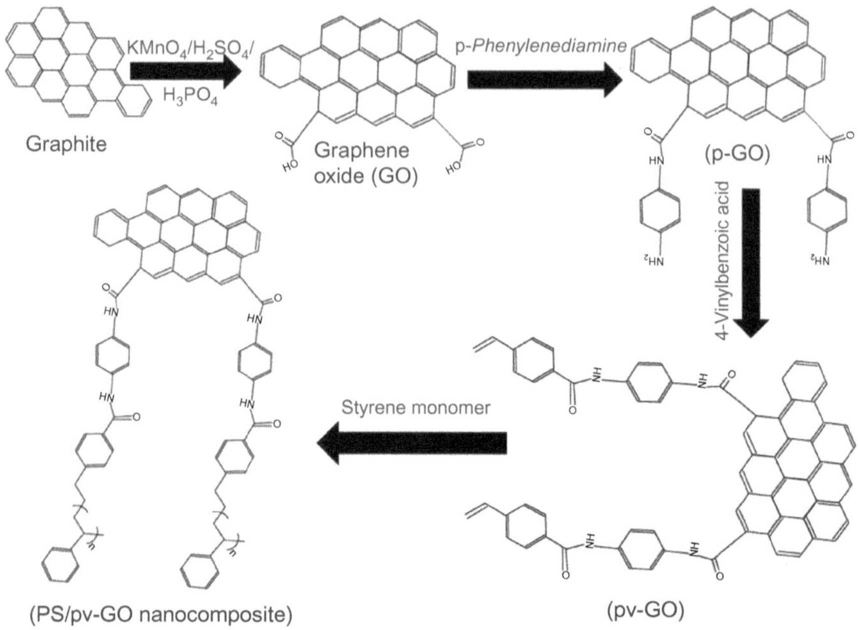

Figure 2 Schematic illustration for the preparation steps of well-dispersed PS/modified GO nanocomposites through *in situ* miniemulsion polymerization.

increased, whereas the corrosion current and the corrosion rate decreased with pv-GO loading. PS/pv-GO nanocomposites provide substantial improvements in corrosion protection efficiency over that provided by pure PS: the PS and PS/pv-GO2 (i.e., 2% nanocomposite content). The protection efficiency percentages were found to be 37.90% and 99.53%, respectively. The superior corrosion protection of the nanocomposite coating was attributed to the improvement of the gas barrier properties of the coated nanocomposite film, which delay the initiation of the corrosion process and consequently prolong the lifetime of the coated metal surface. In this case, the improved barrier properties of the nanocomposite coating were attributed to (1) the presence of graphene nanofiller with high aspect ratio and (2) the effective dispersion of graphene into the PS matrix through grafting methodology that probably allows the exfoliation of grafted graphene nanosheets (GNSs). Thus, the nanocomposite coating was able to enhance the diffusion pathways of O_2 and H_2O, preventing the occurrence of corrosion at the interface between the polymer and the metal electrodes (Figure 3).

Wei et al. [55] evaluated the conductive polyurethane (PU) nanocomposite coatings filled with multiwalled carbon nanotubes (MWNTs) for

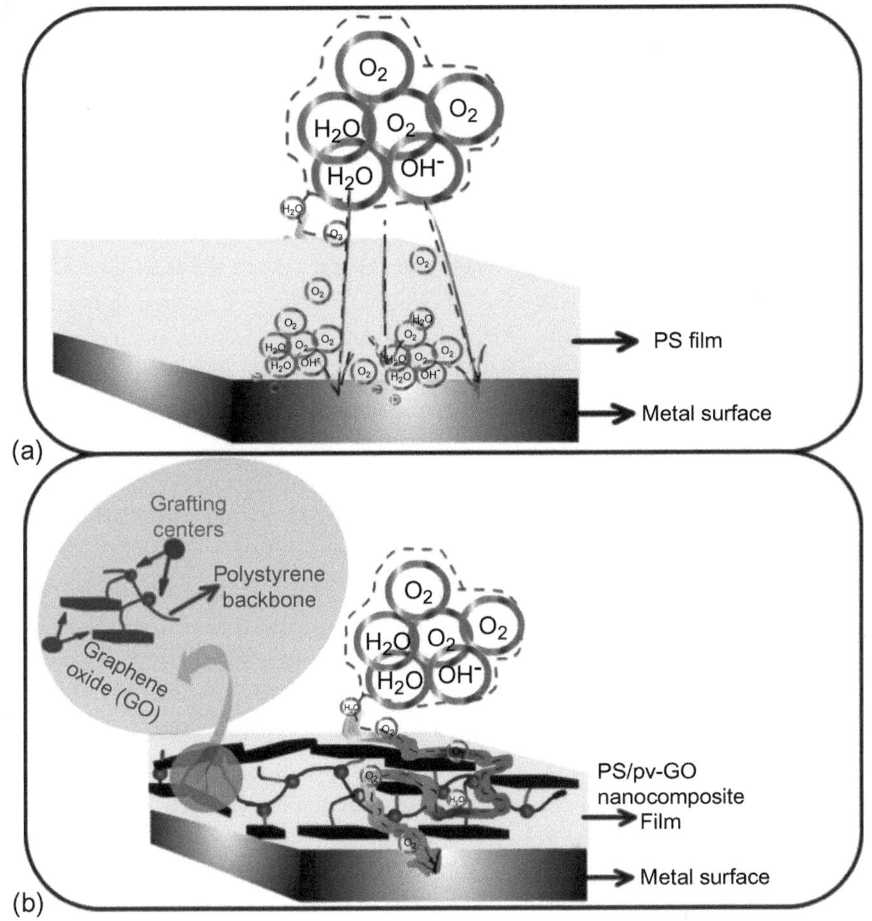

Figure 3 Schematic illustrations for the diffusion pathways of corrosive species through (a) pure PS organic film coating and (b) hybrid PS/pv-GO nanocomposite coating.

corrosion prevention of stainless steel (SS) substrates in 3.0 wt.% NaCl aqueous solution. The conductive PU nanocomposites filled with 5.0 wt.% MWNTs were fabricated using an *in situ* surface-initiated polymerization (SIP) methodology. A significant positive shift of nearly 1.0 V in the open circuit potential (E_{ocp}) was observed. Extrapolation of Tafel plots provided a much more positive corrosion potential (E_{corr}) and a lower corrosion current (I_{corr}). The nanocomposite coating provided protection efficiency as high as 97.70%. The substantial protection was attributed to the improved barrier properties of the PU polymer matrix containing the well-dispersed

MWNTs, which increased the diffusion path of the corrosive ions and rendered their penetration.

Chang et al. [56] prepared PMMA/graphene nanocomposite (PGN) with biomimetic hydrophobic structures by the nanocasting technique utilizing negative patterns of a *Xanthosoma sagittifolium* leaf as transparent soft template for the preparation of PDMS imprinting substrate. In the imprinting step, drops of the precursor solution were spread on the CRS surface and then the PDMS template was pressed against the precursor solution surface. It was then exposed under ultraviolet (UV) light (Fusion UV curing, light intensity $=2000\ mJ/cm^2$) for 300 s. After the PDMS template was peeled off, the nanocomposite coating with hydrophobic imprinted substrates was obtained.

SEM revealed that the surface morphology of as-synthesized hydrophobic PMMA (HP) and PGN coatings exhibited lots of microscaled mastoids, each decorated with many nanoscaled wrinkles. The contact angle of a water droplet on the sample surface was increased from $\approx 80°$ for the PMMA surface to $\approx 150°$ for HP and HPGN and the sliding angle decreased from $\approx 60°$ to 5°. Electrochemical corrosion measurements showed that the hydrophobic PMMA (HP) coatings provided superior corrosion resistance to the CRS electrode compared to that of neat PMMA coatings. The enhanced corrosion protection performance of the HPGN coatings was attributed to (1) the coating hydrophobicity that repelled the moisture/corrosive media, preventing the underlying metals from corrosion attack, (2) the enhanced oxygen barrier properties due to the presence of well-dispersed GNSs in the HPGN matrix, and (3) the relatively higher aspect ratio of GNS compared to clay nanoplatelets.

3. NANOCOMPOSITE COATINGS OF SMART FUNCTIONALITY

Self-healing nanocomposites are a novel class of smart materials that are able to repair the damage caused by mechanical usage over time. The inspiration comes from biological systems, which have the ability to heal after being wounded. Self-healing nanocomposites have been utilized for the fabrication of "smart" functional coatings and modification of different surfaces for various applications such as antifogging [57,58] and anti-icing coatings [59–61], scratch-resistant coatings [62,63], antistain coatings [64,65], moisture-resistant coatings [66,67], oil-repellent coatings [68,69], self-cleaning coatings [70,71], antimicrobial coatings [72–75], UV protection coatings [76,77], adhesive coatings [78], and anticorrosive coatings [79,80].

Self-healing nanocomposite coatings provide superior performance over their traditional coatings counterparts. Traditional coating systems use direct mixing of the corrosion inhibitor in the formulated coating. The direct mixing can result in the loss of inhibitor efficiency, leaching out of inhibitor, or coating degradation. However, in the case of nanocomposite self-healing coatings, the corrosion inhibitor is stored in a variety of micro- or nanostructured materials such as microcapsules, nanocontainers, and nanoreservoirs. Microencapsulation offers advantages such as (1) lower consumption of corrosion inhibitor, (2) prevention of direct contact with the coating, and (3) controlled release of the stored inhibitor at the time.

3.1 Microencapsulation-Based Smart Nanocomposite Coatings

Microcapsules are one of most recent topics for the development of nanocomposite coatings with self-healing functionalities. Microencapsulation is a technique by which solid, liquid, or gaseous active ingredients are packaged within a second material for the purpose of shielding the active ingredient from the surrounding environment. The encapsulated active ingredient is called the core materials, whereas the surrounding material is called the shell. The principle of microencapsulation dates back to 1950s when gelatin and gum arabic were utilized for fabrication of microencapsulated dyes by coacervation for manufacturing of carbonless copying papers [81]. Microencapsulation technology has been utilized for a number of applications such as drug delivery, cosmetics, catalysis, food applications, and preparation of self-healing nanocomposites [82–88].

There are a variety of methods for microencapsulation. Two main categories can be identified for microencapsulation. In the first category, monomers/prepolymers are used as starting materials. Examples of this category include emulsion polymerization, suspension polymerization, and interfacial polycondensation. In the second category, polymers are utilized as the starting material for the preparation of microcapsules as in the case of suspension cross-linking and solvent extraction/solvent evaporation. Other microencapsulation methods include coacervation/phase separation, spray drying, and polymer precipitation.

3.1.1 Emulsion Polymerization

Microcapsules can be prepared by emulsion polymerization (oil dispersed in continuous water phase) or inverse emulsion polymerization (water disperse in continuous oil phase) [89–91]. In emulsion polymerization, stirring or sonication creates droplets that become the core material of the capsule.

The capsule shell wall is created by the polymerization at the aqueous/organic interface around these droplets. Emulsion polymerization can be employed for both step growth (condensation) polymerization and chain growth (addition) polymerization.

3.1.2 Suspension Polymerization

Suspension polymerization also can be employed for the preparation of microcapsules. In this case, a stabilizing agent, such as polyvinyl alcohol, is needed to prevent agglomeration of the formed microcapsules. In suspension polymerization, the oily liquid monomer is dispersed into an agitated stabilizing medium consisting of water and small amounts of stabilizing agent [92,93]. The added initiator is soluble in the liquid monomer. In the proper conditions, the polymer reacts mainly at the interface of the drop forming a shell around the core since this interface is the only locus of polymerization.

3.1.3 Interfacial Polycondensation

Polymers such as polyamides, PUs, polyureas, and polyesters are prepared by interfacial polycondensation techniques [94–97]. In this case, two monomers with symmetrical bifunctional end groups are dissolved in two different phases and the polymerization occurs at the interface. A small amount of a suitable stabilizer is employed to prevent droplet coalescence or particle coagulation during the polymerization and microcapsule formation. Interfacial polycondensation can be used to prepare micrometer- or nanometer-sized particles.

3.1.4 Suspension Cross-linking

Microcapsule formation by this technique involves dispersion of an aqueous solution of the polymer containing core material in immiscible organic solvent in the form of small droplets [98,99]. A suitable stabilizer is employed to prevent droplets coalescence and coagulation. The subsequent step is covalent cross-linking of the formed droplets either thermally (>500 °C) or using a suitable cross-linking agent such as formaldehyde solution.

3.1.5 Solvent Evaporation/Solvent Extraction

The polymer is dissolved in organic solvent, into which the core material is also dissolved or dispersed [100,101]. The resulting solution is added dropwise to a stirred aqueous solution containing suitable water-soluble steric stabilizer such as poly(vinyl alcohol), to form small polymer droplets containing encapsulated material. The microcapsule hardening process is achieved by solvent evaporation or solvent extraction.

3.2 Smart Nanocomposite Coatings Containing Self-Healing Microcapsules

Wang et al. [102] utilized isophorone diisocyanate (IPDI) microcapsules as self-healing component in the alkyd varnish coatings (AVCs), for the corrosion protection of carbon steel Q235 substrates. The performance of the self-healing coating was tested after 100 h immersion in 0.6 M NaCl solution, where the scratched crevice was fully cured by the action of self-healing PU product. The self-healing action was monitored and confirmed by infrared spectroscopy, where new absorption bands for urea in the formed PU appeared at 3339 (N–H), 1644 (–C=O), and 1558 (N–H) cm^{-1}.The self-healing action of IPDI microcapsules was triggered by the presence of water and resulted in formation of water-insoluble PU during the corrosion initiation process.

Huang et al. [103] prepared organic silane microcapsules by condensation emulsion polymerization procedure for corrosion protection of steel substrates. The microcapsule composed of poly(urea-formaldehyde) (PFU) external shell and hydrolyzable organic silane internal core. The hydrolyzable organic silane compound 1H,1H,2H,2H-perfluorooctyl triethoxysilane (POTS) was selected for the following reasons: (a) POTS has the ability to hydrolyze in wet environment to form hybrid organic-inorganic silane-based film without the presence of added catalyst; (b) the newly formed hybrid film has a hydrophobic nature [12] due to the presence of perfluorooctyl chains, indicating such a special wetting property will assist to repel aqueous electrolyte solution away from metal and to provide supplementary corrosion protection to metal substrate; and (c) the commercial availability of POTS enables the possibility of mass production of such anticorrosive self-healing coating. The prepared microcapsules were incorporated into epoxy resin (10 wt.%) and were employed for corrosion protection of steel substrates with a coating thickness of 250-300 μm. The corrosion test in the salt solution demonstrated the good corrosion resistance of the self-healing coating compared to the control sample. The enhanced anticorrosion performance of the microencapsulated coating was attributed to the self-healing action of the embedded organic POTS molecules that were hydrolyzed upon contact with water by polycondensation reaction to form a hybrid hydrophobic protective film on the scratched locations of the tested steel substrates. The key advantage of such self-healing functionality is the absence of any external interventions such as heating or UV exposure, making it promising for the development of catalyst-free, self-healing coatings, which is of considerable technical and commercial importance.

The long-term performance of POTS microcapsule-based self-healing coatings was also reported by the same group recently [104]. The prepared POTS-based anticorrosive silicone elastomer was exposed to 1 M HCl solution for 1 month. A neat silicone elastomer without microcapsules was used as control and treated in the same manner. After 1 month of exposure, the silicone elastomer was peeled away from the steel substrate. In the case of specimen coated with the POTS-based coating, SEM revealed that the steel substrate was less corroded compared with the specimen coated with neat silicone elastomer. The self-healing performance was assessed by EIS measurements, which showed that the scratched coating was autonomously self-healed due to the formation and increment of a new film formed within the scratches during exposure to HCl solution.

3.3 Smart Nanocomposite Coatings Containing Self-Healing Nanocontainers

Different types of nanocontainers such as halloysite nanotubes (HSNs), mesoporous silica nanocontainers (MSNs), hydroxyapatite (HAP) microparticles, and nanocrystalline layered double hydroxides (LDHs) have been utilized as nanocarriers and nanoreservoirs for immobilization, storage, and controlled release of inhibitors that are incorporated into various coatings [16,17,79,80,89,90,105–119].

The incorporation of the active corrosion inhibitor can be achieved by different techniques, including (1) simple physical absorption through the porous nanocontainer structure, (2) various encapsulation processes as discussed in Section 3.1, (3) ion exchange reactions with the available counterions in the respective nanocontainer, and (4) embedding of the inhibitor between the alternating polyelectrolyte multilayers (Section 4) [16,79,80,105,107,115,118–122].

Figure 4 illustrates the general steps for the immobilization of corrosion inhibitor into various types of inorganic nanocontainers. Inorganic nanocontainers (HSNs, MSNs, LDHs, and HAPs) represent the inner core host material that can be loaded with corrosion inhibitor through any of the abovementioned pathways. In the next step, organic polymeric electrolytes (polyelectrolytes) are coated on the external surface of the nanocontainers to form the external organic shell. Based on the specific properties of the employed polyelectrolytes such as acidity, basicity, and the availability of grafted

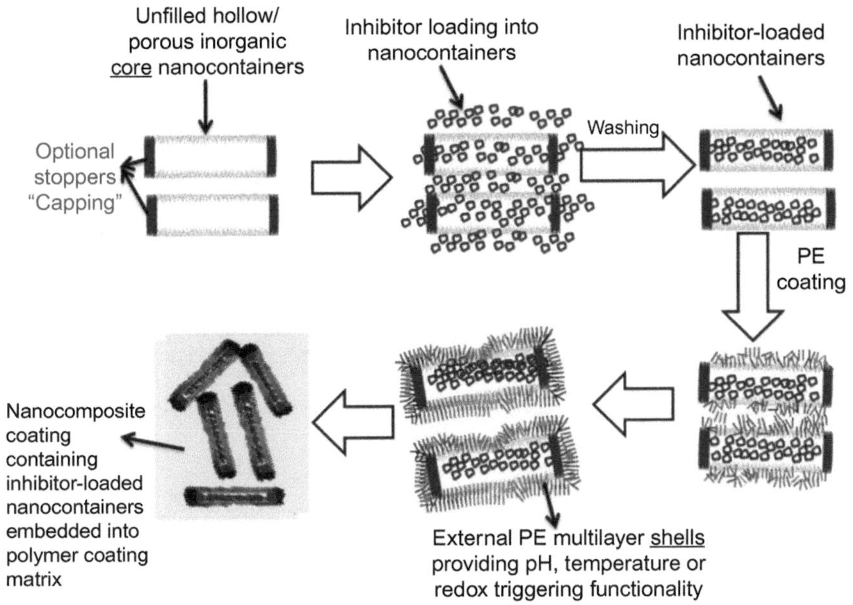

Figure 4 Schematic illustration for the preparation of nanocomposite coating embedded with functional inorganic-organic core-shell nanotubes. Initially, inorganic nanotubes are loaded with suitable corrosion inhibitor (internal core) and then coated with polyelectrolytes multilayers (external shell) bearing specific responsive functionalities for effective and controlled self-healing action through the release of stored inhibitor.

responsive functionalities, polyelectrolyte multilayers can be designed to respond to certain stimuli as described in the previous sections.

3.3.1 Mesoporous Silica Nanoparticles (MSNs)

Chen and Fu [123] synthesized novel pH-responsive nanovalves for on-demand release of stored corrosion inhibitor. The synthesis procedure utilized hollow mesoporous silica nanospheres (HMSs) as nanocontainers for the corrosion inhibitor, benzotriazole (BTA), and the supramolecular nanovalves, consisting of cucurbit[6]uril (CB[6]) rings and the functional stalks attached to the surface of HMSs (Figure 5). The study showed that the HMSs assembled with the nanovalves possess a higher encapsulation capacity for BTA than MCM-41 assembled under the same procedure due to their huge hollow internal structure. The controlled release of the stored BTA corrosion inhibitor was monitored by UV absorption spectra. The results indicated that the nanovalves were closed tightly without evident leak at

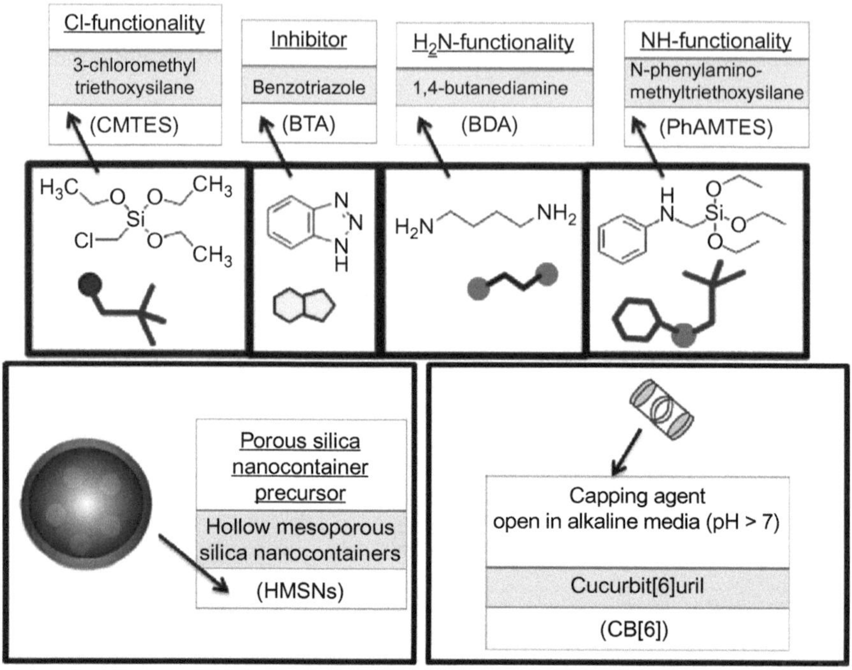

Figure 5 Chemical structures and functionalities for various components utilized in the fabrication process of pH-responsive nanovalves for on-demand release of stored corrosion inhibitor.

neutral pH but opened and released the BTA when the pH was increased. This behavior was attributed to the pH-dependent binding ability of CB[6] with bisammonium stalks that enabled the controlled release of BTA from the assembled HMSs.

In another work, Chen and Fu [113] reported the preparation of intelligent anticorrosion coating based on pH-responsive supramolecular nanocontainers. Hollow HMSNs were utilized as starting porous precursor for the preparation of functionalized supramolecular nanocontainers that are able to release the embedded BTA corrosion inhibitor upon pH change. The preparation process of the intelligent pH-responsive coating is depicted in Figure 6. Undoped hybrid sol-gel coating was prepared by mixing zirconia sol and organosiloxane sol in 1:2 volume ratio. The intelligent corrosion protection coating was prepared following the same steps, except that approximate amount of mixture of alkaline-responsive nanocontainers (1.5 mg/ml) and acid-responsive nanocontainers (1.5 mg ml) was added to the final sol-gel solution and uniformly dispersed by ultrasonication for

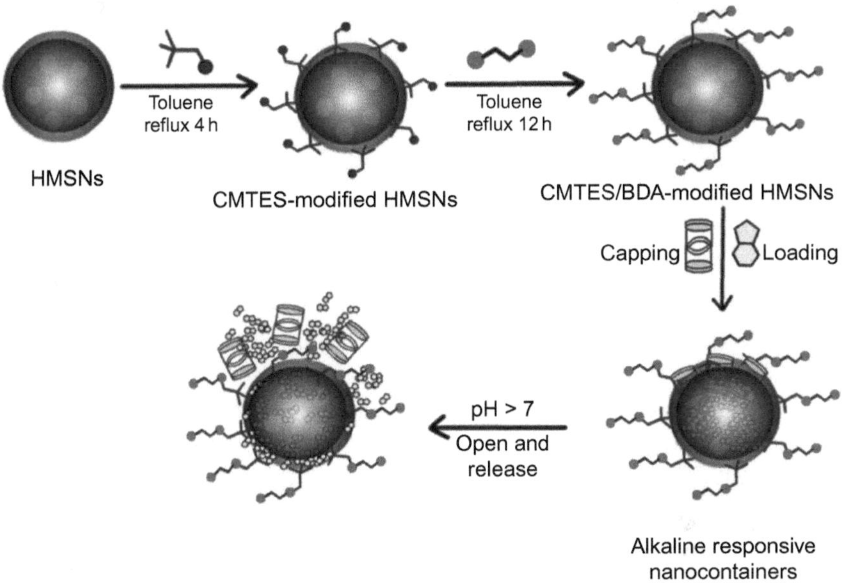

Figure 6 Schematic illustrations for the preparation of alkaline-responsive supramolecular nanocontainers through multistep functionalization of HMSNs.

few minutes. Results showed that the release of encapsulated corrosion inhibitor was quick to respond to environmental pH changes. The employed alkaline and acidic pH-responsive nanocontainers were found to be compatible with hybrid zirconia-silica sol-gel coatings. The self-healing property of the undoped sol-gel coating and the pH-responsive nanocontainer-doped sol-gel coating was assessed by making artificial scratches in the coated panels. After immersion in 0.5 M NaCl for 96 h, the surface morphologies of the scratched samples were observed with a striking difference between the two coatings. The undoped coating failed to provide corrosion protection as the alloy matrix was almost invisible and the corrosive trace was very clear near the artificial scratch. In contrast, an obvious inhibition effect occurred in the intelligent anticorrosion coating. This conclusion was evidenced by the constructed film on the alloy surface, which was attributed to the released corrosion inhibitor deposits originating from the pH-responsive nanocontainers and hence triggered the termination of the corrosion activities.

In a recent study, the same research group prepared stimuli-responsive nanocontainers with self-healing functionality [105]. The intelligent anticorrosion coating utilized mechanized HMSs as smart nanocontainers implanted into the self-assembled nanophase particle (SNAP) coating. The operation of mechanized HMSs as smart nanocontainers mainly depends on the supramolecular nanovalves installed on the external surface of HMSs, which have two vital functions: to prevent any undesirable release under neutral environment and to release inhibitor molecules upon the acid or alkaline stimulus. The pendant supramolecular nanovalves consist of two components: the linear stalks containing the hexylammonium units and the ferrocenecarboxylic acid units linked by amide bonds and the movable gates, CB[7] macrocycles, which are employed to regulate the flow of corrosion inhibitors out of HMSs. The prepared HMSs were loaded with caffeine molecules ($\approx$20 wt.%) as corrosion inhibitor. Caffeine is neutral, nontoxic, and environmentally friendly substance that has been utilized as corrosion inhibitor for a number of metals and alloys owing to its special structure of multiadsorption centers. Smart nanocontainer-doped SNAP coating showed the first sign of corrosion after 4 h. After 16 h immersion, the surface presents no remarkable electrochemical activity, which is in sharp contrast to the pure SNAP coating suggesting that all the corrosion activities were suppressed by self-healing functionality.

3.3.2 Layered Double Hydroxides

LDHs, also known as anionic clays or hydrotalcite-like compounds, are anion exchange materials consisting of stacks of positively charged, mixed-metal hydroxide layers between which anionic species and solvent molecules are intercalated [124,125]. LDHs have been utilized as nanocarriers for drug delivery applications due to their low toxicity, biocompatibility, and ability to release active species under control [126–129]. Therefore, LDHs were proposed as nanocarrier for immobilization and encapsulation of corrosion inhibitors [130–133,122]. LDHs can be loaded with suitable corrosion inhibitors and incorporated into corrosion protective organic coatings to provide self-healing functionality through the release of the stored species. As illustrated in Figure 7, LDHs can provide self-healing functionality through (1) controlled release of the immobilized negatively charged organic/inorganic inhibitors and (2) entrapment of the corrosive anions such as chloride and sulfate anions [114].

Poznyak et al. [134] prepared nanocrystalline LDHs, with platelike morphology, as novel inorganic host materials intercalated with guest organic inhibitors for anticorrosion applications. LDHs nanocontainers were loaded with two different organic inhibitors, 2-mercaptobenzothiazolate (MBT) and quinaldate (QA), through anion exchange methodology with pristine

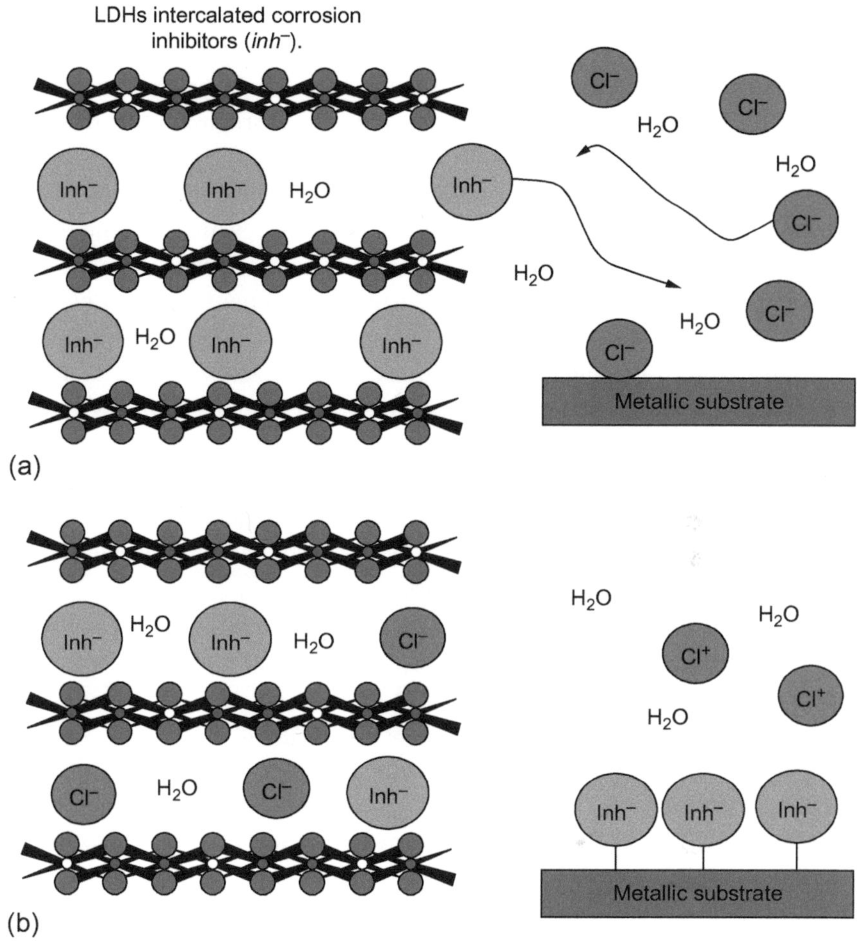

Figure 7 (a) The release of inhibitors (Inh^-) is triggered by the presence of anions in solution (Cl^-). (b) LDHs play a double role, providing inhibitors to protect the metallic substrates and entrapping aggressive species from the environment. *(Reprinted with permission from Ref. [114]. Copyright (2010) American Chemical Society.)*

nitrate-loaded LDH precursor. The release of the negatively charged organic inhibitors was found to occur in two parallel ways (i.e., release of inhibitor and entrapment of the aggressive anions), providing active feedback for corrosion protection control. The AA2024 samples that were placed in contact with LDH suspensions in a 0.05 M NaCl solution for 24 h showed a small anticorrosion activity of these systems. This observation was attributed to the local alkalization of the solution in the presence of hydroxide-rich LDHs particles, which causes a decrease in the inhibition efficiency of MBT and QA inhibitors. However, for longer immersion times, the corrosion processes are efficiently inhibited because of the formation of a protective film, with the MBT-loaded LDHs revealing a better performance than the QA-loaded ones.

The anion exchange property of LDHs was employed for the preparation of Zn-Al LDHs intercalated with phosphate, vanadate, and MBT inhibitors [114]. Both primer and sol-gel formulations were loaded with ≈10 wt.% LDH nanocontainers in dry paint film. The developed coatings showed distinct active corrosion protection toward AA2024 substrates. The best results were obtained with a combination of sol-gel film (pretreatment layer) doped with LDH-MBT and primer doped with LDH-VO_x. The superior performance of the formed coating was attributed to the availability of MBT anions in a very short timescale for the protection of the interface layer between the aluminum substrate and the sol-gel film, whereas the long-term protection was conferred by vanadate anions released from LDH nanocontainers added to the primer.

Hu et al. [135] prepared PANI-grafted-LDH nanocomposite coating for corrosion protection application. The grafting methodology aimed to improve the hydrophilicity of the LDH nanoclays, which will increase the compatibility of the nanoclays with the organic polymeric coating. The preparation procedure of the hybrid coating was carried out in three consecutive steps. First, decavanadate corrosion inhibitor anions were intercalated into LDHs via anion exchange reaction with the originally intercalated nitrate (NO_3^-) anions, and then, the decavanadate intercalated LDH (D-LDH) was treated with APTS, to form a bonding layer on its surface, where PANI was *in situ* polymerized to form PANI/AD-LDH nanocomposite. The corrosion protection performance of the PANI/AD-LDH hybrid coating was superior to that of D-LDH and PANI. This enhanced corrosion protection property was attributed to (1) improved barrier properties of the coating, (2) redox properties of PANI, and (3) the release of decavanadate inhibiting anions.

3.3.3 HAP microparticles

HAP microparticles, which are calcium phosphate-based minerals ($Ca_{10}(PO_4)_6(OH)_2$), possess a number of features that make them suitable carriers for corrosion inhibitors. These features include crystallinity, insolubility in water, submicrometer size, high surface area, and chemical dissolution at acidic pH range. The dissolution of HAP microparticles in the acidic medium provides pH-dependent triggering mechanism for the release of stored inhibitors, whereas their high surface area enables a high loading capacity [136].

Snihirova et al. [136] investigated the incorporation of inhibitor-doped HAP microparticles into a hybrid sol-gel coating as new feedback-active anticorrosion system. Cerium (Ce^{3+}) and lanthanum (La^{3+}) cations, salicylaldoxime (Sal), and 8-hydroxyquinoline (8HQ) were used as dopants in the HAP microparticles. Cation exchange reactions were utilized for the incorporation of the inorganic cationic inhibitors (Ce^{3+} and La^{3+}), whereas the organic inhibitors (Sal and 8HQ) were embedded by physical adsorption on the surface of HAP microparticles (Figure 8). The loading percentages of doped corrosion inhibitors were 22, 20, 9.3, and 0.54 wt.% for (Ce^{3+}), (La^{3+}), (8HQ), and (Sal), respectively. AA2024-T3 was coated with thin films of the formulated coating that contains 0.2 wt.% inhibitor-doped HAP microparticles, by using dip-coating technique. It was found that the release of doped inhibitors occurred in acidic medium ($pH < 6$), with an acceleration of Ce^{3+} and La^{3+} release when the pH falls below 3.1 (Figure 9). The local acidification down to $pH = 3.65$ was attributed to the anodic dissolution of Al, as revealed from localized pH measurements that were conducted using pH-selective microelectrodes. Hence, the HAP microparticles were able to release the stored inhibitors on demand by providing a sensing mechanism for corrosion initiation through local

(a) (b)

Figure 8 Scanning electron microscopy images of (a) blank HAP and (b) HAP loaded with (i) Ce^{3+}, (ii) La^{3+}, (iii) Sal, and (iv) 8HQ. *(Reprinted with permission from Ref. [137]. Copyright (2010) American Chemical Society.)*

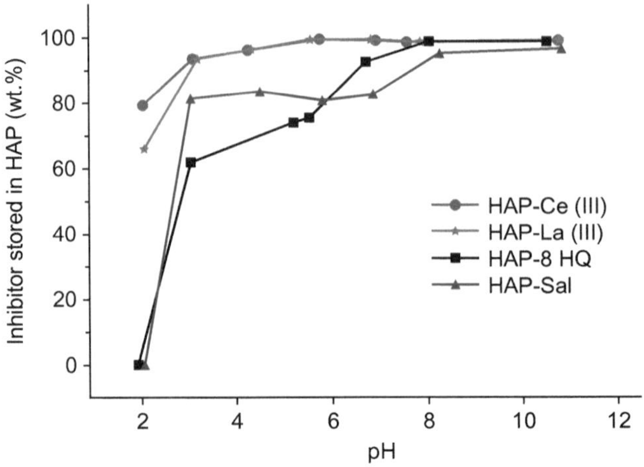

Figure 9 Release of different inhibitors from the HAP as a function of the pH. The curves show the amount of inhibitor remaining in the HAP after 24 h of immersion in a 0.05 M NaCl solution at specified pH. *(Reprinted with permission from Ref. [137]. Copyright (2010) American Chemical Society.)*

acidification process. Also, the dissolution of HAP microparticles themselves contributed to corrosion protection by pH buffering mechanism, as conferred from the delayed corrosion of the AA2404-treated substrates with sol-gel coating embedded with blank HAP microparticles (i.e., undoped with inhibitor).

3.3.4 Ceramic-Based Nanocontainers

Kartsonakis et al. [110] prepared cerium oxide (CeO_2) nanocontainers as carriers for corrosion inhibitor and incorporated them as a part of multilayer hybrid coating for protection of aluminum alloys (AA) 2024-T3. Porous ceria nanocontainers were prepared by a three-step procedure. The first step utilized emulsion polymerization for the preparation of PS latex core particles with diameter of 85 ± 5 nm and specific area of 84.2 m^2/g. The outer shell of PS lattices was coated with a layer of cerium oxide by controlled hydrolysis of cerium (III) acetylacetonate $Ce(acac)_3$ aqueous solution in the presence of PS latex and polyvinylpyrrolidone (PVP) stabilizer. In order to obtain hollow cerium oxide nanospheres, PS cores were removed by calcination process in the last step. Porous ceria nanocontainers were loaded with MBT corrosion inhibitor. MBT-loaded ceria nanocontainers were incorporated with PANI-PPy conductive polymer (CP) coating via electrochemical deposition technique. The incorporation of MBT-loaded ceria nanocontainers into the CP coatings improved the corrosion protective properties of the films, as revealed from EIS measurements, which showed an increase of the total impedance values as well as reduction of both anodic and cathodic currents compared to coatings without nanocontainers.

In a recent study, the former research group investigated the corrosion protection effectiveness of multifunctional epoxy coatings modified with ceramic nanocontainers loaded with corrosion inhibitor, chloride, and water traps, applied on AA2024-T3 [17]. Hollow CeMo nanocontainers were synthesized following the previously reported three-step procedure [110]. The prepared CeMo nanocontainers were loaded with 8-hydroxyquinoline (8-HQ) corrosion inhibitor (8% w/w inhibitor loading). Pigments of polymeric water traps were prepared by distillation precipitation polymerization [137]. Methacrylic acid (MAA), 2,2′ azobis(2-methylpropionitrile) (AIBN), and ethylene dimethacrylate (EGDMA) were employed as monomer, initiator, and cross-linker, respectively (Figure 10). The polymerization of cross-linked MAA-EGDMA microspheres was carried out in acetonitrile solvent. Chloride trap pigments were prepared by two-step procedure. The first step utilized sol-gel chemistry for the preparation of silica microparticles by adapting the Stöber method. In the second step, the surface of silica microparticles was functionalized with positively charged quaternary amine groups as functional traps for chloride anions. The final hybrid organic-inorganic coating (epoxy-ORMOSIL) was prepared by incorporation of 1-2 wt.% of the prepared CeMo pigment nanocontainers and traps into the epoxy-based coating as illustrated in Figure 11. The coatings were deposited via dip-coating process on chemically cleaned coupons of aluminum alloy AA2024-T3, with coating thickness ≈9 μm. It was found that the addition of CeMo pigment nanocontainers, water traps, and chloride traps improved the corrosion resistance. The release of the corrosion inhibitor,

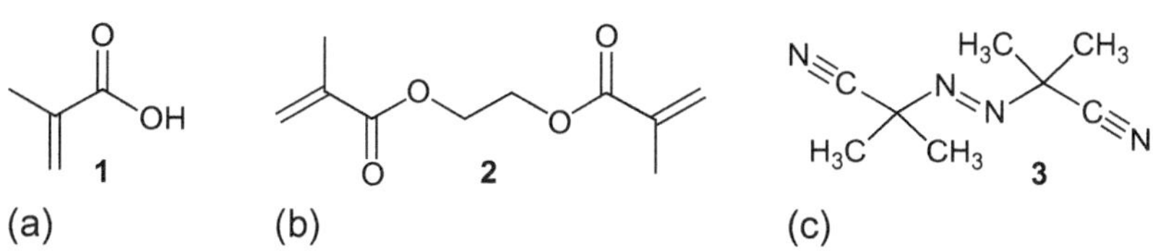

Figure 10 The chemical structures of reagents used in precipitation polymerization carried out in acetonitrile for the preparation of water traps polymeric pigments: (a) MAA monomer, (b) EGDMA cross-linker, and (c) 2,2′ azobis(2-methylpropionitrile) (AIBN) initiator.

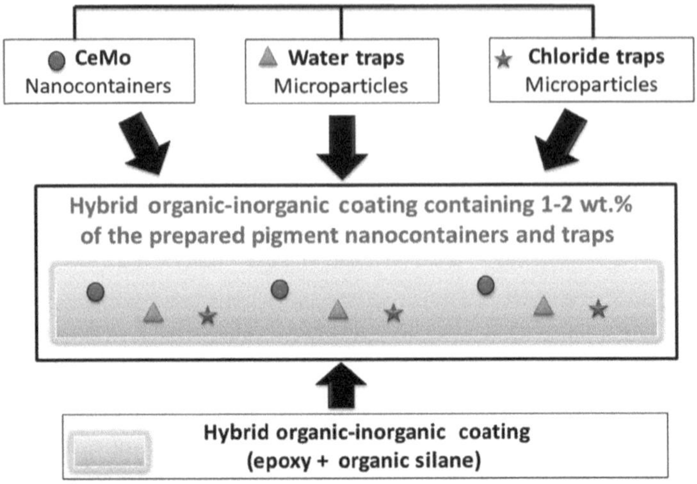

Figure 11 Schematic illustration for the preparation of hybrid organic-inorganic coating containing ceramic CeMo nanocontainers, water traps, and chloride traps.

8-HQ, from the nanocontainers resulted in corrosion inhibition and development of self-healing behavior.

4. NANOCOMPOSITE COATINGS BASED ON POLYELECTROLYTE COMPLEXES

Polyelectrolyte complexes can be prepared using layer-by-layer (LBL) deposition technique. Different types of nanoparticles can be employed as the capsule core material. Polyelectrolyte layers with alternating positive and negative charges are deposited on the capsule shell wall [13,16,18,138]. The thickness of capsule shell wall can be controlled by the number of deposited layers.

Polyelectrolytes can be utilized for the preparation of capsules/nanocontainers by adapting different approaches. In the first approach, polyelectrolyte multilayers are introduced after loading the inhibitor into the inorganic core nanocontainers, as discussed in Section 3.3 (see Figure 4). However, in the second approach, polyelectrolyte capsules are created in a three-stage process as illustrated in Figure 12. Initially, a suitable capsule core material is selected. Then, polyelectrolyte multilayers are deposited on the capsule shell wall. Finally, the metal oxide core material is removed (usually by acid hydrolysis), leaving behind hollow, porous, and semipermeable polyelectrolyte capsules. The porosity of the prepared capsules can be optimized by

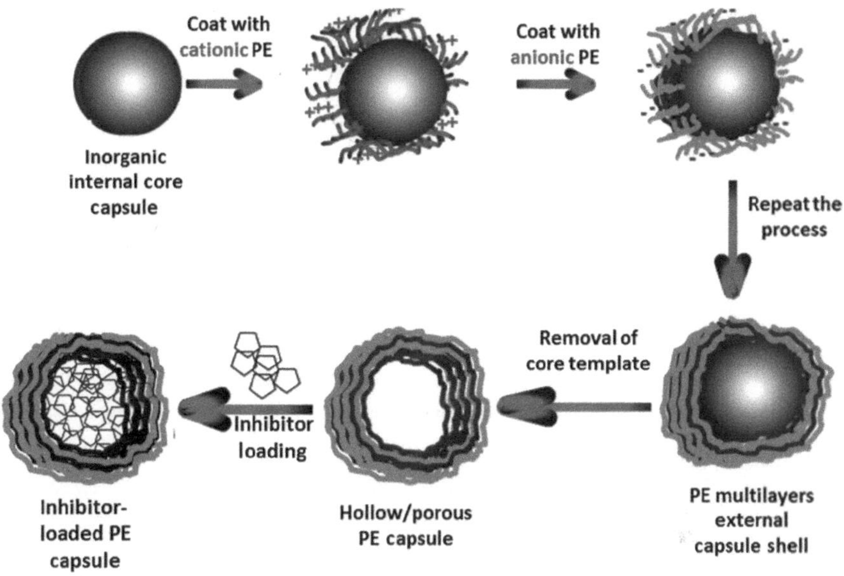

Figure 12 Schematic illustration for the preparation of polyelectrolyte capsules using template inorganic core methodology.

controlling the number of deposited polyelectrolyte layers. This porosity is essential as it allows the capsule core to be loaded with various types of organic and inorganic corrosion inhibitors.

Andreeva et al. [13] noticed that the release of stored corrosion inhibitors from the pH-sensitive polyelectrolyte capsules and their anticorrosion activities follow one or more of these mechanisms: (1) pH buffer formed by polybase (anionic polyelectrolyte) and polyacid (cationic polyelectrolyte) complex suppress pH changes caused by corrosion degradation; (2) relative mobility of polymer chains in swollen state leads to coating regeneration and defect elimination; (3) polyelectrolyte layers form a carrier for inhibitor allowing its release on demand; and (4) polyelectrolyte nanonetwork provides a barrier between surface and environment. For example, when the scratched metal substrate was coated with two weak polyelectrolytes, polyethyleneimine (PEI) and poly(acrylic acid) (PAA), and immersed in 0.1 M NaCl solution, the (PEI/PAA) multilayer system was able to stop the initiation of the corrosion [13]. The self-healing mechanism was attributed to the high mobility of the nanolayers that facilitate the water penetration through the polyelectrolyte multilayers and trigger the release of the embedded inhibitor. However, when the couple of employed polyelectrolytes are strong (strong acidic/strong basic polyelectrolytes), no such self-healing performance was observed. In this case, the high charge density of the two highly ionizable polyelectrolytes cannot be altered by pH change.

He et al. [139] prepared a novel polyelectrolyte nanostructured photocatalytic coating for corrosion protection with self-healing functionality. Initially, TiO_2 nanoparticles (TiO_2 NPs, <5 nm) were prepared and spin-coated on aluminum alloy sheets with coating thickness of ≈ 1.1 μm and followed by heat treatment to 300 °C for 4 h. In the next step, PEI and polystyrene sulfonate (PSS) were deposited alternatively on the TiO_2 NPs coated with Al alloy sheets by spin LBL coating technique. The corrosion inhibitor 8-hydroxyquinoline (8-HQ) was embedded between the polyelectrolyte layers by following the same LBL technique with the following layers sequence: PEI/PSS/8-HQ/PSS/PEI. The photocatalytic action of TiO_2 NPs triggers the release of 8-HQ under UV light resulting in self-healing effect. This conclusion was supported by the real-time EIS results, which showed that the new passivated layer was formed after the release of 8-HQ molecules.

Plawecka et al. [140] prepared self-healing water-based epoxy coating modified with nanocapsules loaded with 2-mercaptobenzothiazole (MBT) for corrosion protection of aluminum alloy and steel. The nanocapsule containers were prepared by direct encapsulation of hydrophobic active agent (surfactant) in emulsion drops assembled with polyelectrolyte shells. Chemical structures of the used reagents are shown in Figure 13, and the preparation

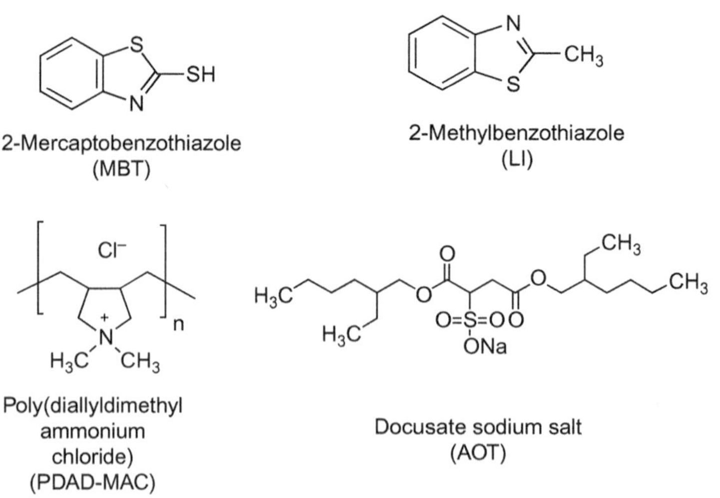

Figure 13 Chemical structures of used reagents for synthesis of inhibitor-loaded polyelectrolyte capsules.

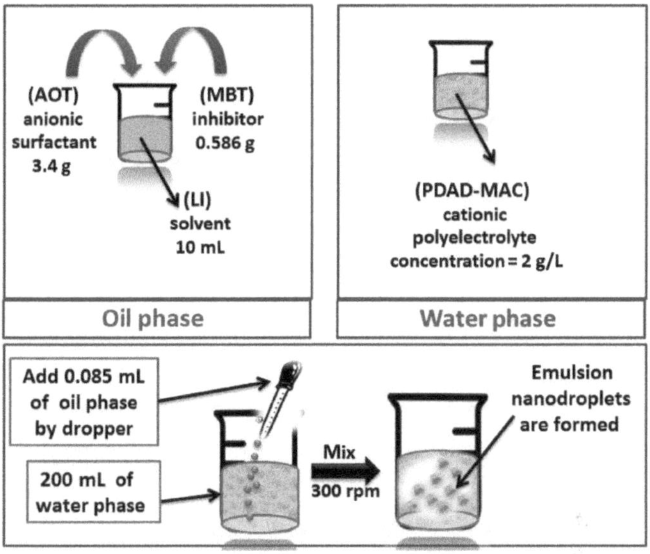

Figure 14 The preparation of inhibitor-loaded polyelectrolyte capsules by emulsion polymerization method. The corrosion inhibitor (oil phase) is encapsulated into the organic water-soluble polyelectrolyte solution in the presence of anionic surfactant emulsifier. Chemical structures of the employed reagents are provided in Figure 13.

procedure for the polyelectrolyte capsules is illustrated in Figure 14. Two sets of water-based epoxy solutions were prepared: a blank system (without inhibitor-loaded nanocapsules) and a solution modified with inhibitor-loaded nanocapsules. The inhibitor content in the final coating solution was 0.175 g/L LI + 0.00875 g/L MBT. The epoxy coating was applied on aluminum and steel samples by dip-coating. It was concluded that the addition of nanocapsules to water-based epoxy primers does not damage the barrier properties of these coatings. The presence of inhibitor-loaded nanocapsules improves the corrosion resistance of aluminum alloys and steel. These results highlighted that the utilization of inhibitor-loaded nanocapsules in water-based epoxy coatings is a promising strategy toward the development of self-healing water-based epoxy coatings.

5. CONCLUSIONS AND FUTURE REMARKS

Nanocomposite coatings can be tailored to provide unique functions for a variety of industrial applications including material corrosion protection. Nanocomposite coatings can be engineered, not only to provide barrier protection features but also to provide autonomous repair and maintenance for the defected sites upon corrosion initiation. The essential components in such smart and self-responsive nanocomposite coatings are the inhibitor-loaded carriers with nano- or micro-scale dimensions that enable them to be dispersed more effectively in the final organic coating matrix. Different nanostructured components such as nanoparticles, nanowires, nanotubes, nanosheets, and nanocontainers have been designed and tested for corrosion protection applications. Also, hybrid organic-inorganic nanocomposite coatings containing a combination of one or more of nanostructured components can be formulated to provide more robust and prolonged protection under different environmental conditions.

Inorganic nanofillers such as nanoparticles and nanowires offer properties such as very fine grain size, high surface area to volume ratio, and improved dispersion in the respective coating. Eventually, this will lead to lower loading percentages of such nanofillers compared to their conventional counterparts. The nanoscale dimensions of these nanofillers enable their incorporation into the various hybrid coatings during the hydrolysis steps without agglomeration. One-dimensional nanowires such as CMNs showed superior corrosion resistance due to the enhanced solubility of the amorphous CMNs compared to crystalline bulk cerium molybdate pigments.

Electrically conductive components with nanostructured identity can be employed to improve the corrosion resistance. Nanostructured carbons such as CNTs and graphene have been incorporated into organic coatings to enhance the corrosion resistance by improving the barrier properties and reducing the gas permeability of the nanocomposite coating. ECPs can be introduced into the nanocomposite coatings to provide superior corrosion protection in multiple mechanisms that can work simultaneously. These mechanisms include barrier protection, anodic protection, and smart self-healing protection.

The recent advancements in corrosion protective nanocomposite coatings have focused on a smart/self-responsive/self-healing new type of coatings. Smart coatings are expected to provide continuous and autonomous corrosion protection performance by sensing the corrosion initiation processes and responding by release of suitable self-healing agents. The basic strategy for designing smart self-responsive coatings is based on establishing trigger and release functionalities in the embedded nanocarrier components. Nanocarriers can be synthesized/fabricated by different means including emulsion polymerization, precipitation polymerization, polyelectrolyte complexes, mesoporous silica nanoparticles, naturally occurring nanotubes/nanocontainers, or a combination of one or more of these methods.

Future studies have to focus on finding more simple, cost-effective, and reproducible procedures for synthesizing various nanoparticles, nanowires, and nanocontainers. One attractive and powerful future research direction related to nanocomposite coatings is multifunctional protective coatings. Multifunctional protective coatings can be designed to serve more than one function, such as corrosion and water-proof superhydrophobic self-cleaning coating. Future research is expected to grow very fast in the area of smart protective coatings. This research area requires more robust test and assessment of the smart protective coatings in the real wet/dry environments and for certain applications.

REFERENCES

[1] K. Liu, Y. Tian, L. Jiang, Bio-inspired superoleophobic and smart materials: design, fabrication, and application, Prog. Mater. Sci. 58 (4) (2013) 503–564.
[2] S.A. Corr, Y.P. Rakovich, Y.K. Gun'ko, Multifunctional magnetic-fluorescent nanocomposites for biomedical applications, Nanoscale Res. Lett. 3 (3) (2008) 87–104.
[3] J. Baur, E. Silverman, Challenges and opportunities in multifunctional nanocomposite structures for aerospace applications, MRS Bull. 32 (04) (2007) 328–334.
[4] X. Zhang, et al., Multifunctional antireflection coatings based on novel hollow silica-silica nanocomposites, ACS Appl. Mater. Interfaces 6 (3) (2014) 1415–1423.
[5] S. Kango, et al., Surface modification of inorganic nanoparticles for development of organic-inorganic nanocomposites—a review, Prog. Polym. Sci. 38 (8) (2013) 1232–1261.
[6] T. Mishra, A. Mohanty, S. Tiwari, Recent development in clay based functional coating for corrosion protection, Key Eng. Mater. 571 (2013) 93–109.
[7] M. Omastová, M. Mičušík, Polypyrrole coating of inorganic and organic materials by chemical oxidative polymerisation, Chem. Pap. 66 (5) (2012) 392–414.
[8] Z. Zheng, et al., Self-healing and antifouling multifunctional coatings based on pH and sulfide ion sensitive nanocontainers, Adv. Funct. Mater. 23 (26) (2013) 3307–3314.
[9] D. Shchukin, Container-based multifunctional self-healing polymer coatings, Polym. Chem. 4 (18) (2013) 4871–4877.
[10] U. Riaz, C. Nwaoha, S. Ashraf, Recent advances in corrosion protective composite coatings based on conducting polymers and natural resource derived polymers, Prog. Org. Coat. 77 (4) (2014) 743–756.
[11] M. Samadzadeh, et al., A review on self-healing coatings based on micro/nanocapsules, Prog. Org. Coat. 68 (2010) 159–164.
[12] Y. Li, L. Li, J. Sun, Bioinspired self-healing superhydrophobic coatings, Angew. Chem. 122 (35) (2010) 6265–6269.
[13] D.V. Andreeva, E.V. Skorb, D.G. Shchukin, Layer-by-layer polyelectrolyte/inhibitor nanostructures for metal corrosion protection, ACS Appl. Mater. Interfaces 2 (7) (2010) 1954–1962.
[14] H. Mohwald, D.G. Shchukin, Self-repairing coatings containing active nanoreservoirs, Small 3 (2007) 926–943.
[15] M. Kendig, M. Hon, L. Warren, 'Smart' corrosion inhibiting coatings, Prog. Org. Coat. 47 (3–4) (2003) 183–189.
[16] D.G. Shchukin, et al., Layer-by-layer assembled nanocontainers for self-healing corrosion protection, Adv. Mater. 18 (13) (2006) 1672–1678.

[17] I. Kartsonakis, et al., Multifunctional epoxy coatings combining a mixture of traps and inhibitor loaded nanocontainers for corrosion protection of AA2024-T3, Corros. Sci. 58 (2014) 147–159.
[18] A. Zhuk, S.A. Sukhishvili, Stimuli-responsive layer-by-layer nanocomposites, Soft Matter 9 (2013) 5149–5154.
[19] C. Sanchez, et al., Applications of hybrid organic-inorganic nanocomposites, J. Mater. Chem. 15 (35–36) (2005) 3559–3592.
[20] T. Gopakumar, N. Patel, M. Xanthos, Effect of nanofillers on the properties of flexible protective polymer coatings, Polym. Compos. 27 (4) (2006) 368–380.
[21] W. Keyoonwong, et al., Corrosion behavior of three nanoclay dispersion methods of epoxy/organoclay nanocomposites, Int. J. Corros. 2012 (2012).
[22] C. Motte, et al., Improvement of corrosion protection offered to galvanized steel by incorporation of lanthanide modified nanoclays in silane layer, Prog. Org. Coat. 74 (2) (2012) 326–333.
[23] E. Huttunen-Saarivirta, et al., Characterization and corrosion protection properties of epoxy powder coatings containing nanoclays, Prog. Org. Coat. 76 (4) (2013) 757–767.
[24] M. Behzadnasab, S.M. Mirabedini, M. Esfandeh, Corrosion protection of steel by epoxy nanocomposite coatings containing various combinations of clay and nanoparticulate zirconia, Corros. Sci. 75 (2013) 134–141.
[25] C. Zhou, et al., Polybenzoxazine/SiO_2 nanocomposite coatings for corrosion protection of mild steel, Corros. Sci. 80 (2014) 269–275.
[26] N. Elhalawany, M.A. Mossad, M.K. Zahran, Novel water based coatings containing some conducting polymers nanoparticles (CPNs) as corrosion inhibitors, Prog. Org. Coat. 77 (3) (2014) 725–732.
[27] N. Elhalawany, M.M. Saleeb, M.K. Zahran, Novel anticorrosive emulsion-type paints containing organic/inorganic nanohybrid particles, Prog. Org. Coat. 77 (2) (2014) 548–556.
[28] A.S. Hamdy, M. Shoeib, H. Hady, The effect of grain refining and phosphides formation on the performance of advanced nanocomposite and ternary alloy coatings on steel, Mater. Lett. 80 (2012) 191–194.
[29] A.S. Hamdy, et al., Electroless deposition of ternary Ni-P alloy coatings containing tungsten or nano-scattered alumina composite on steel, J. Appl. Electrochem. 38 (3) (2008) 385–394.
[30] A.S. Hamdy, et al., Corrosion behavior of electroless Ni-P alloy coatings containing tungsten or nano-scattered alumina composite in 3.5% NaCl solution, Surf. Coat. Technol. 202 (1) (2007) 162–171.
[31] A.S. Hamdy, M. Shoeib, H. Hady, O.F. Abdel Salam, Effect of the experimental parameters on the coating performance of Ni-P nano-composite alloy coatings, in: Processing and Product Manufacturing: Innovative Processing and Synthesis of Ceramics, Glasses and Composites Symposium, Proceedings of MS&T '07, 16-20 September 2007, Michigan, 2007.
[32] A.S. Hamdy, M.A. Shoeib, H. Hady, Nano-composite and ternary-alloy protective coatings for steel, in: Proceedings of 14th NACE Middle East Corrosion Conference & Exhibition, 12-15 February 2012, Bahrain, 2012, Paper: 75-CR-12.
[33] A.S. Hamdy, H. Hady, M. Shoeib, O. Abdel Salam, Electrochemical impedance studies on Ni-P-W and Ni-P-Al_2O_3 nano-composite alloy coatings in 3.5% NaCl, in: Proceedings of European Corrosion Congress 2007, Germany, 2007.
[34] Y. Huang, et al., Directed assembly of one-dimensional nanostructures into functional networks, Science 291 (5504) (2001) 630–633.
[35] Y. Xia, et al., One-dimensional nanostructures: synthesis, characterization, and applications, Adv. Mater. 15 (5) (2003) 353–389.

[36] K. Yasakau, et al., Cerium molybdate nanowires for active corrosion protection of aluminium alloys, Corros. Sci. 58 (2012) 41–51.
[37] K. Yasakau, et al., Active corrosion protection of AA2024 by sol-gel coatings with cerium molybdate nanowires, Electrochim. Acta 112 (2013) 236–246.
[38] K. Yasakau, et al., Influence of inhibitor addition on the corrosion protection performance of sol-gel coatings on AA2024, Prog. Org. Coat. 63 (3) (2008) 352–361.
[39] K.A. Yasakau, et al., Mechanisms of localized corrosion inhibition of AA2024 by cerium molybdate nanowires, J. Phys. Chem. C 117 (11) (2013) 5811–5823.
[40] A. Subathira, R.M. Meyyappan, Inhibition of corrosion of steel alloy using polyaniline conducting polymer coatings, Int. J. Chem. Sci. 8 (4) (2010) 2563–2574.
[41] V.S. Saji, A review on recent patents in corrosion inhibitors, Recent Pat. Corros. Sci. 2 (2010) 6–12.
[42] M. Rohwerder, Conducting polymers for corrosion protection: a review, Int. J. Mater. Res. 100 (10) (2009) 1331–1342.
[43] N.Y. Abu-Thabit, A.S.H. Makhlouf, Recent advances in polyaniline (PANI)-based organic coatings for corrosion protection, in: A.S.H. Makhlouf (Ed.), Handbook of Smart Coatings for Materials Protection, Woodhead Publishing, United Kingdom (UK), 2014, chapter no 17, pp. 459–486.
[44] B. Wessling, Passivation of metals by coating with polyaniline: corrosion potential shift and morphological changes, Adv. Mater. (Weinheim, Ger.) 6 (1994) 226–228.
[45] B. Wessling, Scientific and commercial breakthrough for organic metals, Synth. Met. 85 (1997) 1313–1318.
[46] U. Rammelt, L.M. Duc, W. Plieth, Improvement of protection performance of polypyrrole by dopant anions, J. Appl. Electrochem. 35 (12) (2005) 1225–1230.
[47] M. Łapkowski, G. Bidan, M. Fournier, Synthesis of polypyrrole and polythiophene in aqueous solution of Keggin-type structure heteropolyanions, Synth. Met. 41 (1–2) (1991) 407–410.
[48] P.J. Kulesza, et al., Polyoxometalates as inorganic templates for monolayers and multilayers of ultrathin polyaniline, Electrochem. Commun. 4 (6) (2002) 510–515.
[49] D. Kowalski, M. Ueda, T. Ohtsuka, The effect of ultrasonic irradiation during electropolymerization of polypyrrole on corrosion prevention of the coated steel, Corros. Sci. 50 (2008) 286–291.
[50] D. Kowalski, M. Ueda, T. Ohtsuka, Self-healing ion-permselective conducting polymer coating, J. Mater. Chem. 20 (36) (2010) 7630–7633.
[51] P. Liu, F. Chen, Conducting polyaniline nanoparticles and their dispersion for waterborne corrosion protection coatings, ACS Appl. Mater. Interfaces 3 (2011) 2694–2702.
[52] M. Hosseini, et al., Effect of polypyrrole-montmorillonite nanocomposites powder addition on corrosion performance of epoxy coatings on Al 5000, Prog. Org. Coat. 66 (3) (2009) 321–327.
[53] C.-W. Peng, et al., Nano-casting technique to prepare polyaniline surface with biomimetic superhydrophobic structures for anticorrosion application, Electrochim. Acta 95 (2013) 192–199.
[54] Y.-H. Yu, et al., High-performance polystyrene/graphene-based nanocomposites with excellent anti-corrosion properties, Polym. Chem. 5 (2) (2014) 535–550.
[55] H. Wei, et al., Anticorrosive conductive polyurethane multiwalled carbon nanotube nanocomposites, J. Mater. Chem. A 1 (36) (2013) 10805–10813.
[56] K.-C. Chang, et al., Synergistic effects of hydrophobicity and gas barrier properties on the anticorrosion property of PMMA nanocomposite coatings embedded with graphene nanosheets, Polym. Chem. 5 (3) (2014) 1049–1056.
[57] D.M. Sim, et al., Ultra-high optical transparency of robust, graded-index, and antifogging silica coating derived from si-containing block copolymers, Adv. Opt. Mater. 1 (6) (2013) 428–433.

[58] J. Xiong, et al., A multifunctional nanoporous layer created on glass through a simple alkali corrosion process, J. Mater. Chem. 20 (45) (2010) 10246–10252.

[59] Y. Wang, et al., Verification of icephobic/anti-icing properties of a superhydrophobic surface, ACS Appl. Mater. Interfaces 5 (8) (2013) 3370–3381.

[60] W. Li, et al., Fabrication and anti-icing behavior of superhydrophobic surfaces on aluminum alloy substrates, Langmuir 29 (27) (2013) 8482–8491.

[61] J. Chen, et al., Robust prototypical anti-icing coatings with a self-lubricating liquid water layer between ice and substrate, ACS Appl. Mater. Interfaces 5 (10) (2013) 4026–4030.

[62] C. Xin, et al., A novel route to prepare weather resistant, durable antireflective films for solar glass, Sol. Energy 93 (2013) 121–126.

[63] M. Sangermano, et al., Multifunctional antistatic and scratch resistant UV-cured acrylic coatings, Prog. Org. Coat. 76 (9) (2013) 1191–1196.

[64] S. Nagappan, et al., Polymethylhydrosiloxane-based organic-inorganic hybrids for amphiphobic coatings, Compos. Interfaces 20 (1) (2013) 33–43.

[65] S. Nagappan, et al., Highly transparent, hydrophobic fluorinated polymethylsiloxane/silica organic-inorganic hybrids for anti-stain coating, Macromol. Res. 21 (6) (2013) 669–680.

[66] C.H.T. Vu, K. Won, Novel water-resistant UV-activated oxygen indicator for intelligent food packaging, Food Chem. 140 (1–2) (2013) 52–56.

[67] C.H.T. Vu, K. Won, Bioinspired molecular adhesive for water-resistant oxygen indicator films, Biotechnol. Prog. 29 (2) (2013) 513–519.

[68] L. Zhu, et al., Ice-phobic coatings based on silicon oil infused polydimethylsiloxane, ACS Appl. Mater. Interfaces 5 (10) (2013) 4053–4062.

[69] L.-P. Xu, et al., An ion-induced low-oil-adhesion organic/inorganic hybrid film for stable superoleophobicity in seawater, Adv. Mater. 25 (4) (2013) 606–611.

[70] X. Li, J. He, Synthesis of raspberry-like SiO_2-TiO_2 nanoparticles toward antireflective and self-cleaning coatings, ACS Appl. Mater. Interfaces 5 (11) (2013) 5282–5290.

[71] V.A. Ganesh, et al., A review on self-cleaning coatings, J. Mater. Chem. 21 (41) (2011) 16304–16322.

[72] F. Hui, C. Debiemme-Chouvy, Antimicrobial N-halamine polymers and coatings: a review of their synthesis, characterisation and applications, Biomacromolecules 14 (3) (2013) 585–601.

[73] S. Forbes, et al., Comparative surface antimicrobial properties of synthetic biocides and novel human apolipoprotein E derived antimicrobial peptides, Biomaterials 34 (22) (2013) 5453–5464.

[74] W. Cloete, L. Verwey, B. Klumperman, Permanently antimicrobial waterborne coatings based on the dual role of modified poly (styrene-co-maleic anhydride), Eur. Polym. J. 49 (5) (2013) 1080–1088.

[75] R. Andre, et al., Self-cleaning antimicrobial surfaces by SnO_2 coatings on glass deposited with surface-bound spermine, Nanoscale 5 (8) (2013) 3447–3456.

[76] G. Broasca, et al., Characterization of ZnO coated polyester fabrics for UV protection, Appl. Surf. Sci. 279 (15) (2013) 272–278.

[77] T. Dispinar, C.A.L. Colard, F.E.D. Prez, Polyurea microcapsules with a photocleavable shell: UV-triggered release, Polym. Chem. 4 (3) (2013) 763–772.

[78] A. Ershad-Langroudi, et al., Adhesion enhancement of 316L stainless steel to acrylic bone cement through nanocomposite sol-gel coating system, Plast. Rubber Compos. 42 (3) (2013) 101–107.

[79] D. Borisova, et al., Nanocontainer-based anticorrosive coatings: effect of the container size on the self-healing performance, Adv. Funct. Mater. 23 (30) (2013) 3799–3812.

[80] D. Borisova, H. Moehwald, D. Shchukin, Influence of embedded nanocontainers on the efficiency of active anticorrosive coatings for aluminum alloys. Part II: influence of nanocontainer position, ACS Appl. Mater. Interfaces 5 (1) (2012) 80–87.

[81] B.K. Green, S. Lowell, Oil-containing microscopic capsules and method of making them, US Patent No. 2800457A, 1957.
[82] A.K. Anal, H. Singh, Recent advances in microencapsulation of probiotics for industrial applications and targeted delivery, Trends Food Sci. Technol. 18 (5) (2007) 240–251.
[83] A.M. Borreguero, et al., Characterization of rigid polyurethane foams containing microencapsulated Rubitherm® RT27: catalyst effect. Part II, J. Mater. Sci. 46 (2) (2011) 347–356.
[84] M. Huang, J. Yang, Facile microencapsulation of HDI for self-healing anticorrosion coatings, J. Mater. Chem. 21 (30) (2011) 11123–11130.
[85] M. Serra, et al., Microencapsulation technology: a powerful tool for integrating expansion and cryopreservation of human embryonic stem cells, PLOS ONE 6 (8) (2011) e23212.
[86] G.Y. Zhu, et al., Fragrance and flavor microencapsulation technology, Adv. Mater. Res. 535 (2012) 440–445.
[87] S.V. Bhujbal, P. de Vos, S.P. Niclou, Drug and cell encapsulation: alternative delivery options for the treatment of malignant brain tumors, Adv. Drug Deliv. Rev. 67 (2014) 142–153.
[88] I.M. Martins, et al., Microencapsulation of essential oils with biodegradable polymeric carriers for cosmetic applications, Chem. Eng. J. 245 (2014) 191–200.
[89] D.O. Grigoriev, et al., New method for fabrication of loaded micro-and nanocontainers: emulsion encapsulation by polyelectrolyte layer-by-layer deposition on the liquid core, Langmuir 24 (3) (2008) 999–1004.
[90] D.O. Grigoriev, et al., Emulsion route in fabrication of micro and nanocontainers for biomimetic self-healing and self-protecting functional coatings, Bioinspired Biomimetic Nanobiomater. 1 (2) (2012) 101–116.
[91] D. Yin, et al., Pickering emulsion: a novel template for microencapsulated phase change materials with polymer-silica hybrid shell, Energy 64 (2014) 575–581.
[92] D. Simón, et al., The effect of the dry glass transition temperature on the synthesis of paraffin microcapsules obtained by suspension-like polymerization, Polym. Eng. Sci. 54 (1) (2014) 208–214.
[93] T. Suzuki, A. Osumi, H. Minami, One-step synthesis of "rattle-like" polymer particles via suspension polymerization, Chem. Commun. 50 (69) (2014) 9921–9924.
[94] J.-S. Cho, A. Kwon, C.-G. Cho, Microencapsulation of octadecane as a phase-change material by interfacial polymerization in an emulsion system, Colloid Polym. Sci. 280 (3) (2002) 260–266.
[95] M. Kobašlija, D.T. McQuade, Polyurea microcapsules from oil-in-oil emulsions via interfacial polymerization, Macromolecules 39 (19) (2006) 6371–6375.
[96] D. Liu, X. Jiang, J. Yin, One-step interfacial thiol-ene photopolymerization for metal nanoparticle decorated microcapsules (MNP@ MCs), Langmuir 30 (24) (2014) 7213–7220.
[97] G. Wu, et al., Robust microcapsules with polyurea/silica hybrid shell for one-part self-healing anticorrosion coatings, J. Mater. Chem. A 2 (30) (2014) 11614–11620.
[98] R. Arshady, Microspheres and microcapsules: a survey of manufacturing techniques. Part 1: suspension cross-linking, Polym. Eng. Sci. 29 (24) (1989) 1746–1758.
[99] K.-D. Yao, et al., Microcapsules/microspheres related to chitosan, J. Macromol. Sci. Polym. Rev. 35 (1) (1995) 155–180.
[100] S. Freitas, H.P. Merkle, B. Gander, Microencapsulation by solvent extraction/evaporation: reviewing the state of the art of microsphere preparation process technology, J. Control. Release 102 (2) (2005) 313–332.
[101] R. Dubey, Microencapsulation technology and applications, Def. Sci. J. 59 (1) (2009) 82–95.

[102] W. Wang, et al., Self-healing properties of protective coatings containing isophorone diisocyanate microcapsules on carbon steel surfaces, Corros. Sci. 80 (2014) 528–535.
[103] M. Huang, H. Zhang, J. Yang, Synthesis of organic silane microcapsules for self-healing corrosion resistant polymer coatings, Corros. Sci. 65 (2012) 561–566.
[104] M. Huang, J. Yang, Long-term performance of 1H, 1H′, 2H, 2H′-perfluorooctyl triethoxysilane (POTS) microcapsule-based self-healing anticorrosive coatings, J. Intell. Mater. Syst. Struct. 25 (1) (2014) 98–106.
[105] J. Fu, et al., Acid and alkaline dual stimuli-responsive mechanized hollow mesoporous silica nanoparticles as smart nanocontainers for intelligent anticorrosion coatings, ACS Nano 7 (12) (2013) 11397–11408.
[106] M.L. Zheludkevich, J. Tedim, M.G.S. Ferreira, "Smart" coatings for active corrosion protection based on multi-functional micro and nanocontainers, Electrochim. Acta 82 (1) (2012) 314–323.
[107] S.H. Sonawane, et al., Improved active anticorrosion coatings using layer-by-layer assembled ZnO nanocontainers with benzotriazole, Chem. Eng. J. (Amsterdam, Neth.) 189–190 (2012) 464–472.
[108] M.F. Montemor, et al., Evaluation of self-healing ability in protective coatings modified with combinations of layered double hydroxides and cerium molibdate nanocontainers filled with corrosion inhibitors, Electrochim. Acta 60 (2012) 31–40.
[109] F. Maia, et al., Silica nanocontainers for active corrosion protection, Nanoscale 4 (4) (2012) 1287–1298.
[110] I.A. Kartsonakis, et al., Hybrid organic-inorganic multilayer coatings including nanocontainers for corrosion protection of metal alloys, Corros. Sci. 57 (2012) 56–66.
[111] I.A. Kartsonakis, et al., Incorporation of ceramic nanocontainers into epoxy coatings for the corrosion protection of hot dip galvanized steel, Corros. Sci. 57 (2012) 30–41.
[112] M.F. Haase, et al., Development of nanoparticle stabilized polymer nanocontainers with high content of the encapsulated active agent and their application in water-borne anticorrosive coatings, Adv. Mater. 24 (18) (2012) 2429–2435.
[113] T. Chen, J. Fu, An intelligent anticorrosion coating based on pH-responsive supramolecular nanocontainers, Nanotechnology 23 (50) (2012) 505705.
[114] J. Tedim, et al., Enhancement of active corrosion protection via combination of inhibitor-loaded nanocontainers, ACS Appl. Mater. Interfaces 2 (5) (2010) 1528–1535.
[115] D.G. Shchukin, D.O. Grigoriev, H. Möhwald, Application of smart organic nanocontainers in feedback active coatings, Soft Matter 6 (4) (2010) 720–725.
[116] E. Abdullayev, et al., Halloysite tubes as nanocontainers for anticorrosion coating with benzotriazole, ACS Appl. Mater. Interfaces 1 (7) (2009) 1437–1443.
[117] M.L. Zheludkevich, et al., Anticorrosion coatings with self-healing effect based on nanocontainers impregnated with corrosion inhibitor, Chem. Mater. 19 (3) (2007) 402–411.
[118] M.L. Zheludkevich, et al., Active corrosion protection based on nanocontainers with corrosion inhibitors, Chem. Mater. 19 (2006) 402–411.
[119] D.G. Shchukin, M. Zheludkevich, H. Möhwald, Feedback active coatings based on incorporated nanocontainers, J. Mater. Chem. 16 (47) (2006) 4561–4566.
[120] D.G. Shchukin, et al., Active anticorrosion coatings with halloysite nanocontainers, J. Phys. Chem. C 112 (4) (2008) 958–964.
[121] D.G. Shchukin, H. Möhwald, Smart nanocontainers as depot media for feedback active coatings, Chem. Commun. 47 (31) (2011) 8730–8739.
[122] M.L. Zheludkevich, et al., Active protection coatings with layered double hydroxide nanocontainers of corrosion inhibitor, Corros. Sci. 52 (2) (2010) 602–611.
[123] T. Chen, J. Fu, pH-responsive nanovalves based on hollow mesoporous silica spheres for controlled release of corrosion inhibitor, Nanotechnology 23 (23) (2012) 235605.

[124] S. Newman, Synthesis, characterization and applications of layered double hydroxides containing organic guests, New J. Chem. 22 (2) (1998) 105–115.
[125] M. Shao, et al., Hierarchical structures based on functionalized magnetic cores and layered double-hydroxide shells: concept, controlled synthesis, and applications, Chem. Eur. J. 19 (13) (2013) 4100–4108.
[126] J.-H. Choy, et al., Layered double hydroxide as an efficient drug reservoir for folate derivatives, Biomaterials 25 (15) (2004) 3059–3064.
[127] F.-J. Xu, et al., Functionalized layered double hydroxide nanoparticles conjugated with disulfide-linked polycation brushes for advanced gene delivery, Bioconjug. Chem. 24 (6) (2013) 968–978.
[128] U. Riaz, M. Ashraf, Double layered hydroxides as potential anti-cancer drug delivery agents, Mini Rev. Med. Chem. 13 (4) (2013) 522–529.
[129] L. Li, et al., Co-delivery of siRNAs and anti-cancer drugs using layered double hydroxide nanoparticles, Biomaterials 35 (10) (2014) 3331–3339.
[130] R.G. Buchheit, et al., Active corrosion protection and corrosion sensing in chromate-free organic coatings, Prog. Org. Coat. 47 (3–4) (2003) 174–182.
[131] Z. Liu, et al., Synthesis, anion exchange, and delamination of Co-Al layered double hydroxide: assembly of the exfoliated nanosheet/polyanion composite films and magneto-optical studies, J. Am. Chem. Soc. 128 (14) (2006) 4872–4880.
[132] B. Chico, et al., Anticorrosive behaviour of alkyd paints formulated with ion-exchange pigments, Prog. Org. Coat. 61 (2–4) (2008) 283–290.
[133] F. Zhang, et al., Corrosion resistance of superhydrophobic layered double hydroxide films on aluminum, Angew. Chem. Int. Ed. 47 (13) (2008) 2466–2469.
[134] S.K. Poznyak, et al., Novel inorganic host layered double hydroxides intercalated with guest organic inhibitors for anticorrosion applications, ACS Appl. Mater. Interfaces 1 (10) (2009) 2353–2362.
[135] J. Hu, et al., Synthesis and anticorrosive properties of polymer-clay nanocomposites via chemical grafting of polyaniline onto Zn-Al layered double hydroxides, Surf. Coat. Technol. 240 (2013) 55–62.
[136] D. Snihirova, et al., Hydroxyapatite microparticles as feedback-active reservoirs of corrosion inhibitors, ACS Appl. Mater. Interfaces 2 (11) (2010) 3011–3022.
[137] A. Karatzas, et al., Reversible spherical organic water microtraps, J. Non-Cryst. Solids 358 (2) (2012) 443–445.
[138] K. Skorb, D. Andreeva, Layer-by-Layer approaches for formation of smart self-healing materials, Polym. Chem. 4 (18) (2013) 4834–4845.
[139] X. He, et al., Nanostructured photocatalytic coatings for corrosion protection and surface repair, Surf. Coat. Technol. 237 (2013) 320–327.
[140] M. Plawecka, et al., Self healing ability of inhibitor-containing nanocapsules loaded in epoxy coatings applied on aluminium 5083 and galvanneal substrates, Electrochim. Acta 140 (2014) 282–293.

CHAPTER 25

Synthesis, Characterization, and Development of PZT-Based Composition for Power Harvesting and Sensors Application

Chandrashekhar Madhav Lonkar, Dinkar Kisanrao Kharat, Sahab Prasad, Balasubramanian Kandasubramanian
Department of Materials Engineering, Defence Institute of Advanced Technology (Deemed University), Pune, India

Chapter Contents

Handbook of Nanoceramic and Nanocomposite Coatings and Materials
http://dx.doi.org/10.1016/B978-0-12-799947-0.00025-0

1. INTRODUCTION

Lead zirconate titanate (PZT)-based piezoceramics are of technological interest to design various sensors and actuators as they offer excellent piezoelectric properties [1]. Optimum piezoelectric properties can be obtained when Zr/Ti ratio is about 52/48, which corresponds to the composition of the ferroelectric morphotropic phase boundary (MPB) separating the ferroelectric tetragonal (F_T) and ferroelectric rhombohedral (F_R) phases [2]. Since PZT-based piezoelectric ceramics are excellent electromechanical energy converters, they can produce electrical output in response to ambient energy, vibrations, movements, etc. [3]. It could be stored and subsequently used for operation of low-electric power devices. This can offer the solution to the "limited battery life" problem of the modern low-power devices like embedded structural health monitoring (SHM) system and microelectromechanical systems (MEMSs) where the replacing of exhausted battery is very expensive or impossible sometimes under very specific conditions [4].

The electrical output obtained from piezoceramic is determined mainly by the product of piezoelectric charge constant and voltage constant ($d_{33} \times g_{33}$) known as figure of merit for power harvesting (FoM_{PH}). Material with higher d_{33} and g_{33} leads to higher $d_{33} \times g_{33}$ that offers higher electrical output and hence is the essential criteria for the material intended to be used for energy harvesting applications [5–8]. This condition is satisfied when piezoelectric charge constant (d_{33}) and voltage constant (g_{33}) are on higher side and dielectric constant (K_3^T) is on lower side. Since electromechanical coupling factor (k_p) represents the energy conversion efficiency of the material, its higher value is preferred. Moreover, the higher the piezoelectric voltage constant (g_{33}), the higher is the sensitivity of the material making it suitable for sensor application [9].

The aim of the present work was to synthesize, characterize, and optimize the PZT-based composition with advanced properties suitable for power harvesting and sensor applications.

Compositions represented by stoichiometric formula $Pb_{1-z}La_z(NiSb)_x[(Zr_{1-y}Ti_y)_{1-z/4}]_{1-x}O_3$ were investigated for suitability for the intended purpose. Efforts were taken to improve properties for power harvesting applications. This work focuses on the following aspects:

(i) Processing of the said compositions by mixed oxide route
(ii) Preparation of $NiSb_2O_6$
(iii) Preparation of PZT-based $Pb_{1-z}La_z(NiSb)_x[(Zr_{1-y}Ti_y)_{1-z/4}]_{1-x}O_3$ compositions

(iv) Achieving the intended properties through optimization of
- Zr/Ti ratio,
- NS concentration,
- La concentration

(v) Characterization of compositions for
- crystal structure,
- microstructure, and
- electromechanical properties

(vi) Evaluation of optimized PNS-PZT composition as a power harvester and sensor (hydrophone) and comparison of its performance with commercially available PZT type 5A

2. SYNTHESIS OF COMPOSITIONS

The compositions were synthesized using the oxides of elements in the powder form. The raw materials used were NiO (Acros, 97%), Sb_2O_5 (Loba Chemie, 99%), PbO (Waldies Ltd., Kolkata, 99.5%), ZrO_2 (Loba Chemie, 99.37%), TiO_2 (Travancore Titanium Products, 98.5%), and La_2O_3 (Indian Rare Earths Ltd., 99.99%).

Calcination is one of the important stages of mixed oxide route described below. It is essential to determine the calcination temperature since solid-state reactions takes place at this temperature. Desired compositions were prepared by two-stage calcinations. In the first stage, $NiSb_2O_6$ (NS) was prepared, and in the second stage, Pb(NiSb)(ZrTi) O_3 (i.e., PNS-PZT) was prepared using NS.

2.1 Synthesis of NS

Calcination temperature of *mixed and milled* NS powder was determined through SDTA-TGA studies. Phase formation in calcination was confirmed from X-ray diffraction (XRD) pattern.

As shown in Figure 1, up to 350 °C obtained peaks are associated with slight weight loss, which corresponds to loss of water [10]. Water molecules adsorbed on the surface of the particle evaporate at lower temperatures, while the molecules entrapped, during the milling, require larger temperatures [10,11]. Around 380-400 °C, sharp endothermic peak without any weight loss was observed, which corresponds to melting of Sb_2O_5. Small exothermic peak followed by significant endothermic peak was observed from 700 to 740 °C. At this stage, no significant weight loss was observed. The endothermic peak in DTA may be due to transformations into

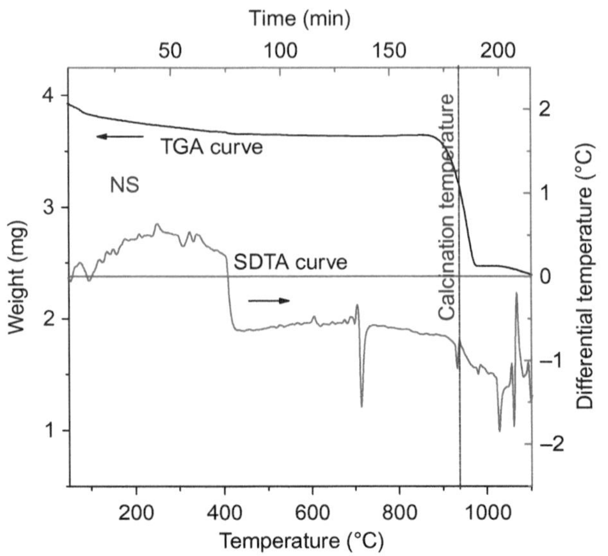

Figure 1 SDTA-TGA profile for NS.

metastable phases, which formed prior to the formation of nickel antimonate [12,13]. The fact was confirmed by the XRD pattern of the powder calcined at 740 °C, which does not show the formation of NS. An endothermic peak along with some weight loss is observed from 920 to 945 °C, which corresponds to the formation of NS as observed in the XRD patterns recorded for the powder calcined at 945 °C. It was confirmed by comparing the XRD patterns with ICCD card no 86-110 (Figure 2a and b).

2.2 Synthesis of PNS-PZT

Calcination temperature of *mixed and milled* powder of composition $Pb(NiSb)_{0.02}[(Zr_{0.52}Ti_{0.48})]_{0.98}O_3$ was determined from SDTA-TGA studies. Desired phase formation in calcination was confirmed from XRD pattern.

As shown in Figure 3, an exothermic peak observed around 600 °C, which is associated without any weight loss, corresponds to the formation of $PbTiO_3$, and small exothermic peak followed by significant endothermic peak seen from 690 to 730 °C without significant weight loss corresponds to the formation of $PbZrO_3$ [2,14]. A long endothermic peak observed from 950 to 1050 °C without significant weight loss corresponds to transformations into metastable phases, which formed prior to the formation PNS-PZT, which took place around 1060 °C [12,13]. The formation of perovskite PNS-PZT at 1060 °C was confirmed from XRD pattern indicated by intense (110) peak and other perovskite peaks shown in Figure 4.

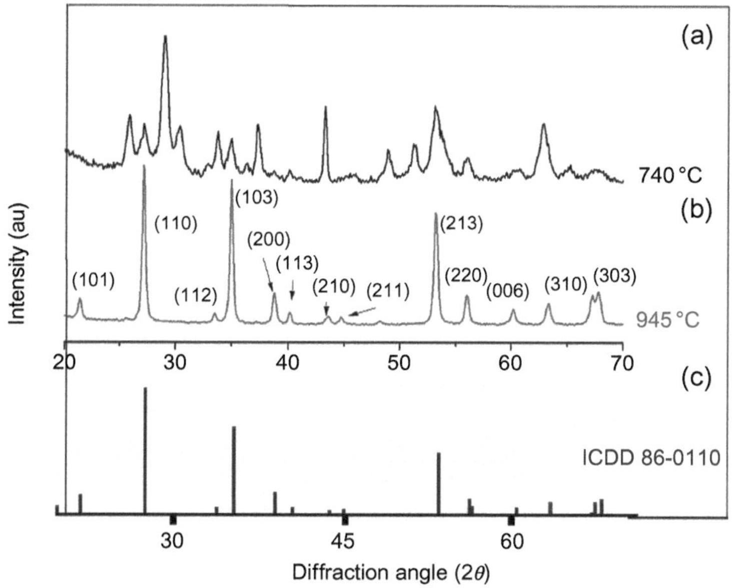

Figure 2 XRD pattern of NS calcined at (a) 740 °C and (b) 945 °C. (c) Standard XRD for NS (ICDD 86-110).

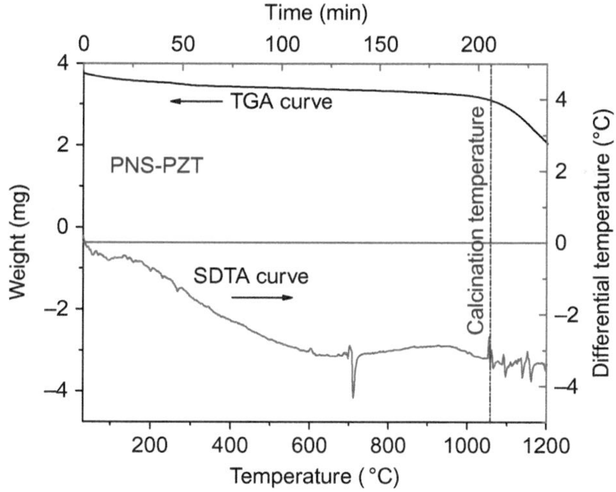

Figure 3 SDTA-TGA profile for PNS-PZT.

PNS-PZT compositions were prepared and processed through mixed oxide route. To compensate the lead loss during calcination, 0.5 mol% excess quantity of PbO was taken. Powders were wet milled for 24 h in pure water medium (Millipore, Elix-10), and zirconia balls were used as grinding

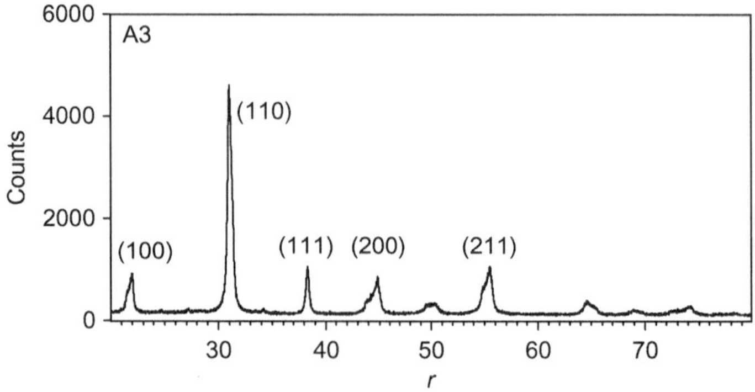

Figure 4 XRD pattern of PNS-PZT calcined at 1060 °C.

media. Solid-state reaction was carried out at 1060 °C, and further, powder was ultrafine ground to 1.0–1.2 μm. The compositional effect on crystal structure was studied from XRD patterns recorded for position 2θ from 20° to 70° using Philips X'pert Pro. The powders were compacted by semi-automatic compaction machine (GMT,India) to form disks of Ø 17.2 mm × 1.7 mm thickness maintaining green density about 4.8 g/cc. About 6-9 specimens were sintered, lapped, and subsequently electroded with silver paste. Poling was performed by applying DC field of 35 kV/cm at 100 °C for 30 min. Specimens were aged for 10 days and characterized for piezoelectric and dielectric properties. Capacitance (C), tan δ, impedance (Z_m), resonance frequency (f_r), and antiresonance frequency (f_a) were measured using variable frequency LCR meter (Hioki Hi Tester model 3532). Piezoelectric charge constant (d_{33}) was measured using d_{33} Berlincourt meter (CPDT 3330). Hysteretic studies and response to cyclic electrical loading of La-doped compositions were assessed using TF Analyzer 2000, aixACCT Systems.

The following standard relations were used to calculate K_3^T, g_{33}, k_p, and Q_m [15].

$$K_3^T = \frac{Ct}{\varepsilon_0 A} \tag{1}$$

$$g_{33} = \frac{d_{33}}{K_3^T \varepsilon_0} \tag{2}$$

$$k_p = \sqrt{2.51 \frac{f_a - f_r}{f_r}} \tag{3}$$

$$Q_m = \frac{1}{2\pi f_r Z_m C}\left(\frac{f_a^2}{f_a^2 - f_r^2}\right) \tag{4}$$

where K_3^T is the dielectric constant, C is the capacitance (F), ε_0 is the permittivity of free space (8.854×10^{-12} F/m), t is the separation of electrodes (m), A is the area of electrode (m^2), g_{33} is the piezoelectric voltage constant (V m/N), d_{33} is piezoelectric charge constant (C/N), Z_m is the impedance, f_r is the resonance frequency (kHz), and f_a is the antiresonance frequency (kHz).

Microstructure analysis of chemically etched polished samples was carried out using a scanning electron microscope (FEI Quanta 200 ESEM model).

To develop the composition of desired properties, following compositions were investigated and optimized for product of piezoelectric charge constant and voltage constant ($d_{33} \times g_{33}$):

(a) *Optimization of Zr/Ti ratio*: $Pb(Ni_{1/3}Sb_{2/3})_{0.02}[(Zr_{1-y}Ti_y)]_{0.98}O_3$ compositions were investigated with y=0.46-0.50, that is, Zr/Ti from 54/46 to 50/50. Samples were sintered at 1280 °C for 60 min.

(b) *Optimization of NS concentration*: $Pb(Ni_{1/3}Sb_{2/3})_x[(Zr_{0.52}Ti_{0.48})]_{1-x}O_3$ compositions were investigated with x=0.03-0.07. Samples were sintered at 1280 °C for 60 min.

(c) *Optimization of La concentration*: $Pb_{1-z}La_z(NiSb)_{0.05}[(Zr_{0.52}Ti_{0.48})_{1-z/4}]_{0.95}O_3$ compositions were investigated with z=0.01-0.05. Samples were sintered at 1270 °C for 60 min.

Compositions were characterized for crystal structure, microstructure, and electromechanical properties. Performance of the optimized composition was evaluated as a power harvester and sensor and compared with PZT type 5A.

3. EFFECT OF Zr/Ti RATIO AND NS CONCENTRATION

Effect of Zr/Ti ratio and NS on crystal structure, microstructure, and electromechanical properties of PNS-PZT compositions and their optimization will be discussed here.

3.1 Effect on Crystal Structure

The effect of Zr/Ti ratio and NS on crystal structure of PNS-PZT compositions was analyzed from XRD patterns of calcined powders. All the patterns show the formation of perovskite phase, indicated by the perovskite planes indexed and particularly by the intense peak for (110) plane [13,15,16]. The splitting in the peak intensity at triplet (200), (210), and

(211) indicates the presence of both ferroelectric tetragonal (F_T) and ferroelectric rhombohedral (F_R) phases [16–20]. When Zr/Ti ratio was reduced, the splitting of peak intensity at (200), (210), and (211) was predominant, indicating the increase in tetragonality, which was confirmed from the calculations of *c*/*a* ratio from XRD data (Table 1). It was also observed that reduction in Zr/Ti ratio reduced the unit cell parameter "*a*" and increased "*c*." This has net effect on increasing the *c*/*a* ratio and thus the tetragonality. As the NS was increased, the splitting of peak intensity at (200), (210), and (211) was reduced, indicating the reduction in tetragonality. This was confirmed from the calculations of *c*/*a* ratio from XRD data (Table 2). It was also observed that increase in NS increased the unit cell parameter "*a*" and decreased "*c*." This has net effect on decreasing the *c*/*a* ratio and thus the tetragonality.

3.2 Effect on Microstructure

Effect of Zr/Ti was analyzed from SEM. Average grain size was measured using the "line intercept method." In case of Zr/Ti ratio 52/48, narrow grain size distribution and dense microstructure with minimum porosity were seen. Most of the grains were polygonal in shape. The average grain size measured was about 1.85 μm. Over all, it was observed that the grain growth was promoted as the Zr/Ti ratio was reduced.

Table 1 Effect of Zr/Ti ratio on lattice parameters

Zr/Ti	*a* (Å)	*c* (Å)	*c*/*a*
54/46	4.0428	4.1127	1.0173
53/47	4.0388	4.1173	1.0194
52/48	4.0357	4.1210	1.0211
51/49	4.0313	4.1267	1.0237
50/50	4.0278	4.1288	1.0251

Table 2 Effect of NS on lattice parameters

NS (mol%)	*a* (Å)	*c* (Å)	*c*/*a*
3	4.0342	4.1157	1.0202
4	4.0374	4.1120	1.0185
5	4.0385	4.1069	1.0169
6	4.0394	4.1017	1.0154
7	4.0388	4.0976	1.0146

As shown in Figure 5a–e, where NS increased from 3 to 5 mol%, there was no remarkable grain growth. Grains increased slightly in size with increased NS concentration. For 3 mol% of NS, average grain size was about 1.71 μm; further, it increased to 1.82 μm for 4 mol% of NS. The microstructure at 5 mol% of NS seems to be denser compared with earlier cases having average grain size of 1.86 μm. On further increasing, the NS grains increased largely in size along with wider grain size distribution, and the presence of few pores was also noticed. The average grain size measured was 2.28 and 3.8 μm, respectively, for 6 and 7 mol% of NS.

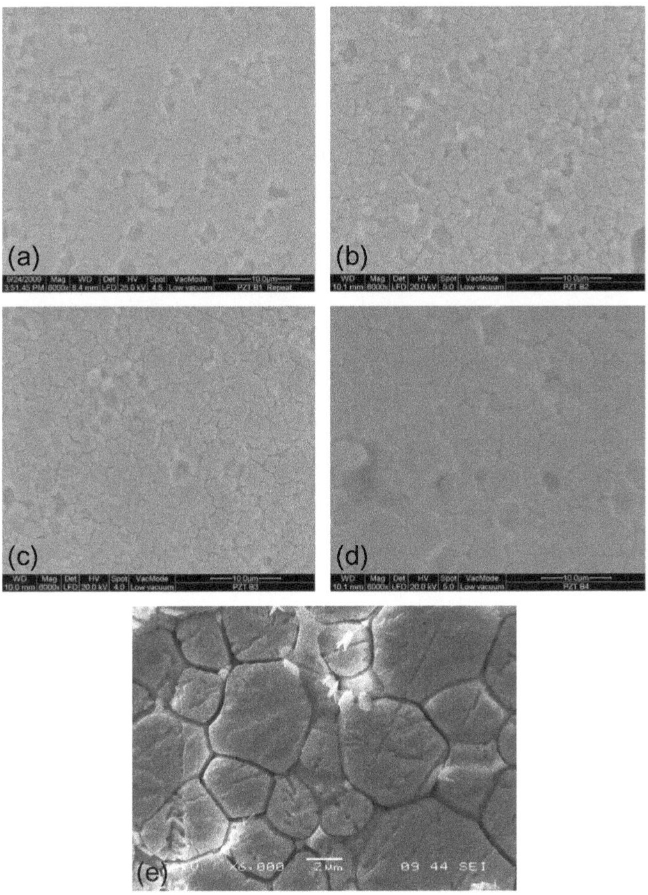

Figure 5 Effect of NS concentration on microstructure (a) 3 mol%, (b) 4 mol%, (c) 5 mol%, (d) 6 mol%, and (e) 7 mol%.

3.3 Effect on Electromechanical Properties

The composition $Pb(Ni_{1/3}Sb_{2/3})_{0.02}[(Zr_{1-y}Ti_y)]_{0.98}O_3$ has maximum polarizability when Zr/Ti ratio was 52/48 at which dense and uniform microstructure was obtained. Maximum polarizability has resulted into minimum Q_m and maximum k_p (Figure 6a) [21]. Addition of NS, up to 5 mol%, shows similar trend like Zr/Ti ratio. Q_m increased from 174 to 200, while k_p decreased from 0.58 to 0.54. A further increase of the NS is shown in Figure 6b. As shown in Figure 6c, with the decrease in the Zr/Ti ratio, d_{33} initially increase till the

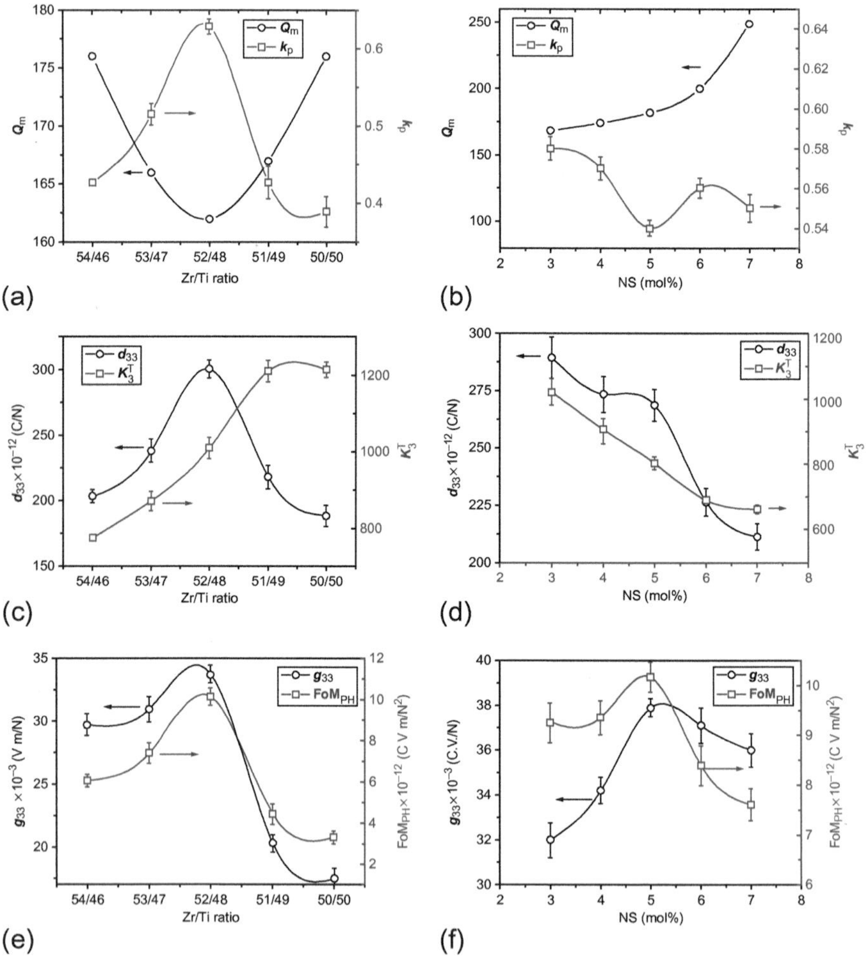

Figure 6 (a) Effect of Zr/Ti on Q_m and k_p, (b) effect of NS on Q_m and k_p, (c) effect of Zr/Ti on d_{33} and K_3^T, (d) effect of NS on d_{33} and K_3^T, (e) effect of Zr/Ti on g_{33} and FoM_{PH}, and (f) effect of NS on g_{33} and FoM_{PH}.

Zr/Ti ratio was 52/48. At this ratio, the optimum value of d_{33} (301×10^{-12} C/N) was obtained. This was attributed to the reduction in the porosity. The presence of pores decreases the polarization per unit volume [20,21], which leads to decrease in d_{33}. On further decreasing the Zr/Ti ratio to 50/50 in steps, d_{33} was decreased attributed to the increase in porosity, while K_3^T increased with decrease in Zr/Ti. As g_{33} is directly proportional to d_{33} and was optimum (33.7×10^{-3} V m/N) for 52/48 ratio (Figure 6e). This indicates that material has higher voltage-producing ability and higher sensitivity for Zr/Ti ratio 52/48. Optimum d_{33} and g_{33} resulted into optimum figure of merit. As shown in Figure 6f, with the increase in NS concentration, tetragonality decrease resulted in decrease in K_3^T. There is lesser decrease in d_{33} from 4 to 5 mol% of NS compared to K_3^T. This led to higher g_{33} and FoM_{PH} for 5 mol% NS.

3.4 Conclusions

1. Perovskite (ABO_3) phase with coexistence of F_T and F_R phases formed in compositions studied for optimization of Zr/Ti ratio and NS.
2. On reducing the Zr/Ti ratio, tetragonality has increased; on increasing NS, it has decreased, which is indicated by splitting of (200) planes.
3. Dielectric constant has shown linear corelationship with tetragonality.
4. Sintering temperature of 1280 °C and sintering time of 60 min are parameters for the optimized compositions:
   - $Pb(Ni_{1/3}Sb_{2/3})_{0.02}[(Zr_{1-y}Ti_y)]_{0.98}O_3$ with $y=0.48$
   - $Pb(Ni_{1/3}Sb_{2/3})_x[(Zr_{0.52}Ti_{0.48})]_{1-x}O_3$ with $x=0.05$

   For these compositions, the obtained microstructure is dense with a minimum porosity and narrow grain size distribution.
5. The optimum grain size for the compositions was about 1.85 μm for the above compositions where optimum results were obtained.
6. The optimum properties obtained for the composition—$Pb(Ni_{1/3}Sb_{2/3})_{0.02}[(Zr_{1-y}Ti_y)]_{0.98}O_3$ with $y=0.48$, that is, Zr/Ti=52/48—are
   - $d_{33}=301 \times 10^{-12}$ C/N,
   - $g_{33}=33.7 \times 10^{-3}$ V m/N,
   - $FoM_{PH}(d_{33} \times g_{33})=12.1 \times 10^{-12}$ C V m/N^2,
   - $k_p=0.63$,
   - $K_3^T=1010$.
7. The optimum properties obtained for the composition—$Pb(Ni_{1/3}Sb_{2/3})_x[(Zr_{0.52}Ti_{0.48})]_{1-x}O_3$ with $x=0.05$—are
   - $d_{33}=269 \times 10^{-12}$ C/N,
   - $g_{33}=37.9 \times 10^{-3}$ V m/N,

- $FoM_{PH}(d_{33} \times g_{33}) = 12.2 \times 10^{-12}$ C V m/N^2,
- $k_p = 0.54$,
- $K_3^T = 801$.

4. EFFECT OF LANTHANUM CONCENTRATIONS

Effect of La concentrations on crystal structure, microstructure electromechanical properties, phase transition, hysteretic behavior, and behavior of materials under electric cyclic loads will be discussed here.

When La^{3+} ($r = 1.22$ Å) is substituted in PZT composition, it replaces Pb^{2+} ($r = 1.32$ Å) [22] ions in the lattice according to the relation $Pb^{2+} \leftrightarrow La^{3+} + e^-$. To maintain the electrical neutrality, lead vacancies are created in the lattice, making domain wall motion comparatively easier, which assists the domains to align in the direction of applied DC field during poling process. This increases dipole moment [23] and in turn results in the enhanced piezoelectric properties of the ceramic [24,25] like d_{33}, k_p, remnant polarization (P_r), and coupling factor (k_p) [24–26].

However, modification of PZT ceramics by lanthanum creates the defects in the lattice, leading to disturbance in the long-range polar order by weakening the ferroelectric interactions causing reduction in grain size, domain size, and tetragonality [27–29]. The normal micrometer-sized ferroelectric domains sustain up to critical value of lanthanum concentration [30]. Above critical value, instead of micrometer-sized, polar nanodomains are stabilized due to strong decoupling effects, and hence, the macroscopic properties change from normal ferroelectric to relaxor-type ferroelectric [29] and shows a diffuse phase transition (DPT) [31,32] resulting in increase in dielectric constant (K_3^T) and lowering piezoelectric constant (d_{33}) [16,18].

4.1 Effect on Crystal Structure

XRD patterns of calcined powder are shown in Figure 7 for the compositions $Pb_{1-z}La_z(NiSb)_{0.05}[(Zr_{0.52}Ti_{0.48})_{1-z/4}]_{0.95}O_3$ with $z = 0.01$-0.05. XRD patterns show the polycrystalline nature. Formation of PZT perovskite phase for all compositions is indicated by intense peaks indexed and particularly by the highest peak intensity at (110) plane [14,16,17]. Coexistence of ferroelectric tetragonal (F_T) and ferroelectric rhombohedral (F_R) phases was indicated by the splitting in the peak intensity at triplet (200), (201), and (211) [17–21]. However, peaks of PbO were also seen [17]. It can be seen that as the La concentration is increased from 1 to 2 mol%, splitting in peak intensity at (200) increases indicating increase in c/a ratio of

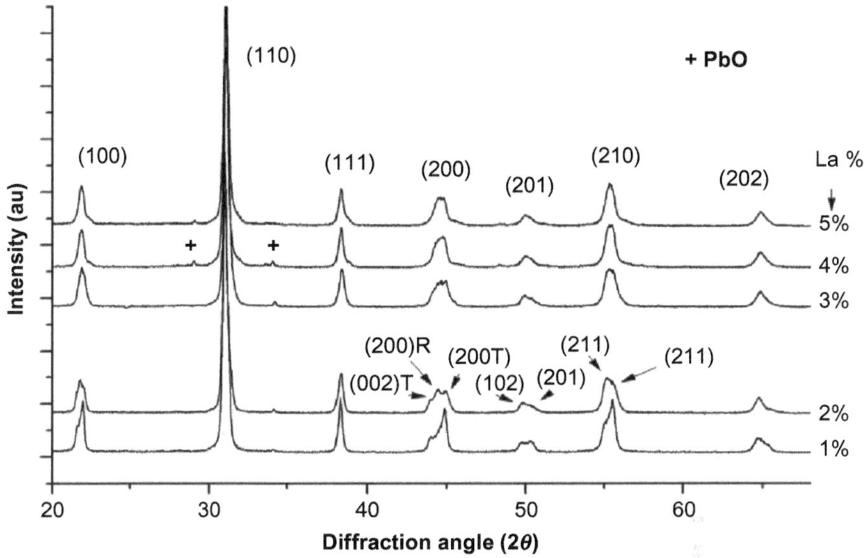

Figure 7 XRD patterns for La-doped PNS-PZT.

tetragonal unit cell, that is, tetragonality. Furthermore, the intensity of the peaks for 2 mol% of lanthanum is approximately the same indicating the approximately equal content of F_T and F_R phases, that is, composition in the vicinity of MPB. At 3 mol% of La, tetragonality is slightly reduced. Further increase in La to 4 and 5 mol% reduces the splitting indicating reduction in c/a ratio and thus reveals that tetragonal phase may be approaching to cubic. Furthermore, patterns show the presence of perovskite phase only. Compositions obtained were free from pyrochlore phase ($Pb_2Sb_2O_7$) as the pyrochlore-related peaks of (222) plane at about $2\theta = 30°$ and (440) and (622) nearly about $2\theta = 49°$ and 58°, respectively, were not observed in the diffraction patterns [33,34].

4.2 Effect on Microstructure

It is known that electromechanical properties are greatly influenced by the grain size [22,35]. Similarly, dense microstructure with optimum grain size results in better piezoelectric properties [36]. Over all, it was observed that the grain growth was retarded as the La concentration was increased (Figure 8). The grain size was measured by linear intercept method at least at four places, and the average was about 7.05 ± 0.1 μm, 6.45 ± 0.1 μm,

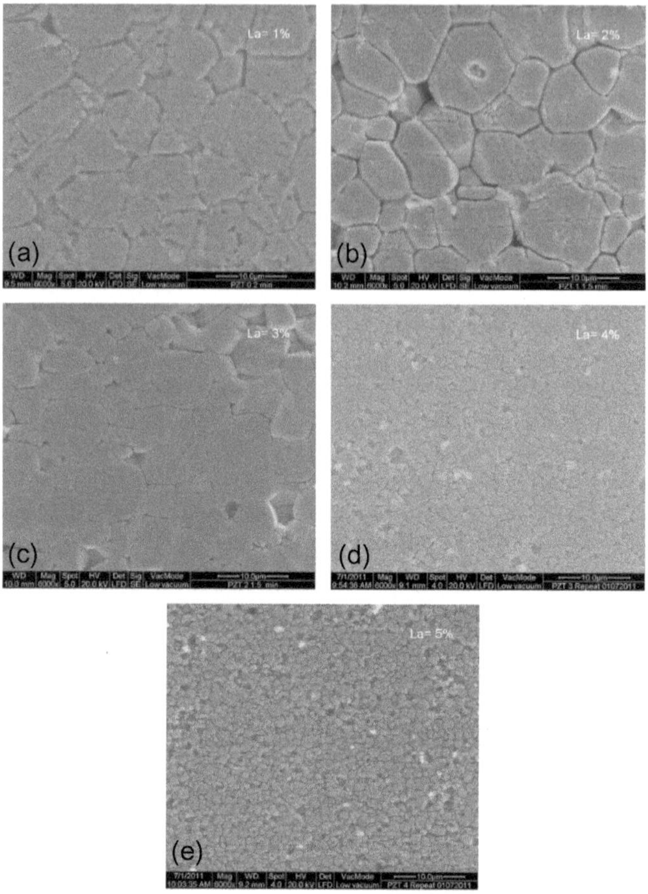

Figure 8 Microstructure of samples sintered at 1270 °C for 60 min with (a) La 1 mol%, (b) La 2 mol%, (c) La 3 mol%, (d) La 4 mol%, and (e) La 5 mol%.

5.35 ± 0.12 μm, 1.2 ± 0.05 μm, and 1 ± 0.05 μm, respectively, for 1-5 mol% of lanthanum concentrations. There is drastic reduction in grain size after 3 mol% of lanthanum.

4.3 Effect on Electromechanical Properties

Influence of La concentration on density and average grain size is shown in Figure 9a.

Density increases while average grain size decreases with increase in La concentration. In fact, decrease in grain size increases grain boundaries, and hence, decrease in density was expected. However, increase in density with reduction in c/a ratio supports the approaching of tetragonal in cubic phase.

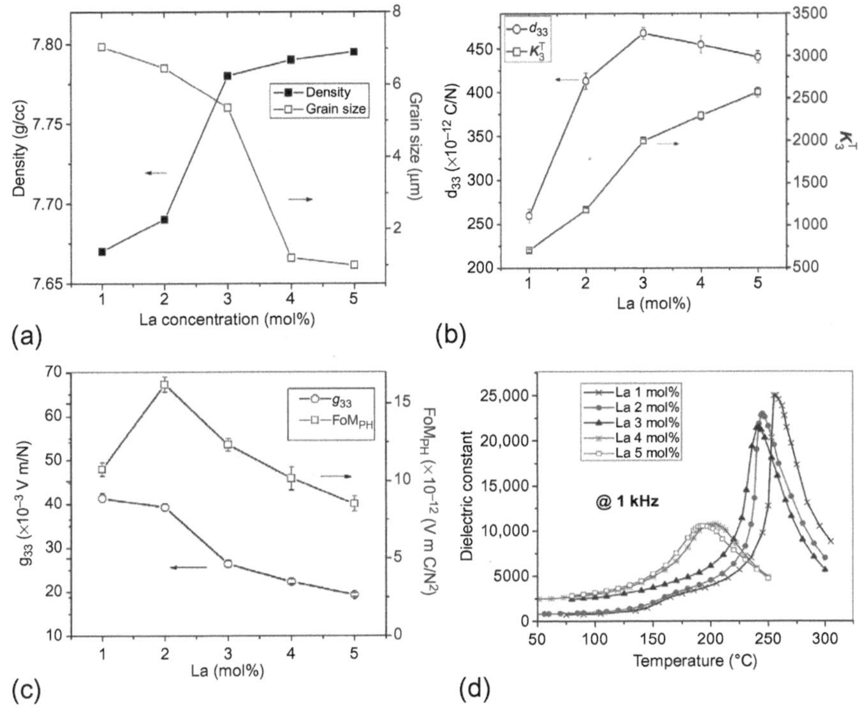

Figure 9 (a) Effect of La on density and grain size, (b) effect of La on d_{33} and K_3^T, (c) effect of La on g_{33} and FoM_{PH}, and (d) effect of La on phase transition.

Effect of La concentration on d_{33} and K_3^T is shown in Figure 9b. Denser microstructure with optimum polarizability for the composition containing 3 mol% of La concentration resulted in optimum d_{33} values. K_3^T value increased with increased La concentrations.

Figure 9c shows the effect of La substitution on piezoelectric voltage constant (g_{33}) and on FoM_{PH}. Voltage constant, g_{33}, is directly proportional to d_{33} and inversely proportional to K_3^T. With increase in La content to 2 mol%, g_{33} slightly decreased from 41.5 ($\times 10^{-3}$ V m/N) to 39.3 ($\times 10^{-3}$ V m/N). Further, it reduced substantially to 26.4 ($\times 10^{-3}$ V m/N) at 3 mol% and to 19.4 ($\times 10^{-3}$ V m/N) at 5 mol% of La. In general, it was observed that increase in La content has reduced the sensitivity and voltage generation ability of the composition. Though the value of g_{33} was slightly higher for the composition containing 1 mol% of La, d_{33} was at minimum. This led to the lesser value of FoM_{PH} (10.8×10^{-12} C V m/N^2). With the increase in the La content to 2 mol%, FoM_{PH} was increased to 16.2×10^{-12} C V m/N^2 showing optimum value. Such composition could be useful for the sensor and power harvesting applications. With further increase in La content, FoM_{PH} was reduced.

Influence of temperature on dielectric constant at 1 kHz is shown in Figure 9d. As the La concentration was increased, Curie temperature was decreased. The dielectric permittivity increased gradually until the transition temperature and then decreased sharply for 1 to 3 mol% of lanthanum. As the La concentration further increased to 4 and 5 mol%, dielectric maxima broaden indicating diffuse phase transition (DPT). In case of 4% and 5%, the region in the neighborhood of the dielectric maxima is highly broadened, which is one of the most important characteristics of a disordered perovskite structure with a DPT [37,38]. This indicates that, for composition $Pb_{1-z}La_z(NiSb)_{0.05}[(Zr_{0.52}Ti_{0.48})_{1-z/4}]_{0.95}O_3$, the critical amount of La is 3 mol% beyond which decoupling effects are stronger [29] and normal micrometer-sized ferroelectric domains do not sustain [30] leading to drastic reduction in grain size, tetragonality, and polarizability, which are supported, respectively, by microstructure, crystal structure, and k_p values [27–29]. Broadening of the dielectric peak is attributed to compositional fluctuations and substitutional disordering in the arrangement of cations in one or more crystallographic sites of the structure [39].

4.4 Hysteretic Response

Figure 10a–e shows the effect of La doping on polarization and displacement or strain on application of electric field. The values for the coercive field, remnant polarization, and displacement obtained are shown in Table 3. Defect produced due to La substitution at "A" site minimizes the local stresses and makes the domains remarkably mobile under the electric field [38]. Such materials show highly symmetric P-E hysteresis loops with low coercivity, higher polarizability, and good squareness [1,40,41]. As La content was increased from 1 to 3 mol%, remnant polarization (P_r) and strain, that is, displacement (D), were improved. This was attributed to a better alignment of domains and enhanced domain wall motion [28,42]. Upon further increasing the La concentration, these properties were reduced attributing to the decrease in domain wall motions.

4.5 Behavior Under Electric Cyclic Load

Piezoelectric materials, when used in converse effect, are very often subjected to the electric cyclic loads, which disturb and reorient the aligned domains leading to the reduction in the properties and thus affect its

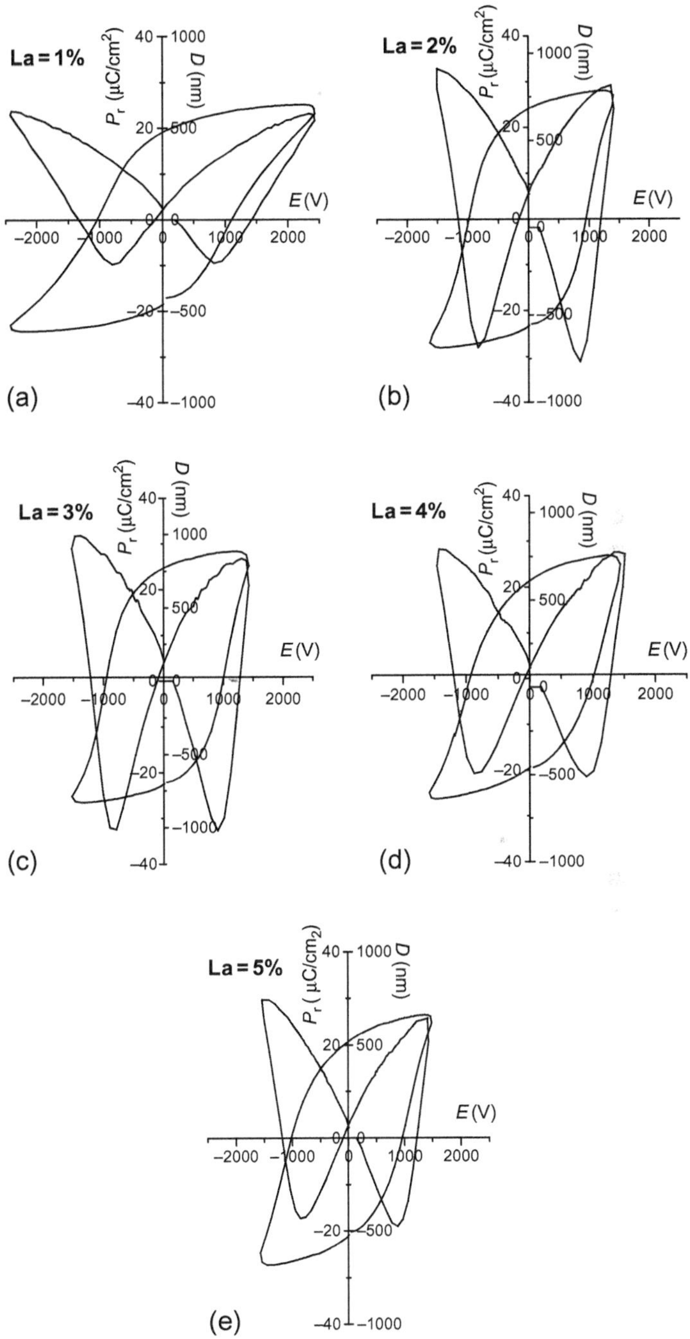

Figure 10 Hysteretic response of PNS-PZT as a function of La content. (a) La 1 mol%, (b) La 2 mol%, (c) La 3 mol%, (d) La 4 mol%, and (e) La 5 mol%.

Table 3 Hysteretic properties of PNS-PZT as function of La concentration

La content (mol%)	Remnant polarization, P_r ($\mu C/cm^2$)	Coercive field, E_c (V/mm)	Displacement/ strain, D (nm)
1	19.01	1040	837
2	24.03	976	1681
3	24.65	956	2012
4	21.33	969	1307
5	20.07	1017	1217

performance. Reorientation is due to switching of domains cyclically with applied cyclic field [43,44]. At lower fields, few ferroelectric domains are switched, and at higher cyclic fields, nearly all domains are switched with applied cyclic field causing heavier decrease in properties [43–45]. Though the degradation of properties at high and low cyclic fields is quantitatively different, the trend in the degradation remains the same [45].

Figure 11 depicts the effect of electric cyclic load on impedance and resonance and degradation of piezoelectric properties of lanthanum-doped PNS-PZT ceramics. 1500 V alternating voltage and 20 V fatigue amplitude were applied at 1000 Hz. After each 10^6 cycles, samples were analyzed for impedance and resonance and other properties.

Figure 11a–e represents change of impedance as a function of frequency, of first radial mode resonance (f_r) and antiresonance (f_a). In electric field cycled samples, the Δf (i.e., $f_a - f_r$) decreases indicating degradation in k_p (Eq. 3), which represents electromechanical energy transduction. Poled sample containing 1 mol% of La (Figure 11a) shows abnormal resonance behavior indicating presence of microcracks induced during poling. This may be due to larger internal stress in the F_T-dominant composition induced by 90° domain switching arising in poling [46], which increases further during electric field cycling. Samples with 2 and 3 mol% of La (Figure 11b and c), which are in the vicinity of MPB, show a lesser amount of degradation due to the well-known fact that both F_T and F_R crystallographic phases coexist in approximate equal amounts, and electrical domain switching is in favorable orientation for maximum strain. Samples with 4 and 5 mol% of La show significant change for 10^6 cycles itself and completely decayed for 2×10^6 cycles (Figure 11d and e) as compositions move away from MPB. Moreover, increasing the lanthanum to 4% and 5%, drastic reduction in grain size (Figure 8) was observed. Smaller grain size is associated with larger grain boundary density where space charge gets accumulated under cyclic electric loading. These trapped space charges reorient domain and further

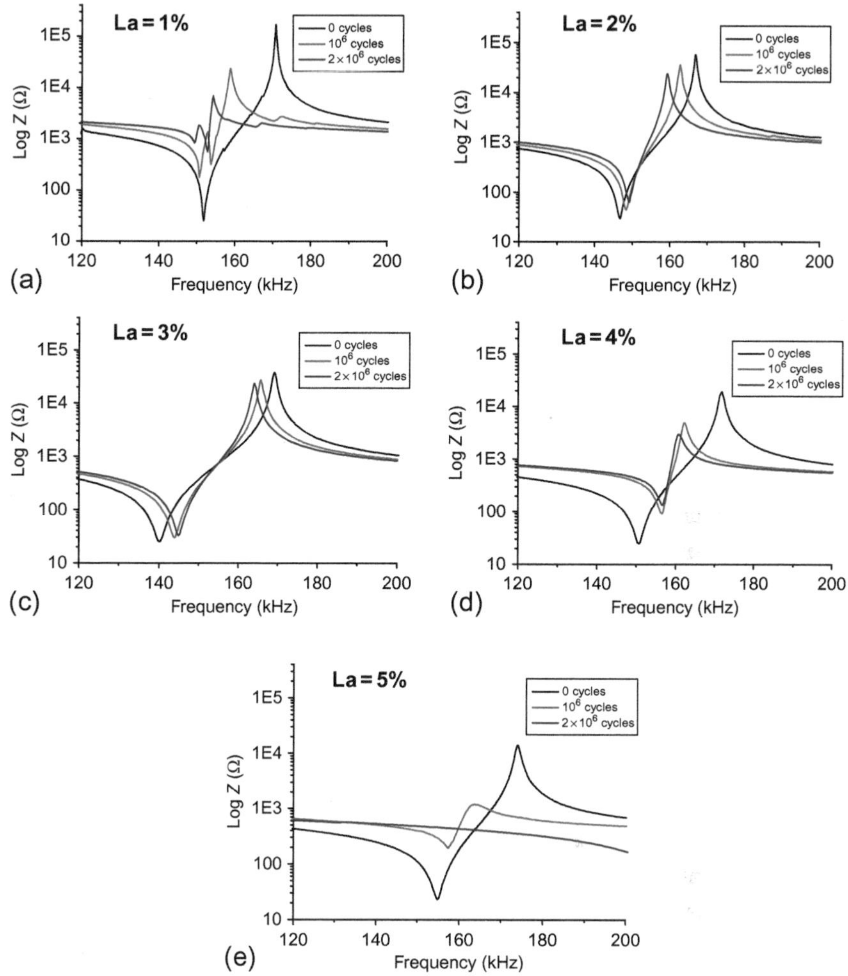

Figure 11 Effect of cyclic electric loading on resonance and impedance.

switching is arrested, resulting in decrease in properties [47]. This implies that cyclic load freezes certain regions of domains at 10^6 cycles, and finally, these nonswitchable domains grow as the number of cyclic electric load increases leading to complete degradation in the degree of poling.

4.6 Conclusions

1. Dense microstructure with narrow grain size distribution with minimal pores is obtained for the composition containing 2% of lanthanum. Most of the grains are polygonal in shape. The average optimum grain size was 6.45-6.8 μm.

2. Dense microstructure obtained for the composition containing 3% of lanthanum. Most of the grains are polygonal in shape and few ovals. The average optimum grain size was 7.05 μm.
3. XRD patterns shows
   - perovskite phase with coexistence of F_T and F_R,
   - nonlinear relationship of lanthanum content and tetragonality,
   - optimum tetragonality for the composition containing 2% of lanthanum indicated by the higher splitting at (002) plane.
4. Dielectric constant and density increased with lanthanum content, while voltage constant decreased.
5. Increase in lanthanum increased dielectric constant and reduced piezoelectric voltage constant.
6. PNS-PZT composition containing 2 mol% of La resulted in moderately higher
   - $d_{33}=413\times10^{-12}$ C/N,
   - $g_{33}=39.3\times10^{-3}$ V m/N,
   - $k_p=0.65$, and
   - optimum $FoM_{PH}=16.2\times10^{-12}$ C V m/N^2.

   Such composition could be useful for sensor and power harvesting applications.
7. Presence of 3 mol% of La in PNS-PZT resulted in optimum
   - $d_{33}=468\times10^{-12}$ C/N,
   - $k_p=0.68$,
   - $P_r=24.65$ μC/cm^2,
   - strain/displacement, $D=2012$ nm.

   Minimum degradation in the properties under cyclic load was attributed to higher strength of poling and presence of F_R on larger scale. This indicates the usefulness of composition for actuator applications.
8. Composition containing 1-3 mol% of lanthanum sharp phase transition is observed.
9. The broadening of the permittivity peak may be considered due to the disorder in the arrangements of rare earth and other atoms, leading to a microscopic heterogeneity for 4% and 5% of lanthanum.

5. EVALUATION OF OPTIMIZED COMPOSITION FOR POWER HARVESTING AND SENSORS APPLICATIONS

Performance of optimized La-doped PNS-PZT composition, that is, $Pb_{0.98}La_{0.02}(NiSb)_{0.05}[(Zr_{0.52}Ti_{0.48})_{0.995}]_{0.95}O_3$, was evaluated as a power harvester and sensor (cymbal hydrophone) and compared with PZT type 5A. This uses disks with diameter of 25 mm and thickness of 1.2 mm.

6. METHODOLOGY OF HARVESTING THE POWER

Piezoelectric effect is direction-dependent. Compressive stress and tensile stress will generate voltage of opposite polarity. As shown in Figure 12, sinusoidal power output may be produced in response to physical movements or ambient vibrations. This output is converted into DC output and is used to charge the capacitor. The energy stored in the capacitor can be utilized to power the low-power electronics.

6.1 Evaluation of La-Doped PNS-PZT as a Power Harvester

In general, impedance is the measure of the opposition to the flow of current through a circuit when an *alternating voltage* is applied. When the impedance of the piezoelectric element and the load is matched, maximum power and thus energy can be obtained [48].

Measurement of electrical output, across the load resistance ranging 15-6000 Ω, in response to impact of freely released stainless steel ball weighing 8.25 g from 150 mm height was carried out. Figure 13a shows voltage output traced by oscilloscope. Maximum power output by positive pulse was calculated from V_{max} and resistance, while maximum energy was calculated from maximum power and full width at half of maximum (FWHM) as shown in Table 4. It is observed that the power and energy output by PNS-PZT was better than PZT 5A across the load resistances under experiment.

As shown in Figure 13b, maximum power output of 5.11 W and energy output of 71.13 μJ were given by PNS-PZT across the load resistance of 4000 Ω and 2.08 W maximum power, and 20.79 μJ energy output is given by PZT type 5A across 1000 Ω resistance.

Compared to PZT type 5A, about 2.5 times the higher electrical output can be obtained from optimized La-doped PNS-PZT composition, which is

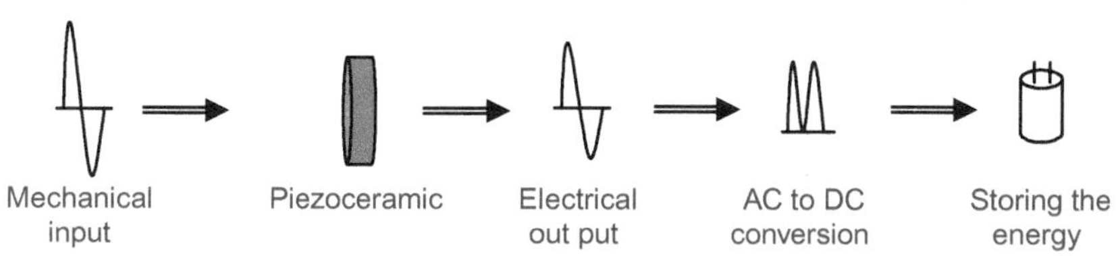

Figure 12 Methodology of harvesting the power using piezoceramics.

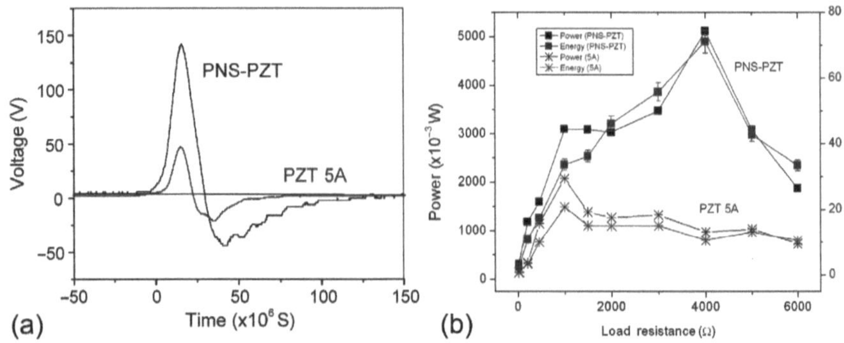

Figure 13 (a) Optimum voltage output recorded on oscilloscope and (b) electrical output at various load.

Table 4 Electrical output by PNS-PZT and PZT type 5A

Parameter	PNS-PZT at 4000 Ω	PZT type 5A at 1000 Ω
V_{max} (V)	143.0	45.6
FWHM (μS)	13.91	10.0
P_{max} (W)	5.11	2.08
Energy (μJ)	71.13	20.79
Energy density (μJ/cc)	120.7	35.3

attributed to its higher power harvesting piezoelectric properties, that is, d_{33}, g_{33} and FoM_{PH}.

6.2 Evaluation of La-Doped PNS-PZT as a Sensor

The sensitivity of the optimized composition was evaluated in the form of cymbal hydrophone and compared with PZT type 5A. Hydrophone is a device used as a passive sensor to listen or pick up the underwater acoustic energy. In response to pressure generated due to acoustic energy, piezoelectric element in the hydrophone produces analog voltage, which is further amplified and processed for detection and analysis [49]. Cymbal hydrophone is class V flextensional transducer, consisting of a "thickness poled ceramic disk" sandwiched between the metallic end caps, which amplify the applied stress due to its geometry [50]. Receiving sensitivity of the hydrophones is highly influenced by the cymbal design and piezoceramic material used to fabricate the cymbal transducers [51]. Considering the hydrostatic mode, the receiving sensitivity (M_0) is given by Equations (5) and (6) [51,52]:

$$M_0 = g_h \times t = (g_{33} + g_{32} + g_{31}) \times t \quad (5)$$

Since $g_{31} = g_{32}$

$$M_0 = (g_{33} + 2g_{31}) \times t \tag{6}$$

Hydrostatic voltage constant (g_h) is related to hydrostatic charge constant (d_h) and dielectric constant (K_3^T) according to Equation (7):

$$g_h = \frac{d_h}{\varepsilon_0 K_3^T} \tag{7}$$

Cymbal transducers were fabricated using piezoceramic disks of compositions $Pb_{0.988}(Zr_{0.52}Ti_{0.48})_{0.976}Nb_{0.024}O_3$ (PZT type 5A) and $Pb_{0.98}La_{0.02}(NiSb)_{0.05}[(Zr_{0.52}Ti_{0.48})_{0.995}]_{0.95}O_3$ (PNS-PZT). Piezoelectric and hydrostatic constants for disks of PNS-PZT composition were better. End caps were made up of brass sheet having thickness of 0.5 mm. Underwater testing of the polyurethane-molded hydrophones were carried out in acoustic tank and compared.

6.3 Fabrication of Cymbal Hydrophones

Cymbal transducers as per the schematic in Figure 14 and physical dimensions in Table 5 were fabricated out of PNS-PZT and PZT type 5A disks and pair of brass metal caps; 3 M adhesive was used to adhere metal end caps with

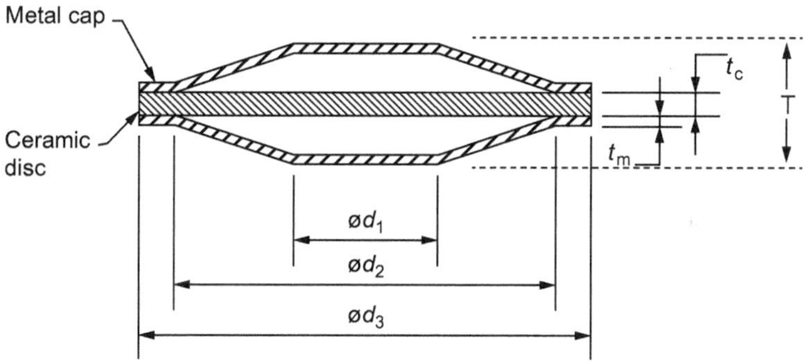

Figure 14 Schematic of the cymbal transducer.

Table 5 Dimensions of cymbal transducer

Dimension	Value in mm
d_1	8.0
d_2	21.0
d_3	25.0
t_c	1.2
t_m	0.5
T	6.0

Figure 15 Molded cymbal hydrophone.

ceramic disk and allowed to cure for 48 h at ambient temperature. Further, they were molded in acoustically transparent polyurethane and cured for 72 h (Figure 15).

Hydrophones were tested in cubical acoustic tank, 8 m × 8 m × 8 m. Calibration of standard hydrophone (Brüel and Kjær model 9718) was carried out by reciprocity method, and the data obtained were used to measure the receiving sensitivity of hydrophones under testing. Standard hydrophone and test hydrophone were mounted at 1 m depth and 1 m apart. Receiving sensitivity was measured from 2 to 40 kHz. The beam patterns were obtained at 10, 20, 30, and 40 kHz.

6.4 Results and Discussions

Receiving response, for both the hydrophones, is shown in Figure 16a. Resonance frequency for both the hydrophones was 10.1 kHz. At resonance frequency, PNS-PZT hydrophone shows sensitivity −183.2 dB re 1 V/μPa, which is higher by 8 dB compared with PZT type 5A (−191.2 dB re 1 V/μ

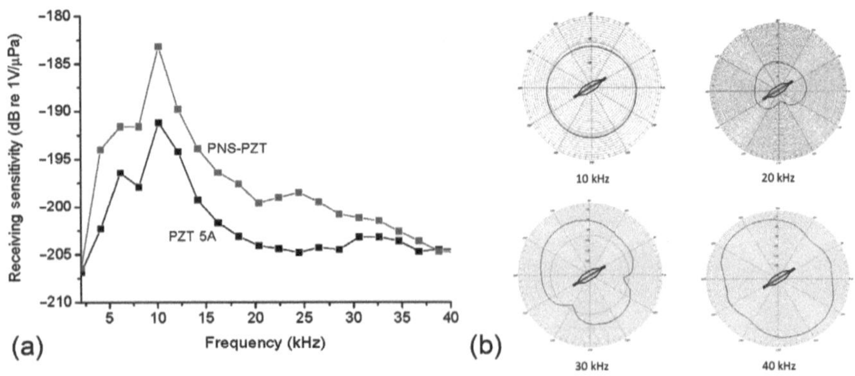

Figure 16 (a) Receiving response of PNS-PZT hydrophone and (b) receiving directivity patterns of PNS-PZT hydrophone.

Pa). It is also higher over the frequency range 2-38 kHz. This is attributed to higher d_h, g_h, and $d_h \times g_h$ for PNS-PZT [53].

Directivity patterns for PNS-PZT composition are shown in Figure 16b. Very close to resonance frequency, that is, at 10 kHz, the pattern is perfectly omnidirectional. Beyond resonance frequency, patterns are distorted. Main lobes are along the flat face of the hydrophone.

6.5 Conclusions

1. Cymbal hydrophones were fabricated and tested for underwater acoustic performance.
2. The resonance frequency for both the hydrophones was 10.1 kHz.
3. At resonance frequency, PNS-PZT hydrophone shows about 2.5 times higher sensitivity (−183.2 dB re 1 V/μPa) compared with PZT 5A (−191.2 dB re 1 V/μPa) attributed to the higher hydrostatic properties (d_h, g_h, $d_h \times g_h$).
4. Directivity of PNS-PZT hydrophone was found to be perfectly omnidirectional at resonance frequency.

REFERENCES

[1] G.H. Haertling, Ferroelectric ceramics: history and technology, J. Am. Ceram. Soc. 82 (1999) 797–818.
[2] B. Jaffe, W.R. Cook Jr., H. Jaffe, Piezoelectric ceramic, Academic Press, New York, 1971.
[3] H.A. Sodano, D.J. Inman, G. Park, Comparison of piezoelectric energy harvesting devices for recharging batteries, J. Intell. Mater. Syst. Struct. 16 (2005) 799–807.
[4] H.A. Sodano, D.J. Inman, G. Park, A review of power harvesting from vibration using piezoelectric, Shock Vib Digest 36 (2004) 197–205.
[5] H.W. Kim, S. Priya, K. Unchino, R. Newnham, Piezoelectric energy harvesting under high pre-stressed cyclic vibrations, J. Electroceram. 15 (2005) 27–34.
[6] F. Mohammadi, A. Khan, R.B. Cass, Power generation from piezoelectric lead zirconate titanate fiber composites, in: Proceedings of the Material Research Society Symposium, vol. 736, 2003, pp. D5.5.1–D5.5.6.
[7] C. Green, K.M. Mossi, R.G. Bryant, Scavenging energy from piezoelectric materials for wireless sensor applications, in: Proceedings of IMECE2005 International Mechanical Engineering Congress and Exposition, Orlando (Florida), 2005, pp.1–7.
[8] A. Mathers, K.S. Moon, Y.A. Jingang, Vibration-based PMN-PT energy harvester, IEEE Sens. J. 9 (2009) 731–739.
[9] J.F. Tressler, A. Sedat, R. Newnham, Piezoelectric sensors and sensor materials, J. Electrocerm. 2 (4) (1998) 257–271.
[10] A.R. James, Characteristics of PLZT Ceramics derived through mechano-chemical processing, in: D.K. Kharat (Ed.), Proceedings of the National Seminar on Advances in Electroceramics, Allied Publishers, New Delhi, May 2006.
[11] Z. Brankovic, G. Brankovic, C. Jovalekic, Y. Maniette, M. Cilense, J.A. Varela, Mechanochemical synthesis of PZT powders, Mater. Sci. Eng. A 345 (2003) 243–248.
[12] M. Zhang, X.D. Wang, F.M. Wang, I.M. Miranda Salvado, P.M. Vilarinho, W.C. Li, Preparation and ferroelectric properties of PZT fibers, Ceram. Int. 31 (2005) 281–286.

[13] B. Praveenkumar, H.H. Kumar, D.K. Kharat, B.S. Murty, Investigation and characterization of la-doped PZT nanocrystalline ceramic prepared by mechanical activation route, Mater. Chem. Phys. 112 (2008) 31–34.
[14] Chandratreya, R.M. Fulrich, J.A. Pask, Reaction mechanisms in the formation of PZT solid solutions, J. Am. Ceram. Soc. 64 (1981) 422–425.
[15] C.M. Lonkar, D.K. Kharat, H.H. Kumar, S. Prasad, K. Balasubramanian, Effect of La on piezoelectric properties of $Pb(Ni_{1/3}Sb_{2/3})O3–Pb(ZrTi)O_3$ ferroelectric ceramics, J. Mater. Sc.: Mater. Electron. 24 (2013) 411–417.
[16] R. Tipakontitikul, S. Anantha, A modified two-stage mixed oxide synthetic route to lead zirconate titanate powders, Mater. Lett 58 (2004) 449–454.
[17] V. Singh, H.H. Kumar, D.K. Kharat, S. Hait, M.P. Kulkarni, Effect of lanthanum substitution on ferroelectric properties of niobium doped PZT ceramics, Mater. Lett. 60 (2006) 2964–2968.
[18] R.S. Nasar, M. Cerqueira, E. Longo, J.A. Varela, A. Beltran, Experimental and theoretical study of the ferroelectric and piezoelectric behavior of strontium doped PZT, J. Eur. Ceram. Soc. 22 (2002) 209–218.
[19] A. Boutarfafia, Investigations of co-existance region in lead zirconate titanate solid solutions: X-ray diffraction studies, Ceram. Int. 26 (2000) 583–587.
[20] M.C. Wang, Sintering and piezoelectric properties of Pb(Ni1/3Sb2/3)O3-PbZrO3-PbTiO3 ceramics, J. Mater. Sc. 3 (7) (2002) 663–668.
[21] C.M. Lonkar, D.K. Kharat, H.H. Kumar, S. Prasad, M. Ghosh, C.S. Kumbhar, Effect of Zr/Ti ratio on piezoelectric properties of $Pb(Ni_{1/3}Sb_{2/3})O_3–Pb(ZrTi)O_3$ ceramics, Ceram. Int. 37 (2011) 3509–3514.
[22] V. Raghvan, Materials Science and Engineering, Sixth ed., Prentice Hall of India Pvt Ltd., New Delhi, 1995.
[23] K. Puja Goel, L. Yadav, A.R. James, Double doping effect on the structural and dielectric properties of PZT ceramics, J. Phys. D: Appl. Phys. 37 (2004) 3174–3179.
[24] A. Govindan, A. Sharma, A.K. Pandey, S.K. Gaur, Piezoelectric and pyroelectric properties of lead lanthanum zirconate titanate (PLZT) ceramics prepared by sol gel derived nano powders, Indian J. Phys. 85 (2011) 1829–1832.
[25] A.R. James, J. Subramanyam, Processing and structure property relation of fine-grained PLZT ceramics derived from mechanochemical synthesis, J. Mater. Sci: Mater. Electron. 17 (2006) 529–535.
[26] R. Laishram, O.P. Thakur, D.K. Bhattacharya, Dielectric and piezoelectric properties of la doped lead zinc niobate–lead zirconium titanate ceramics prepared from mechanochemically activated powders, Mater. Sci. Eng. B. 172 (2010) 172–176.
[27] Li Tao, Li Huiqin, Xu. Youchao, Dielectric behavior and raman spectra of lanthanum-doped lead magnesium niobate ceramics, J. Mater. Sci: Mater. Electron. 22 (2011) 1188–1194.
[28] M.D. Durruthy-Rodríguez, F. Calderón-Piñar, C. Malfatti, L.D. Pérez-Fernández, Sintering kinetics of soft-doped PZT (54/46) systems, J. Mod. Phys. 2 (2011) 416–420.
[29] P.-H. Xiang, N. Zhong, X.-L. Dong, R.-H. Liang, H. Yang, C.-D. Feng, Fabrication and dielectric properties of lanthanum-modified lead zirconate titanate using coprecipitation powder coating, Mater. Lett. 58 (2004) 2675–2678.
[30] J.F. Li, X. Dai, A. Chow, D. Viehland, Polarization switching mechanisms and electromechanical properties of la-modified lead zirconate titanate ceramics, J. Mater. Res. 10 (1995) 926–938.
[31] Q. Tan, D. Viehland, Alternating and direct current field effects on the dielectric response of la-modified lead zirconate titanate, J. Appl. Phys 81 (1997) 361–368.
[32] A. Krumins, T. Shiosaki, S. Koizumi, Jpn. spontaneous trasition between relaxor and ferroelectric states in lanthanum-modified lead zirconate titanate (6–7)/65/35, J. Appl. Phys. 33 (1994) 4940–4945.

[33] M.C. Wang, M.S. Huang, N.C. Wu, Effects of 30B2O3–25BiO3–45CdO glass addition on sintering of 12Pb(Ni1/3 Sb2/3)O3 40PbZrO3 48PbTiO3 piezoelectric ceramics, J. Eur. Ceram. Soc. 21 (2001) 695–701.
[34] M.C. Wang, M.S. Huang, N.C. Wu, Low-temperature sintering of 12Pb(Ni1/3Sb3/3) O3 40PbZrO3 48PbTiO3 with V2 O5 and excess PbO additives, J. Eur. Ceram. Soc. 22 (2002) 697–705.
[35] H. Kungl, J.H. Michael, Effects of sintering temperature on microstructure and high field strain of niobium-strontium doped morphotropic lead zirconate titanate, J. Appl. Phys. 107 (2010) 054111.
[36] M. Chen, X. Yao, L. Zhang, Grain size dependence of dielectric and field–indiced strain properties of chemical prepared (Pb, La)(Zr, Sn, Ti)O3 antiferroelectric ceramics, Ceram. Int. 28 (2002) 201–207.
[37] G.A. Smolenskii, A new framework for understanding relaxor ferroelectric, J. Phys. Sot. Japan. 28 (Suppl. 26) (1970).
[38] K.L. Yadav, R.N.P. Choudhary, Structural and electrical properties of PZT (La, K) ceramics, Mater. Lett. 16 (1993) 291–294.
[39] S. Dutta, R.N.P. Choudhary, P.K. Sinha, Ferroelectric phase transition in bi-doped PLZT ceramics, Mater. Sc. Eng. B 98 (2003) 74–80.
[40] L.E. Cross, Ferroelectric Ceramics: Tailoring Properties for Specific Applications, Birkhauser Verlag, Basel, (1991).
[41] X. Zeng, X. Hea, W. Chenga, X. Zhenga, Pingsun Qiu, Dielectric and ferroelectric properties of PZN–PZT ceramics with lanthanum doping, J. Alloys Compd. 485 (2009) 843–847.
[42] D. Damjanovic, Hysteresis in piezoelectric and ferroelectric materials, in: I. Mayergoyz, G. Bertotti (Eds.), The Science of Hysteresis, vol. 3, Elsevier, 2005, pp. 353–355.
[43] R. Qian, S. Lukasiewicz, Q. Gao, Electrical fatigue for ferroelectric ceramics under electrical cylic load, Solid-State Electron. 44 (2000) 1717–1722.
[44] K.L. Lee, A.K. Soh, Effects of electric fatigue on piezoelectric ceramics, Key Eng. Mater. 306–308 (2006) 91–96.
[45] M.G. Cain, M. Stewart and M. Gee, Degradation of piezoelectric materials, NPL Report CMMT(A) 148, 1999.
[46] W.P. Tai, S.H. Kim, The effect of poling treatment and crystal structure of PZT on fracture toughness and fatigue resistance, J. Mater. Sc. 38 (2003) 1787–1792.
[47] C.R. Bowen, S. Mahon, M.L. Prieto, Non-destructive evaluation of multi layers actuators, Ferroelectrics 241 (2000) 199–206.
[48] H. Kim, Ph.D. thesis, Impedance adaptation methods of the piezoelectric energy harvesting, The Pennsylvania State University, USA, 2006.
[49] Z. Deng, M. Weiland, T.M. Carlson, B. Eppard, Design and instrumentation of a measurement and calibration system for an acoustic telemetry system, Sensors 10 (2010) 3090–3099.
[50] S. Chen, C. Sun, An improved cymbal hydrophone made of lanthanum -modified lead zirconate titanite, J. Southwest Jiaotong University 17 (2009) 148–152.
[51] L. Zhaohui, H. Aigen, L. Guidong, Z. Jinduo, Finite element analyzing of underwater receiving sensitivity of PMN-0.33PT single crystal cymbal hydrophone, Ultrasonics 44 (2006) 759–762.
[52] J.F. Tressler, Piezoelectric Transducer Designs for SONAR Applications, in: A. Safari, E.K. Akdoğan (Eds.), Piezoelectric and acoustic materials for transducer applications, Springer Science+Business Media, New York, 2008, pp. 217–239.
[53] C.M. Lonkar, D.K. Kharat, H.H. Kumar, N.S. Prasad, $Pb(Ni_{1/3}Sb_{2/3})O_3$-$PbZrTiO_3$ ceramic sensors for underwater transducer application, Def. Sci. J. 62 (2012) 269–273.

INDEX

Note: Page numbers followed by *f* indicate figures and *t* indicate tables.

I

L

O

R

U

FSC
www.fsc.org
MIX
Papier | Fördert gute Waldnutzung
FSC® C083411

Kolibri 360 GmbH
Ferdinand-Jühlke-Straße 7
99095 Erfurt, Deutschland
produktsicherheit@kolibri360.de